QMC 364080 5

a30213 0036408056

D1429332

Some Griffin books on statistics

Studies in the history of statistics and probability (first volume)
Selected and edited by E. S. PEARSON & SIR MAURICE KENDALL

The advanced theory of statistics (3 volumes)
SIR MAURICE KENDALL & A. STUART

An introduction to the theory of statistics
G. U. YULE & SIR MAURICE KENDALL

Rank correlation methods   SIR MAURICE KENDALL

Time-series   SIR MAURICE KENDALL

Multivariate analysis   SIR MAURICE KENDALL

Scientific truth and statistical method   M. BOLDRINI

Games, gods, and gambling: A history of
  probability and statistical ideas   F. N. DAVID

Statistical papers of George Udny Yule
Selected by A. STUART & SIR MAURICE KENDALL

Combinatorial chance   F. N. DAVID & D. E. BARTON

Style and vocabulary: Numerical studies   C. B. WILLIAMS

Biomathematics: C. A. B. SMITH
  Volume II: Numerical methods, matrices,
    probability, statistics

Griffin Statistical Monographs

Geometrical probability   SIR MAURICE KENDALL & P. A. P. MORAN

The analysis of categorical data   R. L. PLACKETT

Statistical models and their experimental application   P. OTTESTAD

The linear hypothesis: A general theory   G. A. F. SEBER

Families of bivariate distributions   K. V. MARDIA

Families of frequency distributions   J. K. ORD

Green's function methods in probability theory   J. KEILSON

Statistical tolerance regions: Classical and Bayesian   I. GUTTMAN

Analysis of variance: a basic course   A. HUITSON

WITHDRAWN
FROM STOCK
QMUL LIBRARY

STUDIES IN THE HISTORY

STATISTICS AND PROBABI

VOLUME II

STUDIES IN THE
HISTORY OF STATISTICS
AND PROBABILITY

VOLUME II

A series of papers selected and edited by

SIR MAURICE KENDALL, Sc.D., F.B.A.

R. L. PLACKETT, M.A., Sc.D.

Quis separabit nos

CHARLES GRIFFIN & COMPANY LIMITED
LONDON & HIGH WYCOMBE

195479

Charles Griffin & Company Limited
Registered Office
5A Crendon Street, High Wycombe, Bucks HP13 6LE

Copyright © 1977

260 mm × 184 mm, ix + 488 pages
2 plates
ISBN 0 85264 232 6

PRINTED IN GREAT BRITAIN BY
J. W. ARROWSMITH LTD., BRISTOL

QUEEN MARY
COLLEGE
LIBRARY

CONTENTS

Book Page*

1   **On the possible and probable in Ancient Greece**
S. Sambursky *Osiris*, **12,** 35–48 (1956)   1

2   **Probability in the Talmud**
Studies in the history of probability and statistics XXII
Nachum L. Rabinovitch *Biometrika*, **56,** 437–41 (1969)   15

3   **Combinations and probability in rabbinic literature**
Studies in the history of probability and statistics XXIV
Nachum L. Rabinovitch *Biometrika*, **57,** 203–5 (1970)   21

4   **A budget of paradoxes**
H. L. Seal *J. Inst. Actuaries Students' Soc.*, **13,** 60–65 (1954)   24

5   **An argument for Divine Providence, taken from the constant regularity observ'd in the births of both sexes**
John Arbuthnott *Phil. Trans.*, **27,** 186–90 (1710)   30

6   **Measurement in the study of society**
M. G. Kendall *Man and the Social Sciences*, ed. W. A. Robson, pp. 133–47, Allen & Unwin, London (1972)   35

7   **The early history of index numbers**
Studies in the history of probability and statistics XXI
M. G. Kendall *Rev. Int. Statist. Inst.*, **37,** 1–12 (1969)   51

8   **Abraham De Moivre's 1733 derivation of the normal curve: a bibliographical note**
Studies in the history of probability and statistics XXX
R. H. Daw and E. S. Pearson *Biometrika*, **59,** 677–80 (1972)   63

9   **The historical development of the use of generating functions in probability theory**
H. L. Seal *Bull. de l'Association des Actuaires suisses*, **49,** 209–28 (1949)   67

10   **Boscovich and the combination of observations**
C. Eisenhart *Roger Joseph Boscovich*, ed. L. L. Whyte, pp. 200–12, Allen & Unwin, London (1961)   88

** The original pagination of each paper is retained in order to preserve cross-referencing.
Book page numbers are printed thus*   **7**

11   **Daniel Bernoulli on the normal law**
Studies in the history of probability and statistics XXIII
O. B. Sheynin *Biometrika*, **57,** 199–202 (1970)
101

12   **D. Bernoulli's work on probability**
O. B. Sheynin *RETE Strukturgeschichte der Naturwissenschaften*, **1,** 273–300
(1972)
105

13   **Progress in the middle of the eighteenth century**
Süssmilch and his contemporaries
Estimates and enumerations of population
Progress of theory at the close of the eighteenth century
H. Westergaard *Contributions to the History of Statistics*, pp. 53–99, P. S. King
& Son, London (1932)
133

14   **Leading British statisticians of the nineteenth century**
Paul J. FitzPatrick *J. Amer. Statist. Ass.*, **55,** 38–70 (1960)
180

15   **Notes on the history of quantification in sociology – trends, sources
and problems**
Paul F. Lazarsfeld *Isis*, **52,** 277–333 (1961)
213

16   **Laplace, Fisher, and the discovery of the concept of sufficiency**
Studies in the history of probability and statistics XXXII
Stephen M. Stigler *Biometrika*, **60,** 439–45 (1973)
271

17   **The discovery of the method of least squares**
Studies in the history of probability and statistics XXIX
R. L. Plackett *Biometrika*, **59,** 239–51 (1972)
279

18   **Development of the notion of statistical dependence**
H. O. Lancaster *Math. Chronicle (N.Z.)*, **2,** 1–16 (1972)
293

19   **Florence Nightingale as a statistician**
E. W. Kopf *J. Amer. Statist. Ass.*, **15,** 388–404 (1916)
310

20   **On the history of some statistical laws of distribution**
Studies in the history of probability and statistics XXV
O. B. Sheynin *Biometrika*, **58,** 234–6 (1971)
328

21   **The work of Ernst Abbe**
Studies in the history of probability and statistics XXVI
M. G. Kendall *Biometrika*, **58,** 369–73 (1971)
331

22   **Entropy, probability and information**
M. G. Kendall *Int. Statist. Rev.*, **41,** 59–68 (1973)
337

CONTENTS

23   **A history of random processes. I. Brownian movement from**   347
      Brown to Perrin
      S. G. Brush *Archive for the History of the Exact Sciences*, **5,** 1–36 (1968)

24   **Branching processes since 1873**   383
      David G. Kendall *J. Lond. Math. Soc.*, **41,** 385–406 (1966)

25   **The simple branching process, a turning point test and a**   406
      fundamental inequality: a historical note on I. J. Bienaymé
      Studies in the history of probability and statistics XXXI
      C. C. Heyde, and E. Seneta *Biometrika*, **59,** 680–3 (1972)

26   **Simon Newcomb, Percy Daniell, and the history of robust**   410
      estimation 1885–1920
      Stephen M. Stigler *J. Amer. Statist. Ass.*, **68,** 872–9 (1973)

27   **The hypothesis of elementary errors and the Scandinavian school**   419
      in statistical theory
      Studies in the history of probability and statistics XXVII
      Carl-Erik Särndal *Biometrika*, **58,** 375–92 (1971)

28   **On the history of certain expansions used in mathematical**   437
      statistics
      Studies in the history of probability and statistics XXVIII
      Harald Cramér *Biometrika*, **59,** 205–7 (1972)

29   **Historical survey of the development of sampling theories and**   440
      practice
      You Poh Seng *J.R. Statist. Soc.* A, **114,** 214–31 (1951)

30   **Sir Arthur Lyon Bowley (1869–1957)**   459
      W. F. Maunder. An inaugural lecture delivered in the University of
      Exeter (1972)

31   **Note on the history of sampling methods in Russia**   482
      S. S. Žarković *J.R. Statist. Soc.* A, **119,** 336–8 (1956)

32   **A supplement to "Note on the history of sampling methods**   486
      in Russia"
      S. S. Zarkovich *J.R. Statist. Soc.* A, **125,** 580–2 (1962)

PREFACE

The first selection of articles on the history of statistics and probability, edited by E. S. Pearson and M. G. Kendall, was published in 1970* and received a warm welcome. Interest in the subject on the part of statisticians and other scientists has been increasing in the meantime and we decided, therefore, to assemble a second set, which forms the contents of this volume. As before, the articles are arranged roughly in the chronological order of the period to which they relate. A glance at the list of articles will illustrate both the scope of the subject and the time period over which the collected set extends.

The editors of the first series pointed out that there were few books on the history of statistics and probability, but expressed the hope that the articles would among other things suggest some of the gaps which still remain to be filled in. The present set both extends the field of study and fills in some of those gaps. Taken as a whole the two volumes will, we hope, give a fairly continuous account of the historical development of this fascinating and important branch of scientific methodology.

Most of the articles in the first volume had appeared in *Biometrika* and were therefore in uniform format. The articles in this second volume, although including a number from *Biometrika*, go further afield and are more varied. They are reproduced by a photographic process to preserve their character – though minor enlargement or reduction has sometimes been necessary.

We are indebted to the living authors of the papers, the editors of the publications in which they appeared, and the following for permission to reproduce: George Allen & Unwin Ltd; the American Statistical Association; the Trustees of *Biometrika*; the University of Exeter; the International Statistical Institute; the London School of Economics and Political Science; the Council of the London Mathematical Society; Mathematical Chronicle Committee; the Royal Society; the Royal Statistical Society; the Master and Fellows of Trinity College, Cambridge.

M. G. KENDALL
R. L. PLACKETT

June 1976

---

*Studies in the History of Statistics and Probability (first volume), ed. E. S. Pearson and M. G. Kendall.

On the Possible and the Probable in Ancient Greece

S. SAMBURSKY (1956)

Greek science has been a serious precursor of modern developments in several fields. Geometry and astronomy are, of course, the most conspicuous examples, but there are other instances of anticipation of the basic concepts which have ruled science since the days of GALILEO and NEWTON. The use of some of the methods of the infinitesimal calculus, although in a rudimentary form, is well known, and one might also include methods of inference from the visible to the invisible which were applied by the Greek atomists. The extent to which the Greeks were able to develop these basic concepts or, to put it the other way round, their failure to carry them beyond the first stages of definition and analysis, is characteristic of the fundamental problem of the difference between their way of thinking and ours. In this respect the associated notions of the Possible and the Probable seem especially worthy of study, as they touch equally upon epistemology, mathematics and the inductive sciences. The progress made by the theory of probability in all fields of theoretical and applied science from the time of PASCAL and FERMAT down to our own days has reached spectacular dimensions. Probability concepts were a tool to solve problems in games of chance in the 16th and 17th centuries. From there to the mathematical and philosophical enlightenment of the 18th century, connected with such names as BERNOULLI and LAPLACE, to the theory of errors, the rise of statistics and the penetration of statistical methods and probability concepts into the whole field of natural science, economy and psychology, we witness a development whose significance by far transgresses the boundaries of scientific thought. The psychology and action of society and individuals today is dominated by modes of thought and principles which derive from the philosophy of the Probable, and the rigid concepts of causality in the field of evaluation and prediction of

1

facts have given way to the more complex and flexible notions of statistics and probability. These notions have shaped our mental structure to an extent that makes it difficult for us to appreciate the enormous efforts made by some of the best Greek thinkers to grasp the first elements of inductive logic. It is remarkable that these efforts did not carry them much beyond the analysis of the Possible.

The Greek equivalent for " plausible " or " probable "—εἰκός— was in use from pre-Socratic literature onwards and well into the Hellenistic period, in the same sense as these terms are applied today, i.e. " to be expected with some degree of certainty." In one of the fragments of Antiphon the Sophist says : " whenever anyone does the beginning of anything correctly, *it is likely (εἰκός)* that the end also will be right " (1). Similarly, DEMOCRITUS is quoted as having said " The man who is prevented by law from wrongdoing will *probably* do wrong in secret " (2). PLUTARCH uses ὑπὸ τοῦ εἰκότος in the sense of "plausible" and παρὰ τὸ εἰκός with the opposite meaning when quoting examples given by CHRYSIPPUS to prove that " we are taken in easily and against reason by things which seem plausible and again distrust those which seem improbable " (3). ARISTOTLE, in the " Poetics " (4) contrasts εἰκός and ἀπίθανον (unconvincing), when giving instructions on how to tell a story. Likely impossibilities, he says, are always preferable to unconvincing possibilities.

PLATO comes nearer to the core of the scientific significance of εἰκός when he considers its relation to truth. SOCRATES in his discourse on rhetoric practice says that in the law courts men " care nothing about truth, but only about conviction, and this is based on probability " (5). Later he goes on to define εἰκός as having a " likeness to truth " *(ὁμοιότης τοῦ ἀληθοῦς)* (6) which is exactly the literal meaning of the latin " verisimilitudo " or the German " Wahrscheinlichkeit ". One of the characteristics of truth is that we are forced to accept or to recognize it, whereas

(1) H. DIELS, *Fragmente der Vorsokratiker*, 6 ed. 87B60.
(2) DIELS, *l. c.* 68B181.
(3) PLUT., *Quaest. conviv.* 626F.
(4) ARIST., *Poet.* 1460ª27.
(5) *Phaedr.* 272D.
(6) *Ibid.* 273D.

2

εἰκός only creates the illusion of having this power to convince. It is in this sense that ARISTOTLE regards εἰκός as a generally approved proposition which is not based on any sort of proof, in contradistinction to the sign (σημεῖον), a demonstrative proposition (7). A probable proposition starting from signs possesses a higher degree of certainty because of its reference to the causal nexus. Characteristically, both PLATO and ARISTOTLE frequently point to the lack of demonstrative power in a probability proposition. In PHAEDO, SIMMIAS says to SOCRATES that " the proposition that the soul is a harmony has not been demonstrated at all (ἄνευ ἀποδείξεως) but rests only on probability (μετὰ εἰκότος) " (8). In THEAETETUS, too, SOCRATES finds fault with the lack of a logical compulsion (ἀνάγκη) in a demonstration based on probability and he adds : " a mathematician who argued from probabilities in geometry would not be worth an ace " (9). ARISTOTLE finally, when talking of the idea of reality propounded by certain philosophers, bluntly declares : " While they speak plausibly (εἰκότως) they do not speak what is true (ἀληθῆ) " (10).

The awareness of the fact that empirical truth does not rest on the same safe foundations as does mathematical truth independent of the time factor, led to the coining of a very striking term for the natural sciences. SIMPLICIUS tells us : " Aptly did Plato call natural science (φυσιολογία) the science of the probable (εἰκοτολογία); ARISTOTLE was of the same opinion and postulated that exact evidence must spring from immediate and reliable principles and from exact and essentially primary causes " (11). This remarkable definition anticipates the point of view of HUME, who attributed probability to all knowledge inferred from the law of causality and not from logical relations between ideas.

In view of this characterization of φυσιολογία, a science based largely on prediction, as εἰκοτολογία, it no longer seems farfetched to ask why such advanced epistemological insight did not lead to a more quantitative development of the concept of probability. We know from the history of the beginnings of the theory of

(7) ARIST., Anal. pr., 70ᵃ3.
(8) Phaedo, 92D.
(9) Theaet., 162E.
(10) ARIST., Met., 1010ᵃ4.
(11) SIMPL., ~~in Arist.~~ Phys., ~~325, 24.~~   (Amendments made by Author, 1974)

V 18, 29

In

3

probability in modern times that the way for such a development was paved by the clarification of two elements. On the one hand there was the logical aspect, comprising the elaboration of the concept of equally possible cases, i.e. disjunction and the notion of the Possible, and on the other hand there were the mathematical tools—the elements of combinatorial analysis.

It is well known that hypothetical and disjunctive propositions were for the first time considered by THEOPHRASTUS and EUDEMUS (12) and were later developed in detail by the Stoics (13). As prerequisites for the logic of the Possible only the so-called exclusive disjunctions are of interest to us, i.e. the disjunctions which are true if just one of their component members is true: " Either A or B or C ... is true ". From the falsity of B, C, ... one can then argue to the truth of A. The majority of the examples used in literature on this point are two-membered disjunctions, such as " A number is either odd or even " (14), and only very few instances of tri-membered ones are found, including an example taken from geometry : " One section is either equal to or greater or smaller than another section " (14). As we are interested in the notion of the empirically possible rather than the logically possible, mention might be made of one of the very rare examples of this kind which takes the form of a tri-membered disjunction (15) : CHRYSIPPUS who maintained that some animals have the faculty of drawing logical conclusions, gave the following instance : A hunting dog chasing an animal arrives at a tri-furcation of a road; he sniffs along two of the continuations and finds no scent. After that, without sniffing, he starts chasing along the third way. CHRYSIPPUS concludes that the dog is acting like a human being endowed with logic : if there are n possibilities and (n-1) of them are proved wrong, the nth remains as the true one.

But what about the category of the Possible in those cases where no immediate decision about right or wrong can be taken, such as in the case of an event taking place only at some future date ? Where is the place of the Possible between the Necessary and the Impossible ? It is well known that even before the disjunctive

(12) C. PRANTL, *Geschichte der Logik im Abendlande*, Vol. 1, p. 375 (1855).
(13) B. MATES, *Stoic logic*, Univ. of Calif. Press, 1953.
(14) PHILOP., *ad anal. pr.*, quoted in PRANTL, *l. c.*, p. 388.
(15) SEXT. EMP., *pyrrh. hyp.*, I, 69.

245, 27 and 36

4

propositions were brought up, the concept of the Possible was discussed by ARISTOTLE. Taking into account the relevant passages, especially in Metaphysics and *De Interpretatione*, one must agree with ROSS' conclusion (16) that ARISTOTLE was no absolute determinist. He believed in the existence of a real contingency, particularly in everything pertaining to the field of human action. His famous analysis of the sentence " tomorrow there will be a naval battle " (17) hardly leaves any room for doubt that he only recognized the necessity of the following alternative : " tomorrow it will either be true or it will not be true that a naval battle has taken place ". But he rejected the idea that today it is already either true or not true that a naval battle will take place tomorrow. This is one of the cases where the attributes " true " and " not true " do not make sense until the realization of the event (18). Only predictions relating to a completely determined future can be true or false before their verification. This applies in the first place to the recurrent series of the heavenly motions or the succession of seasons (19). It seems, therefore, that for ARISTOTLE the Possible in part coincides with the Necessary which is going to happen, and in part comprises those cases which, like human actions, contain an element of genuine contingency and cannot be predicted even by the Laplacean Intelligence. This ambiguity of the Possible was not conducive to the development of the concept of the Probable. We know from the elements of the theory of probability that probability statements about any event are based on the assumption that a whole group of cases, of which the one considered is a member, are " equally possible ", i.e. that each case has full equality among the collective of causally conceivable cases. Any idea of arbitrariness or the non-determination of certain events, any thought that causes of a non-physical nature could ever influence them will necessarily overthrow that assumption of equal possibilities.

In the post-Aristotelian period we are confronted by two developments. On the one hand there is the idea of indeterminacy, brought to a culmination by the Epicurean School, and lack of

---

(16) W. D. Ross, *Aristotle*, 5th ed., 1949, p. 80, p. 201.
(17) ARIST., *De Interpr.*, 19ᵃ28.
(18) H. SCHOLZ, *Hauptgestalten der Logik*, Münster, 1934 (typescript).
(19) ARIST., *de Gen. et Corr.*, II, 11

determination is extended into the realm of purely physical nature. The " inclinamen " of the atoms, their uncaused deviation from their course " at quite indeterminate times and places " (20), excludes any idea of a distinct definition of " equally possible cases " or their pre-determination. The Stoic School, however, developed an extreme concept of rigid determinism and of the causal nexus of the whole of existence. Quite naturally, therefore, it was the Stoics who came to grasp most clearly and to define with the greatest precision the notion of the Possible. The reference to experience in their polemics against the Epicurean hypothesis of contingency best characterizes the inductive approach of the Stoics to the problem of causality. " Chrysippus confuted those who would impose lack of causality upon nature by pointing to dice and scales and many other things which can never fall or swerve without some internal or external cause. For there is no such thing as lack of cause or chance. In the impulses mentioned which some have arbitrarily called accidental, there are causes hidden from our sight which determine movement in a certain direction " (21). Two points should be noted here : the example of the dice whose throws served as stimulus for the first probability calculations in the 17th Century, and the definition of chance as an event whose causes are hidden from us. Here the Stoics followed the steps of the pre-Socratic mechanists (22), who had already anticipated the idea of general causality to a certain degree; only, as we shall see presently, they regarded the Possible, too, as a consequence of the imperfection of human knowledge as well as the concept of chance.

A definition of the Possible widely accepted in the post-Aristotelian era and down to the late Hellenistic period was that of DIODORUS THE MEGARIAN who said : " The Possible is that which either is or will be true " (23). This identification of the Possible with actual or potential happenings was challenged by the Stoics who defined the Possible as " that which is not prevented by anything from happening even if it does not happen " (24). At the same time the Stoics maintained that the Possible is part and parcel

(20) LUCRETIUS, de rerum natura, II, 217.
(21) PLUTARCH, de Stoic. repugn., 1045B.
(22) AETIUS, I, 29.
(23) PLUT., de Stoic. repugn., 1055E.
(24) PLUT., l. c. and ALEX. APHR., de Fato, ch. 10.

6

of " things happening according to fate ", and they regarded the Possible as an integral part of the causal set-up of the world. According to the meager information available on the subject (mainly ALEXANDER OF APHRODISIAS, *de fato*, the Stoics argued as follows : Let us suppose that there are two mutually exclusive possibilities, A and B; then the non-realization of A means the realization of B.   The same causal nexus, therefore, which led to the happening of B was the reason which prevented A from happening, and both A and B have their place in the causal scheme. The very fact, however, that A could be assumed as possible, although it did not happen, is attributed to our ignorance of the future, i.e. of the complete causal nexus.   It is this ignorance which gives meaning to the category of the Possible and which for a consistent determinist is the prerequisite of the existence of " equally possible cases ", of which one alone is going to take place. Here we see the Stoics taking the same attitude as the most consequential determinist of the modern era—LAPLACE, who defined equally possible cases as cases about whose existence we are in equal ignorance (25).

But the Stoics had still another approach to the problem, starting from the classification of propositions (26).   According to them, a proposition, even if it was true, need not be necessary.   The attribute " necessary " should be reserved for propositions which are *always* true, irrespective of time, i.e. logical (or mathematical) statements.   Propositions about future events, however, which lose their significance after the realization of the event (or after they have been proved false by developments) are possible propositions.   " Tomorrow there will be a naval battle ", therefore, is not a necessary proposition, even should it prove to be true, and the conclusion drawn by the Stoics is that the event itself, although completely determined, is not a necessary one but belongs, like all empirical facts, to the category of the Possible.

All these attempts of the Stoic School to show the compatibility of the Possible with determinism are one aspect of their tendency to point out the inductive character of our comprehension of the causal nexus.   A few words more might be added about this very

---

(25) LAPLACE, *Théorie analytique des probabilités*, (1812), p. 178.
(26) ALEXANDER APHROD., *l. c.*

important feature of Stoic epistemology. Although the universal
validity of causal law was axiomatic for the Stoics, they tried also
to give it experimental support. In this respect, their belief in
divination is of great interest. On the one hand they justified
divination by their hypothesis of universal causality, and on the
other hand every successful diviner gave additional support to
their belief in fate. The critical remarks of EUSEBIUS confirm
this attitude : " Chrysippus based his proof on the mutual interdependence of things. For he wants to show by the truth of
divination that everything happens in accordance with fate; but
he cannot prove the truth of divination without first assuming
that everything happens in accordance with fate " (27). This
indeed is the method of natural science whose epistemological
foundation was correctly grasped by the Stoics : every new instance
of inductive evidence strengthens the law of causality, while the
postulation of this law increases our confidence that any given
chain of events is not arbitrary. This empirical approach of the
Stoics is very well stated by CICERO (28), and we are confronted
with its implications in another context (29). Here we learn that
CHRYSIPPUS avoided giving conditional form to a molecular proposition related to the future. The instance discussed is a proposition based on the observation of astrologers : " If anyone was
born at the rising of the dog star, he will not die at sea ". Clearly
such a conditional proposition could be used for the formation of
the following argument : " If anyone was born at the rising of the
dog star, he will not die at sea; FABIUS was born at the rising of the
dog star; therefore, Fabius will not die at sea ". CHRYSIPPUS, in
order to avoid the commitment involved in such an apodictic
statement, gave the proposition the weaker form of a negated conjunction : " Not both—someone is born at the rising dog star and
he will die at sea ". CICERO makes sarcastic comments on this
careful approach by CHRYSIPPUS : " What proposition is there that
cannot in this way be changed from a conditional to a negated
conjunction ? " CICERO adds a few more examples of his own,
e.g. " not both—a person's pulse is high and he has not got a
fever " and " not both—two given circles on the sphere are the

---

(27) EUSEBIUS, *praepar. evang.*, IV, 3.
(28) CICERO, *De divinatione*, I, 110, 118.
(29) CICERO, *De fato*, 12-16.

greatest and they will not bisect each other ".   The last one proves that CICERO did not grasp the real motive of CHRYSIPPUS' formulation; in this case of a geometrical proposition CHRYSIPPUS himself would have stuck to the conditional form.

The development outlined above laid some of the essential foundations for the theory of probability.   Once it became evident that several mutually exclusive results are possible under certain conditions, the concept of the probability of a result could emerge as the quotient of the possibility of a certain result and the sum of all possibilities.   The mathematical definition would have followed through coordination of numbers characterizing the different results.   We know that the first probability calculations were carried out on problems of games of chance in the 16th Century by Tartaglia and Cardano in Italy and in the 17th Century by PASCAL and FERMAT in France.   TARTAGLIA was interested in the question of how many different throws (different sums of pips) are possible with 2, 3, etc. dice, and so combinatorial analysis was introduced into probability calculations.   CARDANO determined the probability of such a throw by calculating how many times a certain number, such as 7, can originate from the throw of two dice.   A hundred years later PASCAL determined the probability of throwing 6 with a single die at least once in n throws.   Obviously games of dice are especially suitable for probability calculations, and it is well known that games of chance, with the symmetrical die as well as with the asymmetrical astragalus, were in common usage throughout the whole period of Greek and Roman antiquity, in all strata of society.   SOCRATES relates (30) that upon entering the Palaestra on the festival of Hermes after the sacrifice he found the young people busy playing dice.   Many Roman emperors were passionate gamblers, and CLAUDIUS ~~even~~ wrote a book on the game of dice (31).   It is therefore surprising to discover that apparently games of chance did not lead to any probability calculations in Greek antiquity.

A term for mathematical combination (συμπλοκή) is occasionally used, but never in connection with dice.   The available information, though fragmentary, does not leave any doubt that the com-

---

(30) PLATO, *Lysis*, 206D.
(31) PAULY-WISSOWA, *Real-Encyclopaedie der klassischen Altertumswissenschaft:* " Lusoria tabula " and " Astragalomanteia ".

binatorial analysis of the Greeks was faulty. We learn (32) that "Xenokrates determined the number of syllables which are produced through mixing the letters of the alphabet up to 1,002,000 millions". We cannot check how Xenokrates arrived at this figure, as nothing is said about the maximum number of letters in a syllable or any other details relevant to the calculation. As there are few syllables with more than four letters, the figure given seems to be a gross exaggeration. Another example · is quoted by the same source and repeated, with small variations, elsewhere (33). We are told that CHRYSIPPUS calculated the number of molecular propositions which can be formed out of all the possible combinations of 10 atomic propositions. CHRYSIPPUS' figure (over a million) is rejected as grossly exaggerated and it is maintained that "the arithmeticians, among them Hipparchus" proved that the correct figure for positive propositions is 10I,049 and for negative ones 3I0,952. Again we cannot check the figures as we do not know how many molecular propositions were taken into account. But it is of interest to note that the whole story is told in connection with the discussion of the enormous possibilities of illnesses which may arise out of the combination of the many faculties of the human body with all the qualities introduced into it by various foodstuffs. Here we have an example with an empirical background hinting at the low probability of a correct diagnosis of an illness.

The fact that combinatorial analysis was not developed beyond these primitive beginnings can be partly explained by the lack of an algebraic notation which was a severe handicap to the whole of Greek arithmetic. However, the apparent absence of any theoretical approach to games of chance in Greek antiquity probably had other and more important reasons. Information on rules used for games of dice is scarce (34), but some conclusions may be drawn from it. Besides the dice whose six faces had the same notation as today, the astragalus (knuckle-bones taken from the hind-feet of sheep) was frequently used. Because of its peculiar oblong shape the astragalus can rest only on four faces, two of them broad and two narrow. Although the chance of the astra-

(32) PLUT., *Quaest., conv.*, 733A.
(33) PLUT., *De Stoic. repugn.*, 1047C.
(34) PAULY-WISSOWA, *l. c.*

galus falling on one of the broad faces is much greater than for its falling on one of the narrow ones, the values attributed to the narrow faces were 1 and 6 and to the broad ones 3 and 4. No account was apparently taken, in fixing the values of the faces, of the relative frequencies with which they turn up, otherwise (even disregarding asymmetries) the narrow faces should have counted 4 and 6 and the broad ones 1 and 3. However, no conclusions should be drawn from this arbitrary arrangement as to the lack of understanding of the probability nexus. But there seems to be significance in this fact if it is taken together with the arbitrariness which prevailed in the evaluation of the simultaneous throws of several knuckle-bones. Usually four bones were thrown together, and the most widely used rule attributed the highest value to the throw which showed different faces for each of the four bones. This throw (1, 3, 4, 6) which was called "Aphrodite" or "Venus" was therefore valued higher than all the other, less frequent throws out of the remaining 34 combinations, including the throw 6, 6, 6, 6, which has a much lower probability. There were also games where the throw with the highest pips won and there were apparently also rules fixing the value of the throw according to a scale which had nothing to do with the sum of the pips. But the rule giving the "Venus" throw the highest value was by far the most common, and the significance of this fact can hardly be overlooked in our present discussion.

We know nothing whatsoever about any probability assessment with regard to the results of future throws. More than anywhere else the belief in "personal factors", in good luck or bad luck seems to have prevailed in these games, or else the reliance on manual skill in influencing the result of the throw—cheating was very popular at all times. But one looks in vain for any references indicating that regularities in the recurrence of throws were observed or noted. Nothing seems to have been known of the law of large numbers, and there is only occasional mention of the obviously higher frequency of small recurrences and the practical impossibility of long series of identical throws. "To succeed in many things, or many times, is difficult; for instance, to repeat the same throw ten thousand times with the dice would be impossible, whereas to make it once or twice is comparatively easy"

11

says ARISTOTLE (35). And 300 years later CICERO gives two similar examples : " Can anything be an accident which bears upon itself every mark of truth ? Four dice are cast and a Venus throw results—that is chance; but do you think it would be chance, too, if in one hundred casts you made one hundred Venus throws ? " (36) And in another place he says : " Nothing is so uncertain as a cast of dice, and yet there is no one who plays often who does not sometimes make a Venus-throw and occasionally twice and thrice in succession. Then are we, like fools, to prefer to say that it happened by the direction of Venus rather than by chance ? " (37) Apparently in a period of 300 years no progress was made in probability considerations beyond these very vague and qualitative statements. This failure of Greek scientific thought to discover the mathematical elements of the regularity of chance is astonishing for two reasons which we can sum up now. Certain theoretical foundations had already been laid for a theory of probability, particularly by the Stoic teachings on disjunctions and the concept of the Possible. Furthermore, the great popularity of games of dice in all strata of society had created favourable conditions for the accumulation of observations which could have led to quantitative results.

One is tempted to look for more deep-rooted reasons for this failure to define the concept of mathematical probability, reasons connected with and depending on the cosmic outlook of the Greeks. Let us return for a moment to the passage in ARISTOTLE's *De Caelo* quoted above (35). The parallel of the dice appears there in connection with his reflections on the degrees of regularity in the heavenly motions. The nearer a thing is to perfection, the simpler and more regular its movements, as is the case with the outermost sphere which carries all the fixed stars. Things which are farther away from the perfect state need to perform many actions in order to reach perfection, and the more numerous the intermediate steps, the likelier they are to fail. In the sphere of human action success is usually granted only to less complex endeavours. Here ARISTOTLE introduces the quoted parallel. To repeat the same cast ten thousand times would be an event completely outside

---

(35) ARIST., *De Caelo*, 292ª29.
(36) CICERO, *De divinat.*, I, 23.
(37) *l. c.*, II, 121.

the frame of terrestrial happenings, because it would represent a perfection existing only in the sphere of celestial movements which reproduce themselves accurately over and over again. In contradistinction to these, terrestrial events are marked by a complex irregularity and by unpredictable vacillations. The rift created between heaven and earth by the philosophy of PLATO and ARISTOTLE firmly established in the consciousness of the Greeks the identification of every mathematically defined repetition or every recurrent series with the rigid regularity of the heavenly movements. " Things whose substance is imperishable will be numerically the same in their recurrence " (38). True, there are recurrent sequences in the sublunar region and on earth, such as the seasons, but they exist only in direct dependence on the eternal cyclical movements of the heavens. The idea, however, of regular sequences produced by man seemed a contradiction in terms, and the notion of a certain number or sequence of numbers returning with a frequency exhibiting some sort of regularity in repeated throws of the dice seemed an absurdity.

This inability to discern the laws of chance behind the seemingly chaotic series of random events is linked to another conspicuous feature of Greek science : the absence—with few exceptions—of systematic experimentation. An essential element of each experiment is its reproducibility, and since GALILEO the artificial production of reproducible series of phenomena, whereby certain factors are kept constant and others varied, has become the basis for the inductive comprehension of the laws of nature. Every experimental repetition of a sequence of phenomena increases our certainty that the same cause always produces the same effect. As a matter of fact, the astronomical phenomena which to a large extent formed the Greek's picture of the universe display all the ideal qualities of a laboratory experiment, because the astronomer is again and again presented by Nature with an unending sequence of the same initial conditions. But, with a few exceptions, it never occurred to the Greek to devise systematic repetitions in imitation of nature, to restore, by a " dissection of nature ", the same situation A which will give rise again to the same situation B. The unnatural device of a series of identical, man-made

---

(38) ARIST., *De Gen. et Corr.*, 338*b*14.

13

events, artificially isolated from a complex of interwoven processes, was foreign to his mind which conceived the Cosmos as an organic entity. The Stoics, although the first to postulate the idea of universal causality, did not—apart from heavenly phenomena— connect it with the notion of reproducibility. What they had in mind was a rigidly determined chain of events whose every link was connected with its neighbours both in the past and in the future. " Fate is an eternal and unalterable series of circumstances and a chain rolling and entangling itself through an unbroken series of consequences from which it is fashioned and made up " (39). There is no notion whatsoever of any verification of this postulate by repetition of artificially constructed series of phenomena (40).

It was not chance that the beginnings of systematic experimentation coincided with the inception of the theory of probability in the 16th and 17th centuries. Both have their origins in the new approach of Man to Nature which developed after the Renaissance. This approach involved the breakdown of the barriers between " heavenly " and " terrestrial " phenomena and laws. Once the " dissection of nature " had started, the results of experiments led to the conclusion that " terrestrial " series, too, exhibit regularities and that natural phenomena can be reproduced ad libitum and that they are in principle no different from the " heavenly " occurrences. By the same change of attitude the observation of repeated, man-made sequences of figures in games of dice brought about the discovery of the laws of chance. Here the development started from the observation of " artificial " sequences and expanded to that of natural ones; the drawing up and study of " mortality bills " led to the establishment of statistics.

Neither the technique of experimentation nor the theory of probability, two cornerstones of modern science, could develop in ancient Greece, because Greek mentality saw precision restricted to the heavens and failed to see laws in the fluctuations and irregularities on earth.

Hebrew University, Jerusalem.   S. SAMBURSKY.

---

(39) GELLIUS, *Noct. Att.*, VII, II, 1.

(40) Cf. S. SAMBURSKY, *The Physical World of the Greeks*, Routledge & Kegan Paul, London, 1956 (p. 233 ff.).

Note added in March 1975. Attention is drawn to my *Physics of the Stoics* (Routledge and Kegan Paul, London, 1959, pp. 71–80) in which some relevant points are further elaborated.

Biometrika (1969), **56**, 2, *p.* 437

Printed in Great Britain

Studies in the History of Probability and Statistics. XXII

Probability in the Talmud

By NACHUM L. RABINOVITCH

University of Toronto

SUMMARY

Typical examples of probability arguments are adduced from the Talmud to show that the ancient Rabbis formulated rules for adding, multiplying and comparing probabilities as well as criteria for sampling and estimating populations.

1. INTRODUCTION

Hasofer (1966) has reviewed some occurrences in the Talmud of the use of random mechanisms to secure a 'fair' outcome. In the present paper, probabilistic notions in the Talmud are investigated particularly as they occur in situations other than those involving a deliberate use of chance. Questions involving probability considerations occur frequently in the ancient rabbinic literature, and we shall show that the Rabbis computed probabilities in accordance with certain preconceived logical principles.

There are, in fact, two Talmuds. They consist of (*a*) a common basic text, the *Mishnah*, substantially completed in the second century, and (*b*) the *Gemara*, the discussions of the Rabbis on the text of the *Mishnah*. The *Jerusalem Talmud* was brought to essentially its present form by the end of the fourth century, the *Babylonian Talmud* about a century later. References to Talmudic texts are given in the standard form, i.e. tractate, folio number and side for the Babylonian Talmud, and, for the Jerusalem Talmud, tractate, chapter and section, just as for the Mishnah.

For a general introduction to the Talmuds, see Strack (1945).

2. 'FOLLOW THE MAJORITY'

A typical question involves objects whose identity is not known and reference is made to the likelihood that they derive from a specific type of source in order to determine their legal status, i.e. whether they be permitted or forbidden, ritually clean or unclean, etc. Thus, only meat which has been slaughtered in the prescribed manner is *kasher*, permitted for food. If it is known that most of the meat available in a town is *kasher*, there being, say, nine shops selling *kasher* meat and only one that sells non-*kasher* meat, then it can be assumed when an unidentified piece of meat is found in the street that it came from the majority and is therefore permitted.

However, the Talmud distinguishes between an enumerated majority and one that is just taken for granted *a priori*.

(Hullin 11*a*): Whence is derived the rule which the Rabbis stated: 'Follow the majority'?...As for a majority that is enumerated as in the case of the Nine Shops or the Sanhedrin we do not ask the question. Our question relates to cases where the majority is not explicit, as in the case of the Boy and Girl [who we assume will grow up to be fertile and are therefore not to be considered eunuchs with

15

respect to levirate marriage, although this cannot now be ascertained. The reason is that we follow the majority, and the majority of people are not sterile. Yet, this majority is not an explicit one since we cannot count all the people].

Not all the sages agreed to follow a majority when that is not actually enumerated. Rabbi Meir and others 'are concerned for the minority' (Gittin 2b), except where the uncounted majority is overwhelming. Such is the presumption that most of 'the scribes of the courts know the law' (Gittin 2b), for an ignorant scribe must be so rare that a legal document issued by the court may be assumed to have been properly executed.

Even for explicit majorities, it is important to determine whether the majority is relevant to the case in question.

(Kethuboth 15a): All that is stationary (fixed) is considered as half and half...If nine shops sell ritually slaughtered meat and one sells meat that is not ritually slaughtered and he bought in one of them and does not know which one; it is prohibited because of the doubt; but if meat was found [in the street], one goes after the majority.

The reasoning seems to be that when the question arises at the source, the chances are not really nine to one. For the other nine shops do not enter into the picture at all, since the piece of meat in question certainly does not come from any of them. Therefore there are only two possibilities and the chances of its being *kasher* or not must be considered even. However, if meat is found in the town at large, the chances that it comes from any one of the ten shops are equal and therefore the probability that it is *kasher* is 9/10. This is expressed in the rule: 'that which is detached, is detached from the majority'.

3. Addition and multiplication of probabilities

If a husband dies, leaving his wife pregnant, the probability that she will bear a live male child is a factor in the application of certain laws of inheritance. This probability must be less than one-half. For,

(Yevamoth 119a): A minority [of pregnant women] miscarry and of all the live births half are male and half female. Add the minority of those who miscarry to the half who bear females and the males are in a minority.

A 'minority of a minority' is defined in a discussion concerning cheese made with rennet derived from the stomachs of cattle.

(Avodah Zarah 34b): Since there would be a majority of calves that are not slaughtered for idolatry and then there are the other [adult] cattle [of whom none are slaughtered for idolatry and they are a majority as against all calves] it would be a minority of a minority and even Rabbi Meir is not concerned for a minority of a minority.

An interesting example is the following, where equal probabilities are multiplied to obtain a minority. This is called a 'doubt of a doubt' or 'double doubt', and is distinguished from a minority of a minority.

An adulteress is forbidden to live with her husband, so that her husband may divorce her without penalty. But this is not so if he claims that his bride committed adultery before the consummation of the marriage, even though it be established that she is not a virgin.

(Kethuboth 9a): It is a double doubt. It is a doubt: whether under him [i.e. during the period—usually one year—between formal betrothal and consummation] or not under him [i.e. prior to the betrothal]. And if you say that it was under him there is the doubt whether it was by violence or by her free will.

It is assumed that the chances are at most even that the incident occurred during the period of betrothal. But even if such were the case, since a woman violated by another man does not become forbidden to her husband, it can be argued that there is at least an even chance that she did not submit of her own free will. Thus the probability that the husband's claim is correct is only $\frac{1}{2} \times \frac{1}{2} = \frac{1}{4}$.

The commentators distinguish between a 'reversible' double doubt such as the above, where the probabilities are independent, i.e. there are four possible cases, and a 'non-reversible' one where the second doubt is conditional on the first so that there are only three distinct possibilities.

4. COMPARING PROBABILITIES

Ordinarily, when a forbidden object falls into a majority of similar objects that are permitted, the prohibition is annulled, so that if one draws any object it is permitted (Zevahim 72 *a*).

However, the prohibition of objects used in idolatrous worship is so grave that 'even one in ten thousand' makes them all prohibited (Zevahim 74 *a*). From that discussion we quote:

If an idol's ring was mixed up with [others making a total of] one hundred rings and forty of them were separated to one place and sixty to another...the forty detached to one place do not render others forbidden: the sixty in one place do render others forbidden.

Although the odds are 3:2 that the idol's ring is in the sixty, this is not sufficient reason to permit drawing from the forty, since on an individual draw from either group the probability is the same, 1/100, that the forbidden ring is chosen. Thus the remark in the same source:

Why is one from the forty different? [Presumably] because we say, The forbidden one is among the majority. Then in the case of one from sixty too we must say, The forbidden one is in the majority [i.e. the remaining fifty-nine]!

However, when at least one additional ring is added to each group the probabilities are no longer equal for an individual draw from each set. This is the meaning of the rule 'the forty... do not render others forbidden'.

While no computations are given in the text, it is remarkable that the division of 40:60 is chosen rather than 49:51 which would make a stronger point (cf. Pesahim 79 *b* ff.). However, a calculation of the different probabilities is revealing, and we can attempt to reconstruct what might be a plausible line of Talmudic reasoning.

Suppose to the set of X rings ($X < 50$), separated from the original mixture, is added at least one more ring (for the worst case) to make the set A. Similarly to those remaining from the original mixture add at least one to form the set B.

The probability p_A of drawing the idol's ring from the set A is

$$p_A = \frac{X}{100} \frac{1}{X+1}, \quad \text{while} \quad p_B = \frac{100-X}{100} \frac{1}{101-X}.$$

Clearly $p_B > p_A$.

Elsewhere, the Rabbis specify for an effective majority that it exceed the minority by a certain minimum (e.g. Berakhoth 48 *a*, Sanhedrin 2 *a*, Hullin 28 *a*, etc.). Thus, $(p_B - p_A)/p_A$ can be considered as an indicator of the greater likelihood that a single draw from B, rather than from A, gives the forbidden ring, and this must be of significant magnitude. We seek a lower bound for this ratio.

Now for large X, both p_A and p_B are close to 1/100; and, since 1 in 10,000 renders a mixture unfit, a probability of 1×10^{-4} relative to 1×10^{-2} is still decisive, i.e. if $(p_B - p_A)/p_A$ is near $1 \times 10^{-4}/1 \times 10^{-2}$, the set A can be permitted. In fact, the minimum significant probability must be somewhat less than 1×10^{-4}. Therefore the upper bound for X should not be the least value such that

$$L(X) = \frac{p_B - p_A}{p_A} \frac{1}{100} < \frac{1}{10,000}.$$

It suffices to choose the next larger X.

We get the following values:

X	38	39	40
$L(X)$	1/9975	$11/120,900 \simeq 1/10,991$	1/12,200

Therefore $X = 40$ is chosen as the maximum admissible value for which the set A can be permitted.

5. SAMPLING

The problem of drawing a typical example from a population is raised in several contexts. In general, there are two views; one maintains that two instances indicate a pattern, while according to another view, three are required, but there are some variations.

Tefillin are ritual objects which must be prepared according to exacting stipulations. One is worn on the arm and another on the head.

(Eiruvin 97a): If one acquires [bundles] of Tefillin from one who is not certified, he checks two for the arm and one for the head or two for the head and one for the arm...similarly, for the second and third bundles...Since each bundle was made by a [different] person...the third one is mentioned to show that there is no presumption with respect to the bundles...and even the fourth and fifth bundles [require sampling].

Thus it appears that a sampling is required to establish the reliability of each individual source. However, in the Jerusalem Talmud the question is left open;

(Eiruvin X—1): If one found two or three bundles, he checks a pair from the first bundle and similarly from the second and third. Isaac ben Elazar asked: Is there one pattern for all of them or must it be for each one separately? If you say one pattern for all of them, he checks the first pair from the first bundle, etc.; if you say each one has its own pattern, he checks three pairs in each bundle.

In dealing with major physical abnormalities, the Talmud rules that two recurrences of a phenomenon are to be considered a significant indication of variation. Rabbi Judah the Prince (2nd cent.) ruled

(Yevamoth 64b): [A mother] had one child circumcised and he died; a second one and he died; one must not circumcise the third.

The same rule applies if the children were not born to the same woman but the mothers are sisters for 'sisters establish a pattern' (ibid.). However, if there is only one unusual occurrence, it may be mere coincidence (cf. Kethuboth 87b).

To this day, in the absence of medical proof that the deaths are unrelated, Rabbi Judah's rule applies.

The size of the sample relative to the population was specified in the following instance.

(Taanith 21a): A town bringing forth five hundred foot-soldiers like Kfar Amiqo, and three died there in three consecutive days; it is a plague...A town bringing forth one thousand five hundred foot-soldiers like Kfar Akko, and nine died there in three consecutive days; it is a plague.

There are, however, no indications what was the proportion of foot-soldiers to the total population.

6. INCONCLUSIVE PROBABILITIES

Legal principles are applied when the probabilities are inconclusive.

An offering called *Terumah* is to be set aside from the grain crop. This is sacred and may be eaten only by priests, so that a mixture of *Terumah* and ordinary grain becomes forbidden to all except priests. However,

(Mishnah Terumoth VII–5): Two bins, one containing *Terumah* and one ordinary grain, and a bushel of *Terumah* fell into one of them but it is not known into which one; I say that it fell into the *Terumah*.

Thus the ordinary grain remains permitted. None the less there is a difference of opinion whether the same rule applies in the following circumstances:

(Yevamoth 82a): Two bins, one containing *Terumah* and one ordinary grain, and before them are two bushels one of ordinary grain and one of Terumah and the latter fell into the former, one into each.

Since it is known that each bushel fell into a different bin, the events are not independent and the probability is just 1/2 that the Terumah fell into the ordinary grain. The commentators explain that the rule permitting the grain in the first case is based on the legal principle that we assume the *status quo ante* unchanged if there is not sufficient contrary evidence. Since the chances are even that nothing fell into the bin of ordinary grain, our assumption stands. However, in the second case, one bushel did fall into each bin, and therefore one can take a more stringent view since the chances are even that it was the *Terumah* that fell into the ordinary grain, making it forbidden.

7. CONCLUSION

In the framework of Jewish Law, it is natural that probability considerations appear. Because the law requires making manifold decisions every day respecting duties and rights, it deals of necessity with empirical data which are by their very nature often incomplete. The essential question therefore is how convincing are the available data. This accounts for the very many instances of probabilistic reasoning in the Talmud.

In fact some of the rules cited above, and similar ones, originated well before the Talmudic era. On the other hand, most of them have been incorporated into the subsequent codifications of Jewish law up to our own times. However, in practice, their application is often modified by other considerations.

Thanks are due to Dr R. M. Fischler, who suggested the subject of this inquiry and made valuable comments.

REFERENCES

EPSTEIN, I. (1961), (Editor). *Babylonian Talmud in English Translation.* 18 vols. London: Soncino Press.
HASOFER, A. M. (1966). Random mechanisms in Talmudic literature. *Biometrika* **54**, 316–21.
STRACK, H. L. (1945). *Introduction to the Talmud and Midrash.* Philadelphia: Jewish Publication Society.

[*Received December* 1968. *Revised February* 1969]

Biometrika (1970), **57**, 1, *p.* 203
Printed in Great Britain

Miscellanea

Studies in the History of Probability and Statistics. XXIV
Combinations and probability in rabbinic literature

By NACHUM L. RABINOVITCH

University of Toronto

SUMMARY

Examples are cited from the rabbinic literature up to the fourteenth century illustrating computation of combinations, permutations and probabilities. It is shown that the rabbis had some awareness of the different conceptions of probability as a measure of relative frequencies or of a state of ignorance.

1. INTRODUCTION

In a previous paper (Rabinovitch, 1969) probabilistic notions occurring in the Talmud were described. The present paper explores the development and application of some of these concepts as well as the closely related subject of permutations and combinations in the rabbinic literature.

Of particular interest are the *Sefer Yetsirah* or *Book of Creation* and its commentaries. Containing less than 1600 words, much of the *Sefer Yetsirah* is devoted to enumerating various permutations of the letters of the alphabet. Critical opinion is divided as to when it was written. Gandz (1943, p. 159) dates it before the third century, i.e. early in the Talmudic period. Among the major commentaries are those by R. Saadya Gaon (892–942) and R. Shabbatai Donnolo (913–970). References to *Sefer Yetsirah* are to chapter and paragraph and those to the Talmud-tractate are to folio number and side and for the Babylonian, chapter and section for the Jerusalem Talmud.

2. COMBINATIONS AND PERMUTATIONS

For the author of *Sefer Yetsirah*, the twenty-two letters of the Hebrew alphabet represent the building blocks of creation. Thus

(I–13): He selected three [letters] and fixed them in His great Name—JHV, and sealed with them the six directions.

All six permutations of the three letters are then listed and each arrangement is made to correspond to one of the directions: up, down, east, west, south and north.

The number of combinations of the twenty-two letters taken two at a time is given as follows:

(II–4): Twenty-two element letters are set in a cycle in two hundred and thirty-one gates—and the cycle turns forward and backwards... How is that?... Combine A with the others, the others with A, B with the others, and the others with B, until the cycle is completed.

The actual combinations are tabulated by the commentators; for example, Donnolo, who points out that the description of one arrangement as 'forwards' and the other as 'backwards' refers to the fact that in 231 permutations the letters appear in the order of the alphabet, the earlier one in the alphabet followed by the later one, while in the 'backward' 231 permutations the order is reversed; also that the computation rule—'A with the others', etc., means that with A in the first place there are twenty-one pairs and similarly for every one of the twenty-two letters of the alphabet, so that the total number of arrangements is $22 \times 21 = 462$.

A more general rule is

(IV–12): Two stones [letters] build two houses [words], three build six houses, four build twenty-four houses, five build one hundred and twenty houses, six build seven hundred and twenty houses, seven build five thousand and forty houses; henceforward go and calculate...

To illustrate, Saadya actually computes the number of permutations of eleven letters, since the longest word in the scripture consists of eleven letters.

21

Donnolo gives the following proof for the rule that n letters can be arranged in $n!$ ways:

A single letter stands alone but does not form a word. Two form a word—the one preceding the other and vice versa—give two words, for twice one is two. Three letters form three times two—that is six. Four letters form four times six—that is twenty-four...and in this way continue for more letters as far as you can count. The first letter of a two-letter word can be interchanged twice, and for each initial letter of a three-letter word the other letters can be interchanged to form two two-letter words—for each of three times. And all the arrangements there are of three-letter words correspond to each one of the four letters that can be placed first in a four-letter word: a three-letter word can be formed in six ways, and so for every initial letter of a four-letter word there are six ways—altogether four times six making twenty-four words...and so on.

The problem of determining the number of permutations when some of the letters are identical is dealt with in *Sefer Yetsirah* commentaries only in connexion with the tetragrammaton, where two out of the four letters are the same. The number of arrangements is correctly given as 12 and all of them are listed.

Later, as is well known, R. Abraham Ibn Ezra (d. 1167) investigated combinations of n things taken r at a time and obtained the result $C_{n,r} = C_{n,n-r}$ (Sarton, 1931, vol. 2, p. 124).

3. Application to probability

The idea that in the absence of contrary evidence all permutations may be considered to be equally probable was applied in his Code by Maimonides (1135–1204) in a ruling dealing with the Biblical Numb. xviii. 15) obligation of the father to redeem his wife's first born male child by giving five pieces of silver to the priest. The male first born of donkeys too must be redeemed.

Rabinovitch (1969) showed that the Talmud assumes that for a live birth the chances are even that the child will be a boy or a girl. For each birth is an independent event. Suppose one or more women have given birth to a number of children and the order of the birth is unknown, nor is it known how many children each mother bore nor which child belongs to each. What is the probability that a particular woman bore boys and girls in a specified sequence? This hypothetical case is dealt with in the Talmud (Bekhoroth 49 a), as a parallel to a similar case referring to cattle, where it is not unusual for such a situation to arise.

Maimonides rules as follows (Laws of First Fruits XI–29, 30):

Two wives of different husbands, one primiparous and the other not, who gave birth to two males [and they were mixed up]—he whose wife is primiparous gives five pieces of silver to the priest, but if they bore a male and a female, the priest has no claim. [Since the probability is only $\frac{1}{2}$ that the mother giving birth for the first time had the boy.] If they bore two males and a female, the husband of the primipara, must give five pieces of silver, because he would be free in only two cases [1, if his wife bore the daughter only, and 2, if she bore twins; the daughter first followed by a son], while if his wife bore only male offspring [i.e. two cases for one male child and one case for both male children] he is obligated, and if she bore a male and a female he is obligated as well, unless the female was first. Since the chance is remote [two cases versus four], he must pay redemption.

4. Empirical probabilities

The question whether a probability measure can be assigned in case of ignorance seems to be at issue in the following discussion in the Jerusalem Talmud.

(Bava Kama IV, 1): We learnt: Three put [coins] into a money-bag, and part of it was stolen [the remainder is divided between them proportionately].

Is that how they divide it? Have we not learnt concerning these stones [of a collapsed two-storey house and each storey belonged to another] that if [some of the building stones] were stolen—half is of the one and half of the other [regardless of the proportions of ownership]? Said R. Shammai (4th cent.), Building stones are large [and when one takes them, it is not at random] and it is not known whether he took from this one's or from that one's, and because of doubt—half is of the one and half of the other. But coins are small and it is possible to mix them up—to do justice to all, each one takes according to his investment.

On what grounds do you say that we consider the stolen ones [and each owner suffers half the loss in the case of building stones]; perhaps we should consider the remainder [and each will get a half of that]?

R. Shammai's proposal sounds like the modern 'Principle of Indifference' by virtue of which each proposition of whose correctness we know nothing is regarded as having probability $\frac{1}{2}$, for the proposition and its contradiction can be considered to be two equally probable cases. To counter R. Shammai's 'Principle of Indifference', the Talmud shows that it leads to a paradox reminiscent of Bertrand's Paradox.

In the case of a bride accused of adultery before the consummation of the marriage, the Talmud (Kethuboth 9a) argues that the chances are at most even that the incident occurred during the period of betrothal and even in that case there is an even chance that she was violated against her will. One of the early commentators points out that rape is known to occur infrequently. However, this is countered by the objection that even so the probability of adultery is still less than half in view of the first doubt. In discussing this question R. Isaac bar Sheshet (1326–1408) makes some interesting observations on the difference between probabilities based upon *a priori* laws or observed relative frequency and those which reflect ignorance.

> (Responsum 372): [When the Talmud says that] half of all [children born] are male and half female, it is certain and necessary, for thus did the King of the Universe establish it for the preservation of the species. Therefore of necessity, of all pregnant women those who bear males are a minority, for some abort, and this is inescapable. But, here we cannot say that of all those who have illicit relations—half do so after betrothal and half before...for whence do we know that it is half and half? It is only that we say the matter is in doubt, for the one or the other is possible...therefore the willing adulteress after betrothal is not certainly in the minority, and we can still say the one or the other is possible...and the doubt still exists.

He is prepared to accept that the probability of rape is less than $\frac{1}{2}$. In fact much smaller, because that is a conclusion based on past experience. As for the equal distribution of boys and girls, this is indicated not only by observation alone, but also seems to reflect natural law. These probabilities, then, represent observed relative frequencies. However, with respect to the question of when the incident occurred, R. Isaac bar Sheshet makes it clear that he regards any assignment of a probability measure as merely tentative, since it is based upon our ignorance.

REFERENCES

EPSTEIN, I. (ed.) (1961). *Babylonian Talmud in English Translation*, 18 vols. London: Soncino Press.

GANDZ, S. (1943). *Saadia Gaon as a Mathematician*, A.A.J.R. Texts and Studies, vol. 2, 141–193.

RABINOVITCH, N. L. (1969). Probability in the Talmud. *Biometrika* **56**, 437–41.

SARTON, G. (1931). *Introduction to the History of Science*, 3 vols.

[*Received June* 1969. *Revised September* 1969]

CORRESPONDENCE

The Joint Editors
The Journal of the Institute of   10 *March* 1954
  Actuaries Students' Society

A Budget of Paradoxes

Sirs,

It is quite common nowadays for actuarial and statistical authors to embellish their publications with the *cachet* of a historical reference. A harmless enough foible, to be sure, but it has its dangers. For all too often it turns out that a widely quoted statement about early work can be traced back to a single author's misinterpretation of the original source.

This contention may be illustrated by two frequently quoted 'inventions': (i) the measurement of mortality by contemporaries of Ulpianus, the Roman jurist who died A.D. 228, and (ii) the (continuous) Normal curve of errors by Abraham de Moivre. Neither of these claims stands up too well under close scrutiny.

Let us first consider the so-called *lex Falcidia* which was passed in Rome in 40 B.C. According to Braun (1921) it was intended to exercise some rein on the growing practice of disinheriting the heirs of estates. It stipulated that no testator could leave less than one-quarter of his estate to his legal heir.

In his A.D. 230 commentary on the law (see, for example, Charond, 1575), Aemilius Macer reported that it had been the custom to value a life tenant's annuity by multiplying it by a certain number of years m_x, where

$$m_x = \begin{cases} 30 & x < 30 \\ 60 - [x] & 30 \leqslant x \leqslant 60. \end{cases}$$

If the legatee were a municipal community the number of years to be used was 30. (Those who enjoy 'writing history backwards' might argue that this indicates that the underlying interest rate was 3·3%—cf. Hodge *J.I.A.* **6**, 315.)

However, Macer reports, some years earlier the jurist Ulpianus had laid down a new rule. His multiplicative factor n_x was defined by

$$n_x = \begin{cases} 30 & x < 20 \\ 28 & 20 \leqslant x < 25 \\ 25 & 25 \leqslant x < 30 \\ 22 & 30 \leqslant x < 35 \\ 20 & 35 \leqslant x < 40 \\ 59 - [x] & 40 \leqslant x < 50 \\ 9 & 50 \leqslant x < 55 \\ 7 & 55 \leqslant x < 60 \\ 5 & 60 \leqslant x. \end{cases}$$

This fragmentary commentary by Macer* is the only reference that has ever been found to mortality estimation in Roman times (van Haaften, 1943). And although Hodge (*loc. cit.*) states that the legal rate of interest in Rome was 4% in A.D. 230, there is no evidence that the Romans could calculate the present value of a number of future monetary payments. It is even more doubtful whether the crude arithmetic of Roman times would have extended to the calculation of an expectation of life from mortality data. Remember that the decimal system was only introduced into Europe in the twelfth century, that a proper theory of compound interest was not available until the fifteenth (Braun, 1925), and that the first published evidence of the 'concept of the mortality table'—l_x at integral ages—is to be found in Graunt (1662) (see, for example, Westergaard, 1901†).

Major Greenwood (*J.R. statist. Soc.* **103**, 1940, pp. 246–8) has referred to Ulpian's table as a 'statistical mare's nest'. A year

* Most of the Latin text is to be found in Trenerry (1926, p. 151). Although this doctoral thesis is very useful as a source-book, many of its conclusions are of the since-A-knew-the-meaning-of-x-he-must-have-understood-how-to-calculate-y variety.

† 'It should not be wondered at that this first attempt to find a mortality table gave only a most imperfect representation of mortality. But in spite of its imperfection it is right to designate this investigation as pioneering, not only in mortality but in the whole field of statistics. Graunt was the first who attempted to draw conclusions from statistical data....'

later he wrote (*Biometrika*, **32**, 101–27): 'There is not, I think, any reason to believe that the practical Romans had anticipated Graunt and Petty'.

We turn now to Simpson who is sometimes referred to as the author of a triangular curve of errors.

Simpson's work (1756,* 1757) should be particularly interesting to statisticians since it (1) is the first example of the application of probability theory to errors of measurement, and (2) contains the first reference to a continuous probability law. With the exception of some introductory sentences, the whole of the 1756 article is repeated in the 1757 publication. I have described most of the mathematical content of these papers elsewhere (1949) but the following additional remarks are relevant to the discussion, later in this letter, of the invention of the Normal curve of errors.

Simpson starts his 1757 paper by making two 'suppositions' about the 'errors arising from the imperfection of instruments and of the organs of sense':

'1. That there is nothing in the construction, or position of the instrument whereby the errors are constantly made to tend the same way, but that the respective chances for their happening in excess, and in defect, are either accurately, or nearly, the same.

2. That there are certain assignable limits between which all these errors may be supposed to fall; which limits depend on the goodness of the instrument and the skill of the observer.'†

The articles then consist of two propositions. In modern terminology the first of these derives the probability distribution of a sample mean of *n* observations from a *discrete* rectangular universe; the second provides the distribution of a sample mean from a discrete symmetric triangular universe—'much better adapted than if all the terms were to be equal'.

* This article was in the form of a letter to the Earl of Macclesfield and was dated 4 March 1755. There is no reference in it to a 'curve' of errors nor to a triangle.

† The copy of Simpson's *Tracts* in the library of Yale University is bound up with some of that author's other publications and, by a coincidence, bears on the flyleaf the ink signature: Francis Baily 1803. A pencil note indicates that it was purchased on 9 June of that year for £1. 11s. 6d.

The 1756 article concludes with some numerical examples on the discrete triangular law. However, in his second paper Simpson goes on to 'show how the chances may be computed, when the error admits of any value whatever, whole or broken, within the proposed limits, or when the result of each observation is supposed to be *accurately* known'. To do this he allows the number of discrete ordinates of the distribution law to increase without limit, the range of the abscissae remaining unchanged. This limiting argument is illustrated by reference to a geometric figure—an isosceles triangle.

The 1757 article concludes with a problem which clearly shows that Simpson was aware that his limiting procedures had resulted in continuous error laws. He asks, in fact, what is the probability that $\bar{x}$, the mean of n observations from a continuous symmetric triangular law with (true) mean equal to zero (and a range of two units), numerically exceeds the value of a single observation, x.

Simpson writes (Seal, 1949), in effect,

$$Pr\left(|\bar{x}-x|>0\right)$$

$$=2\int_0^1 Pr(x=\xi)\,Pr(|\bar{x}|>\xi)\,d\xi$$

$$=2\int_0^1 (1-\xi)\,\frac{2n^{2n}}{(2n)!}\sum_{j=0}^{[n-n\xi]}(-1)^j\binom{2n}{j}\left(1-\xi-\frac{j}{n}\right)^{2n}d\xi$$

$$=\frac{4n^{2n}}{(2n)!}\sum_{j=0}^{n}(-1)^j\binom{2n}{j}\int_0^1 z\left(z-\frac{j}{n}\right)^{2n}dz.$$

and proceeds to carry out the definite integration. Readers of this *Journal* will, no doubt, quickly perceive the grave errors involved in *two* of these three supposed identities.

Our statement that Simpson was the originator of the continuous probability distribution is in conflict with the oft-repeated statement that De Moivre invented the Normal curve of errors.

The theorem on which these allegations are based is to be found in the second (1738) and third (1756) editions of De Moivre's *The Doctrine of Chances*. As Pearson (1924) points out, it was first published in 1733 as a second Supplement to De Moivre's *Miscellanea Analytica* (1730).

This theorem provides an approximation to the sum of a

27

number of terms of the discrete binomial distribution law. It is derived by De Moivre in the following steps:

$$\sum_{r=np-l}^{np+l} \binom{n}{r} p^r (1-p)^{n-r} \sim \sum_{k=-l}^{l} \{2\pi np(1-p)\}^{-\frac{1}{2}} \quad \exp[-k^2/2np(1-p)]$$

$$= \{2\pi np(1-p)\}^{-\frac{1}{2}} \sum_{k=-l}^{l} \sum_{j=0}^{\infty} (-1)^j k^{2j}/2^j n^j p^j (1-p)^j j!$$

$$\doteq 2\{2\pi np(1-p)\}^{-\frac{1}{2}} \sum_{j=0}^{\infty} (-1)^j l^{2j+1}/(2j+1) 2^j n^j p^j (1-p)^j j!,$$

where summation has been replaced by (approximate) integration but only *after* the exponential function has been expanded. De Moivre works out some numerical illustrations of this approximation but there is nowhere any indication that, at the first step above, the result is a continuous probability law with an overall integral of unity. (In fact, the integral $\int_{-\infty}^{\infty} e^{-x^2} dx$ seems to have been evaluated by Laplace for the first time in 1778.)

In my opinion it is unjustified to see in the above series of approximations, the discovery of the Normal *curve of errors*. Neither 'error' nor 'curve' is involved in De Moivre's theorem.

Yours faithfully,

HILARY L. SEAL

Yale University
New Haven
Connecticut

REFERENCES

BRAUN, H. (1921). Vom Rentenwesen im Mittelalter bis zur Berechnung des Rentenbarwertes. *Verzekerings-Archief*, **2**, 209–37.

BRAUN, H. (1925). *Geschichte der Lebensversicherung und der Lebensversicherungs-technik*. Nurenberg.

CHAROND, L. (1575). *Dn. Sacratissimi Principis Justianiani PP.A. Juris enucleati ex omni vetere iure collecti Digestorum seu Pandectarum*. Antwerp.

GRAUNT, J. (1662). *Natural and Political Observations upon the Bills of Mortality....* London.

PEARSON, K. (1924). Historical note on the origin of the normal curve of errors. *Biometrika*, **16**, 402–4.

SEAL, H. L. (1949). The historical development of the use of generating functions in probability theory. *Mitt. Ver. schweiz.-Math.* **49**, 209–28.

SIMPSON, T. (1756). On the Advantage of taking the Mean of a Number of Observations, in practical Astronomy. *Phil. Trans.* **49**, 82–93.

SIMPSON, T. (1757). An Attempt to shew the Advantage arising by Taking the Mean of a Number of Observations, in practical Astronomy. *Miscellaneous Tracts on Some curious, and very interesting Subjects in Mechanics, Physical-Astronomy, and Speculative Mathematics*. London.

TRENERRY, C. F. (1926). *The Origin and Early History of Insurance, including the Contract of Bottomry*. London.

VAN HAAFTEN, M. (1943). Lijfrentekoopsommen bij de Romeinen in de tweede eeuw na Christus. *De Verzek. bode*, **62**, no. 24.

WESTERGAARD, H. (1901). *Die Lehre von der Mortalität und Morbilität*. Jena.

From Phil. Trans. (1710) 27, 186-90.

II. *An Argument for Divine Providence, taken from the constant Regularity observ'd in the Births of both Sexes. By Dr.* John Arbuthnott, *Physitian in Ordinary to Her Majesty, and Fellow of the College of Physitians and the Royal Society.*

AMong innumerable Footsteps of Divine Providence to be found in the Works of Nature, there is a very remarkable one to be observed in the exact Ballance that is maintained, between the Numbers of Men and Women ; for by this means it is provided, that the Species may never fail, nor perish, since every Male may have its Female, and of a proportionable Age. This Equality of Males and Females is not the Effect of Chance but Divine Providence, working for a good End, which I thus demonstrate :

Let there be a Die of Two sides, M and F, (which denote Cross and Pile), now to find all the Chances of any determinate Number of such Dice, let the Binome M + F be raised to the Power, whose Exponent is the Number of Dice given ; the Coefficients of the Terms will shew all the Chances sought. For Example, in Two Dice of Two sides M + F the Chances are $M^2 + 2 MF + F^2$, that is, One Chance for M double, One for F double, and Two for M single and F single; in Four such Dice there are Chances $M^4 + 4 M^3 F + 6 M^2 F^2 + 4 MF^3 + F^4$, that is, One Chance for M quadruple, One for F quadruple, Four for triple M and single F, Four for single M and triple F, and Six for M double and F double; and universally, if the Number of Dice be *n*, all their Chances will be expressed in this Series

$$M^n +$$

$$M^0 + \frac{n}{1} \times M^{n-1}F + \frac{n}{1} \times \frac{n-1}{2} \times M^{n-2}F^2 + \frac{n}{1} \times \frac{n-1}{2} \times \frac{n-1}{3} \times M^{n-3}F^3 +, \text{ \&c.}$$

It appears plainly, that when the Number of Dice is even there are as many M's as F's in the middle Term of this Series, and in all the other Terms there are moſt M's or moſt F's.

If therefore a Man undertake with an even Number of Dice to throw as many M's as F's, he has all the Terms but the middle Term againſt him ; and his Lot is to the Sum of all the Chances, as the coefficient of the middle Term is to the power of 2 raiſed to an exponent equal to the Number of Dice : ſo in Two Dice his Lot is $\frac{2}{4}$ or $\frac{1}{2}$, in Three Dice $\frac{6}{16}$ or $\frac{3}{8}$, in Six Dice $\frac{20}{64}$ or $\frac{5}{16}$, in Eight $\frac{70}{256}$ or $\frac{35}{128}$, &c.

To find this middle Term in any given Power or Number of Dice, continue the Series $\frac{n}{1} \times \frac{r-1}{2} \times \frac{n-2}{3}$, &c. till the number of terms are equal to $\frac{1}{2}n$. For Example, the coefficient of the middle Term of the tenth Power is $\frac{10}{1} \times \frac{9}{2} \times \frac{8}{3} \times \frac{7}{4} \times \frac{6}{5} = 252$, the tenth Power uf 2 is 1024, if therefore A undertakes to throw with Ten Dice in one throw an equal Number of M's and F's, he has 252 Chances out of 1024 for him, that is his Lot is $\frac{252}{1024}$ or $\frac{63}{256}$, which is leſs than $\frac{1}{4}$.

It will be eaſy by the help of Logarithms, to extend this Calculation to a very great Number, but that is not my preſent Deſign. It is viſible from what has been ſaid, that with a very great Number of Dice, A's Lot would become very ſmall ; and conſequently (ſuppoſing M to denote Male and F Female) that in the vaſt Number of Mortals, there would be but a ſmall part of all the poſſible Chances, ior its happening at any aſſignable time, that an equal Number of Males and Females ſhould be born.

It is indeed to be confeſſed that this Equality of Males and Females is not Mathematical but Phyſical, which alters much the foregoing Calculation ; for in this Caſe

the

the middle Term will not exactly give A's Chances, but his Chances will take in some of the Terms next the middle one, and will lean to one side or the other. But it is very improbable (if mere Chance govern'd) that they would never reach as far as the Extremities: But this Event is wisely prevented by the wise Oeconomy of Nature ; and to judge of the wisdom of the Contrivance, we must observe that the external Accidents to which Males are subject (who must seek their Food with danger) do make a great havock of them, and that this loss exceeds far that of the other Sex, occasioned by Diseases incident to it, as Experience convinces us. To repair that Loss, provident Nature, by the Disposal of its wise Creator, brings forth more Males than Females ; and that in almost a constant proportion. This appears from the annexed Tables, which contain Observations for 82 Years of the Births in *London*. Now, to reduce the Whole to a Calculation, I propose this

Problem. A lays against B, that every Year there shall be born more Males than Females : To find A's Lot, or the Value of his Expectation.

It is evident from what has been said, that A's Lot for each Year is less than $\frac{1}{2}$; (but that the Argument may be stronger) let his Lot be equal to $\frac{1}{2}$ for one Year. If he undertakes to do the same thing 82 times running, his Lot will be $\overline{\frac{1}{2}}|^{82}$, which will be found easily by the Table of Logarithms to be $\dfrac{1}{4\,8360\,0000\,0000\,0000\,0000\,0000}$

But if A wager with B, not only that the Number of Males shall exceed that of Females, every Year, but that this Excess shall happen in a constant Proportion, and the Difference lye within fix'd limits ; and this not only for 82 Years, but for Ages of Ages, and not only at *London*, but all over the World ; (which 'tis highly probable is Fact, and designed that every Male may have a Female of the same Country and suitable Age) then A's Chance will be near an infinitely small Quantity, at least

less

lefs than any affignable Fraction. From whence it follows, that it is Art, not Chance, that governs.

There feems no more probable Caufe to be affigned in
Phyficks for this Equality of the Births, than that in
our firft Parents Seed there were at firft formed an equal
Number of both Sexes.

Scholium. From hence it follows, that Polygamy is
contrary to the Law of Nature and Juftice, and to the
Propagation of Human Race; for where Males and
and Females are in equal number, if one *Man* takes
Twenty Wives, Nineteen Men muft live in Celibacy,
which is repugnant to the Defign of Nature; nor is it
probable that Twenty Women will be fo well impreg-
nated by one Man as by Twenty.

Christened.				Christened.		
Anno.	*Males*	*Females.*		*Anno.*	*Males.*	*Females.*
1629	5218	4683		1648	3363	3181
30	4858	4457		49	3079	2746
31	4422	4102		50	2890	2722
32	4994	4590		51	3231	2840
33	5158	4839		52	3220	2908
34	5035	4820		53	3196	2959
35	5106	4928		54	3441	3179
36	4917	4605		55	3655	3349
37	4703	4457		56	3668	3382
38	5359	4952		57	3396	3289
39	5366	4784		58	3157	3013
40	5518	5332		59	3209	2781
41	5470	5200		60	3724	3247
42	5460	4910		61	4748	4107
43	4793	4617		62	5216	4803
44	4107	3997		63	5411	4881
45	4047	3919		64	6041	5681
46	3768	3395		65	5114	4858
47	3796	3536		66	4678	4319

B b

Christened.

(190)

Anno.	Christened. Males.	Females.	Anno.	Christened. Males.	Females.
1667	5616	5322	1689	7604	7167
68	6073	5560	90	7909	7302
69	6506	5829	91	7662	7392
70	6278	5719	92	7602	7316
71	6449	6061	93	7676	7483
72	6443	6120	94	6985	6647
73	6073	5822	95	7263	6713
74	6113	5738	96	7632	7229
75	6058	5717	97	8062	7767
76	6552	5847	98	8426	7626
77	6423	6203	99	7911	7452
78	6568	6033	1700	7578	7061
79	6247	6041	1701	8102	7514
80	6548	6299	1702	8031	7656
81	6822	6533	1703	7765	7683
82	6909	6744	1704	6113	5738
83	7577	7158	1705	8366	7779
84	7575	7127	1706	7952	7417
85	7484	7246	1707	8379	7687
86	7575	7119	1708	8239	7623
87	7737	7214	1709	7840	7380
88	7487	7101	1710	7640	7288

III.

(Reproduced by permission of the Royal Society)

34

MEASUREMENT IN THE STUDY OF SOCIETY

Maurice G. Kendall
(1972)

It is in the spirit of these lectures to look back over the past in order to appreciate the present. That I shall do. But since the subject of this lecture, 'Measurement in the Study of Society', has been revolutionized in almost every respect in the last twenty years and therefore lies mostly in the future, I shall devote a considerable part of the lecture to the work which still remains to be done. The break with tradition in the study of societal organization in the last two decades has been extensive and profound, and somewhat paradoxically perhaps, I must, in order to illustrate the nature of that break, go further back in time and take a longer run at my subject than some of my colleagues in this series.

History never seems to begin in a clean and satisfactory way, but we may perhaps start from a lively figure of the early eighteenth century, John Arbuthnot, mathematician, Doctor of Medicine, Fellow of the Royal Society, satirist (he was the inventor of the character of John Bull), and friend of Pope and Swift, a man of whom Swift said that if there were a dozen like him he could burn *Gulliver's Travels*. Arbuthnot became interested in one of the first sociological laws to be discussed in scientific terms, the constancy of the sex ratio of births and its slight but significant departure from equality. Among other things he was physician to Queen Anne, which unfortunate woman had seventeen children, all of whom died before her accession to the throne. Whether or not this fact stimulated Arbuthnot's interest in the subject I do not know; but he published in 1710 an essay entitled 'An Argument for Divine Providence, taken from the constant regularity observed in the births of both sexes'.

The demographic laws of birth and death are so familiar to us nowadays that we find it hard to imagine how ignorant our fore-fathers were only two or three centuries ago. In the Middle Ages there were, in general, only two reasons for counting anything relating to human society; one was to find out how many men could bear arms, and the other was to ascertain how much money could be levied by way of tax. Apparently it was fourteenth-century Venice which has the credit of first recognizing women and children as possessing human souls which were worth count-ing as members of her community. Even in Elizabethan times there seems to have been doubt whether males and females were born in equal numbers. Arbuthnot's work was therefore far from being the rather slight contribution to knowledge that it may now appear. He was one of the early statisticians, a man who not only observed quantitative phenomena, but applied the laws of chance to them and, in fact, he has an honourable place in the memory of statisticians as the first man to set up a formal test of a statistical hypothesis.

However, it is not his statistical work which I wish to empha-size, but the explanation he gave for his phenomena. He proved that there were more males born than females, in London at least, but that as males were prone to risk in their early days, the sex ratio at the marriageable age was about equal. As the title of his pamphlet indicates, he regarded this admittedly very remarkable fact as due to Divine Providence. This same explanation persisted through a great part of the eighteenth century. Whether one approved of the state of society or not, whether the Supreme Power took an interest in each individual event or merely wound up the universe like a clock and left it to run under predetermined laws, the ultimate outcome was that the arithmetical features of the world were laid firmly at the door of the Almighty. The concept was expanded and developed by a number of people, notably a German pastor, one J. P. Suessmilch, whose book, the *Göttliche Ordnung*, 'Divine Order as proven by birth, death and fertility of the human species', published in 1741, is an early landmark in the sociological interpretation of statistical observa-tion. Suessmilch also was impressed by the fact that the sexes are in balance at the marriageable age; surely a demonstration that

36

the Creator thoroughly approves of a monogamistic reproduction of the species.

The development of political arithmetic during the eighteenth century brought into prominence more and more statistical regularities in human behaviour, in population, in trade, the incidence of disease, and so forth. The statistical accounts of the time read more like a cross between a Baedeker guide and *Whitaker's Almanack* than our current *Statistical Digests*. But more and more material became available in tabular form and social comparisons were possible over an ever-widening geographical domain. Population census were systematically conducted; the first in Britain occurred in 1801, although I think Iceland had one somewhat earlier. Life assurance became a science. Costs of living were measured by means of index numbers. The kind of enumeration and measurement which we should nowadays call descriptive statistics became a recognized part of government duties. But at the same time doubts were raised whether all this could be or should be blamed on to the Almighty. It seemed that social conditions could be altered by man himself, either by slow evolution or by revolution. The degree to which they should be changed and the manner of changing them have generated acute conflicts ever since.

We now enter on a long period during which a succession of men and women were drawn by their sociological interests into the measurement of social phenomena. They came from all kinds of disciplines. Lagrange, who suggested the first sample social survey round about the year 1790, was a mathematician. Quetelet, one of the greatest names in the subject, after a few false starts as painter, poet, and literary critic, succumbed to mathematics and became an astronomer. Le Play was a mining engineer. Charles Booth was a shipowner. Florence Nightingale was a nurse. Patrick Geddes was a botanist. I mention only a few of the formative individuals; it is no part of my present purpose to catalogue them all. But whatever their origins, whatever class of society from which they were drawn, however they acquired an interest in sociological matters, they were all, I think, motivated by two things; one was a profound dissatisfaction with things-as-they-were and a determination to change them; the other was a realization that it was necessary to measure social phenomena in some scientific way in order to clinch their sociological arguments.

Now statistics are well-known, in O. Henry's phrase, to be the lowest grade of information known to exist, and the compilation of statistical facts is often regarded as a necessary kind of drudgery but a drudgery none the less. In its practitioner it requires a degree of dedication that only another statistician could admire or wish to emulate. Down the years there have been continual protests from humanitarians that statistics can be twisted and misinterpreted and in any case give no true insight into the problems of humanity. The statistician, in fact, like the physical scientist, is pictured as a person devoid of human emotion. It may come as a surprise, then, to find that the early social scientists were dominated by an intensity of emotion which puts our current angry brigades in the shade. Florence Nightingale was described by her biographer as a passionate statistician. She had good reason to be, when the mortality of the British Army at the Crimea was 60 per cent per annum and that due to disease alone, apart from casualties due to enemy action. But she was far from being alone in her horror at the appalling social conditions which confronted her or in her determination to study them numerically.

The situation in which the social reformers of the nineteenth century found themselves differs radically from the one which confronts us today. They had no need to set up elaborate sociological hypotheses. All around them were poverty, slums, crime, disease, illiteracy, drunkenness, immorality. As they saw it these things were caused by man and it was man's duty to get rid of them. Social evil was so manifest to the Victorians that diagnosis was hardly necessary; it was the remedies that must be urgently sought. A few dissonant voices, of course, expressed the view that the poor had only themselves to blame, or that they were a regrettable but essential part of society, but the least altruistic men of the time could hardly avoid the mixture of guilt and compassion which was aroused by social studies. Nor, as it seems to me, was there much difference of opinion about the relative priorities of these various aspects of social degradation. The primary target was poverty. If that could be abolished much if not all social evil, crime and disease in particular, would disappear with it. 'I hate the poor,' said Bernard Shaw, 'I hope to live to see them abolished.' It was a holy simplicity of social aim which we can only envy in the light of our current problems.

There were two problems of present relevance which faced the sociologists of the nineteenth century. They are both methodological. One was to decide what to measure; the other to find some way of measuring it.

In all statistical work there is an important distinction between counting and measuring. It is relatively easy to count births, deaths, the number of tubercular patients, the number of indictments for murder, the numbers of persons paying income tax, and so forth. A great deal of sociological data is still based on head-counting, and rightly so. But measurement is a different thing, especially in sociology. We must remember that whereas in the physical sciences everything can be expressed in terms of a few fundamental units, mass, length, time, and electric charge (which, whatever the philosophical problems involved, can in practice be measured with a high degree of precision) the behavioural sciences are constantly struggling to measure, or at least to express in quantifiable form, concepts which are not directly observable such as utility, welfare, health, moral standards, opinions, and attitudes of mind.

These problems are still with us. They are of three kinds.

1. First of all, it is not obvious that a particular word which we currently use to describe social circumstances corresponds to something which is even theoretically measurable. The fact that we live in a welfare state and spend a lot of effort in maximizing welfare does not mean that we can measure a quantity of welfare.

2. It is therefore often necessary to measure quantities which are known to be associated with these indefinite concepts and are in fact observable. But this raises problems of choosing the right observables, and very often there are too many of them. We are, in fact, only just beginning to learn how to handle these multivariate complexes.

3. Thirdly our units of measurement have the embarrassing property of being impermanent. The standard case is that of prices. One of the few things that the variables of economics have in common is that of value as expressed in price, and we

know only too well what can happen to prices even over a relatively short period.

The early sociologists saw these difficulties clearly enough and it is to their credit that they were not thereby deterred from a quantitative assault on their problems. An early example will illustrate the nature of the problem and the lines along which they attempted a solution. Pierre Le Play (1806–82), engineer, economist, and senator, was one of the earliest sociologists to insist on empirical studies of the human environment. It was he who was mainly responsible for the family budget as a social instrument, and he was in a position to devote considerable resources to budgetary analysis. The budget, in fact, was to him a supreme measure of social effects. 'There is nothing,' he says, 'in the existence of a worker, no sentiment and no action worth mentioning which does not leave a clear market trace in the budget.' But Le Play was what might be described as a statistician *malgré lui*. He distrusted any kind of generalization which lost sight of the individual human being. He quantified case studies but was flatly opposed to statistical summarization. It was Ernst Engel, another mining engineer by training, who, basing himself in part on Le Play's data, formulated the well-known law that the proportion of income spent on food increases as the family income decreases.

Le Play is, in fact, an interesting example of how a man can evolve important research techniques on completely false hypotheses about the phenomena which he is studying. He was deeply concerned with social conditions, but none of the evils which were so evident to him was attributed to social evolution. As he saw it some men were evil because they were born that way. Others were, he believed, corrupted by false ideas, especially those of Rousseau that man was born good and of Thomas Jefferson that they had equal rights. Teachers, he maintained, do harm if they do not consider themselves as supplementary agents of parental authority; otherwise they disseminate dangerously novel ideas. It is no surprise to find that Le Play does not figure very prominently in the sociological syllabus of the L.S.E.[1] And yet

1. I am indebted for this information about Le Play to a comprehensive and scholarly account by Paul Lazarfeld, 'Notes on the history of quantification in sociology – trends, sciences, and problems', *Isis* 52 (1961), 277.

the man inspired a very great deal of quantified studies in the sociological field, and he is fairly to be regarded as one of the founding fathers of measurement in the behavioural sciences.

We have now to notice the monumental work of Charles Booth (1840–1916) on the *Life and Labour of the People of London*, the first volume being published in 1889 and the final, seventeenth volume, in 1903. The impact of this work was enormous. It may be fairly regarded as forming the evidence on which a long train of social legislation was to be based. Some of his early work was done in collaboration with Beatrice Potter, who later became Mrs. Sidney Webb, although he was, if anything, a conservative in politics. For our present purposes the importance of his work is not what he revealed but how he revealed it.

Booth's importance in empirical sociology rests on two grounds, his system of classification of the amorphous material which he had to reduce to order, and the fact that his social survey was conducted by sampling. I daresay that a modern purist would be able to fault him on both counts; certainly his sample would not attain the standards laid down by the Market Research Society. But in retrospect none of this matters against the overwhelming force of the numerical evidence which he produced. And all this, by the way, was carried out largely by his own efforts and at his own expense. It is typical that private enterprise led official statistics until quite recently. Like Florence Nightingale, Booth was highly critical of official statistics and his influence did much to improve them.

Booth's work on London was extended and supplemented by Seebohm Rowntree's work at York. Rowntree's report entitled *Poverty: A Study of Town Life*, published in 1901, rivals Booth's in thoroughness. He extended Booth's work by obtaining information more directly from the families themselves and can be regarded as the inventor of the Poverty Line, the standard minimum below which no individual should be allowed to fall if his health and well-being were not to suffer.

There are many distinguished names between Booth and Rowntree and the present day, Sidney Webb, Graham Wallas, Patrick Geddes, Harold Laski, to name only those most closely associated with the L.S.E. It was characteristic of their work, whether in sociology or in economics, that they set up their hypotheses

41

without much statistical help. It was always pleasant, perhaps essential, to be able to adduce statistical information to support their arguments or to rebut those of their opponents. Indeed none of them would have denied the necessity for scientific evidence to support their theses. But I think it is true to say that few lines of sociological thought were derived by an inspection of data which the modern statisticians would regard as a representative sample. They came from the heart, not the head. Measurement was a rather lowly handmaiden whose duty it was to provide corroborative detail, not to throw up laws of behaviour or to discover new social relationships; least of all to suggest social remedies. This attitude is still quite common. In fact only recently have we begun to realize that intuitively based social policies are not only without scientific foundation but may frustrate the objectives of the very people who advocate them.

To the nineteenth-century sociologist, then, statistics as a subject was almost entirely a matter of data collection and validation. The upsurge of interest in statistics as a branch of scientific method occurred elsewhere in a biological context, deriving largely from the work of Galton and Karl Pearson. It was round about 1890 that it first dawned on the world that chance had laws as inexorable as the deterministic laws of physics, although earlier writers, and notably Quetelet, had drawn attention to the fact in the demographic context. This was exciting but not particularly shocking. There was no particular emotional resistance to the discovery that things like height of human beings or the lengths of claws in crabs or barometric heights at Kew had definite frequency distributions which could conveniently be summarized in mathematical form. But there was a good deal of emotional resistance to the assertion that social phenomena were equally subject to law. Although I do not know that anyone put it quite in this way, the general feeling seemed to be that, having taken the control of society out of the realm of Divine Providence and put it into human hands, humanity was not going to see it taken out of human hands and put at the mercy of the laws of chance. The feeling still exists. In consequence statistical theory developed in biology, meteorology, and latterly in psychology alongside the advances in social descriptive statistics, with little more than a distant nodding acquaintance with social theories.

42

There were exceptions to this general statement, but they serve only to emphasize its general truth. Pareto, for example, called attention to the distribution of incomes which bears his name; and incidentally it is remarkable how stable the shape of the Pareto distribution has been over the past eighty years in spite of all the egalitarian measures which have been introduced to undermine it. Edgeworth, although working mainly as an economist, produced observational patterns in voting behaviour and contributed a good deal to the mathematical theory of frequency distributions. Udny Yule applied correlation theory to studies of pauperism. It was left for one of the most honoured names in the rolls of the L.S.E. to bring together here the mathematical and the socio-economic approach. I refer to Arthur Lyon Bowley, whose work was lineally descended from that of Booth and Rowntree.

As long ago as 1860 Florence Nightingale wanted to found a chair in applied statistics at Oxford. Apparently she discussed the project at some length with Benjamin Jowett, Arthur Balfour, and Francis Galton. However, it seems that our senior universities were still whispering from their towers the last enchantments of the Middle Ages; only in the last few years have they recognized the enchantments of stochastic processes, and after thirty years of effort Florence gave it up. London was more receptive. Karl Pearson became Professor of Statistics at University College in 1911, having held a chair in applied mathematics since 1884. Bowley joined the staff of the L.S.E. when its first session began in 1895 and became its first Professor of Statistics in 1915, retiring in 1936 after forty years of continuous service.

Bowley was by training a mathematician, graduating as a wrangler at Cambridge in 1891. But he was not a creative mathematical mind and never used his mathematics to any effective extent. He was, however, an outstanding practical statistician. His work on livelihood and poverty, on foreign trade, and on wages and incomes formed models of statistical pertinacity and genuine insight which have had a profound effect on both private and official statistics. He wrote extensively on index numbers and helped to make concepts like national income useful by painstaking steps to show exactly how they should be measured. It has always seemed to me that one of his greatest contributions to the behavioural sciences was his insistence on the importance of

43

measurement. Without going as far as Kelvin, who said that if you cannot measure what you are talking about your knowledge is of a meagre and unsatisfactory kind, Bowley made it plain that he had very little use for concepts which were not operationally meaningful. There was never a man over whose eyes one could pull so little wool.

Bowley's second great contribution was his advocacy of the sampling method. He may be fairly regarded as one of the founding fathers of the social survey. Round about the turn of the present century there were some vigorous arguments about the relative merits of complete census and what was then called 'the representative method', namely inquiry by representative sample. Official statisticians were in favour of complete enumeration. Bowley, who was always highly critical of official statistics and had an extensive knowledge of their shortcomings, argued for the representative method. He had no difficulty in convincing theoretical statisticians but it was uphill work in official quarters. And indeed, it was not until the Second World War that the British Government was forced to recognize sample surveys officially. The theory of sampling as we now have it in the social sciences is a much more sophisticated and mathematical subject than when Bowley considered it. But it still rests on the principles on which he insisted, the greater care which could be taken to secure accurate responses with trained interviewers confining their efforts to a moderately sized sample, the importance of avoiding bias, and the ascription of precision in quantified terms to the results.

Under Bowley's successors the statistical work at the L.S.E. has continued and flourished. It may be invidious to mention living individuals by name but I don't mind being invidious enough to call attention to the academic and public work of Sir Roy Allen and the fact that our professor of social statistics, Claus Moser, was chosen to become head of the Central Statistical Office. Wherever I go in the world I encounter students who recollect gratefully the statistical instruction they received here. In fact, from Bowley's time I think we can claim that we have at the L.S.E. one of the most powerful groups in the world of statisticians working in the behavioural sciences. Years ago I myself had some say in what they did. In the meantime the whole situation has changed radically with the advent of the computer. And so, with a complete lack of responsibility and possibly with their cordial disagreement, I want to go and describe what I think they should do next.

As I have already remarked, until a few years ago there were in existence social evils so glaring that no over-all social strategy was required to justify efforts to remove them. But sooner or later it becomes necessary to define what our social objectives are and in some way to quantify benefit or measure of success in attaining them. The nineteenth and the first half of the twentieth centuries paid little attention to this aspect of social organization, and when they did, usually came up with some vague and unsatisfactory criterion. Blake never explained just what kind of Jerusalem he wanted to build in England's green and pleasant land. Bentham's 'greatest happiness of the greatest number', in a period of expanding population, is nonsense. Phrases like 'from each according to his means, to each according to his needs' are not much better as operational criteria. We are now a little more sophisticated but not much. Not long ago I heard a trade unionist declare that his objective was to bring all wages up to the average. We have, in fact, a large set of social objectives, to eliminate infant mortality, to take care of the aged, to provide comprehensive medical care, to provide adequate housing, to ensure ease of transport, to extend educational facilities, and so on. It is obvious enough that limitation of resource prevents our carrying out comprehensive programmes for all of these simultaneously. It is almost equally obvious that a government tends to minimize electoral dissatisfaction rather than maximize social benefit in reconciling demands on those resources. But these are not points which I wish to make. The more important point is that if we pursue these social aims independently we may frustrate even the limited objectives which we set ourselves. In modern terminology, individual sub-optimization in a cybernetic system may be the reverse of optimization for the system as a whole.

An excellent example of this was provided two or three years ago by Professor Jay Forrester of the Massachusetts Institute of Technology in his book on *Urban Dynamics* (1969). A model structure was developed to represent the fundamental urban

45

processes, showing how industry, housing, and people interact with one another as the city grows and decays. It is only by constructing models which reflect the highly interactive nature and the dynamic nature of the system that the consequences of any particular social measure can be followed through. Forrester, by the way, had as his collaborator a former mayor of Boston, Professor Collins, so he can hardly be accused of an uninformed approach. The two examined four common programmes for improving the depressed nature of the central city. One is the creation of jobs by transporting the unemployed by bus to the suburbs, or by governmental jobs as employer of last resort. The second was a training programme to increase the skills of the lowest income group. The third was financial aid from federal funds. The fourth was the construction of low-cost housing. 'All of these', says Forrester, 'are shown to lie between detrimental and neutral almost irrespective of the criteria used for judgement. They range from ineffective to harmful judged either by their effect on the economic health of the city or by their long range effect on the low-income population of the city.'

These conclusions, especially those concerning housing, are so unpalatable that I doubt whether any politician who hopes to retain power dare adopt them or even mention them. But the reasoning is quite clear.

In fact [says Forrester], it emerges that the fundamental cause of depressed areas in the cities comes from *excess* housing in the low-income category rather than the commonly presumed housing shortage. The legal and tax structures have combined to give incentives for keeping old buildings in place. As industrial buildings age the employment opportunities decline. As residential buildings age they are used by the lower income groups who are found to use them at a higher density population. . . . Housing, at the higher population densities, accommodates more low-income urban population than can find jobs. A social trap is created where excess low cost housing beckons low-income people inward. They continue coming to the city until their numbers so far exceed the available income opportunities that the standard of living declines far enough to stop further inflow. Income to the area is then too low to maintain all of the housing. Excess housing falls into disrepair and is abandoned. One can simultaneously have extreme crowding in those buildings that are occupied

46

while other buildings become excess and are abandoned because the economy of the area cannot support all of the residential structures.

Many of you, I suppose, would instinctively repudiate Forrester's conclusions, or would claim that they do not apply to British conditions, although you may perhaps think that some of what he says strikes unpleasant notes of recognition. I adduce the example to show that under analysis by modern methods, intuition, and common sense are not enough. What I do claim is that in a highly interactive social organism simple and obvious remedies for evils in one domain may be entirely deceptive, and that it is essential to consider the organism as a whole.

Let us consider another example of a similar kind, the problem of transport by road. It goes without saying that a large city has traffic problems which will increase in intensity over the next decade at least. It is obvious, then, that we ought to widen our streets where possible, provide good parking facilities and build good access highways from the suburbs. No highway authority would have difficulty in getting approval for a programme along these lines and possibly financial assistance to implement it. But what then happens? The middle income groups move out of the city, to which they can now commute more easily, and abandon the centre to office blocks, expensive flats, and ghettos, and another cycle of social malaise is set up.

I have deviated somewhat from my main line of argument in order to broaden my terms of reference. The word 'measurement' assigned to the title of this lecture is only a part of the quantification which we can now bring to bear on social problems. We must consider it in the broader context of systems analysis, model-building, and simulation on the computer. Technology and political evolution have put at our disposal some new instruments of different orders of magnitude from those of thirty or forty years ago. The theory of social sampling; the generation of social data on regional, national, and international scale; and, of course, the computer itself. At the same time we have social problems of both old and new varieties. We are therefore in a position to consider anew the whole problem of sociological inquiry and to examine the consequences of social progress.

Admittedly the problem of constructing a model of our highly complex society is not an easy one. But it is essential if we are to attain our social goals. There is, after all, nothing new about model-building, and nothing new about building social models. The fact that we think about social organization at all in a logical way is equivalent to building a mental model, however ill-defined and incomplete it may be. It is nearly twenty years since Professor Phillips here constructed a physical model of the financial system, and although nowadays we do our modelling on a digital computer and not by analogue, it is still an instructive teaching instrument. The plain fact is that the unaided human mind is not capable of thinking about a highly interactive system in a quantifiable way. We can only think about it consciously a piece at a time and it is precisely this limitation which renders thinking about the system in its totality something in which we have to call on help from the logician, the mathematician, and the computer.

I take it that I need not describe at any length what the computer can do, but let me remove one or two misconceptions which still linger. The computer cannot think in any constructive way. It merely thinks strictly along lines which have been laid down for it in meticulous detail by a human being. Nor, in our present context, is it to be regarded as an arithmetic engine, a sort of electronic magnification of the desk calculators with which we are all familiar. It can, of course, do arithmetic and indeed all its operations proceed at a speed which even its high priests find astonishing. What makes it important for our purposes is the size of its memory which, if you add peripheral records on magnetic tape, is practically unlimited. This implies that we can write a program for the simulation of a model, however complex (within reason) and, having fed that program of instructions into memory, run the model as often as we like under all kinds of initial assumptions. We are at last able to experiment in the behavioural sciences.

One implication of this is that economics can no longer be left to the economists or sociology to the sociologists. The study of complex interactive systems requires an interactive team, behavioural scientists, mathematicians, statisticians, psychologists, systems analysts, and numerical data handlers. This is what we have, or are capable of having at the L.S.E. here. It is not for me, at least on this occasion, to suggest how such work should be put in hand, what its implication would be on research; or for that matter on the undergraduate syllabus. I am content to call attention to the enormous potential now available for new behavioural studies of an interdisciplinary kind, something which the L.S.E. has successfully done in the past and, I am confident, can equally successfully do in the future.

(Reproduced by permission of the London School of Economics and Political Science, and George Allen & Unwin Ltd.)

REVIEW OF THE INTERNATIONAL
STATISTICAL INSTITUTE
Volume 37 : 1, 1969

I

STUDIES IN THE HISTORY OF PROBABILITY AND STATISTICS, XXI. THE EARLY HISTORY OF INDEX NUMBERS

by

M. G. Kendall

C-E-I-R Limited

1. A complete history of index numbers has never been written. There is an excellent and scholarly summary in the book by C. M. Walsh (1901) and a shorter one in that by Irving Fisher (1922). The article by Edgeworth in Palgrave's *Dictionary of Political Economy* is also useful, and refers to his own monumental memoranda published in the Reports of the British Association for 1887–89. Sir Roy Allen and Dr. W. R. Buckland prepared a fairly comprehensive bibliography for the International Statistical Institute (1956) and this is being revised by Mr. W. F. Maunder of the London School of Economics and Political Science. The story behind these references is, perhaps, worth telling and in this article I propose to review the progress of the subject up to the time of Edgeworth's British Association memoranda.

2. We begin with Bishop Fleetwood's well known *Chronicon Preciosum* (1707, 2nd ed. 1745). The theory of statistics has many quaint origins, but this is the only one I know which began from a matter of conscience. A certain college (Fleetwood avoids saying which, but Edgeworth later said that it was All Souls, Oxford, and as a Fellow he should know) was founded between 1440 and 1460, and one of its original statutes required a person, when admitted to Fellowship, to swear to vacate it if coming into possession of a personal estate of more than £5 per annum. The question was whether, in the year 1700, a man might conscientiously take this oath even if he possessed a larger estate, seeing that the value of money had fallen in the meantime.

3. To answer this Fleetwood did an extensive inquiry into the course of prices over the previous 600 years. In particular he considers how much money would be required to buy £5 worth (at 1440/60 prices) of four commodities, Corn, Meat, Drink and Cloth, these being then, as they still are, the necessities of academic life. He comes to the conclusion that for these four, respectively, the present value of £5 was £30, £30, "somewhat above £25" and "somewhat less than £25".

"And therefore I can see no cause why £28 or £30 per annum should now be accounted a greater estate than £5 was heretofore betwixt 1440 and 1460". The inference is that an income of £30 or less "may be enjoyed, with the same innocence and honesty, together with a Fellowship, according to the Founder's Will".

4. The bishop considers briefly two questions which, over the next two centuries, did not come to the surface in discussions of index numbers though they are not without some topical interest: (a) whether laws which condemn to death for stealing things to the value of 12 pence (equivalent in 1700 to about forty shillings) do not automatically become too harsh – "tis certain that many die for stealing things of less value than 20 shillings", and (b) whether financial qualifications for Parliamentary electors, a minimum of 40 shillings net per annum in 1430, need bringing up to date – "But in these affairs, it is not fit for private people to meddle".

5. Fleetwood thus chose four items for his basket-of-goods. As he found, in each case, a decrease in the purchasing power of money of about the same extent, he was

relieved of the necessity of averaging his four price-relatives, or of considering their weights. This, perhaps, was fortunate for him, but unfortunate for us. One would like to have seen the bishop tackling the problem of amalgamating four different figures into a single measure. I do not think it was beyond him.

6. Fleetwood's book, which included an interesting study of early English coins, was fairly well known in the eighteenth century, and in particular was quoted by Sir George Evelyn, to whom I refer below. Our next author, Dutot (1738), was also concerned with the fall in the value of money, but on a grander scale; whether, for example, Louis XV, with a revenue of 100 million livres in 1735, was better off than Louis XII had been with 7,650,000 livres in 1515. He was, concludes Dutot, considerably worse off. To decide this question, he lists the prices at the two relevant dates, or thereabouts, of a large number of items, a goat, a chicken, a rabbit, a pigeon, a rick of hay, a day's labour for man and for woman, and so forth. He adds these prices together, unweighted, and divides one set by the other, concluding that in the period concerned the value of money had declined by a factor of 22 to 1.

If p_0 represents prices in the base year and p_1 those in the current year, Dutot's measure is then

$$I_1 = \Sigma\, p_1 / \Sigma\, p_0 \tag{1}$$

where the summation is over the articles. We may regard this as a rather arbitrary basket of goods. It is interesting, however, that Dutot includes services as well as commodities in his calculations.

7. Dutot's work was noticed by Dupré de Saint Maur (1746) who attempts some similar work of his own, but in a somewhat obscure way. He objects to some of Dutot's data, but not to his method. Like Dutot and Fleetwood, he is interested in comparing the value of money at two points of time, not in compiling a current index. I do not know of any further work by a French author for the next hundred years, but I should be surprised if some of the Encyclopedists did not have something to say on the subject. Comments on inflation during the Revolution and the Napoleonic period were probably discouraged.

8. We next have to notice Gian Rinaldo Carli (1720–1795) who was professor of astronomy at Padua before settling in Milan, where he established a considerable reputation as antiquary, historian and economist. His tract of 1764 was a discussion of the decline in the value of money since the discovery of America. He takes the prices of grain, wine and oil about 1500 and 1750 and constructs a price-relative for each commodity. He then strikes an arithmetic average of the price relatives, so that his index may be written

$$I_2 = \frac{1}{n} \Sigma\, (p_1/p_0) \tag{2}$$

where n is the number of commodities. He also does not attempt any weighting; nor does he include services among his commodities.

9. It would seem that Fleetwood, Dutot and Carli were independent pioneers and two of them at least were led to their researches by practical problems of some urgency. Even more urgent was the problem of the depreciation of paper money in the New England states throughout the eighteenth century. W. C. Fisher (1913) has given some account of experience in Massachusetts. Public bills of credit were first issued there in 1690 but soon depreciated in value. A series of enactments failed, as they have always failed, to check inflation. An Act of 1747 stated that

"when any valuation shall be made of the bills of public credit on this province . . .

regard shall be had not only to silver and bills of exchange, but to the prices of provisions and other necessaries of life".

10. According to Fisher, this was a requirement "in which the principle of the tabular standard is given recognition, perhaps one might say, full recognition". Well, perhaps one might, but I think that Fisher is being a trifle chauvinistic. To pass a law saying that regard shall be had to the cost of living is one thing; to explain how it should be done is something else. Even Fisher admits that the law had little effect.

11. During the War of Independence, Massachusetts, like other colonies, issued paper money which suffered depreciation at an alarming rate. Attempts to fix prices, as usual, failed. The main problem was to pay the soldiers who, by 1780, were nearly destitute. An Act of that year required that the depreciation of money from 1st January, 1777 to 1st January, 1780 was to be "calculated upon an average of the rates of depreciation as computed by the prices of Beef, Indian Corn, Sheep's Wool and Sole Leather". Actual calculation revealed that prices over the three years had increased by a factor of no less than $32\frac{1}{2}$. It was enacted that for the rest of the war the army was to be paid in notes maturing at various dates up to 1788 and, at maturity, were to be valued in current money according as "five bushels of corn, sixty eight pounds and four sevenths part of a pound of beef, ten pounds of sheep's wool and sixteen pounds of sole leather shall then cost, more or less than one hundred and thirty pounds current money, at the current prices of the said articles."

12. Just how this sytem worked is not known. Payment, if made at all, seems to have been very grudging. Fisher was able to trace only one valuation by justices under this Act, in March 1781. I have little doubt that the notes were treated rather like the British Post-War Credits of the Second World War. However, the principle of updating the value of a declining currency is clear enough. One wonders whether the authors of the Act were acquainted with Fleetwood's book.

13. These early efforts at correcting for a fall in the value of money, or a rise in prices (which were generally thought of as the same thing), like most early work which is revealed by the historian's retrospective disinterment, present the appearance of isolated studies with no continuity of development. We now arrive at a pioneer who may be said to have launched a movement which can be traced in direct line of descent to the present day. Sir George Shuckburgh Evelyn[1] was primarily interested in establishing "an invariable and unperishable standard of weights and measures." Nearly the whole of his paper (1798) is concerned with apparatus for the measurement of length and weight. But right at the end of the paper he says (p. 175)

"Before I close this paper, after having said so much on the subject of weights and measures, it may not be improper to add a few words upon a topic that, although not immediately connected, has some affinity to it; I mean the subject of the prices of provisions, and of the necessaries of life, etc. at different periods of our history, and, in consequence, the depreciation of money. Several authors have touched incidentally upon this question, and some few have written professedly upon it; but they do not appear to me to have drawn a distinct conclusion from their own documents. It would carry me infinitely too wide, to give a detail of all the facts I have collected; I shall therefore content myself with a general table of their results, deduced from taking a mean rate of the price of each article, at the particular periods, and afterwards combining these means, to obtain a general mean for the depreciation at that period; and lastly, by interpolation, reducing the whole into more regular periods, from the

---

[1] Sir George Shuckburgh (1751–1801), the sixth baronet, married as his second wife in 1785 the daughter of James Evelyn and took the additional name of Evelyn in 1791.

Conquest to the present time: and, however I may appear to descend below the dignity of philosophy, in such oeconomical researches, I trust I shall find favour with the historian, at least, and the antiquary."

14. It is interesting to observe the note of apology which he feels it necessary to introduce into any such utilitarian subject as measuring the value of money. Evidently, however, he had given a good deal of time and thought to the subject and he produces a large table "exhibiting the prices of various necessities of life, together with that of day labour, at different periods from the Conquest to the present time."

Evelyn's basket of goods consists of wheat, cattle in husbandry (horse, ox, cow, sheep, hog), poultry (goose, hen, cock), butter, cheese, ale, small beer, beef and mutton, and labour in husbandry. He takes 1550 as his base year and expresses the prices at other periods on the basis of 1550 as 100. He then takes a simple mean of the price relatives to obtain an index at various points of time, and interpolates for inter-mediate dates. His index is thus of the Carli type. His results are of some melancholy interest:

Date	Index		Date	Index		Date	Index
1050	26		1450	88		1740	287
1100	34		1500	94		1750	314
1150	43		1550	100		1760	342
1200	51		1600	144		1770	384
1250	60		1650	188		1780	427
1300	68		1675	210		1790	496
1350	77		1700	238		1795	531
1400	83		1720	257		1800	562

As a pioneer effort, it seems to me, this is highly commendable. Evelyn has the notion of base year, price relatives, the inclusion of services in the basket of goods, and the taking of an arithmetic mean. The index (and he did not use the word) is, however, open to criticism in regard to weights; for he has averaged the price relatives of an ox and a hen, of a day's labour and a pound of butter, without regard to their relative importance.

15. However inaccurate Evelyn's figures may be, they emphasize one point of importance in our present context. Up to 1550 the decline in the value of money was relatively slow. At that point the rate of decline began to accelerate and by the middle of the eighteenth century was rapid enough to be serious from decade to decade. The effects on trade and agriculture were sufficiently marked to force national attention to the hardships caused in fixed-price contracts and other obligations which could not be quickly adapted to meet rising prices. This is undoubtedly the reason why the theory of index numbers emerged at that point of time.

16. Evelyn's paper evidently attracted great attention. Arthur Young (1812, p. 67) says that his Table of Appreciation "has been referred to, and appears to have been entirely trusted, by every writer who has published upon any branch of political arithmetic during the last ten years. It is much to be lamented that such a Table should have been printed without that attendant Essay, which it is said, he drew up, and left in manuscript at the time of his death."

It is indeed to be lamented, but I think we can make a fair guess at the way in which Evelyn approached the subject. His main paper was concerned with fixed physical standards. He thought of depreciation in the value of money as having the same reality as the length of the standard rod. Price relatives were estimates of this central quantity; and it was reasonable to average these estimates to get at the "true" value, just as it was reasonable to average experimental observations to reduce estimational errors.

17. This point of view was strongly attacked by Arthur Young in the pamphlet already noticed (1812). Young also seems to feel it necessary to excuse the amount of work which he put into this study. "At my advanced age, and labouring under a great personal calamity, (I hope and trust from the mercy of God), my voluntary pursuits should be directed to higher purposes than wordly objects, whatever temporary importance may attend them . . . and I could not but ask myself, why engage in such an enquiry?"

18. Young's criticisms are, for the most part, justified, although he fails to give Evelyn (whom he always refers to as Shuckburgh) any credit for originality. The primary data, says Young, were suspect; the objects included in the table are imperfectly defined and may have altered in nature; a number of important items, and particularly manufactures, are entirely omitted; and there is no weighting.

Young then proceeds to examine price changes for a number of commodities – Wheat, Barley and Oats, Beans, Provisions (beef, mutton, pork, bacon, butter, cheese), Labour, Artisan's pay, Wool, Land, Rents, Houses, Timber, Coals, Metals, and Manufactures (among others, shoes, stockings, hats and mops).

19. Price changes, however, he remarks, are not solely, or even mainly due to changes in the value of money. "Have there been no improvements in the breed of horses in the last fifty years? . . . If the animal, upon the whole, is improved, it would be a strange mode of showing a depreciation of money, by discovering that a better horse sells at a higher price; and this remark is applicable to many other objects."

In commenting on Evelyn's figures, Young gives some of his own for the 17th and 18th century and says "Repeating wheat five times on account of its importance, barley and oats twice, provisions four times, labour five times and reckoning wool, coals and iron, each but once, and considering iron as the general representative of all manufactures, the rise per cent from the prices of one century to those of the other, will amount to no more than 22½ per cent; or only the tenth part of the rise stated by Sir George."

This has led some writers to attribute to Young the idea of a weighted index. He does indeed clearly express the concept, but only to criticise Sir George, not to commend it. And indeed he is resolutely opposed to a Table of Appreciation. "I consider such tables as tissues of error and deception; they take for granted, the question in dispute, and assign every rise in price, whatever may be the real cause, to depreciation, even in cases when none whatever is to be discovered" (l.c. p. 134).

20. From this point onwards the theory of index numbers is confused by arguments about economic principles and financial practices; whether bullion provides a standard of value; how much inflation is due to paper money; whether the country should revert to bimetallic standards; what is the basis of value. Man had not yet arrived at the conclusion that there are many ways of constructing index numbers, according to the use to be made of them or the economic theory underlying them. But the idea of a "general" movement in prices persisted and attempts to measure it provided a good deal of the motivation for studies of the subject in increasing depth.

21. The father of index numbers is generally considered to be Joseph Lowe whose book (1822) certainly contains the first discussion of many of the problems involved in their construction. Lowe was concerned with the disturbance caused by the Napoleonic wars. The war period itself was one of prosperity; the ensuing peace brought a recession, especially in agriculture. Among Lowe's objects was "the discussion of propositions for the relief of our suffering classes, founded, partly on the

evident tendencies of our resources to increase, partly on a plan of aiding individuals
to correct the existing disproportion in wages, salaries and other contracts formed
when money was of far less power in the purchase of commodities."

22. He addresses himself specifically to the fluctuations in the value of money in
his chapter 9. Substantial variation in purchasing power was an evil to be avoided if
possible. Previous attempts to link cost of living with corn prices were inadequate
because of the very different amounts spent by different classes on different commodities. He exemplifies this by a table

	% expenditure	
	Family of agricultural labourer (about six persons)	London middle class family – six persons, including two maid-servants
Provisions	74	33⅓
Clothes and washing	13	18⅓
Rent	4½	11⅔
Fuel and light	7	6
Contingencies	1½	—
Assessed taxes and poor rate	—	5
Servants wages	—	3⅔
Education, charity, repair of furniture and all contingencies	—	22
	100	100

In our age of such varied and refined expenditure, says Lowe, a standard of more
comprehensive character ought, if possible, to be adopted. "Now, the progress of
statistics and the multiplication of returns within the last half-century have supplied
data in a great measure (previously unknown) and have suggested to us the practic-
ability of framing a standard from material" (set out in an Appendix).

"What, it may be asked, would be the consequence of our possessing a table such
as we have supposed? The ascertaining on grounds that would admit of no doubt or
dispute, the power in purchase of any given sum in one year, compared to its power
of purchase in another. And what would be the practical application of this knowledge? The correction of a long list of anomalies in regard to rents, salaries, wages etc.,
arising out of the unfortunate fluctuations of our currency."

23. Lowe introduced the term Tabular Standard, which came into common use in

Commodity	£m expended by the public
Wheat	30
Barley	9
Oats (for human food)	10
Meat	35
Woollens	20
Cottons	12
Linen	15
Silk	8
Leather	15
Hardware	9
Sugar	9
Tea	8
Sundry specified articles	100
A multiplicity of unspecified articles	70
	350

the nineteenth century, to denote the array of articles, their weights and prices, entering into an index number. He lists estimates of the total expenditure in the U.K. on commodities entering into domestic consumption (see previous page).

His index for one year, as compared with the previous year, is constructed by evaluating the expenditure *on the same quantities* at the new prices. If, for example, this fixed basket of goods were revalued at £367.5 (to take his actual illustration) "£105 are required to effect the purchases for which £100 sufficed in the previous year".

Thus, with weights in the first year of q_0, Lowe's index is equivalent to

$$I_3 = \Sigma (p_1 q_0) / \Sigma (p_0 q_0), \tag{3}$$

a form which later became known as Laspeyres'.

24. Lowe recognises some of the standard difficulties of index number construction, especially that of getting reliable data. He also has a clear idea of weighting, for he goes on to consider whether different tabular standards are required for different social groups. "Were it intended to compute from such tables the consumption of the *nation at large*, the obvious course would be, to form a product by multiplying the sums in each column by the number of families in each class."

25. It is not entirely clear whether Lowe would compare prices in two years on the basis of weights from some other year; whether, in fact, he would use formula (3). This has led some writers to attribute to him an index of the form

$$I_4 = \Sigma (p_0 q) / \Sigma (p_1 q) \tag{4}$$

where q are some set of weights not necessarily related to either year 0 or year 1. This degree of generality rather flatters his line of thought. He contemplates no appreciable change in consumption over the periods for which he wanted to make comparisons (five years), in which case weights would remain constant and need not be attributed to any particular year.

26. Ideas of fluctuations in the value of money continued to dominate the Tabular Standard concept. A certain G. Poulett Scrope, combining Membership of Parliament with Fellowship of the Royal Society, published in 1833 a book on *Principles of Political Economy*. When reading the work of our statistical forefathers I nearly always get a sense of affection for them and of admiration for their valiant assaults on the problems of their time. Mr. Poulett Scrope is an exception. He was a violent anti-Malthusian and commits himself to statements like "Food can easily be made to increase faster than population" and "Misery is the result of crime and folly, not of any natural law". It is no surprise, then, to find that his book contains practically no statistics. It does, however, argue for a Tabular Standard:

"Take, for instance, a price-current, containing the prices of one hundred articles in general request, in quantities determined by the proportionate consumption of each article – and estimated (as they are under the standard of this country) in gold. Any variations from time to time in the sum or mean of these prices will measure, with sufficient accuracy for all practical purposes, the variations which have occurred in the general exchangeable value of gold."

The details of the method he would employ have to be gleaned from a footnote (*Political Economy*, p. 408):

"From tables of average prices drawn up by the Board of Trade, and printed in the Appendix to the Report of the Select Committee on the Bank Charter, 1832, it appears that on a comparison of the prices of the principal necessaries of home production, viz. wheat, meat, coals, iron, cheese, and butter, in the years 1819 and 1830, the average fall in that interval had been thirty-five per cent. In the principal articles of foreign importation, viz. sugar, coffee, hemp, cotton, tallow, oil, timber and tobacco, the fall had been near forty per cent. This relates to raw produce only. But the reduction in manufactured goods has been much greater; on the average, certainly not less than sixty per cent. On the whole, therefore, the gross average fall in the prices of the principal articles of consumption, raw and wrought, can scarcely have been less than fifty per cent. In other words the 'purchasing power' on money has doubled between 1819 and 1832."

A pamphlet of 1833 argues the same case. He remarks in it that "after the greater part of the above was written I have found in Mr. Lowe's valuable work a proposal to the same effect". So far as I can see, Scrope added nothing to Lowe's work and falls far behind him in understanding. Lowe, in fact, was careful to disclaim all political bias. Scrope is full of it.

27. There are occasional references in the literature to a book or pamphlet by W. Cross (1856) entitled *Standard pound versus pound sterling*. I have been unable to trace this work, but while looking for it came across a much earlier publication (1837) called *Monetary Reform* consisting of three letters published in the *Aberdeen Herald* in that year.

Cross wishes to maintain a steady price level by issuing or withdrawing paper currency. His second letter deals with the state of prices, and is the first I have seen which uses the word "index".

A list should be carefully made up of all or a number of the principal articles of trade (wheat, sugar, silver, gold etc. but not labour or services). "Let the proportions of the several commodities to each other be determined and fixed in the list, as nearly as can be found practicable, according to their relative importance in the market – that is, in proportion to the gross value of the entire quantity consumed of each kind of article respectively within a given territory, for instance the United Kingdom."

He then estimates as follows the annual consumption:

	£ million
Wheat	44
Coal	12
Wood	12
Cotton	12
Sugar	6.5
Iron	6
Tea	3.2
Gold and silver	2
	97.7

He appears to have intended to include flax in the list (about £m2.3).

Like Lowe, he arrives at weights for 1835 by dividing these sums by representative prices.

This procedure is thus the same as Lowe's: the use of national expenditures to arrive at weights and hence the determination of a fixed basket of goods, and then the calculation of the value of this basket at new prices. Cross was arguing an impossible economic thesis and his work seems to have been lost to sight, by statisticians at least.

28. About the same time G. R. Porter began to publish his three volume work on the *Progress of the Nation*, in the second volume of which (1838) he, from January 1833 to December 1837, calculates an index based on the average of the price relatives of

50 commodities. This is thus an index of the Carli type, but Porter does not specify which commodities, except that he says they are the most important 50 in the *export* trade. He calls the unit price of the base month an "index price". Porter has had some rather rough handling by later writers, unjustly in my opinion. His statistical survey of the nation, covering its activities from crime to colonisation, was a work of encyclopedic scope. Whether he drew on Lowe's work is not known; he certainly does not mention it. His index was submitted to a Select Committee of the House of Commons on Banks of Issue in 1840 and was severely criticised on the grounds that it was unweighted.

29. An interesting variant occurs in a book on *The State of the Nation* (1835) attributed to Henry James. This author finds his work done for him, so to speak, in Customs figures for exports from the U.K. These aggregate figures were of two kinds – the "official" value, based on prices of 1694, and the "declared" value, supposedly at current prices. The ratio of the second to the first is taken by James as an index of the value of British produce and manufactures. The ratios are available for the run of years from 1798 to 1833. There is no fixed basket of goods here. The index is essentially of the Paasche type (paragraph 36 below)[1].

30. For the next thirty years little seems to have been done to carry the subject further. There are occasional references to the Tabular Standard and a few attempts at comparing prices over a period, but nothing new in the way of theory. As late as 1859 Newmarch published the wholesale prices of 41 articles for 1851–9 and gave price relatives for 20 of them based on January 1851 = 100. But he attempted no amalgamation of the price relatives into an index for groups or for commodities as a whole. Later in 1864 he began the famous series which were annually published in "The Economist". Not until 1869 was a total index published. It was then a simple average of the Carli type. Unless there is work lying undiscovered in the libraries of Europe, it seems that the subject lay fallow for a whole generation.

31. As so often happens, there then occurred a renewal of interest in several different places, beginning with a famous book by Jevons (1863). For the first time something more than simple arithmetic made its appearance. Sophistication appeared in discussions of the type of average to be employed, the writing down of algebraic formulae, and definitions of the nature and purpose of indices.

32. Jevons introduced a new note into the forthcoming discussions by advocating the use of the geometric mean. He was primarily interested in the price of gold, and price-movements, to him, were significant in their proportional increases, not their absolute increases. With one or two exceptions the geometric mean has never come into general use, partly because of arithmetic inconvenience, partly because of the difficulty of explaining its nature to laymen, partly because it gives much the same answer as the arithmetic mean when prices are moving relatively slowly. Jevons' work sparked off a good deal of argument about the best kind of average to use. Edgeworth, in particular, wrote a good deal on the question. But the history of measures of location is really a separate subject and to dwell on controversy about the geometric mean tends to obscure the history of index numbers themselves.

33. To follow the course of the next twenty-five years we have to examine the various different objects which index numbers were supposed to serve. Edgeworth (1889) distinguishes seven:

[1] The Customs valuation was a rather complicated procedure based on "sworn values" where "official values" did not exist. A new basis of "real" values was introduced in 1854.

1) The *capital* standard. Intended to measure changes in the monetary value of a fixed set of articles, these articles being representative of all purchasable items in existence.
2) The *consumption* standard. Ditto, except that the population of articles consists of all commodities *consumed* by the community.
3) The *currency* standard. Ditto, except that the population of articles consists of all commodities *sold*.
4) The *income* standard. Intended to measure the change in the monetary value of the average consumption by individuals, or in *income per head*, of the community.
5) The *indefinite* standard, which simply takes a Carli type average of unweighted price-relatives. (The object seems to be to measure the "intrinsic value" of money, as compared with the consumption standard, which measures its "power". We are here, in my view, approaching the dangerous situation where the analytical procedure defines the concept, as in factor analysis.)
6) The *production* standard. Intended to measure changes in the monetary value of all commodities *produced* in the community yearly.
7) The *wages* standard. Intended to measure changes in the pecuniary remuneration of a certain set of services, such set being all those rendered in the course of production during a year.

34. Items (1), (2), (3) and (6) are indices of prices of commodities. Item (7) is an index of the price of service. Item (4) I reproduce as Edgeworth gave it, but I do not perceive the equality of an index of personal budgets and an index of income, unless we assume that all the income is spent on goods and services, or include savings, etc. in the personal budget. Item (5) seems to reify a rather nebulous concept.

35. It is convenient to consider next a group of German writers: Laspeyres (1864, 1871), Drobisch (1871), Paasche (1874, 1878). They knew of Jevons' work but do not seem to have followed his references backwards to Lowe. In their terminology they reject both the arithmetic and the geometric average; but what they really mean is that they reject *unweighted* averages such as Jevons used. Their methods are essentially those of Lowe and Cross, except that they have direct information about quantities consumed. They are concerned only with material goods, not with services, and work on Hamburg prices.

Laspeyres uses weights in a base year to obtain an index of the form

$$I_3 = \Sigma\,(p_1\,q_0)\,/\,\Sigma\,(p_0\,q_0) \tag{3 bis}$$

Paasche prefers the weights in the second period and hence obtains an index of type

$$I_5 = \Sigma\,(p_1 q_1)\,/\,\Sigma\,(p_0 q_1). \tag{5}$$

Drobisch mentions both and suggests an arithmetic mean of the two weights.

36. I_3 and I_5 are now known as indices of the Laspeyres and Paasche types. They are obviously particular cases of Lowe's index I_T, but perhaps it is fair for the German authors' names to be attached to them, for the earlier writers were not much concerned with weights moving through time, and it was not until consumption began to expand fairly rapidly that detailed thought had to be given to the question of changing weights. None of the authors, it would seem, contemplated that his index number would be valued over more than a five years' stretch. At a later point of time (1927) Irving Fisher was to propose a geometric mean of I_3 and I_5 as an "ideal" index number

$$I_6 = \left\{ \frac{\Sigma(p_1 q_0)}{\Sigma(p_0 q_0)} \frac{\Sigma(p_1 q_1)}{\Sigma(p_0 q_1)} \right\}^{\frac{1}{2}} \tag{6}$$

In the meantime Marshall and Edgeworth proposed, and Bowley endorsed, a method of averaging weights, giving

$$I_7 = \frac{\Sigma\{p_1(q_1 + q_0)\}}{\Sigma\{p_0(q_1 + q_0)\}} \tag{7}$$

37. Most theoretical elaborations of the simplest types of index numbers have foundered on the difficulty of acquiring the necessary information or the complicated calculations required. The Paasche index, for example, requires up-to-date figures of the weights q_1, which are usually not available from official sources in time to permit of the construction of an up-to-date index. Without having attempted an enumeration of the many indices now in use, I should guess that by far the biggest proportion of them are of the Laspeyres type, with an occasional up-dating of the weights.

38. Indices of baskets of physical commodities usually allow of the determination of weights in physical and comparable terms. Where intangibles such as services are involved, and physical units are not comparable, the only common unit is money and some such procedure as Lowe's (paragraph 23), based on total values, is required. But there are cases where weighting has to be decided on *ad hoc* grounds. For example, an index of stock exchange prices of equities is sometimes weighted, not by the number of outstanding shares, but by the total value of the company (shares being only one way of raising money to run a company, and often quite unrelated to its real capital). Palgrave (1886) weighted price-relatives by the current total value of the commodity concerned, with the result that the price appears as a quadratic term:

$$I_8 = \frac{\Sigma\left\{ p_1 q_1 \left(\frac{p_1}{p_0} \right) \right\}}{\Sigma(p_1 q_1)} \tag{8}$$

39. By 1900 what we may consider as the elementary theory of index numbers had been fairly well established; and for practical reasons most of the official and semi-official indices which have since been published have, from the theoretical viewpoint, been of the elementary kind. The subject has, however, been cultivated in a more sophisticated manner by economists and econometricians who have been concerned with changes in consumption standards or with the measurement of economic variables other than price, such as productivity. The reader who wishes to pursue these topics should begin with Sir Roy Allen's article on Index Numbers in Chambers Encyclopedia and the review article by Frisch (1939).

REFERENCES

Only those works actually mentioned in the text are given here. The comprehensive bibliography by Allen and Buckland (1956) has been mentioned in section 1.

[1] Walsh, C. M. (1901) *The Measurement of General Exchange Value.* New York, Macmillan.
[2] Fisher, Irving (1922) *The Making of Index Numbers.* Boston and New York, Houfton Mifflin.
[3] Fleetwood, W. (1707) *Chronicon Preciosum,* London. (Second edition 1745).
[4] Dutot (1738) *Réflexions politiques sur les finances et le commerce.* The Hague.
[5] Dupré de St. Maur (1746) *Essai sur les monnaies, ou réflexions sur le rapport entre l'argent et les denrées.* Paris.

[6] Carli, G. R. (1764) Del valore etc. (Opere scelte di Carli, ed. Custodi, vol. 1, p. 299).

[7] Fisher, W. C. (1913) The Tabular Standard in Massachusetts. *Quarterly Journal Economics*, May.

[8] Evelyn, Sir George Shuckburgh (1798) An account of some endeavours to ascertain a standard of weight and measure. *Phil. Trans.*, 113.

[9] Young, A. (1812) *An inquiry into the progressive value of money in England etc.* London.

[10] Lowe, J. (1822) *The present state of England etc.* London.

[11] Scrope, G. Poulett (1833) *Principles of Political Economy*. London.

[12] Scrope, G. Poulett (1833) *An Examination of the Bank Charter Question, with an inquiry into the nature of a just standard of value etc.* London.

[13] Cross, W. (1837) *Monetary Reform*. Edinburgh.

[14] Porter, G. R. (1838) *The Progress of the Nation*. London (2nd ed. 1847).

[15] James, Henry (1835) *The State of the Nation*, London.

[16] Newmarch, W. (1859) Mercantile Reports of the character and results of the trade of the U.K. during the year 1858; with reference to the progress and prices 1851-9. *J. Roy. Statist. Soc.*, 22, 76.

[17] Jevons, W. S. (1863) *A serious fall in the value of gold ascertained and its social effects set forth.* London.

[18] Edgeworth, F. Y. (1889) Third report of the British Association Committee for the purpose of investigating the best methods of ascertaining and measuring variation in the value of the monetary standard.

[19] Laspeyres, E. (1864) Hamburger Warenpreise 1850-1863 und die Kalifornisch-australischen Gold-entdeckung seit 1848. *Jahrbücher für Nationaloekonomie und Statistik*, 3, 81.

[20] Laspeyres, E. (1871) Die Berechnung einer mittleren Warenpreissteigerung *Jahrbücher für Nationaloekonomie und Statistik*, 16, 296.

[21] Drobisch, M. W. (1871) Über Mittelgrössen und die Anwendbarkeit derselben auf die Berechnung des Steigens und Sinkens des Geldwerts. *Berichte über die Verhandlungen der König. Sachs. Ges. Wiss. Leipzig. Math-Phy. Klasse 23, 25.*

[22] Paasche, H. (1874) Über die Preisentwicklung der letzten Jahre, nach den Hamburger Börsennotierungen. *Jahrbücher für Nat-oek. und Stat.* 23, 168.

[23] Palgrave, R. H. I. (1886) Currency and standard of value in England, France and India etc. *Memorandum to the Royal Commission on Depression of Trade and Industry, Third Report, Appendix B.*

[24] Frisch, R. (1936) Annual Survey of Economic Theory: The Problem of Index Numbers. *Econometrica*, 4, 1.

Note added in proof: Since the above article was written I located a copy of the pamphlet by Cross, referred to in section 27. It adds little to his earlier letters so far as concerns index numbers.

RESUME

l'Auteur retrace le développement des nombres indices depuis le début du dix-huitième siècle jusqu'à la fin du dix-neuvième siècle. Les premières oeuvres, de Fleetwood (1707), Dutot (1738), et Carli (1764) apparaissent comme des travaux précurseurs indépendants. Il est évident que les nombres indices utilisés maintenant, et spécialement ceux de Laspeyres (1864), Paasche (1874) et Palgrave (1886) trouvent leur origine dans les travaux d'Evelyn (1798), de Young (1812) et de Lowe (1822).

Biometrika (1972), **59**, 3, *p.* 677
Printed in Great Britain

Miscellanea

Studies in the History of Probability and Statistics. XXX. Abraham De Moivre's 1733 derivation of the normal curve: A bibliographical note

By R. H. DAW

R. Watson and Sons, London

AND

E. S. PEARSON

University College London

SUMMARY

De Moivre's original derivation of the normal curve in a 7 page Note in Latin and its English translations in the 1738 and 1756 editions of his *The Doctrine of Chances* are described. Karl Pearson's and R. C. Archibald's views on it are discussed. The finding of three further copies of the Note is reported.

Some key words: History of normal distribution; History of binomial distribution.

1. De Moivre's derivation

The first known derivation of the normal curve was given by Abraham De Moivre in a 7 pp. printed Note entitled:

*Approximatio ad Summam Terminorum
Binomii $\overline{a+b}|^n$ in Seriem expansi
Autore A.D.M.R.S.S.**
November 12, 1733
[*i.e. Abraham De Moivre, Socius of the Royal Society]

To quote Smith (1929, p. 566): 'This paper gave the first statement of the formula for the 'normal curve', the first method of finding the probability of the occurrence of an error of a given size when that error is expressed in terms of the variability of the distribution as a unit, and the first recognition of that value later termed the *probable error*. It shows also that before Stirling, De Moivre had been approaching a solution of the value of factorial *n*.'

In the second edition (1738) of his *The Doctrine of Chances* (pp. 235–43) De Moivre included an English translation of the *Approximatio* Note, which was prefaced by the remark ' I shall here translate a Paper of mine which was printed November 12, 1733, and communicated to some Friends, but never yet made public....' This translation, apart from a few minor alterations and additions and a substantial addition after the last corollary, follows the original Latin text even to the extent that Corollary VII, or at any rate the heading as such, is missing in both! In the third edition (1756) of *The Doctrine of Chances* (pp. 243–54) published two years after De Moivre's death, there is essentially the same translation of the *Approximatio* (with Corollary VII still missing) but the addition made in the 1738 edition after the last Corollary is now headed 'Remark I' while a further three pages of comment are added under the heading 'Remark II'.

A Fascimile of the 7 pp. *Approximatio* Note of 1733 in Latin was given by Archibald (1926c). Smith (1929) gives the English version from the 1738 edition of *The Doctrine of Chances* and David (1962) included as Appendix 5 the version from the 1756 edition, including Remarks I and II. Facsimile reprints of the whole of the second and third editions of *The Doctrine of Chances* were published in 1967 by Frank Cass and Co. Ltd, London and The Chelsea Publishing Company, New York, respectively.

2. Views of Karl Pearson and R. C. Archibald on the 'Approximatio' note

During the years 1921–32 Karl Pearson gave a series of lecture courses at University College London on the History of Statistics in the 17th and 18th centuries; see Appendix to Pearson (E. S.) & Kendall (1970). When in 1922 he began to examine the work of De Moivre, he found that one of the three copies in the University College Library of that author's *Miscellanea Analytica de Seriebus et Quadratures* (1730) had bound up at the end a copy of the *Approximatio* Note. Immediately preceding this Note and following the main part of the volume was a Supplement entitled *Miscellaneis Analyticis Supplementum*, separately paged but with the printer's signatures continuous with those in the *Miscellanea Analytica*. This Supplement seems to have been written as a result of comments which De Moivre had received from James Stirling shortly after the original publication date, and it was probably included with the later-issued copies of the *Miscellanea*. The *Approximatio* Note, however, was not only separately paged but had a new set of printer's signatures and was not described as a ' Supplementum '.

In his 1922 lectures and in a more detailed article published in *Biometrika* (1924), Pearson described the Note as a ' second supplement' to the *Miscellanea Analytica* which, he suggested, had been included with copies of the latter sold after 1733. Enquiries which he made of some of the more important libraries in the United Kingdom, namely those of the British Museum, the Royal Society and the Universities of Cambridge, Oxford and Edinburgh, showed that of the 12 copies of the *Miscellanea Analytica* located, the first Supplement was included in seven only and the *Approximatio* Note appeared in only one copy, that which he had first discovered in the University College Graves collection.

Following Pearson's 1924 paper, R. C. Archibald of Brown University, Providence, R.I., a well-known authority on the history of mathematics, wrote a letter to *Nature* (1926a) contesting Pearson's description of the *Approximatio* Note as a second supplement to the *Miscellanea Analytica*, principally on the basis of De Moivre's remarks quoted above from the 1738 edition of *The Doctrine of Chances*. Finally, after remarking that Pearson had found only a single copy of the Note, Archibald said that it would be interesting to learn if any other copies of this Latin, 1733 original existed. Pearson (1926) in his reply seems to admit that the term ' second supplement' was perhaps not appropriate, but adds that were he to search for other copies of the Note, he would seek for them *first* at the back of the *Miscellanea Analytica*. His letter is, however, mainly concerned with emphasising the importance from the viewpoint of the history of science of the long sections headed ' Remarks' which De Moivre had included on pp. 250–4 of the 1756 edition of *The Doctrine of Chances*, after his translation of the *Approximatio*.

Writing from Berlin, Archibald (1926b) replied to Pearson, maintaining his contention that the *Approximatio* was not a second supplement to *Miscellanea Analytica*. He ends his letter by saying: ' I have here found another copy [of the *Approximatio*] in the Preussische Staatsbibliothek bound in with the *Miscellanea*.' Curiously he does not seem to realise either in (1926b) or (1926c) that the discovery of a second copy also bound with the *Miscellanea* might be looked on as weakening his argument against the Supplement theory.

In spite of this correspondence in *Nature*, or perhaps in ignorance of it, many modern references to De Moivre's derivation of the normal distribution continue to describe it as a Supplement to the *Miscellanea Analytica* or merely to quote that book; also, on another matter dealt with in the correspondence, it has sometimes been implied or definitely stated that the 1738 edition of *The Doctrine of Chances* did not include a translation of the *Approximatio*.

3. Results of some further enquiries

The notes below are the result of a joint search by the authors of the present article following the chance coincidence that, for different reasons and quite independently, we were both wanting at the same time to examine the University College copy of the *Miscellanea Analytica* in which Karl Pearson had made his 1922 discovery. When one of us found that the book was not available, he was told by a library assistant that the Graves collection held a second copy of the *Approximatio* Note. This copy was not attached to the *Miscellanea*, but bound up in front of an apparently unconnected 108 pp. Italian work:

Principii di Geografia, Astronomico-Geometrica
di Anton-Mario Lorgna,

published in Verona in 1789. On the spine of this book was printed LORGNA GEOGRAFIA-ASTRONOMICO GEOMETRICA, with no reference to the 7 pp. Note.

Neither this second copy of the *Approximatio*, nor the first copy had been indexed in the original Graves MS catalogue, nor were they entered in the printed College Library catalogue prepared towards the end of the last century, though a reference to the first copy had been added in ink, probably after Pearson's 1922 discovery. Later some diligent librarian going more carefully through the Graves collection had found the second copy and made a pencil note on the catalogue. We mention this incident to show how likely it is that other copies of the 7 pp. Note, besides those referred to below, may be bound up, undiscovered, elsewhere.

An intriguing discovery about this second copy of the *Approximatio* was that after reading to the FINIS on p. 7, one of us turned over the page to what might be expected to be a blank p. 8. But there was inscribed on it in faded ink the words:

'for Mr Stirling'
'The above is an autograph of A. De Moivre'

There is no signature of De Moivre, but comparing the two copies of the *Approximatio* it is seen that while the first had been clipped along the bottom margins of its pages, this treatment had been given to the top margins of the second copy, and so may have cut off a signature; alternatively the word 'autograph' may simply have meant that the words 'for Mr Stirling' were in De Moivre's hand. In any case, here seems to be a genuine copy of one of the Notes which De Moivre says he had 'communicated to some Friends'.

This discovery put us on to a further hunt, which may be summarized as follows:

(*a*) *The Berlin copy.* The copy which Archibald discovered in 1926 could not now be located in either the Deutsche Staatsbibliothek (East Berlin) or the Staatsbibliothek Preussischer Kulturbesitz at Marburg (West Germany), although the Humboldt University Library in Berlin has a copy of the *Miscellanea Analytica*, but neither the Supplement nor the *Approximatio* are with it. As a result of the War, there has evidently been some dispersal or loss of books which were in the Preussische Staatsbibliothek.

(*b*) Through the efforts of Mr Oscar Sheynin we have been able to locate *three* further copies of the *Approximatio*, two of which do not seem to have been reported before, all of which are bound up with the *Miscellanea Analytica* and the *Supplementum*.

(i) In *Basel*. As reported by Schneider (1968), the University Library contains a copy of the *Miscellanea Analytica* bound with the *Supplementum* and the *Approximatio*. Dr R. Reinle of the Science Department now reports that there is no hand-written inscription or dedication on the last; however the *Miscellanea*, the *Supplementum* and the 1733 Note show 'uniform edges and the original leather binding of the publisher. Apparently it (the volume) has not been put together later by any book collector or librarian.'

(ii) In *Moscow*. The Institute for the History of Natural Sciences and Technology contains all three items bound together. The only MS inscriptions are on the front cover and title page of the *Miscellanea*, where are the names of Benjamin Robins, an original subscriber to the *Miscellanea*, of Monsieur L'Abbé Vallier and of Mauduit, presumably the French writer on conic sections, spherical astronomy, etc. active in the years 1750–70.

(iii) In *Leningrad*. The Library of the Academy of Sciences has a copy of the *Approximatio* bound up with the *Miscellanea*, and the *Supplementum*.

(*c*) The five Libraries approached by Karl Pearson 50 years ago had no more information to give us.

(*d*) The name of Robins on the Moscow copy, made us examine the printed list of 'subscribers' given at the beginning of the *Miscellanea Analytica*. From this we identified six Cambridge Colleges and the University of Uppsala. While all but one of their Libraries still possessed copies of the *Miscellanea* and three had also the *Supplementum*, none could find a copy of the *Approximatio*, either bound with the *Miscellanea* or separate.

(*e*) We also approached the Libraries of the Universities of Keele and London, the Science Museum, the Royal Statistical Society and the Institute of Actuaries in this country and, in Paris, the Bibliothèque Nationale and the Institut de France Bibliothèque. In no case was a copy of the *Approximatio* found.

To sum up, six copies of the 1733 Note have been located, including the now missing Berlin copy, five of which are or were bound up with the *Miscellanea Analytica* and only one is unambiguously an independent pamphlet. It is possible that some of the former copies were originally presented to De Moivre's friends and later reached the Libraries, where they were quite naturally bound up with the *Miscellanea* since this book is referred to on the first page of the Note. On the other hand in view of the statement from Basel, and in spite of De Moivre's remarks of 1738, it is just possible that a few copies of the Note were left over after the original private circulation and, if the printer was the same, were bound up with copies of the *Miscellanea Analytica* sold after 1733. That copies should appear in (i) Basel, (ii) Berlin and (iii) Leningrad is not surprising since (i) De Moivre was in frequent correspondence with John Bernoulli (Walker, 1934) who was at the University of Basel, (ii) he was elected to the Berlin

Academy in 1735, and (iii) we should expect him to be in touch with Euler who was appointed Professor of Mathematics in St Petersburg (Leningrad) in 1733.

REFERENCES

ARCHIBALD, R. C. (1926*a*). Abraham de Moivre (Letter). *Nature, Lond.* **117**, p. 551.
ARCHIBALD, R. C. (1926*b*). Abraham de Moivre (Letter). *Nature, Lond.* **117**, p. 894.
ARCHIBALD, R. C. (1926*c*). A rare pamphlet of Moivre and some of his discoveries. *Isis* **8**, 671–6.
DAVID, F. N. (1962). *Games, Gods and Gambling.* London: Griffin.
PEARSON, E. S. & KENDALL, M. G. (Editors) (1970). *Studies in the History of Statistics and Probability.* London: Griffin.
PEARSON, K. (1924). Historical note on the origin of normal curve of errors. *Biometrika*, **16**, 402–4.
PEARSON, K. (1926). Abraham De Moivre. Reply to Professor Archibald. *Nature, Lond.* **117**, 551–2.
SCHNEIDER, I. (1968). Der Mathematiker Abraham de Moivre, 1667–1754. *Archs. Hist. exact. Sci.* **5**, 177–317.
SMITH, D. E. (1929). *A Source Book in Mathematics.* New York: McGraw-Hill.
WALKER, H. M. (1934). Abraham de Moivre. *Scr. math.* **2**, 316–33.

[*Received November* 1971. *Revised June* 1972]

The Historical Development of the Use of Generating Functions in Probability Theory

by *H. L. Seal*, Toronto
(1949)

Introduction

An important part of probability theory consists of the derivation of the probability distribution of the sum of n random variables, each of which obeys a given probability law, and the development of asymptotic forms of these distributions valid for increasing n. Probability generating functions owe their dominant position to the simplifications they permit in both problems. Their employment to obtain the successive moments of a probability distribution and to solve the difference equations of probability theory is ancillary to their chief use in connexion with sums of random variables.

A didactic exposition of the use of generating functions in probability theory might easily be made to parallel the historical development of these functions. This circumstance will be clearly perceived in the following historical sketch of the use of probability generating functions from their origin with De Moivre to their present-day wide application under different guises.

The generating function of a discrete law

Although the theoretical frequencies of the various possible totals obtained in throws with two and three ‚perfect' dice had been known from at least the time of Cardan (Todhunter, 1865), it was De Moivre who gave the problem its first algebraic solution. Considering with De Moivre (1730) a generalized die with k faces any one of which is equally likely to appear when the die is thrown, we ask the probability that with n such dice a total of x points will be thrown.

Without any essential change in the problem we may restate it in modern terminology thus: to derive the probability distribution of the random variable formed by adding n random variables each subject to the same discrete rectangular probability law with k equidistant variate values.

De Moivre expresses his solution in terms of the favourable «chances» (combinations) and the total possible «chances». The latter number k^n in the problem under consideration. On the other hand, the total array of chances on any one of the dice may be represented algebraically by $t + t^2 + t^3 + \ldots + t^k$ the index of any t representing the number of points on the corresponding ‚face'. As De Moivre says this progression «may represent all the Chances of one Die: this being supposed, it is very plain that in order to have all the Chances of two such Dice, this Progression ought to be raised to its Square, and that to have all the Chances of three Dice, the same Progression ought to be raised to its Cube; and universally, that if the number of Dice be expressed by n, that Progression ought accordingly to be raised to the Power of n» [1]). With this preliminary De Moivre proceeds to the algebraic solution mentioned.

The words quoted show that De Moivre attached to the discrete rectangular probability law a function designed to represent that probability distribution (namely, «all the Chances of one Die»). Although the title is quite modern we refer to this type of ‚image' function as a probability generating function and write it generally as $\psi(t)$. De Moivre, then, in effect defined $\psi(t)$, the probability generating function of the discrete rectangular law, by

$$\psi(t) \equiv \sum_{x=1}^{k} \frac{1}{k} t^r = \frac{t}{k} \cdot \frac{1 - t^k}{1 - t} \tag{1}$$

and stated it to be «very plain» [2]) that $\psi_n(t)$ the probability generating function of the sum of n such discrete rectangular variables, should

---

[1]) Quoted from the third edition of *The Doctrine of Chances*, 1756.

[2]) In 1777 Euler (1788) presented a paper to the Academy with the sole purpose of proving inductively that the probability of a variate x appearing as the sum of n random variables each distributed according to the law

$$p(x), \ x = x_1, \ x_2, \ \ldots \ x_k,$$

is given by the coefficient of t^x in $\left(\sum_{x_j} p(x_j) \, t^{x_j} \right)^n$.

be given by

$$\psi_n(t) \equiv \sum_{x=n}^{nk} p_n(x)\, t^x = [\psi(t)]^n = \frac{t^n}{k^n}\,(1-t^k)^n\,(1-t)^{-n}$$

$$= \frac{t^n}{k^n}\left[\sum_{\lambda=0}^{n}\binom{n}{\lambda}(-1)^\lambda\, t^{k\lambda}\right]\left[\sum_{\nu=0}^{\infty}\binom{n+\nu-1}{\nu}t^\nu\right]$$

$$= \frac{1}{k^n}\sum_{x=n}^{nk} t^x \sum_{j=0}^{m_1}(-1)^j\binom{n}{j}\binom{x-kj-1}{x-n-kj},$$

$$m_1 = \min.\left(n,\left[\frac{x-n}{k}\right]\right)$$

where $p_n(x)$ is the required probability of a total x with n dice, and may be read off from the last expression written, viz.

$$p_n(x) = \frac{1}{k^n}\sum_{j=0}^{m_1}(-1)^j\binom{n}{j}\binom{x-kj-1}{x-n-kj} \tag{2}$$

De Moivre's contemporaries were fully seized of the value of the artifice thus introduced. An important application of it was made by Simpson in one of the essays of his *Miscellaneous Tracts*. Simpson's object was to consider mathematically the method «practised by *Astronomers*» of taking the mean of several observational readings «in order to diminish the errors arising from the imperfection of instruments and of the organs of sense». He supposes that at any one reading errors in excess or defect are symmetrically disposed and have assignable upper and lower limits.

In the first of two propositions Simpson considers a discrete rectangular law with an odd number, $2h+1$, of variates centred about zero. The problem of obtaining the probability of a total error x arising as the sum of n individual errors is solved precisely in the manner of De Moivre though now $\psi(t)$ appears with a factor t^{-h} representing a removal of the origin to the centre of the distribution. Simpson finally obtains the probability that an error lies between $-z$ and z by summing the appropriate terms of the expression thus obtained.

In his second proposition Simpson assumes a discrete symmetric triangular probability law for the individual errors. He writes, effectively,

$$p(x) = \begin{cases} \dfrac{h+1-|x|}{(h+1)^2} & x = -h, -h+1, \ldots, -1, 0, 1, \ldots h \\[2mm] 0 & x = \text{any other value} \end{cases}$$

and

$$\psi(t) \equiv \sum_{n=-h}^{h} p(x)\, t^x = \frac{t^{-h}}{(h+1)^2} \cdot \frac{(1-t^{h+1})^2}{(1-t)^2}$$

Hence

$$\psi_n(t) = \frac{t^{-nh}}{(h+1)^{2n}} (1-t^{h+1})^{2n} (1-t)^{-2n}$$

$$= \frac{1}{(h+1)^{2n}} \sum_{n=-nh}^{nh} t^x \sum_{j=0}^{m_2} (-1)^j \binom{2n}{j} \binom{2n+nh+x-\overline{h+1}\,j-1}{nh+x-\overline{h+1}\,j} \quad (3)$$

$$m_2 = \min.\left(2n, \left[\frac{nh+x}{h+1}\right]\right)$$

Simpson notices that with the exception of the displacement factor [1] t^{-nh} and the constant, the form of the generating function in the second proposition is similar to that of the first with $2n$ replacing n.

The probability of a total error lying between $\pm z$ (the limits included) when individual triangular variates have been combined is thus

$$1 - \frac{2}{(h+1)^{2n}} \sum_{x=-nh}^{-z-1} \sum_{j=0}^{m_2} (-1)^j \binom{2n}{j} \binom{2n+nh+x-\overline{h+1}\,j-1}{2n-1}, \quad (4)$$

$$z \leqslant nh$$

since, as Simpson points out, the distribution of the mean error, x/n, is symmetrical about zero and it is simpler to work with $-z$ than with z. Simpson carries out the summation of x arriving at

---

[1] Our terminology not Simpson's.

$$1 - \frac{2}{(h+1)^{2n}} \sum_{j=0}^{m_3} (-1)^j \binom{2n}{j} \binom{2n + nh - z - \overline{h+1}\,j - 1}{2n} \quad (5)$$

$$m_3 = \min.\left(2n, \left[\frac{nh - z - 1}{h+1}\right]\right)$$

and illustrates this result by putting $h = 5$, $n = 6$ and $z = 1$ and 2, respectively. He concludes that «the taking of the mean of a number of observations, greatly diminishes the chances for all the smaller errors, and cuts off almost all possibility of any large ones.»

Up to this point Simpson's analysis has been a straightforward application of De Moivre's technique to a new problem. The real advance due to Simpson lies in the corollary to the second proposition. Here for the first time a continuous (symmetric triangular) probability law is introduced. By making $h \to \infty$ in such a way that the range of variation of an individual error x remains within ± 1, the probability of a total error between $\pm y (0 \leqslant y \leqslant n)$ arising from the addition of n errors each subject to a continous triangular law centered on zero and extending one unit to the right and to the left, is given by [1]

$$1 - \lim_{\substack{h \to \infty \\ \frac{z}{h} \to y}} \frac{2}{(h+1)^{2n}} \sum_{j=0}^{m_3} (-1)^j \binom{2n}{j} \frac{1}{(2n)!} (n - 1 + \overline{n-j}\,\overline{h+1} - z)^{(2n)}$$

$$= 1 - \frac{2}{(2n)!} \sum_{j=0}^{m_4} (-1)^j \binom{2n}{j} (n - y - j)^{2n} \quad (6)$$

$$m_4 = \min.\left(2n, [n - y]\right)$$

Simpson did not attach a similar corollary to his first proposition but his observation on the interchangeability of $2n$ and n indicates that he could have written this down without further calculation. It is noted here that the probability of a total error $< |y|$ arising from the sum of n random variables each subject to a continuons rectangular probability law centered on zero and extending one unit in either direction, is

$$1 - \frac{2}{2^n n!} \sum_{j=0}^{m_5} (-1)^j \binom{n}{j} (n - y - 2j)^n \quad (7)$$

$$m_5 = \min.\left(n, [n - y]\right)$$

---

[1] Although Simpson's formula is correctly stated he mistakenly wrote $nz/h = y$ in his three numerical examples.

The generating function of a continuous law

In an article published (apparently) in 1773 Lagrange restated Simpson's two propositions except that he placed his origins away from the centres of the distributions of the individual errors, and directed his attention to the probability of a total variate (error) lying between $-z_1$, and z_2 instead of between $\pm z$. He also used Simpson's limiting process to obtain the distribution laws of the sums of n variates from continuous rectangular and triangular laws of individual error. As Todhunter (*l. c.*) points out, there are a number of algebraic slips in Lagrange's work, but from the modern point of view his mathematical technique is considerably in advance of Simpson's more pedestrian approach.

However, the real advance of Lagrange's memoir consists in his generalization of generating functions to apply directly to continuous probability distributions. Lagrange argued, by direct analogy with the discrete case, that if

$$\psi(t) = \int_{-\infty}^{\infty} t^x p(x)\, dx \tag{8}$$

then

$$\psi_n(t) \equiv \int_{-\infty}^{\infty} t^x p_n(x)\, dx = \left[\int_{-\infty}^{\infty} t^x p(x)\, dx \right]^n \tag{9}$$

where the meaning to be attributed to $\psi_n(t)$ and $p_n(x)$ will be clear from the similar relations of the discrete case. In order to carry out the calculations indicated in these relations Lagrange provided what is, in effect, the first dictionary of Laplace transforms. This dictionary was effectively as follows, where $P_m(x)$ represents an arbitrary polynominal in x of the mth degree.

No.	$p(x)$	$\psi(t) = \int_{-\infty}^{\infty} t^x p(x)\,dx$
1	$\displaystyle\sum_{j=1}^{k} p_j(x)$	$\displaystyle\sum_{j=1}^{k} \psi_j(t)$
2	$P_m(x) \quad 0 \leqq x \leqq a$ $0 \quad -\infty < x < 0,\ x > a$	$\displaystyle\sum_{j=0}^{m} (-1)^j \frac{t^a P_m^{(j)}(a) - P_m^{(j)}(0)}{(\log_e t)^{j+1}}$
3	$P_m(x)\, e^{-\alpha x} \quad 0 \leqq x \leqq a$ $0 \quad -\infty < x < 0,\ x > a$	$\displaystyle\sum_{j=0}^{m} (-1)^j \frac{t^a e^{-\alpha a} P_m^{(j)}(a) - P_m^{(j)}(0)}{(\log_e t - \alpha)^{j+1}}$
4	$x^m \quad 0 \leqq x \leqq a$ $0 \quad -\infty < x < 0$	$\dfrac{m!}{(-\log_e t)^{m+1}} \quad 0 < t < 1;\ m = 1, 2, 3, \ldots$
5	$0 \quad 0 < x < \infty$ $(-x)^m \quad -\infty < x \leqq 0$	$\dfrac{m!}{(\log_e t)^{m+1}} \quad t > 1;\ m = 1, 2, 3, \ldots$

No.	$p(x)$	$\psi(t) = \int_{-\infty}^{\infty} t^x p(x)\,dx$
6	$\sum_{j=0}^{m_6} A_j (x-j)^m \quad 0 \leqq x < \infty,\ m_6 = \min.(h,[x])$ $0 \qquad\qquad\qquad\qquad -\infty < x < 0$	$\dfrac{m!}{(-\log_e t)^{m+1}} \sum_{j=0}^{h} A_j t^j \quad \begin{array}{l} 0 < t < 1; \\ h,\ m = 1,2,3,\ldots \end{array}$
7	$0 \qquad\qquad\qquad\qquad\qquad 0 < x < \infty$ $\sum_{j=0}^{m_7} A_j (-x-j)^m \quad -\infty < x \leqq 0,\ m_7 = \min.(h,[-x])$	$\dfrac{m!}{(\log_e t)^{m+1}} \sum_{j=0}^{h} A_j t^{-j} \quad \begin{array}{l} t > 1; \\ h,\ m = 1,2,3,\ldots \end{array}$
8	$\sum_{j=0}^{m_6} A_j e^{\beta(x-i)} (x-j)^m \quad 0 \leqq x < \infty$ $0 \qquad\qquad\qquad\qquad\quad -\infty < x < 0$	$\dfrac{m!}{(-\log_e t - \beta)^{m+1}} \sum_{j=0}^{h} A_j t^j \quad \begin{array}{l} 0 < t < 1; \\ h,\ m = 1,2,3,\ldots \end{array}$
9	$0 \qquad\qquad\qquad\qquad\qquad 0 < x < \infty$ $\sum_{j=0}^{m_7} A_j e^{\beta(x-i)} (-x-j)^m \quad -\infty < x \leqq 0$	$\dfrac{m!}{(\log_e t - \beta)^{m+1}} \sum_{j=0}^{h} A_j t^{-j} \quad \begin{array}{l} t > 1; \\ h,\ m = 1,2,3,\ldots \end{array}$

Lagrange points out that α of No. 3 may be purely imaginary so that the cases $P_m(x) \cos bx$ and $P_m(x) \sin bx$ are included. He stated that other functions $p(x)$ than those included above have generating functions $\psi(t)$ obtainable only by approximation.

Four examples of the use of the new type of probability generating function close the memoir. They are based, respectively, on the following continuous probability laws:

$$\frac{1}{h_1 + h_2} \qquad -h_1 \leqslant x \leqslant h_2; \qquad 0 \quad x < -h_1,\ x > h_2 \qquad \text{(I)}$$

$$\frac{l - |x|}{l^2} \qquad -l \leqslant x \leqslant l; \qquad 0 \quad |x| > l \qquad \text{(II)}$$

$$\frac{3}{4}\, a^{-3}(a^2 - x^2) \quad -a \leqslant x \leqslant a; \qquad 0 \quad |x| > a \qquad \text{(III)}$$

$$\frac{1}{2} \cos x \qquad -\frac{\pi}{2} \leqslant x \leqslant \frac{\pi}{2}\ ; \quad 0 \quad |x| > \frac{\pi}{2} \qquad \text{(IV)}$$

We illustrate Lagrange's procedure by considering (III) which he calls the law «la plus simple et la plus naturelle qu'on puisse imaginer».

We have in this case

$$\psi(t) = \int_{-a}^{a} t^x\, \frac{3}{4}\, a^{-3}(a^2 - x^2)\, dx = \int_{0}^{a} t^x\, \frac{3}{4}\, a^{-3}(a^2 - x^2)\, dx + \int_{0}^{a} t^{-x}\, \frac{3}{4}\, a^{-3}(a^2 - x^2)\, dx$$

$$= \frac{3}{2}\, a^{-2} \left\{ \frac{t^a + t^{-a}}{(\log_e t)^2} - \frac{a^{-1}(t^a - t^{-a})}{(\log_e t)^3} \right\}$$

by no. 2 of the dictionary. Hence

$$\psi_n(t) = [\psi(t)]^n = \frac{3^n}{2^n}\, a^{-2n} \sum_{j=0}^{n} (-1)^j \binom{n}{j} a^{-j}\, \frac{(t^a + t^{-a})^{n-j}(t^a - t^{-a})^j}{(\log_e t)^{2n+j}}$$

$$= \frac{3^n}{2^n}\, a^{-2n} \sum_{j=0}^{n} (-1)^j \binom{n}{j} a^{-j} \sum_{\lambda=n-2\left[\frac{n}{2}\right]}^{n} {}_{(2)} \left\{ \frac{A_{\lambda j}}{(\log_e t)^{2n+j}}\, t^{\lambda a} + \frac{A_{\lambda j}}{(-\log_e t)^{2n+j}}\, t^{\lambda a} \right\}$$

where

$$A_{\lambda j} = \sum_{\nu=0}^{\frac{n-\lambda}{2}} (-1)^{j+\nu} \begin{pmatrix} n-j \\ \dfrac{n-\lambda}{2} - \nu \end{pmatrix} \begin{pmatrix} j \\ \nu \end{pmatrix},$$

except that A_{0j}, if it occurs, is to be given half the value ascribed by this formula. The «(2)» affixed to the last summation sign on p. 217 indicates that every second term is to be summed. Using the dictionary inversely (Nos. 6 and 7)

$$p_n(|x|) = \frac{3^n}{2^n} a^{-2n} \sum_{j=0}^{n} (-1)^j \binom{n}{j} a^{-j} \frac{1}{(2n+j-1)!} \sum_{\lambda=n-2\left[\frac{n}{2}\right]}^{m_8} {}^{(2)} A_{\lambda j}(|x| - \lambda a)^{2n+j-1} \quad (10)$$

$$m_8 = \min.\left(n, \left[\frac{|x|}{a}\right]\right)$$

Although this is not mentioned by Lagrange, according to his own dictionary the above transformations hold only when a is a positive integer: actually this limitation is unnecessary and results from the primitive methods of integration utilized by that author.

It will be noticed that the inversion illustrated above involves an interesting device which Lagrange himself did not justify. It was assumed that the integral representing $\psi_n(t)$ had an infinite upper limit whereas in fact the distribution law involved has a finite upper and lower bound. The justification is that since we are not concerned with values of x outside the limits $\pm na$ the integral may be completed arbitrarily.

Inversion in the discrete case

At the commencement of Ch. IV of Book II of his text book on probability theory Laplace (1812) provides a new treatment of De Moivre's problem of the addition of n random variables each rectangularly distributed over a finite number of equally spaced discrete points. He supposes that the individual probability laws consist of $2h + 1$ points (h integral) of equal probability, the range extending from $-h$ to h and thus centering on zero. Replacing the t of De Moivre's generating function by e^{iu} the probability generating function of the distribution law for the sum of n discrete rectangular variates is

$$\psi_n(t) = \psi_n(e^{iu}) = \left[\sum_{x=-h}^{h} \frac{1}{2h+1} e^{ixu} \right]^n = \left(\frac{1}{2h+1} \right)^n \left[\sum_{j=-h}^{h} e^{iju} \right]^n$$

Laplace observes that $p_n(x)$, the coefficient of e^{ixu} in the expansion of the expression last written, is the probability of a total value x; it is obtained as the coefficient independent of x in

$$\left(\frac{1}{2h+1}\right)^n e^{-ixu}\left[\sum_{j=-h}^{h} e^{iju}\right]^n$$

Now since

$$\int_0^\pi e^{i(j-x)u}\,du = \begin{cases} 0 & j \neq x \\ 2\pi & j = x \end{cases} \qquad \Big/-\pi$$

this probability may be written

$$p_n(x) = \frac{1}{2\pi}\left(\frac{1}{2h+1}\right)^n \int_0^\pi e^{-ixu}\left[\sum_{j=-h}^{h} e^{iju}\right]^n du \qquad \Big/-\pi$$

$$= \frac{1}{2\pi}\left(\frac{1}{2h+1}\right)^n \int_0^\pi e^{-ixu}\left[\frac{e^{\left(h+\frac{1}{2}\right)iu} - e^{-\left(h+\frac{1}{2}\right)iu}}{e^{\frac{iu}{2}} - e^{-\frac{iu}{2}}}\right]^n du \qquad \Big/-\pi$$

$$= \frac{1}{2\pi}\left(\frac{1}{2h+1}\right)^n \int_0^\pi e^{-ixu}\left[\frac{\sin\left(h+\frac{1}{2}\right)u}{\sin\frac{u}{2}}\right]^n du \qquad \Big/-\pi$$

$$= \frac{1}{2\pi}\left(\frac{1}{2h+1}\right)^n \int_0^\pi \cos xu\left[\frac{\sin\left(h+\frac{1}{2}\right)u}{\sin\frac{u}{2}}\right]^n du \qquad (11) \qquad \Big/-\pi$$

The reasoning used by Laplace in this example is quite general and he has thus derived an inversion formula for the generating function of a discrete probability law, viz.

if

$$\varphi(u) = \sum_x e^{iux} p(x) \qquad (12)$$

then

$$p(x) = \frac{1}{2\pi} \int_0^\pi e^{-ixu} \varphi(u)\,du \qquad (13) \qquad \Big/-\pi$$

Although the preceding development was new in Laplace's 1812 text book this was not the first time he had used a somewhat similar device. In a paper (1810) published two years before his book appeared Laplace used much the same approach based on a continuous rectangular distribution; owing to the somewhat dubious limiting processes involved the derivation employed would not be acceptable nowadays. The argument was extended to cover the addition of n variates from any continuous probability law and an improved version of this development appears in Ch. IV (pp. 329—333) of his text book [1]).

The interesting fact emerges from these references that in no case did Laplace use a probability generating function to derive an explicit form of probability law for the sum of n specified random variables. The result (11), for instance, was not intended to be integrated to arrive at the exact answer (given by (2) when $2h + 1$ has replaced k and $x + nh$ written instead of x) but was deliberately left in the form of an integral because Laplace had previously (1785) obtained asymptotic forms for such an integral with increasing n. When Laplace required an explicit form for the probability law of the sum of n specified random variables he used an inductive method which he established in an earlier paper (1781) [2]). Comparing his method and the generating function method of Lagrange's (1773) article Laplace wrote: «sa méthode est très-ingénieuse et digne de son illustre auteur; mais la précédente a ... l'avantage d'être plus directe et plus générale, en ce qu'elle réduit la solution du Problème aux quadratures des courbes, quelle que soit la loi de facilité des erreurs des observations.» These remarks were made before Laplace had developed the artifice resulting in (12) and (13) and it is perhaps significant that apart from four articles on the application of probability theory to natural philosophy, astronomy, and geodesy (three of which were reproduced as Supplements in the third, and final, edition of Laplace's book, 1820) he made no further theoretical advances in this subject after the ·publication of his *Théorie analytique*.

---

[1]) In the course of the demonstration Laplace in effect discovered the «moment generating» property of generating functions with $t = e^{iu}$.

[2]) This method is reproduced in the Appendix, *post*.

It should be mentioned that the expression «fonctions génératrices» originated with Laplace in 1782 to denote such functions as

$$\psi(t) = \sum_x t^x f(x) \quad \text{and} \quad \mu(v) = \int x^v f(x)\, dx$$

These generating functions were used with great effect in the solution of difference and differential equations but Laplace never used the term in connexion with the synthesis of a probability distribution.

Inversion in the continuous case

The first quarter of the nineteenth century saw a number of contributions to the theory of functions which were of considerable importance in the establishment of an inversion formula for the generating function of a continuous probability law. It was possibly Fourier's prize paper of 1811 (not published until 1819/20) which led Gauss to the establishment of the pair of reciprocal relations

$$F(u) = \frac{1}{\sqrt{2\pi}} \int_{-\infty}^{\infty} e^{itu} f(t)\, dt$$

and

$$f(x) = \frac{1}{\sqrt{2\pi}} \int_{-\infty}^{\infty} e^{-ixu} F(u)\, du$$

Unfortunately these formulae lay undiscovered until the end of the century in one of Gauss's notebooks which he completed in 1813. Their entry without comment under the title «Schönes Theorem der Wahrscheinlichkeitsrechnung» is indeed provocative.

However, substantially the same relations, namely the sine and cosine pairs of transforms, were published by Cauchy in 1817 a year after Fourier's integral theorem, viz.

$$f(x) = \frac{1}{\pi} \int_{-\infty}^{\infty} f(u)\, du \int_{0}^{\infty} \cos v(u-x)\, dv \ ^1)$$

had first appeared in print.

---

¹) The inner integral is actually divergent and nowadays the theorem is written as

$$\frac{1}{2} \{f(x+0) + f(x-0)\} = \frac{1}{\pi} \int_{0}^{\infty} dv \int_{-\infty}^{\infty} f(u) \cos v(u-x)\, du$$

With the stage thus set it was not long before Poisson (1824) derived similar formulae for continuous probability laws. Writing

$$\varphi(u) = \int_a^b e^{iux} p(x) \, dx \tag{14}$$

where a or b may be infinite, he obtained, by what would now be called non-rigorous methods, the result

$$\int_{c-x}^{c+x} p(z) \, dz = \frac{1}{\pi} \int_{-\infty}^{\infty} \varphi(u) \, e^{-icu} \frac{\sin xu}{u} \, du \tag{15}$$

This relation may be obtained by formally integrating, between $c \pm x$, $(\sqrt{2\pi})^{-1}$ times the second of Gauss's two reciprocal relations given above: it is thus formally equivalent to the following relation which was not, however, written down explicitly by Poisson.

$$p(x) = \frac{1}{2\pi} \int_{-\infty}^{\infty} e^{-ixu} \varphi(u) \, du \tag{16}$$

Subsequent history

It is an extraordinary fact that although the theory of probability generating functions had achieved a considerable development by the end of the first quarter of the nineteenth century it was almost a hundred years before a synthesis of these results was made and further contributions to the subject published. The cause of this seems to have originated with Laplace's (perhaps personal) dislike of this artifice and the great weight of his authority with nineteenth century mathematicians.

In the analysis of pp. 10 and 11 *ante* — which is a pattern of the only uses he made of e^{iu} — Laplace is never far from his own invention (see Appendix), the discontinuity factor. In fact Bessel (1838) re-wrote Laplace's general (Central Limit) theorem in a form which, when applied to the particular case we considered, would run as follows:

80

$$p_n(x) = \sum_{x_1, x_2, \ldots, x_n}' \left(\frac{1}{2h+1}\right)^n \quad \text{where } \sum' \text{represents a summation over the}$$

variables $x_1, x_2, \ldots x_n$ subject to $\sum_{j=1}^{n} x_j = x$

$$= \sum_{x_1=-h}^{h} \sum_{x_2=-h}^{h} \cdots \sum_{x_n=-h}^{h} \left(\frac{1}{2h+1}\right)^n \frac{\sin \pi(x_1 + x_2 + \cdots + x_n - x)}{\pi(x_1 + x_2 + \cdots + x_n - x)}$$

$$= \frac{1}{\pi} \int_0^\pi \prod_{j=1}^{n} \left\{ \sum_{x_j=-h}^{h} \frac{e^{iux_j}}{2h+1} \right\} e^{-iux} du, \quad \text{etc.}$$

A similar procedure was followed by Ellis (1849) and the close link between the discontinuity factor and generating function approaches was emphasized by Cauchy's four 1853 articles [1]).

Developments of the Central Limit theorem in articles and text-books written between Laplace's discovery of it and the first world war were almost the only occasions when techniques at all resembling probability generating functions were utilized. Some of these writers followed Laplace's introduction of $p_n(x)$ closely, others preferred the Bessel approach. Without giving exact references we mention Poisson, Galloway, De Morgan, Jullien, Laurent and Charlier as favouring the generating function approach, and Glaisher, Tchebychef, Sleshinski, Pizzetti, Liapounoff and Markoff, the discontinuity factor

---

[1]) Cauchy's contributions to the development of the technique of probability generating functions and even to the discovery of the properties of the probability law which bears his name, have often been exaggerated. In the first and second of the four papers cited Poisson's relation (15) is derived and used to find the probability distribution of a linear function of n equal variables each distributed as

(I) $\quad \sqrt{\frac{k}{\pi}} \, e^{-kx^2}$ \qquad\qquad (II) $\quad \frac{k}{2} \, e^{-k|x|}$ \qquad ($n = 2$ only)

It is further shown that the probability generating function (with $t = e^{-iu}$) of the law $\dfrac{k}{\pi} \dfrac{1}{1 + k^2 x^2}$ is $e^{-\frac{|u|}{k}}$ but no deductions are drawn about the sum of n such variables. Cauchy is thus less discerning than Poisson (1824) was before him for that author had derived the probability distribution of the sum of n variables each distributed according to $\dfrac{1}{\pi} \dfrac{1}{1 + x^2}$ and had pointed out that however great n may be the probability of the mean of n such variables lying between given limits is the same as that of an individual variable. Bienaymé (1853) made a similar observation. The last two of Cauchy's four papers are devoted to improving Laplace's «proof» of the Central Limit theorem for n equal components.

introduction. Strictly speaking the probability generating function did not reappear as an independent entity in probability theory until Poincaré (1896) devoted just over two pages of his text book to «characteristic functions», namely probability generating functions with $t = e^z$; and he added nothing new except the title.

The publication of Gauss's posthumous note mentioned earlier received notice by Hausdorff (1901) who made explicit use of probability generating functions with $t = e^{iu}$ without, however, giving them a title. The only novelty is the derivation of the distribution law of the linear function

$$\sum_{j=1}^{n} \left\{ \left(j - \frac{1}{2} \right) \pi \right\}^{-1} x_j$$

of the n variables $x_1, x_2, \ldots x_n$ each distributed according to the law

$$p(x) = \frac{1}{2} e^{-|x|} \qquad\qquad -\infty < x < \infty$$

The result is

$$p_n(x) = \left(e^{\frac{\pi x}{2}} + e^{-\frac{\pi x}{2}} \right)^{-1} \qquad\qquad -\infty < x < \infty$$

The first author to attempt the development of an independent theory of probability generating functions was Kameda (1916, 1925). To him is due the disinterment of the title ‚generating function' and the theorem which assimilates discrete and continuous laws for the purpose of inversion. Kameda was closely followed in time by von Mises (1919) who used Stieltjes integrals in the representation of the probability distribution, Soper (1922) who threw some of the results of the English statistical school into generating function form without, however, using an inversion formula, and Lévy (1925) whose work on the asymptotic behaviour of various types of probability law has become classic. None of these successive authors was aware of the work of the others; all their papers are readily accessible and need not be mentioned further here.

Appendix

Laplace's method of determining the probability distribution of the sum of n random variables

As mentioned in the body of this article Laplace (1781) devised a direct method of determining the probability distribution of the sum of n random variables each subject to the same probability distribution. Essentially his method is to write

$$p_n(x) = \int_{-\infty}^{\infty} p_{n-1}(x-z)\, p(z)\, dz$$

in the continuous case and he later (1810) extended this to include discrete distributions by means of the relation

$$p_n(x) = \sum_{j=0}^{\infty} p_{n-1}(x-j)\, p(j)$$

These inductive formulae are now one of the standard methods of deriving probability distributions of the sums of random variables (Kendall, 1945, Ch. 10) and of themselves, perhaps, of little interest. However, in the application of these relations to probability distributions of limited range Laplace made brilliant use of a discontinuity factor nearly fifty years before the supposed introduction of such factors by Libri in 1827 (Burckhardt, 1915).

The simplest example of Laplace's procedure is to be found in his 1810 article. The individual continuous random variables are assumed to be rectangularly distributed over the interval $(0, h)$ and a discontinuity factor ζ is introduced by writing

$$p(x) = \begin{cases} \dfrac{1}{h} & 0 \leqslant x \leqslant h \\[2ex] \dfrac{1-\zeta^h}{h} & h < x < \infty \end{cases}$$

so that the range of the variable has become infinite; the final stage of the procedure is to write $\zeta = 1$.

Now

$$p_n(x) = \int\limits_0^x p_{n-1}(x - z_1)\, p(z_1)\, dz_1$$

$$= \int\limits_0^x p(z_1)\, dz_1 \int\limits_0^{x-z_1} p_{n-2}(x - z_1 - z_2)\, p(z_2)\, dz_2$$

$$= \int\limits_0^x p(z_1)\, dz_1 \int\limits_0^{x-z_1} p(z_2)\, dz_2 \ldots \int\limits_0^{x-z_1-z_2-\ldots-z_{n-2}} p(x - z_1 - z_2 - \ldots - z_{n-1})\, p(z_{n-1})\, dz_{n-1}$$

$$= \frac{1}{h^n}(1 - \zeta^h)^n \int\limits_0^x dz_1 \int\limits_0^{x-z_1} dz_2 \ldots \int\limits_0^{x-z_1-z_2-\ldots-z_{n-2}} dz_{n-1} \tag{a}$$

$$= \frac{1}{h^n}(1 - \zeta^h)^n \int\limits_0^x dz_1 \int\limits_0^{x-z_1} dz_2 \ldots \int\limits_0^{x-z_1-z_2-\ldots-z_{n-3}} (x - z_1 - z_2 - \ldots - z_{n-2})\, dz_{n-2}$$

$$= \frac{1}{h^n}(1 - \zeta^h)^n \int\limits_0^x dz_1 \int\limits_0^{x-z_1} dz_2 \ldots \int\limits_0^{x-z_1-z_2-\ldots-z_{n-4}} 1/2!\,(x - z_1 - z_2 - \ldots - z_{n-3})^2\, dz_{n-3}$$

$$= \frac{1}{h^n}(1 - \zeta^h)^n \frac{1}{(n-1)!}\, x^{n-1}$$

$$= \frac{1}{h^n} \frac{1}{(n-1)!} \sum_{j=0}^n (-1)^j \binom{n}{j} \zeta^{hj}\, x^{n-1} \tag{b}$$

$$= \frac{1}{h^n} \frac{1}{(n-1)!} \sum_{j=0}^{\min.\,(n,\,[x])} (-1)^j \binom{n}{j} (x - hj)^{n-1} \qquad 0 \leqslant x \leqslant nh$$

Two steps in this derivation need explanation. At (a) it has been assumed that each of the n variables is making a contribution to x which exceeds h: that is to say all n variables are now measured from h instead of from zero. In fact either none, one, two, ... n of these variables falls in the $(0, h)$ portion of the range and the respective frequencies of these possibilities are provided by the numerical coefficients of the expansion of $(1 - \zeta^h)^n$. The term in ζ^{hj}, for instance, denotes that j of the variables have been given a variate value h in excess of the truth; ζ^{hj} thus acts on x in (b) to reduce it by hj.

The discrete analogue of the preceding development appears on pp. 253—256 of the *Théorie analytique*.

84

References

Bessel, F. W. (1838) Untersuchungen über die Wahrscheinlichkeit der Beobachtungsfehler. *Astron. Nachr.* xv, 369.

Bienaymé, J. (1853) Considérations à l'appui de la découverte de Laplace sur la loi de probabilité dans la méthode des moindres carrés. *Comptes Rendus,* xxxvii, 309.

Burckhardt, H (1915) Trigonometrische Reihen und Integrale (bis etwa 1850). *Encyk. Math. Wissens.* Bd. II, I. Teil, 2. Hälfte, 825.

Cauchy, A. L. (1817) Sur une loi de reciprocité qui existe entre certaines fonctions. *Bull. philomat.* Paris, 121.

— (1853) Mémoire sur les coefficients limitateurs ou restricteurs.
Sur les résultats moyens d'observations de même nature, et sur les résultats les plus probables.
Sur la probabilité des erreurs qui affectent des résultats moyens d'observations de même nature.
Mémoire sur les résultats moyens d'un très grand nombre d'observations. *Comptes Rendus,* xxxvii, 150, 198, 264, 381.

De Moivre, A. (1711) De mensura sortis, seu, de probabilitate eventuum in ludis a casu fortuito pendentibus. *Philos. Trans.* xxvii, 213.

— (1730) *Miscellanea analytica de seriebus et quadraturis.* London.

Euler, L. (1788) Éclaircissemens sur le mémoire de Mr. de La Grange inseré dans le Ve volume des Mélanges de Turin, concernant la méthode de prendre le milieu entre les résultats de plusieurs observations etc. *Nova acta acad. sci. Petropol.* iii (1785) 289.

Fourier, J. (1816) Théorie de la chaleur. *Ann. chim. phys.* iii, 350.

Gauss, C. F. (1900) *Werke,* viii, 88.

Hausdorff, F. (1901) Beiträge zur Wahrscheinlichkeitsrechnung. *Ber. Verhand. könig. sächs. Gesell. Wissens. Leipzig, Math.-Phys. Cl.,* liii, 152.

Kameda, T. (1916) Theorie der erzeugenden Funktionen und ihre Anwendung auf die Wahrscheinlichkeitsrechnung. *Proc. Tôkyô. Math.-Phys. Soc.,* Ser 2, viii, 262, 336.

— (1925) Theory of generating functions and its application to the theory of probability. *Jour. Fac. Sci. Univ. Tokyo,* Sec. I, i, 1.

Kendall, M. G. (1945) *The advanced theory of statistics.* London.

Lagrange (1773) Mémoire sur l'utilité de la méthode de prendre le milieu entre les résultats de plusieurs observations; dans le quel on examine les avantages de cette méthode par le calcul des probabilités; et où l'on résoud différens problèmes relatifs à cette matière. *Miscellanea Taurinensia,* v, 167.

Laplace, P. S. de (1812) *Théorie analytique des probabilités.* Paris.

— (1781) Mémoire sur les probabilités. *Hist. Acad. Roy. Sci.* (1778), 227.

— (1810) Mémoire sur les approximations des formules qui sont fonctions de très-grands nombres, et sur leur application aux probabilités. *Mém. Sci. Math. Phys. Inst. France* (1809), 353.

— (1785) Mémoire sur les approximations des formules qui sont fonctions de très-grands nombres. *Hist. Acad. Roy. Sci.* (1782), 169.

Lévy, P. (1925) *Calcul des probabilités*. Paris.

Mises, R. von (1919) Die Fundamentalsätze der Wahrscheinlichkeitsrechnung. *Math. Zeitschr.* iv, 1.

Poincaré, H. (1896) *Calcul des probabilités*. Paris.

Poisson, S. D. (1824) Sur la probabilité des résultats moyens des observations. *Connaissance des Temps* (1827), 273.

Simpson, T. (1757) An attempt to show the advantage arising by taking the mean of a number of observations, in practical astronomy. *Miscellaneous Tracts* ... London.

Soper, H. E. (1922) *Frequency arrays*. Cambridge.

Todhunter, I. (1865) *A history of the mathematical theory of probability from the time of Pascal to that of Laplace*. Cambridge.

Boscovich and the Combination of Observations

CHURCHILL EISENHART
(1961)

Boscovich was the first to devise a completely objective procedure for uniquely determining the coefficients of a two-parameter line $y = \alpha + \beta x$ from a set of three or more observational points. Like the *median*, Boscovich's procedure is comparatively insensitive to the more extreme of a set of observations, and is especially well suited to summarizing the linear trend evidenced by a more or less heterogeneous set of data compiled from various sources, or obtained by a measurement procedure that has a tendency to yield occasional discordant values. It is most fitting that an opportunity to review the work of the forerunner should arise at the present time, just as renewed attention is being given to methods of the type proposed by Boscovich, both in statistical analysis[30], and in the solution of problems of operations analysis by the technique of linear programming.[28]

1. BOSCOVICH'S CRITERIA FOR DETERMINING THE STRAIGHT LINE OF BEST FIT TO A SET OF OBSERVATIONAL POINTS

Nearly half a century before Legendre announced (1805) his Principle of Least Squares, and twenty years before the birth of Gauss (1777), Roger Joseph Boscovich had formulated and applied[5] (pp. 391–2) the principle that, given more than two pairs of observed values of variables x and y connected by a linear functional relationship of the form $y = \alpha + \beta x$, then the values (a and b) that one should adopt for α and β in order to obtain the line ($y = a + bx$) that is most nearly in accord with all of the observations should be those determined jointly by the two conditions:

(I) *The sums of the positive and negative corrections (to the y-values) shall be equal.*
(II) *The sum of (the absolute values of) all of the corrections, positive and negative, shall be as small as possible.*

He justified these two conditions as follows: the first he considered to be required by the traditional assumption that positive and negative errors are of equal probability; and the second, to be necessary in order to bring the solution into closest possible agreement with the observations.

Condition (I) states that a and b, the intercept and slope of the best-fitting line $y = a + bx$, must satisfy the equation

$$(1) \qquad \sum_{i=1}^{n} (y_i - a - bx_i) = 0$$

which can also be written in the form

$$(2) \qquad \bar{y} - a - b\bar{x} = 0$$

where

$$\bar{x} = \frac{1}{n} \sum_{i=1}^{n} x_i \text{ and } \bar{y} = \frac{1}{n} \sum_{i=1}^{n} y_i$$

are the arithmetic means of the n observed values of x and y, respectively. In other words, Condition (I) states that the best-fitting line $y = a+bx$ shall pass through the centroid $(\bar{x}, \bar{y})$ of the observational points.

Condition (II) states that a and b must satisfy the equation

$$(3) \qquad \sum_{i=1}^{n} |y_i - a - bx_i| = \text{minimum.}$$

Replacing a in equation (3) by its value

$$(4) \qquad a = \bar{y} - b\bar{x}$$

implied by equation (2), it is seen that Condition (II) *in conjunction with Condition (I)* requires that the slope b shall satisfy the equation

$$(5) \qquad \sum_{i=1}^{n} |(y_i - \bar{y}) - b(x_i - \bar{x})| = \text{minimum.}$$

Consequently, determination of the 'Boscovich line' corresponding to a given set of observational points reduces to determining its slope b from equation (5) and then evaluating a from equation (4).

Boscovich seems to have evolved these criteria for determining a line of best fit to observational data sometime between 1755 and 1757. Near the end of his joint treatise with Maire[4] on determination of the Figure of the Earth from measurements of the lengths of meridian arcs and of seconds pendulums at different latitudes, published in 1755, Boscovich examines (pp. 499–501) the lengths of five meridian arcs measured at five different latitudes, including the arc measured by Maire and himself in the vicinity of Rome, and finds that they yield mutually inconsistent values for the ellipticity of the Earth when considered in pairs. He makes no attempt to educe a consensus from this conflicting evidence, and seems to consider (p. 513) such an objective to be impossible to attain from these data. In 1757 he published a summary[5] of his work with Maire, and it is in this summary that he states for the first time his two criteria for a line of best fit, and reports the result of analysing the aforementioned meridian-arc lengths by this method (pp. 391–2). In this first pronouncement of his method he gives no indication of how he solved equation (5) above to obtain the 'best' value of the slope b. On p. 392, however, he seems to say that he has now solved the problem of how best to combine such observations for the purpose at hand, which he had not considered at the time of publication (1755) of the work under review. Are we to infer from this that he was already in possession of a complete solution?

2. BOSCOVICH'S ALGORITHM FOR DETERMINING THE SLOPE OF HIS LINE OF BEST FIT

In 1760, in a prose *Supplement* to the second volume of a three-volume treatise on Natural Philosophy in Latin hexameters* by Benedict Stay,[6] Boscovich restated his two conditions for determining the line of best fit to observational data, and gave a very useful procedure of his own invention for solving equation (5) above, followed by a step-by-step illustration of its application in terms of the five meridian-arc lengths that he had considered previously. 'Boscovich's exposition of his method takes a geometrical form: it is simple, clear and instructive'[17] (Todhunter, p. 332). It is contained in Sections 385–96 of the *Supplement*[4] (pp. 420–5). A French translation of these sections is included in the *Note* appended to the French edition[7] of his joint treatise with Maire. His reasoning, in outline, is as follows:

In Fig. 1, let A, B, C, D, and E denote the five observational points (x_1, y_1), (x_2, y_2), (x_3, y_3), (x_4, y_4), and (x_5, y_5), respectively; and let G denote their centroid $(\bar{x}, \bar{y})$. Consider a line LL' through G and rotated clockwise through a very small angle from the vertical. At this 'moment' the vertical *distances* of the observational points from the line LL' (i.e. the absolute values of the *residuals*) will all be 'enormous', and so will their sum Σ, i.e. the left-hand side of equation (5). If now the line LL' be rotated slowly in a clockwise direction, all of the vertical distances will decrease until the line intersects the 'first' of the points, E in Fig. 1, after which the distance from the 'first' point (E) to the line will steadily increase, but the distances from the other points (A, B, C, and D) will continue

* Commenting on this treatise in 1873, Isaac Todhunter remarked[17] (p. 322): 'The number of students interested both in Natural Philosophy and in Latin Verse could scarcely ever have been large; and is probably less now than formerly.'

to decrease until the line 'reaches' the 'second' point, A in Fig. 1, after which the distance from the 'second' point (A) to the line will steadily increase, and so forth. By virtue of the similarity of the triangles formed at any 'instant' by the horizontal line through G, the vertical lines through the respective observational points, and the rotating line LL$'$, it is evident that the change in the distance of

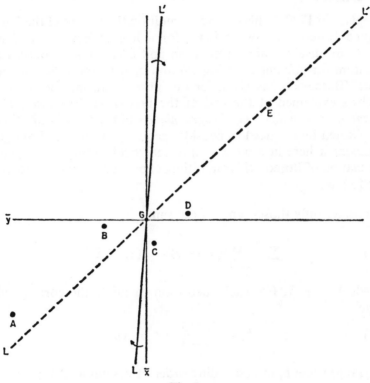

Fig. 1

the line from any observational point (x_i, y_i) produced by a given rotation of the line LL$'$ in the clockwise direction is proportional to the horizontal distance $|x_i - \bar{x}|$ of the point from the vertical line through G; and the 'change' will be positive for points 'already passed' and negative for points 'not yet reached'. Hence the corresponding change in the sum Σ will be proportional to the difference of the sums of the 'horizontal distances' $|x_i - \bar{x}|$ of (a) the points 'already passed' and (b) the points 'not yet reached'; and the change in Σ will be negative until the first of these sums equals or exceeds the second. Consequently, the sum Σ, i.e. the left-hand side of equation (5), achieves its minimum value at the 'moment' when the line LL$'$, rotating clockwise about G, 'reaches' the first observational point such that the sum of the 'horizontal distances' $|x_i - \bar{x}|$

of all points 'reached or passed' exceeds one-half of the sum of the 'horizontal distances' of all of the points, i.e.

$$\sum_{i=1}^{n} |x_i - \bar{x}|;$$

and the desired value of b in equation (5) is the slope of the line at this 'moment'.

Laplace in 1789, in his second memoir on the Figure of the Earth,[8] adopted Boscovich's two criteria for a line of best fit, and gave (pp. 32–6) an algebraic formulation and derivation of Boscovich's algorithm for solving equation (5) above, with the following comments: 'Boscovich has given for this purpose an ingenious method which is explained at the end of the first (French) edition of his *Voyage Astronomique et Geographique*, but as its utilization is complicated by the need to consider geometrical figures, I am going to present it here in a most simple analytical form.' This algebraic formulation of Boscovich's algorithm, expressed in modern notation, is as follows:

Consider only those terms of the sum

(6) $$\Sigma = \sum_{i=1}^{n} |(y_i - \bar{y}) - b(x_i - \bar{x})|$$

for which $x_i \neq \bar{x}$. For each such term calculate the corresponding 'slope'

(7) $$b_i = \frac{y_i - \bar{y}}{x_i - \bar{x}}, \ (x_i \neq \bar{x}).$$

Arrange these b_i in descending order of magnitude, thus

(8) $$b_{(1)} \geqslant b_{(2)} \geqslant b_{(3)} \geqslant \ldots \geqslant b_{(m)}, \ (m \leqslant n).$$

Arrange the absolute values of their denominators in the same order, thus

(9) $$|x_{(1)} - \bar{x}|, \ |x_{(2)} - \bar{x}|, \ \ldots, \ |x_m - \bar{x}|$$

Compute the sum of all of the 'absolute denominators',

(10) $$D = \sum_{j=1}^{m} |x_{(j)} - \bar{x}|,$$

Then the sum Σ of (6) will be a minimum for $b = b_{(r)}$, the rth term of the sequence (8), where r is the smallest integer for which the partial sum of the first r terms of the sequence (9) equals or

exceeds one-half of the sum of the entire sequence, i.e. the smallest
r for which

(11)
$$\sum_{j=1}^{r} |x_{(j)} - \bar{x}| > \frac{1}{2}D.$$

If in determining r the inequality sign holds, then the solution
$b = b_{(r)}$ is unique; but if the equality sign holds, then the solution
is not unique and Σ attains the same minimum for any value of b
between $b_{(r)}$ and $b_{(r+1)}$ *inclusive*.

It should be noticed that this algorithm is of greater generality
than may seem to be the case at first sight. Its derivation does not
depend in any way upon the fact that the first point $(\bar{x}, \bar{y})$ is the
centroid of the observational points (x_i, y_i). Consequently, if it is
desired to determine the coefficients a and b of the line $y = a + bx$
that passes through some particular point (x_0, y_0) *and satisfies Condition* (II), then the required value of the slope b can be found
directly from Boscovich's algorithm with $\bar{x}$ and $\bar{y}$ replaced by x_0 and
y_0, respectively; and the corresponding value of a then found from
the relation $a = y_0 - bx_0$.

3. DISCUSSION

To appreciate fully the significance of Boscovich's contribution one
needs to consider 'the state of the art' of Combination of Observations at the middle of the eighteenth century.

'For a long time', according to Laplace[12] (Book II, Chapter IV,
Article 24, Nat'l ed., p. 351), it had been the custom 'to choose the
circumstances the most favourable' to the determination of magnitudes of principal interest, that is, 'those for which errors of the
observations would alter as little as possible' the values assigned to
the magnitudes of interest, and then to take the arithmetic mean of
the resulting values. Thus, if y were known to be proportional to x,
to determine the constant of proportionality, i.e. the slope β of the
line $y = \beta x$, one would choose values $x_1, x_2, \ldots, x_n$ for x that were
as large as possible, observe the corresponding values $y_1, y_2, \ldots, y_n$
of y, and then take the arithmetic mean of the 'observed ratios'
$b_i = y_i/x_i$, that is, take

(12)
$$\bar{b} = \frac{1}{n} \sum_{1}^{n} b_i = \frac{1}{n} \sum_{1}^{n} \left(\frac{y_i}{x_i}\right) = \overline{\left(\frac{y_i}{x_i}\right)},$$

as the value of β indicated by the data. Similarly, in the case of the
two-parameter line $y = \alpha + \beta x$, one would choose x's as widely
spaced as possible, and then take as the value of β the arithmetic
mean of either all of the 'observed slopes'

$$b_{ij} = (y_j - y_i)/(x_j - x_i), \; i = 1, 2, \ldots, n-1; \; j = i+1, i+2, \ldots, n),$$

or perhaps only of those 'observed slopes' that corresponded to large differences in abscissae. Such was the traditional approach, which Boscovich followed in his original analysis[4,7] (pp. 497–503; pp. 479–84) of the five meridian-arc lengths: first, he took the arithmetic mean of all ten 'observed slopes'; then, he rejected two of these 'which are so different from the others and correspond to such small differences in latitude', and adopted the arithmetic mean of the remaining eight. He was dissatisfied with both answers[4,7] (p. 503; p. 484).

Roger Cotes (1682–1716), in the final paragraph of one[1] of his *Opera Miscellanea* published posthumously in 1722, suggested that, in place of the simple arithmetic mean, one should take the *weighted* arithmetic mean of individual determinations of a single magnitude, the weights to be inversely proportional to the 'influences' of errors in the respective directly observed quantities on the corresponding determinations of the magnitude in question. In other words, if the observational points are to be fitted by a straight line *through the origin, $y = \beta x$*, and the observational points are believed to be affected by errors in the y-direction only, then the value to take for β is the weighted arithmetic mean of the individual determinations $b_i = y_i/x_i$, with weights *inversely* proportional to $1/x_i$, that is,

$$(13) \qquad \bar{b} = \frac{\sum_{1}^{n} x_i \dfrac{y_i}{x_i}}{\sum_{1}^{n} x_i} = \frac{\sum y_i}{\sum x_i} = \bar{y}/\bar{x},$$

the ratio of the means, which is the solution of the equation

$$(14) \qquad \sum_{i=1}^{n} (y_i - bx_i) = 0.$$

In other words, if a set of observational points are to be fitted by a line of the form $y = \beta x$, then Cotes's modification of the traditional procedure implies that the 'best' value to take for β is the solution of equation (14), that is, the value determined by the condition of *zero sum of residuals*. This may be regarded as an extension of the Principle of the Arithmetic Mean, since the arithmetic mean a of a set of numbers $y_1, y_2, \ldots, y_n$ is the solution of the equation

$$\sum_{1}^{n} (y_i - a) = 0.$$

The requirement of zero sum of residuals alone is not sufficient to determine the 'best' values for all of the coefficients of a two- or multi-parameter line or curve from a set of observational points. Consequently, in 1748 Euler[2] and Mayer[3] independently devised the so-called Method of Averages; this consists of subdividing the observational points into as many subsets as there are coefficients to be determined, the subdivision being in terms of the values of (one of) the independent variable(s), and then applying the condition of zero sum of residuals to the points of each subset. Thus, to determine the values for the coefficients of the two-parameter line $y = \alpha + \beta x$ by the Method of Averages one divides the observational points into two 'equal' subsets according to the values of x, and takes as the values of α and β the values a and b determined simultaneously by the two equations

$$(15) \qquad \begin{cases} \sum_{x \leqslant x_0} (y - a - bx) = 0 \\ \sum_{x > x_0} (y - a - bx) = 0 \end{cases}$$

that is by the equations

$$(16) \qquad \begin{cases} \bar{y}' - a - b\bar{x}' = 0 \\ \bar{y}'' - a - b\bar{x}'' = 0 \end{cases}$$

where $(\bar{x}', \bar{y}')$ and $(\bar{x}'', \bar{y}'')$ are the centroids of the points for which $x \leqslant x_0$ and $x > x_0$, respectively.

Provided that there are at least as many x-wise distinct observational points as there are unknown parameters to be determined, the Method of Averages will always come up with a value for each parameter, but there is some arbitrariness and room for subjective choice in the formation of the subgroups, with consequent effect on the answers obtained: If the x's are more or less equally spaced throughout their range, then it is customary to choose x_0 so that there are an equal *number* of x's greater and less than x_0; but if the number of observational points is odd, the outcome will depend on whether the middlemost point is included in the left-hand or right-hand group. If the x's are not even symmetrically dispersed within their range, then the choice of the 'best' subdivision becomes highly subjective, and the end results correspondingly 'arbitrary'.

Laplace clearly recognized the great value of Boscovich's procedure as an objective and unique method of combination of observations subject to a linear law, and employed it not only in his 1789 memoir[8] on the determination of the Figure of the Earth from measured lengths of meridian arcs and seconds pendulums at different latitudes, to which we have already referred, but also in his expanded discussion of this same subject in Book III, Chapter 5, Sections 40–42, of his *Mécanique Céleste*[9], published in 1799. In Section 40 he provides the natural extension of Boscovich's technique to the case of observational points of unequal weight, expressing Conditions (I) and (II) in terms of *weighted residuals* and appropriately modifying his own previous algebraic formulation and derivation of Boscovich's algorithm. But he fails to make any mention of Boscovich! This omission was corrected by Nathaniel Bowditch in his English translation[16] of Laplace's *Mécanique Céleste*, in a lengthy footnote to Book III, Section 40, which begins and concludes as follows:

'This method, proposed by Boscovich, and peculiarly well adapted to the present problem [of deducing the ellipticity of the Earth from meridian-arc measurements], is not now so much used as it ought to be; instead of it, the principle of making the sum of the squares of the errors a minimum, is generally adopted. This method of least squares . . . is extremely well adapted to a set of observations . . . subject to the same degree of uncertainty. . . . But if the measure of one . . . should differ very much from the rest, the method of least squares, applied in the usual manner, would give by far too great an influence to this defective observation . . . (p. 434).

'We shall hereafter find, in several instances, that the method of least squares, when applied to a system of observations, in which one of the extreme errors is very great, does not generally give so correct a result as the method proposed by Boscovich. The reason is, that in the former method, this extreme error affects the result in proportion to the *second* power of the error; but in the other method, it is as the *first* power, and must therefore be less' (p. 438).

Half a century later we find Edgeworth recommending use of the *median** instead of the arithmetic mean, and Boscovich's slope estimate $b_{(r)}$ (which he termed a 'weighted median') instead of the least squares estimate, for the very same purpose—to reduce the influence of 'discordant' observations.[18-23] Edgeworth, however, attributed Boscovich's algorithm to Laplace. Dropping Boscovich's Condition (I), he devised[18] a 'double median' method for finding values of a and b that correspond to the minimum minimorum of the sum

$$\sum_1^n |y_i - a - bx_i|;$$

and a 'multiple median' technique[22,23] for extending the procedure to the fitting of curves and surfaces involving more than two unknown parameters. Bowley has provided a brief summary[24] (pp. 103–9) of Edgeworth's work along these lines, with additional

---

* The median of an odd number of observations $y_1, y_2, \ldots, y_n$, defined to be the middle value in order of size, is the solution of the equation

$$\sum_1^n |y_i - a| = \text{min.}$$

references. More recent work on 'minimizing the sum of absolute deviations' will be found in papers by Rhodes,[25] Singleton,[26] and Bejar.[28,29] In the first of his two papers, Bejar develops the connection between minimizing sums of absolute deviations and some of the techniques of 'linear programming', noted earlier by Harris.[27] None of these writers, I regret to say, mentions Boscovich.

4. CONCLUDING REMARKS

Up to this point our discussion has been completely independent of probability considerations, except for mention of the fact that Boscovich seems to have regarded his Condition (I) as necessary for conformity with the traditional assumption that positive and negative errors are equally probable. We have not found it necessary to mention, or even allude to, any particular probability distribution or law of errors. This is because Boscovich's contribution to the Combination of Observations belongs to a chain of developments, starting with the Principle of Arithmetic Mean and culminating in Legendre's advocacy of the Method of Least Squares, in which the merit of any particular method of adjustment of observations was to be judged in terms of the resulting pattern of the *residuals*.* The actual *errors* of particular observations were regarded as unknown and unknowable. Consequently, one was limited to consideration of their manifestations as evidenced by the *residuals* corresponding to any particular *adjustment* of the observations, and the 'best' adjustment was the one that made the residuals as small as possible. If one seeks to minimize the apparent inconsistency of a set of observations as measured by some simple function of their *residuals*, then the practical requirements of objectivity, general applicability, unique solutions, and computational simplicity lead to adoption of the principle of LEAST SUM OF SQUARED RESIDUALS[10,11] (Preface, p. viii, and Appendix, pp. 72–5. Book II, Section 3, Article 186, *Werke*, Vol VII, pp. 241–3; Eng. Trans., pp. 269–71). Gauss explicitly mentions Boscovich's earlier proposal in this connection[11] (*loc. cit.*).

It is largely for these reasons, I believe, that the Method of Least Squares rapidly pushed Boscovich's method into the background. The ascendency and ultimate supremacy of the Method of Least Squares, to the almost complete exclusion of all other procedures, was greatly aided, of course, by the exceptionally thorough reformulation and development of the Method of Least Squares by

---

* If $y_1, y_2, \ldots, y_n$ are observed values of a magnitude α then $y_1 - \alpha = e_1$, $y_2 - \alpha = e_2, \ldots, y_n - \alpha = e_n$ are the *errors* of the respective observations. If, the value of α being unknown, one adopts some particular value for it, say a, on the basis of the observations, then $y_1 - a = r_1, y_2 - a = r_2, \ldots, y_n - a = r_n$ are the *residuals* of the observations corresponding to the *adjusted value a*.

Gauss[11,14,15] (Book II, Section 3, pp. 205–24, *Werke*, Vol VII, pp. 225–45; Eng. Trans., pp. 249–73) and Laplace[12] (Book II, Chapter 4, Nat'l ed., pp. 309–54) in terms of the mathematical theory of probabilities and the concept of a *probability distribution* (or *law*) *of errors* introduced by Thomas Simpson (1755).

In Section 2 of the Second Supplement[13] dated February 1818, to the third edition of his *Théorie Analytique des Probabilités*, Laplace reproduces (Nat'l ed., pp. 571–3) his own previous algebraic formulation and derivation of Boscovich's algorithm, again without reference to Boscovich. Later in the same section, he refers (Nat'l ed., p. 576) to this procedure as the *method of situation*. He then proceeds to show (Nat'l ed., pp. 576–7) that the *method of situation* may be expected to yield more precise results than his own *most advantageous method* (considered in Section 23 of Chapter IV) whenever the central ordinate $\phi(0)$ of the law of error involved, assumed to be of the form $\phi(x^2)$, is greater than $1/\sigma$, where σ is the root-mean-square error; and, by implication, more precise results than the *method of least squares* whenever $\phi(0) > 1/\sigma\sqrt{2\pi}$, the central ordinate of the so-called 'normal distribution'. He next seeks (Nat'l ed., pp. 577–80) to determine the most precise (or, as we would say, *minimum variance unbiased*) weighted average of the results obtained by the method of situation and his most advantageous method. He obtains very complex expressions for the respective optimum weights, involving the first and second derivations of the law of error at its centre, comments that owing to our usual ignorance of the exact form of the law of error involved these expressions are of no practical value, but notes (Nat'l ed., p. 580) that if the law of error is the error function $ce^{-h^2x^2}$, then the optimum weights are 1 and 0 for the most advantageous method and method of situation, respectively, so that the *method of situation* adds nothing to the result of the *most advantageous method*, which when the law of error is the error function is equivalent to the Method of Least Squares. It remained for Edgeworth to show in 1888[22] that the converse is true, i.e. Boscovich's method receives the unit weight, and is the 'best possible' method, when the law of error is $ce^{-h|x|}$, the law of error Laplace first considered (1774).

NOTES

1. Roger Cotes (1682–1716), 'Aestimatio errorum in mixta mathesi, per variationes partium trianguli plani et spherici', *Opera Miscellania* (appended to his *Harmonia Mensurarum*, Cantabrigiae, 1722), pp. 1–22.
2. Leonhard Euler (1707–83), 'Pièce qui a remporté le prix de L'Académie royale des sciences en 1748, sur les inégalities du mouvement de Saturne et de Jupiter'. Paris, 1749.
3. Johann Tobias Mayer (1723–62), 'Abhandlung über die Umwälzung des Mondes um seine Axe', *Kosmographische Nachrichten und Sammlungen*, Vol I (1748), pp. 52–183 (published 1750).

4. Christopher Maire (1697–1767) and Roger Joseph Boscovich (1711–87), *De Litteraria Expeditione per Pontificiam ditionem ad dimetiendas duas Meridiani gradus, et corrigendam mappam geographicam, jussu, et auspiciis Benedicti XIV Pont. Max. suscepta.* Romae, 1755. French translation.[7]

5. Boscovich, 'De Litteraria Expeditione per Pontificiam ditionem, et Synopsis amplioris Operis, ac habentur plura ejus ex exemplaria etiam sensorum impressa', *Bononiensi Scientiarum et Artum Instituto Atque Academia Commentarii,** Tomus IV, pp. 353–96. 1757.

6. Benedict Stay (1714–1801), *Philosophiae Recentioris, a Benedicto Stay in Romano Archigynasis Publico Eloquentare Professore, versibus traditae, Libri X, cum adnotationibus et Supplementis P. Rogerii Josephi Boscovich S.J.,* Tomus II. Romae, 1760.

7. Maire and Boscovich, *Voyage Astronomique et Géographique dans l'État de l'Église, entrepris par l'Ordre et sous les Auspices du Pape Benoît XIV, pour mesurer deux degrés du méridien, et corriger la Carte de l'État ecclesiastique.* Paris, 1770.

8. Pierre Simon, Marquis de Laplace (1749–1827), 'Sur les degrés mesurés des méridiens, et sur les longueurs observées sur pendule', *Histoire de l'Académie royale des inscriptions et belles lettres, avec les Memoires de littérature tirez des registres de cette académie, Année 1789,* 18–43 of the *Mémoires.* Paris, 1792.

9. Laplace, *Traité de Mécanique Céleste,* Vol II. Paris, 1799. (Reprinted as Vol II of *Œuvres de Laplace,* Paris, 1843, National Edition, Gauthier-Villars. Paris, 1878.)

10. Adrien Marie Legendre (1752–1833), *Nouvelles méthodes pour la détermination des orbites des comètes.* Paris, 1805. (Appendix, 'Sur la Méthode des moindres quarrés', pp. 72–80).

11. Carl Friedrich Gauss (1777–1855), *Theoria Motus Corporum Coelestium in Sectionibus Conicis Solem Ambientium,* Hamburg, 1809. (Reprinted as Vol VII of *Werke,* Gotha, 1871; English translation by Charles Henry Davis. Little, Brown and Company, Publishers, Boston, 1857.)

12. Laplace, *Théorie Analytique des Probabilités.* Paris, 1812. (Third edition, with new Introduction and three Supplements. Paris, 1820; reprinted as Vol VII of *Œuvres de Laplace.* Paris, 1847. National Edition, Gauthier-Villars. Paris, 1886.)

13. Laplace, 'Application du calcul des probabilités aux opérations géodésiques', Second Supplement (February 1818) to third edition (1820) of *Théorie Analytique des Probabilités.*[12] (National edition, pp. 531–80.)

14. Gauss, *Theoria Combinationis Observationum Erroribus Minimis Obnoxiae.* Göttingen, 1823 (Vol IV of *Werke,* pp. 1–53. Göttingen, 1873).

15. Gauss, *Supplementum Theoriae Combinationis Observationum Erroribus Minimis Obnoxiae.* Göttingen, 1828 (Vol IV of *Werke,* pp. 54–108. Göttingen, 1873, pp. 54–108).

16. Laplace, *Mécanique Céleste,* Vol II, English translation, with notes and commentary, by Nathaniel Bowditch. Hilliard, Gray, Little, and Wilkins, Publishers, Boston, 1832.

17. Isaac Todhunter (1820–84), *A History of the Mathematical Theories of Attraction and the Figure of the Earth, from the Time of Newton to that of Laplace,* Vol. I. Macmillan and Company, London, 1873.

18. Francis Ysidro Edgeworth (1845–1926), 'On observations relating to several quantities', *Hermathena,* 1887, No. XIII, Vol VI, pp. 279–85. 1888.

* An early serial publication (Vols 1–7, 1731–91) of what is now referred to as *R. Accademia delle scienze dell'Instituto de Bologna.*

19. Edgeworth, 'On discordant observations', *London, Edinburgh, and Dublin Philosophical Magazine*, Series 5, Vol 23, pp. 364–75. 1887.
20. Edgeworth. Letter calling attention to article in *Hermathena, Phil. Mag.*, Series 5, Vol 24, pp. 222–3. 1887.
21. Edgeworth, 'The Choice of Means', *Phil. Mag.*, Series 5, Vol 24, pp. 268–71. 1887.
22. Edgeworth, 'On a new method of reducing observations relating to several quantities', *Phil. Mag.*, Series 5, Vol 25, pp. 184–91. 1888.
23. Edgeworth, 'On the use of medians for reducing observations relating to several quantities', *Phil. Mag.*, Series 6, Vol. 46, pp. 1074–88. 1923.
24. A. L. Bowley, *F. Y. Edgeworth's Contributions to Mathematical Statistics*. Royal Statistical Society, London, 1928.
25. E. C. Rhodes, 'Reducing observations by the method of minimum deviations', *Phil. Mag.*, Series 7, Vol 9, pp. 974–92. 1930.
26. Robert R. Singleton, 'A method for minimizing the sum of absolute values of deviations', *Annals of Mathematical Statistics*, Vol II, pp. 301–10. 1940.
27. T. E. Harris, 'Regression using minimum absolute deviations' (Answer to Question 25), *American Statistician*, Vol 4, pp. 14–15. 1950.
28. Juan Bejar, 'Regresión en mediana y la programación lineal', *Trabajos de Estadistica*, Vol. 7, pp. 141–58. 1956.
29. Bejar, 'Cálculo práctico de la regresión en mediana', *Trabajos de Estadistica*, Vol 8, pp. 157–73. 1957.
30. John W. Tukey, 'A Survey of Sampling from Contaminated Distributions', Chapter 39 (pp. 448–85) of *Contributions to Probability and Statistics: Essays in Honor of Harold Hotelling* (edited by Ingram Olkin et al.). Stanford University Press, Stanford, California, 1960.

Biometrika (1970), **57**, 1, *p.* 199
Printed in Great Britain

Studies in the History of Probability and Statistics. XXIII
Daniel Bernoulli on the normal law

By O. B. SHEYNIN

Institute for the History of Natural Sciences and Technology,
Academy of Sciences, Moscow

Summary

This paper discusses one of D. Bernoulli's memoirs (1770–1) in which he deduced the 'De Moivre–Laplace' limit theorems, nevertheless credited to De Moivre. The memoir is described in §2 while §1 attempts to sum up Bernoulli's contributions more generally.

1. General

Between 1738 and 1778 D. Bernoulli (1700–82) published seven probabilistic memoirs. The essence of these memoirs, except the memoir to be described in §2, is given by Todhunter (1865). The memoirs contain solutions of important problems in demographic statistics (political arithmetic) and astronomy obtained with the help of probabilistic ideas and methods. As to probability and mathematical statistics proper, Bernoulli was the first to use systematically differential equations for deducing a number of formulae, one of the first to raise the problem of testing statistical hypotheses and the first to introduce 'moral expectation' (due to Cramer) and to study random processes. He is also to be credited, after Lambert, for the second introduction of the maximum likelihood principle (Bernoulli, 1961). In summary, it may be argued that D. Bernoulli's influence upon Laplace, especially concerning applications of probability, was comparable to that of De Moivre.

The account of Bernoulli's memoirs given by Todhunter could well be modernized but the present paper is restricted to the description of the 1770–1 memoir, the second part of which remained unnoticed by Todhunter. For this and other reasons, Todhunter's account of the memoir is unsatisfactory and until now no one has remarked on the appearance in this memoir of the 'De Moivre–Laplace' limit theorems and of the first published small table of the normal distribution. Had these limit theorems been noticed in Bernoulli before, they possibly would not now be called only after De Moivre and Laplace.

2. The normal distribution and the De Moivre–Laplace limit theorems

2·1. *The formula of Wallis*

Bernoulli's 1770–1 memoir was published in two parts with separate numbering of its paragraphs. Preserving this numbering, I shall add digits 1 or 2 to refer to the corresponding parts of the memoir.

Considering the chances of births of both sexes to be equal, Bernoulli in §1·2 deduces the probability of the birth of m males out of $2N$ infants to be

$$\frac{2N(2N-1)\dots(2N-m+1)}{2^{2N}m!} = \frac{P_{2N}}{2^{2N}P_m P_{2N-m}},$$

where $P_k = k!$. In particular, taking $m = N$ (§§ 1·3–1·4), he arrives at

$$\frac{1\,.\,3\,.\,5\ldots(2N-1)}{2\,.\,4\,.\,6\ldots 2N} = q(N).$$

In § 1·5 he calculates q:

$$q(N)\frac{2N+1}{2N+2} = q(N) - \frac{q(N)}{2N+2}, \quad q(N+1) = q(N) - \frac{q(N)}{2N+2}, \quad \frac{dq}{dN} = -\frac{q}{2N+2},$$

$$q(N)\frac{2N}{2N-1} = q(N) + \frac{q(N)}{2N-1}, \quad q(N-1) = q(N) + \frac{q(N)}{2N-1}, \quad \frac{dq}{dN} = -\frac{q}{2N-1},$$

and 'in the mean'

$$\frac{dq}{dN} = -\frac{q}{2N+\frac{1}{2}}, \quad q = A\left(\frac{4f+1}{4N+1}\right)^{\frac{1}{2}},$$

where

$$A = \frac{1\,.\,3\,.\,5\ldots(2f-1)}{2\,.\,4\,.\,6\ldots 2f} = q_0;$$

or, after calculating A for $f = N_0 = 12$,

$$q = \frac{1\cdot 12826}{\sqrt{(4N+1)}}. \tag{1}$$

He could have arrived at

$$q = \frac{1\cdot 12838}{\sqrt{(4N)}}$$

(and, of course, an equivalent result is furnished by the local De Moivre–Laplace limit theorem) using the formula of Wallis. This formula was known, Euler having used it in 1748. It may be inferred that Bernoulli had forgotten the existence of the formula. It would be more difficult to explain the total lack of references to De Moivre in the memoir under consideration. The title of the memoir includes the expression *Mensura Sortis*, which coincides with the title of De Moivre's memoir published in 1712. Furthermore, Bernoulli arrives at the normal distribution (see below), already known to De Moivre, using the same formula as the latter as the origin of his deductions. And if the practice of Laplace is remembered one really could infer that the lack of references to predecessors was characteristic of those times. However, it is also possible that Bernoulli, while reading De Moivre, perhaps some thirty years before 1770, just did not pay due attention to the essence of his work.

2·2. *The Normal Law*

In the 1770–1 memoir Bernoulli deduces the probability of $m = N \pm \mu$ and, for μ of the order of $\sqrt{N}$, asserts (§ 1·18) that

$$\text{prob}\,(m = N \pm \mu) = q e^{-\mu^2/N}. \tag{2}$$

He could have arrived at (2) using Stirling's formula or following De Moivre, who had found the log ratio of the middle term of $(1+1)^{2m}$ to the term whose distance from the middle term is l; in fact

$$\log\left\{\frac{\binom{2m}{m}}{\binom{2m}{m+l}}\right\} = \log\left\{\frac{(m+1)\,(m+2)\ldots(m+l)}{(m-1)\,(m-2)\ldots(m-l)}\right\} \tag{3}$$

is equivalent to l^2/m. If all the factors in the denominator of (3) are increased by unity, this having no influence on the mentioned order of equivalence, the reciprocal value of the fraction (3) will coincide with

$$\text{prob}\,(m = N \pm \mu) = \frac{N(N-1)\dots(N-\mu+1)}{(N+1)(N+2)\dots(N+\mu)}, \tag{4}$$

deduced by Bernoulli in § 1·10. But we shall see that he arrived at (2) by different reasoning.

In the second part Bernoulli generalized his account by considering different probabilities of the births of males and females. Designating the relative frequency of these births by $a:b$, he arrived in §§ 2·9–2·10 at the mathematical expectation of m

$$M = \frac{2Na}{a+b} = 10{,}268, \tag{5}$$

with $a/b = 1·055$ and in § 2·13 deduces that

$$\text{prob}\,(m = M \pm \mu) \equiv \pi = Q \exp\left\{-\frac{(a+b)\,\mu^2}{2bM}\right\} \tag{6}$$

where

$$Q = \text{prob}\,(m = M).$$

Of course, (2) is a special case of (6) because N is a specific value of M corresponding to $a = b$. The deduction of (5) is as follows. If the probabilities of the births of m and $m+1$ males out of $2N$ infants are equal,

$$\frac{2N(2N-1)\dots(2N-m+1)}{m!}\left(\frac{a}{b}\right)^m\left(\frac{b}{a+b}\right)^{2N}$$

$$= \frac{(2N-m)\,2N(2N-1)\dots(2N-m+1)}{(m+1)\,m!}\left(\frac{a}{b}\right)^{m+1}\left(\frac{b}{a+b}\right)^{2N},$$

$$\frac{(2N-m)\,a}{(m+1)\,b} = 1, \quad m = \frac{2Na-b}{a+b} \backsimeq \frac{2Na}{a+b}.$$

In other words, Bernoulli calculated the maximum term of the development of $(a+b)^{2N}$. His calculation presupposes a symmetrical decrease of his function with respect to its maximum value located in $[m, m+1]$, this being the apparent reason for his remark (§ 2·14) that the calculation becomes more accurate with a approaching b. This, of course, is a known property of the local De Moivre–Laplace limit theorem.

Now (6) is deduced from the differential equation

$$\text{prob}\,(m = M+\mu+1) - \text{prob}\,(m = M+\mu) \equiv d\pi = \pi - \frac{2N-M-\mu}{M+\mu+1}\frac{a}{b}\pi\,d\mu,$$

$$-\frac{d\pi}{\pi} = \frac{\mu+1+\mu a/b}{m+\mu+1}\,d\mu,$$

etc. The subsequent transformations are formal and include the development of $\log\{(M+1+\mu)/(M+1)\}$ into a power series.

Bernoulli also proves (§ 2·15) that Q from (6) coincides with q from (2) up to and including terms of order $(a-b)^3$ and notes (§ 2·19) that the total probability (6) from $\mu = 0$ to 47, just as in § 1·12 with $a = b$, is equal to $\frac{1}{2}$ and, of course, notices that the interval $[-47, 47]$ in one case is biased relative to the other case, the centre of the interval coinciding in both cases with the mathematical expectation of m. In § 1·12 Bernoulli starts from (4), not from (2).

In other words, Bernoulli had used the integral De Moivre–Laplace theorem with summation instead of integration. He does not note that with a larger value of a/b the form of the curve (6) would have changed, i.e. that this curve is specified by an important parameter (standard deviation) and, as is also the case with De Moivre, pays little attention to the curve itself.

2·3. *Table of normal distribution*

In contrast with De Moivre, Bernoulli computed a small table of the normal curve, the first ever published. This table is in § 1·19 (p. 43) and is compiled for $\exp(-\mu^2/100)$, $\mu = 1\,(1)\,5$ and $10\,(5)\,30$ with four significant digits. I have checked these ten tabulated values and in three cases the error of the last digit is unity and in one case the error is equal to two; see Table 1.

Table 1. *Bernoulli's table for* $y = \exp(-\mu^2/100)$

μ	y	Corrected y	μ	y	Corrected y
1	0·9901	.	10	0·3679	.
2	0·9608	.	15	0·1054	.
3	0·9141	0·9139	20	0·01832	.
4	0·8522	0·8521	25	0·001931	.
5	0·7789	0·7788	30	0·0001235	0·0001234

2·4. *Testing statistical hypotheses*

Bernoulli concludes his memoir (§ 2·22) with a table of male and female births for London, 1721–30, comparing it with values computed with $a/b = 1\cdot055$ and $1\cdot040$. Although he doubted the constancy of a/b in time and space, his goal was to find the 'real' value of this ratio.

This means that he again raised the question of choosing one or another value of a statistical parameter. He noted the signs of the deviations between computed and observed m's, singled out deviations with absolute values less than 47, noticed the prevalence of deviations of one sign with $a/b = 1\cdot055$ and of the opposite sign with $a/b = 1\cdot040$ but failed to make a definite selection. However, already the raising of the question of testing statistical hypotheses seems to be very important.

It should be emphasized that he returned to this question in 1778 (Bernoulli, 1961) and that the method of differential equations, twice used in this (1770–1) memoir, had been extensively used by him elsewhere.

Acknowledgement is due to Dr L. N. Bolshev and to Dr A. A. Jushkevich for advice and corrections on the first version of this paper.

REFERENCES

BERNOULLI, D. (1770–1). Mensura sortis ad fortuitam successionem rerum naturaliter contingentium applicata. *Novi Comm. Acad. Scient. Imp. Petrop.* **14**, 26–45; **15**, 3–28.

BERNOULLI, D. (1961). The most probable choice between several discrepant observations and the formation therefrom of the most likely induction. *Biometrika* **48**, 3–13.

TODHUNTER, I. (1865). *History of the Mathematical Theory of Probability.* Cambridge and London: Macmillan.

[*Received January* 1969. *Revised August* 1969]

D. BERNOULLI'S WORK ON PROBABILITY

by
O. B. Sheynin
(1972)

Introduction

DANIEL BERNOULLI's (1700–1782) main contributions to probability are contained in eight memoirs which he published between 1738 and 1780. The essence of these memoirs is given by several authors[1,2,3] but in our opinion their description is not sufficient. The memoirs contain solutions of important problems in demographic statistics (political arithmetic), astronomy, and the theory of errors obtained with the help of probabilistic ideas and methods. As to probability and mathematical statistics proper, BERNOULLI was the first to use systematically differential equations for deducing a number of formulae, one of the first to raise the problem of testing statistical hypotheses and the first to introduce ‚moral expectation' (due to G. CRAMER, 1704–1752) and to study random processes. He is also to be credited, after LAMBERT, for the second introduction of the maximum likelihood principle. Of special interest is the appearance in BERNOULLI of the DE MOIVRE-LAPLACE limit theorems[4] which are now named only after DE MOIVRE and LAPLACE. In summary, it may be argued that BERNOULLI's influence upon LAPLACE, especially concerning applications of probability, was comparable to that of DE MOIVRE.

Our paper is presented in sections dealing primarily with BERNOULLI's different applications of probability: games of chance, astronomy, political arithmetic (investigations of smallpox and inoculation, duration of marriages and relative frequency of male and female births) and theory of errors. Besides this two more sections deal with the so-called moral expectation and urn problems respectively. Though the subject-matter of these sections also has a strong bias towards applications of probability, the first towards

1. I. TODHUNTER, *History of the mathematical theory of probability*. Cambridge, 1865 (= TODHUNTER).
2. M. CANTOR, *Vorlesungen über Geschichte der Mathematik*, Bd. 3. Leipzig, 1901, Kap. 108 (pp. 624–641); Bd. 4. Leipzig, 1908, Abschnitt 21, by E. NETTO, section *Wahrscheinlichkeitsrechnung* on pp. 221–257.
3. V. I. SMIRNOV, *D. Bernoulli*. Contained in the Russian translation of D. BERNOULLI's *Hydrodynamica*, Leningrad, 1959 (pp. 433–501). The probabilistic works of BERNOULLI are discussed on pp. 461–470 (in Russian).
4. O. B. SHEYNIN, *D. Bernoulli on the normal law*. *Biometrika* vol. 57 No. 1, 1970, pp. 199–202.

games of chance, freight insurance, etc. and the second towards political arithmetic, it did not seem possible to adopt a completely uniform system for describing BERNOULLI's works. In particular, it did not seem possible to arrange his contributions under purely mathematical (probabilistic) heads, cutting across almost each of these contributions.

An addendum is devoted to D'ALEMBERT and BUFFON.

1. Early interest in probability: a game of chance

Referring to a memoir by J. RIZZETTI, TODHUNTER (§ 1055) mentions a controversy between D. BERNOULLI on the one hand and RIZZETTI and RICCATI on the other hand, relating to some games of chance. RIZZETTI, continues TODHUNTER, cites the *Exercitationes Mathematicae*[5] which TODHUNTER had been unable to see.

In the *Praefatio* to this book BERNOULLI mentions his discussion of a game of chance with RIZZETTI. The game, somewhat unusual in that it contains an escape clause, is described by TODHUNTER, and, of course, in the body of BERNOULLI's book. It had been again discussed by BERNOULLI in his correspondence with CH. GOLDBACH (1690–1764)[6].

2. Probabilistic reasoning in astronomy

Probabilistic reasonings connected with the uniformity the planetary system repeatedly occurred in works of different scholars (NEWTON[7], BERNOULLI, LAPLACE etc). In particular, BERNOULLI devoted several pages of his *Recherches physiques*[8] to this topic, wishing to show that the small mutual inclinations of the five planetary orbits relative to the earth's orbit could not be attributed to chance.

---

5. DANIELIS BERNOULLI / Basileensis / Joh. Fil. / *Exercitationes quaedam / mathematicae /* Venetiis, 1724 / Apud Dominicum Lovisam / . A copy of this book from the Saltykov- Shchedrin library (Leningrad) carries an inscription ,donum Auctoris'.
6. P. N. FUSS, *Correspondance mathématique et physique de quelques célèbres géomètres du xviii siècle.* St. Petersbourg, 1843, t. 2. See letters 10–15 (1724) and 20 (1725) of their correspondence (pp. 199–226, 240–241). Their correspondence lasted from 1723 to 1730 and included 71 letters which did not prevent BERNOULLI from a semicontemptuous reference to GOLDBACH, see letter 22 of the D. BERNOULLI – EULER correspondence (1742), – Ibidem, pp. 479–483.
7. O. B. SHEYNIN, *Newton and the classical theory of probability. Archive for history of exact sciences* vol. 7 No. 3, 1971, pp. 217–243.
8. *Recherches physiques et astronomiques sur le problème . . . Quelle est la cause physique de l'inclinaison des plans des orbites des planètes par rapport au plan de l'Equateur de la révolution du Soleil autour de son axe; Et d'où vient que les inclinaisons de ces Orbites sont différentes entre elles.* In: *Pièces qui ont remporté le prix de l'Acad. Roy. des Sci. en 1734.* Paris, 1735, pp. 95–122. The original Latin version is on pp. 125–144.

Let the inclination of an orbit be equal to a, $0° < a < 90°$. If all the values of a are independent and uniformly distributed, the probability of $a_i < A$ ($A < 90°$) for each of the five planets would be equal to $A/90$; and if $a_i < A$ for $i = 1, 2, \ldots, 5$ (i. e., for the five planets simultaneously), the probability of the whole series of inequalities would be $(A/90)^5$ and, with a small A, becomes insignificantly small. Consequently, the hypothesis of the independence of all the permissible values of a is rejected in favour of an inference that there exists a definite reason for the minuteness of the a's.

Exactly this problem of differentiating between chance and law (divine law) was the main goal of the early theory of probability[9]. Such a differentiation is not yet a test of a statistical hypothesis proper, but this latter is also present in the works of BERNOULLI (see our §§ 4.3, 6). TODHUNTER remarks in his § 396:

> It would seem, that he (BERNOULLI) should rather have considered the poles of the orbits than the planes of the orbits, and have found the chance that all the other poles should lie within a given distance from one of them.

It would have been completely foreign to TODHUNTER's mind to imply that because different modes of calculation may well furnish different results the theory of probability as such is at fault. On the other hand, such assertions were made by D'ALEMBERT (see Addendum).

As we see it, the choice of the correct course of action lies here in astronomy proper, beyond the theory of probability.

‚On trace au hasard une corde dans un cercle', writes BERTRAND[10]. ‚Quelle est la probabilité pour qu'elle soit plus petite que le côté du triangle équilatéral inscrit?'

Au hasard being an insufficient description it turns out that the problem has three different answers. And BERTRAND concludes (p. 5):

> Entre ces trois réponses, quelle est la véritable? Aucune des trois n'est fausse, aucune n'est exacte, la question est mal posée.

To give a modern example from statistical mechanics: depending upon the choice of one or another physical model, probabilistic reasonings lead to three different statistics: the Maxwell-Boltzmann, the Bose-Einstein, and the Fermi-Dirac ones. Clearly, the choice of the model lies beyond the theory of probability.

---

9. See O. B. SHEYNIN, Note 7.
10. J. BERTRAND, *Calcul des probabilités.* Paris 1888. See § 5, p. 4.

It seems that EULER unfavourably referred to BERNOULLI's reasonings. A letter from BERNOULLI to EULER[11] says:

> Ew. Meinung über meine pièce, so den prix erhalten, würde mich sehr mortificiren, wenn ich nicht gesehen hätte, daß Sie dieselbe nur obenhin und in höchster Eil müssen gelesen haben. Es ist mir niemals in den Sinn gekommen das planum aequatoris solis zu verändern, damit die Inclinationen in der Ordnung fortgehen, wie die excentricitates, sondern ich habe nur die Anmerkung gemacht, daß weil das planum aequatoris noch incertum ist, es nicht unfüglich sey zu untersuchen, wie es müsse placirt werden, damit das medium arithmeticum von allen Inclinationen minimum sey, welches ich auch gethan, und gethan zu haben nicht bereue.

The end of this quotation bears upon the theory of errors and we shall return to it in our § S. 5.

3. Moral expectation

Among various games of chance repeatedly considered by mathematicians of the XVIIth and XVIIIth centuries one game devised by N. BERNOULLI and described by MONTMORT[12] merited special attention. The Petersburg Problem, as it came to be known, possibly because of the publication of D. BERNOULLI's *Specimen*[13], which considered this problem, in St. Petersburg (TODHUNTER, § 389), is a problem of solving a paradox of common sense being opposed to mathematical treatment.

If A throws a coin and heads appears at once he is to receive a ducat from B, if heads does not appear until throw number k, he is to receive 2^{k-1} ducats, etc. These are the conditions of the game and the expectation of A is equal to

$$\frac{1}{2} + \frac{2}{2^2} + \frac{4}{2^3} + \ldots = \frac{1}{2} + \frac{1}{2} + \frac{1}{2} + \ldots = \infty$$

On the other hand, none would be willing to pay even a moderate sum for the advantage to be gained.

---

11. P.N. FUSS, *Correspondance mathématique et physique,* letter 9 of their correspondence (1737), pp. 436–437. See also letter 10 (1737), pp. 438–439.
12. P. R. MONTMORT, *Essay d'analyse sur les jeux de hazard.* Paris, 1713. (Published anonymously)
13. *Specimen theoriae novae de mensura sortis. Commentarii academiae scientiarum imperialis Petropolitanae* t. 5 for 1730–1731 (1738), pp. 175–192. Translations: *Versuch einer neuen Theorie der Wertbestimmung von Glücksfällen.* Leipzig, 1896; *Exposition of a new theory on the measurement of risk. Econometrica* vol. 22 No. 1, 1954, pp. 23–36. The quotation given below is from the English translation.

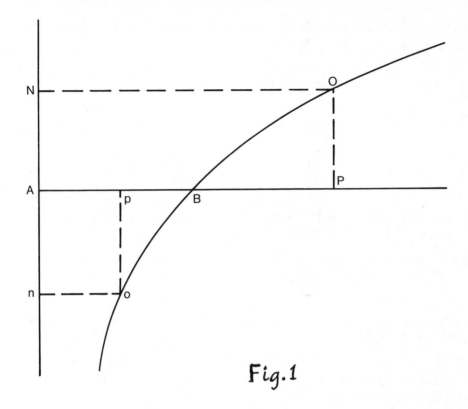

Fig.1

BERNOULLI's *Specimen* was most possibly conceived as a contribution towards solving this paradox although, as we shall see, it contains much more.

BERNOULLI holds (§ 3) that the same (small) gain in a game of chance is more important for a poor than for a rich participant and that the advantage from a (small) gain is inversely proportional to the whole property of the participant. And, measuring gains and losses along the x-axis, and the corresponding advantages (disadvantages) along the y-axis ,BERNOULLI constructs a continuous curve (§ 7) assuming it to be derived from a differential equation (§ 10), possibly the first one to be used in the theory of probability,

$$dy = \frac{Cdx}{x} , \quad C > 0 .$$

Thus, the equation of the curve is

$$y \equiv f(x) = C \log \frac{x}{\alpha}$$

($\alpha = AB$, the initial total fortune of the player, see fig. 1)

109

Calculation of advantages along the y-axis means that the classical mathematical expectation BP (or Bp), or

$$\frac{m_1 x_1 + m_2 x_2 + \ldots + m_n x_n}{m_1 + m_2 + \ldots + m_n}$$

is to be replaced by the moral expectation, or mean advantage PO (or po)

$$PO = \frac{m_1 f(x_1) + m_2 f(x_2) + \ldots + m_n f(x_n)}{m_1 + m_2 + \ldots + m_n} \tag{1}$$

Here, the m's are the frequencies of the occurrences of the corresponding gains (the x's) and the gains and losses are measured along the x-axis, to the right and to the left of B respectively.

In ‚fair‘ games with a zero mathematical expectation of winnings $BP = Bp$ and, for a log curve, the moral expectation of winning becomes negative $(PO < po)$. This is why (§ 13)

> . . . in many games, even those that are absolutely fair, both of the players may expect to suffer a loss; indeed this is Nature's admonition to avoid the dice altogether.

BERNOULLI does not here distinguish between a random quantity and its expectation. And we shall see that this was his usual error, most possibly characteristic for his time.

The subsequent part of the memoir is devoted to the use of the moral expectation in the Petersburg game, freight insurance, etc. The author then describes N. BERNOULLI's comments (1732) on the manuscript of his memoir and appends CRAMER's letter to N. BERNOULLI (1728). CRAMER uses the term ‚moral expectation‘ (not to be found in D. BERNOULLI!) and implicitly gives formula (1). He resolves the paradox of the Petersburg game by using

$$f(x) = \min(x; 2^{24})$$

or, alternatively,

$$f(x) = \sqrt{x}.$$

The moral expectation of A's gain in the Petersburg game will be

$$z = C \frac{\dfrac{1}{2} \log \dfrac{\alpha+1}{\alpha} + \dfrac{1}{4} \log \dfrac{\alpha+2}{\alpha} + \dfrac{1}{8} \log \dfrac{\alpha+4}{\alpha} + \ldots}{\dfrac{1}{2} + \dfrac{1}{4} + \dfrac{1}{8} + \ldots} =$$

$$= C \left[\log\left(\sqrt{\alpha+1}\ \sqrt[4]{\alpha+2}\ \sqrt[8]{\alpha+4}\ \ldots\right) - \log \alpha\right],$$

the corresponding expectation of B's capital will be

$$x = \sqrt{\alpha + 1} \ \sqrt[4]{\alpha + 2} \ \sqrt[8]{\alpha + 4} \ \dots ,$$

the expected increment of capital, $\Delta x = x - \alpha$. Therefore, the payment (s) which A shall have to make so as to equalize the game, is to be found from the equation

$$\sqrt{\alpha - s + 1} \ \sqrt[4]{\alpha - s + 2} \ \sqrt[8]{\alpha - s + 4} \ \dots - \alpha = 0 .$$

With large α's $s \approx \Delta x$ and is therefore finite.

The inequality $E f(x) \geqslant f(Ex)$ where E is the symbol of the mathematical expectation holds for any convex function $f(x)$ situated ,below' its chord[14]. In our case, $E(-\log x) \geqslant \log(Ex)$ or $E(\log x) \leqslant \log(Ex)$ and, with $x > 0$, $E(\log x) < Ex$. But of course, with $Ex = \infty$ this is of little value.

As applied to the shipping of freight Bernoulli's reasoning is as follows. Let the probability of a ship's arrival be p, the value of freight A, the freightowners cash (safe money) α. If all the freight is shipped on board a single ship, the moral expectation of all the freightowner's property is ($q = 1 - p$)

$$z = \frac{p \, C \log [(A + \alpha)/\alpha] + q \, C \log \alpha/\alpha}{p + q} = p \, C \log \frac{A + \alpha}{\alpha}$$

and the property itself is

$$x = \alpha e^{z/C} = \alpha \left(\frac{A + \alpha}{\alpha} \right)^p .$$

The freight having been equally distributed on n ships,

$$z(n) = \sum_{k=0}^{n} \binom{n}{k} p^{n-k} q^k \log \left[\frac{A}{n} (n - k) + \alpha \right] .$$

A numerical example for $n = 1$ and 2 only is given by BERNOULLI with an assertion that z increases monotonically with n, being bounded above by $Ap + \alpha$, the mathematical expectation. I shall now investigate this assertion.

---

14. C. R. RAO, *Linear statistical inference and its applications.* New York, 1965. See § 1 e. 5.

1. The mathematical expectation is (regardless of n)

$$\sum_{k=o}^{n} \binom{n}{k} p^{n-k} q^k \left[\frac{A}{n} (n-k) + \alpha \right] =$$

$$\alpha + A \sum_{\overline{k}=o}^{n} \frac{n-k}{n} p^{n-k} q^k = \alpha + A - Aq = \alpha + Ap .$$

2. The moral expectation is bounded for any increasing function used:

$$z(n) = \sum_{k=o}^{n} \binom{n}{k} p^{n-k} q^k \, f \left[\frac{A}{n} (n-k) + \alpha \right] < f(A + \alpha)(p+q)^n$$
$$= f(A + \alpha)$$

3. For any ~~uniform~~ *continuous*, increasing $f(x)$ $(f'(x) > 0)$ with a decreasing derivative $f'(x)$ (or, otherwise, for any concave function) $z(n)$ monotonically increases with n. Following BUNIAKOVSKII[15] who restricted himself to BERNOULLI's example and only proved that $z(2) > z(1)$, and adding to his conditions written above one more, $f(\alpha) = O$, we have

$$z(n) = (p+q) z(n) = \sum_{k=o}^{n} \binom{n}{k} p^{n-k+1} q^k \, f \left[\frac{A}{n} (n-k) + \alpha \right] +$$

$$+ \sum_{k=1}^{n+1} \binom{n}{k-1} p^{n-k+1} q^k \, f \left[\frac{A}{n} (n-k+1) + \alpha \right],$$

$$z(n+1) = \sum_{k=o}^{n+1} \binom{n+1}{k} p^{n-k+1} q^k \, f \left[\frac{A}{n+1} (n-k+1) + \alpha \right].$$

Now every term of $z(n+1)$, less the first and the last ones, equal to the corresponding terms of $z(n)$, may be divided into two according to the identity

$$\binom{n+1}{k} = \binom{n}{k} + \binom{n}{k-1}$$

15. V. YA. BUNIAKOVSKII, *Principles of the mathematical theory of probability*. St. Petersburg, 1846. (In Russian). See § 44.

so that it is sufficient to show that

$$
\binom{n}{k} \left\{ f\left[\frac{A}{n+1}(n-k+1)+\alpha \right] - f\left[\frac{A}{n}(n-k)+\alpha \right] \right\} > \binom{n}{k-1} \cdot
$$

$$
\cdot \left\{ f\left[\frac{A}{n}(n-k+1)+\alpha \right] - f\left[\frac{A}{n+1}(n-k+1)+\alpha \right] \right\}.
$$

Using the LA GRANGE theorem it is possible to present this inequality in the form

$$
f'\left[\frac{A}{n}(n-k)+\frac{\Xi_1 k}{n+1}+\alpha \right] > f'\left[\frac{A(n-k+1)}{n+1}(1+\frac{\Xi_2}{n})+\alpha \right], 0 < \Xi_i < 1
$$

which becomes an equality if and only if $\Xi_1 = 1$ and $\Xi_2 = 0$ (simultaneously), neither of which is possible.

4. The limit of $z(n)$. This function being a BERNSTEIN polynomial of

$$
f\left[\frac{A}{n}(n-k)+\alpha \right] = \varphi\left(\frac{k}{n} \right),
$$

a function ~~uniform~~ continuous over $0 \leqslant k/n < 1$,

$$
\lim z(n) = \varphi(q) = f(A(1-q)+\alpha) = f(Ap+\alpha) \text{ as } n \to \infty
$$

In particular, for the log curve,

$$
z = \lim z(n) = \log(Ap+\alpha) < Ap+\alpha,
$$

so that $z(n)$ is indeed bounded by the mathematical expectation.

> La théorie de l'espérance morale est devenue classique, jamais le mot ne put être plus exactement employé: on l'étudie, on l'enseigne, on la développe dans des livres justement célèbres. Le succès s'arrête là, on n'en a jamais fait et n'en pourra faire aucun usage.

Strictly speaking, this opinion of BERTRAND[16] is hardly fair: various measures of precision of astronomical and geodetic observations repeatedly

16.   J. BERTRAND, *Calcul des probabilités*, § 53, p. 66.

introduced by LAPLACE, GAUSS and other scholars are of the same nature as the moral expectation (1) in that they employ one or another loss function $f(x)$. And loss functions of various types are of course used in mathematical statistics[16a].

HUBER[17] noticed that the moral expectation is similar to the celebrated WEBER-FECHNER law while SMIRNOV[18] pointed out that the log function is widely used in information theory. But it is our opinion that in both cases the similarity is due not to the essence of the problem — the use of the general formula (1) — but to the highly appropriate choice of the log function as the $f(x)$.

The moral expectation is the subject-matter of LAPLACE's *Théorie analytique des probabilités*, livre 2, chap. 10, described at some length by TODHUNTER. Having given above our own demonstrations pertaining to BERNOULLI's problem of the shipping of freight we shall leave it at that.

4. Political arithmetic

4.1. Smallpox and inoculation

In the middle of the XVIIIth century, the smallpox vaccination being yet unknown, epidemics of smallpox were not unusual. The history of these epidemics and of inoculation with smallpox (communication of a mild form of smallpox to one person from another) is described in various sources[19, 20, 21, 22].

16a Highly relevant are the ideas of FOURIER and OSTROGRADSKY: J.B.J. FOURIER, *Extrait d'un mémoire sur la théorie analytique des assurances. Annales de chimie et de physique*, t. 10, 1819, pp. 177−189; P.N. FUSS, *Compte rendu de l'Académie pour l'année 1835*. With separate pagination in *Recueil des actes de la séance publique de l'Acad. imp. sci. St. Pétersbourg tenue le 29 Dec. 1835*. St. Pétersbourg, 1836. (See pp. 24−25 of the report; no other trace of OSTRO-GRADSKY's report seems to have survived.
     Each of these scholars strived to generalize D. BERNOULLI's idea of the moral expectation.
17. F. HUBER, *D. Bernoulli als Physiologe und Statistiker*. Basel, 1959 (*Basler Veröff. zur Geschichte der Medizin und der Biologie* No. 8).
18. See footnote 3. SMIRNOV refers to a private communication by J. W. LINNIK.
19. C. M. CONDAMINE, *Mémoire sur l'inoculation de la petite vérole. Hist. Acad. Roy. des Sci. 1754 avec mém. math. et phys. pour la même année* (1759), pp. 615−670 of the memoirs.
20. C. M. CONDAMINE, *Second mémoire sur l'inoculation de la petite vérole contenant la suite de l'histoire de cette méthode et de ses progrès, de 1754 à 1758 Hist. Acad. Roy. des Sci.*, 1763, pp. 439−482 of the memoirs.
21. M. N. KARN, *An inquiry into various death-rates and the comparative influence of certain diseases on the duration of life. Annals of eugenics* vol. 4 for 1930−1931 (1931), No. 3−4, pp. 279−326.
22. F. HUBER, Note 17.

In his first memoir CONDAMINE lists the objections against inoculation (physical, i. e. medical, and moral, including religious, ones). He ends his memoir thus:

> l'usage de l'inoculation étoit devenu général en France depuis que la famille royale d'Angleterre sut inoculée, on eût déjà sauve le vie à près d'un million d'hommes (!) sans y comprendre leur postérité.

In his second memoir CONDAMINE mentions the BERNOULLI family (p. 464):

> A Bâle, Mrs Bernoulli, dont le nom seul pourroit à plusieurs titres autoriser une opinione douteuse, ne se sont pas contentés de se déclarer ouvertement en faveur de l'inoculation, et d'obtenir pour les premières épreuves l'approbation des facultés de médecine et de théologie de Bâle: le cadet des deux frères, M. Jean Bernoulli et le seul marié, voulut y joindre son exemple. Il fit inoculer en 1756 les deux plus jeunes de ses fils; et l'année dernière leur frère aîné . . .

KARN who gives statistical data pertaining to the subject and studies BERNOULLI's contribution and the work of subsequent scholars, begins her article by explicitly stating that

> the method used in this paper (her article) for determining the influence of the death-rates from some particular diseases on the duration of life is based on suggestions which were made in the first place by D. Bernoulli.

BERNOULLI's main contribution to this subject is his *Essai*[23] which he begins with statistical data. Then (§ 3) he formulates hypotheses about smallpox epidemics and describes various mortality tables which, without the causes of death having been distinguished, were of little value.

Denoting (§ 5) the age expressed in years by x, the number who survive at that age out of a given number who were born by ξ, the number of these survivors who have not had the smallpox by s BERNOULLI derives a differential equation among these quantities. The integral of this equation is

$$s = \frac{m\,\xi}{1 + (m - 1)\,e^{x/n}} \qquad (2)$$

---

23. *Essai d'une nouvelle analyse de la mortalité causée par la petite vérole, et des avantages de l'inoculation pour la prévenir*. Hist. Acad. Roy. des Sci. 1760 avec les mém. math. et phys. pour la même année (1766), pp. 1—45 of the memoirs. This has been preceeded by an article for a broader circle of readers: *Réflexions sur les avantages de l'inoculation*, published in the *Mercure de France*, Juin 1760, pp. 173—190. On p. 178 BERNOULLI writes:
   Les deux grands motifs pour l'inoculation sont l'humanité et l'intérêt de l'Etat.
   The *Refléxions* end with a *Nota*:
   Le mémoire qu'annonce M. Bernoulli dans ces réflexions (the *Essai*) *a été envoyé à l'Académie des Sciences de Paris.*

where $1/n$ is the annual rate of the occurrence of smallpox in those who have not had the disease and $1/m$ is the corresponding mortality. According to statistical data BERNOULLI assumes $m = n = 8$ (which means an annual probability of death from smallpox equal to $1/64$). All this as well as BERNOULLI's further elaborations are described in detail by TODHUNTER (§§ 398—407).

In particular, beginning from § 11 of his memoir, BERNOULLI studies the problem of inoculation. Employing the same ideas and methods, he calculates the mean duration of life allowing for inoculation which excludes smallpox but in a small number of cases (frequency = $1/200$) proves fatal. His final inference is that inoculation prolongs the mean duration of life from 26 years 7 months to 29 years 9 months.

As to BERNOULLI's mathematical assumptions, he was satisfied to use the approximate equality

$$\frac{\mu}{n} = p$$

(μ = actual number of occurrences of an event in n independent trials, p probability of the occurrence of this event in each trial) which, of course, is a corollary of the law of large numbers. This was the practice of all statisticians in those times, this is still their practice in many instances.
This means that BERNOULLI derived (and solved) differential equations among statistically averaged quantities.

4.2. Duration of marriages

BERNOULLI studied the duration of marriages in his *De duratione*[24] which appeared simultaneously with another of his memoirs, *De usu*[25]. This is no coincidence, because the first memoir is based on the second one, where urn problems are considered (see our § 6). The relevant problem is the problem of extracting strips of two different colours from an urn, with the same or different likelihood of extraction for strips of different colours (= the same or different laws of mortality of both sexes), an unbroken marriage being

---

24. *De duratione media matrimoniorum, pro quacunque conjugum aetate, aliisque quaestionibus affinibus. Novi Comm. Acad. scient. imp. Petrop.*, t. 12 for 1766—1767 (1768), pp. 99—126. A Russian translation is in M. V. PTUCHA, *Essays on the history of statistics in the USSR*, vol. 1. Moscow, 1955, pp. 453—464.
25. *De usu algorithmi infinitesimalis in arte coniectandi specimen. Novi Comm. Acad. scient. imp. Petrop.*, t. 12, pp. 87—98.

equivalent with a pair of strips being still left in the urn. An inaccuracy involved in this is that the duration of lives of man and wife living presumably in similar conditions are hardly independent.

Important as this problem is for demographers and government agencies we shall not deal with it, being content with analysing its mathematical background (§ 6).

We shall also notice that the problems related to the duration of marriages were studied by N. STRUYCK (1687–1769) in his *Hypothèses*[26]. This is STRUYCK's reasoning (p. 229): out of 444 men aged from 30 to 34 years only 262 live to become 20 years older; out of 471 women of the same age-group, only 308, the data being that of life-insurance statistics. Correspondingly, out of 100 men there remains 59, and out of 59 women 39 so that out of 100 marriages in this age-group only 39 remain unbroken.

Such calculations are extremely simple and could have been actually used if only reliable statistical data had been available (but of course statisticians would have been ill-advised to extend life-insurance statistics to cover the population at large). On the other hand, such calculations are completely irrelevant to the formation of probabilistic ideas and methods. Only at the hands of real masters such as BERNOULLI were statistical computations founded upon new, fruitful ideas and methods. To mention another master: somewhat prior to this memoir A. DE MOIVRE, drawing upon HALLEY's statistical data, introduced the continuous uniform distribution as a law of mortality[27].

4.3. Relative frequency of births of both sexes

Problems of political arithmetic naturally included the separate enumeration of annual male and female births and relevant statistical calculations. The classical work of ARBUTHNOT in this field is well known. It could be thought that these statistical calculations led DE MOIVRE to the DE MOIVRE-LAPLACE limit theorems. It could be also noticed that a local limit theorem and, though not explicitly, the exponential function of the

26. N. STRUYCK, *Hypothèses sur l'espèce humaine.* In: *Oeuvres qui se rapportent au calcul des chances* etc. Trad. du holl. Amsterdam, 1912, pp. 165–249. On page VI of the foreword the date of the original publication is given as 1740.

27. O. B. SHEYNIN, *On the early history of the law of large numbers. Biometrika,* vol. 55, No. 3, 1968, pp. 459–467. Reproduced with corrections in *Studies in the history of statistics and Probability.* E. S. PEARSON and M. G. KENDALL, editors. London, 1970, pp. 231–239. On the life and works of De Moivre see I. SCHNEIDER, *Der Mathematiker A. de Moivre. Archive for history of exact sciences* vol. 5, No. 3/4, 1968, pp. 177–317, whose contribution is the most serious general source of information about this scientist.

negative square is to be found in N. BERNOULLI (1713), also in connection with similar statistical calculations[28].

The relevant contribution of D. BERNOULLI is his *Mensura sortis*[29]. Until my recent publication[30], no one had remarked on the appearance in this memoir of the DE MOIVRE-LAPLACE limit theorems (still to be credited to DE MOIVRE).

We shall not here reproduce our remarks. It is, however, necessary to repeat that BERNOULLI published the first (small) table of the normal distribution compiled for exp $(-x^2/100)$, $x = 1\,(1)\,5$ and $10\,(5)\,30$ with four significant digits and gave a quantitative approach to the selection of the appropriate ,real' value of the ratio of male and female births. This in fact was an attempt to test a statistical hypothesis.

We have noted elsewhere[31] the Aristotelian qualitative chance explanation of the production of both sexes in animals[32]. As the approximate equality of male and female births was noticed in the XVIIth century (the permanent predominance of male births having been observed somewhat later) it possibly would not be too far-fetched to propose that this fact could have contributed to the generally prevalent idea that by chance all things are equally probable[33].

We shall see (§ 5.2) that in his last years BERNOULLI used the normal distribution in the theory of errors.

5. Theory of errors

In the middle of the eighteenth century the problem of calculating the ellipticity and dimensions of the earth, one of the main problems of natural science of this century, together with related mathematical problems in astronomy proper, led to the foundation of the theory of errors. From the point of view of the theory of probability, for which the theory of errors soon became the main field of its application, the goals of the theory of errors were:

---

28. See O. B. SHEYNIN, Note 27. Also a related Russian article: *On the history of the De Moivre-Laplace limit theorems. History and methodology of the natural science,* vol. 9. Moscow, 1970, pp. 199–211.

29. *Mensura sortis ad fortuitam successionem rerum naturaliter contingentium applicata. Novi Comm. Acad. scient. imp. Petrop.* t. 14, pars 1 for 1769 (1770), pp. 26–45; *Continuatio argumenti de Mensura sortis ad fortuitam* etc. Ibidem, t. 15 for 1770 (1771), pp. 3–28.

30. See O. B. SHEYNIN, Note 4.

31. See O. B. SHEYNIN, Note 7, § 3.3.

32. *De generatione animalium* iv, 3, 767 b, 5.

33. See O. B. SHEYNIN, note 7.

(i) Given a series of somewhat discrepant observations $x_1, x_2, \ldots, x_n$ of a measured quantity ξ and assuming a certain law of distribution for the errors of these observations, to find the ‚real', ‚most plausible' value of ξ; also, to estimate the error of this value. (The case of *direct observations* of the unknown quantity).

(ii) Given a series of observations $l_1, l_2, \ldots, l_n$ and assuming (usually tacitly) a certain law of distribution for the errors of these observations to find the ‚real', ‚most plausible' values of m $(m < n)$ unknown quantities $x, y, z, \ldots$ from a redundant system of algebraic equations

$$a_i x + b_i y + c_i z + \ldots + l_i = 0, \ i = 1, 2, \ldots, n$$

(usually linear ones as just written); also, to estimate the errors of these values. (The case of *indirect observations* of the unknown quantities).

In particular, the so-called arc measurements (measurements of the length of one degree of a certain meridian at a certain latitude) led to such equations in two unknowns, the parameters of the earth's spheroid.

Each absolute term l_i being derived independently (in the physical sense) such systems of equations were of course inconsistent so that strictly speaking their solution was impossible. For this reason additional conditions were introduced their essence being qualitatively (and later quantitatively) explained in terms of probability. Thus it actually happened that the method of least squares was introduced in the early nineteenth century, essentially by LE GENDRE and GAUSS, its condition being

$$\Delta l_1^2 + \Delta l_2^2 + \ldots + \Delta l_n^2 = \min$$

where the Δl_i stand for the residual absolute terms.

It is not difficult to see that the modern problem of linear (and non-linear) programming could be directly linked with such problems of *adjusting* indirect observations.

We shall see that BERNOULLI's contributions belong to the adjustment of both direct and indirect observations.

5.1. Adjustment of indirect observations

In my § 2 I have quoted one of BERNOULLI's letters to EULER, which contains an interesting point pertaining to the theory of errors. Following are quotations from BERNOULLI's relevant memoir, *Recherches physiques*[34].

---

34. D. BERNOULLI, *Recherches physiques*, note 8.

1. From § 4, p. 98 (inserted in the French version of the memoir):

> . . . la meilleure manière de calculer le degré de probabilité seroit de considérer le plan au milieu des Orbites (qui, selon toutes les apparences, est le plan même de l'Equateur solaire) . . . .

2. From § 6, p. 100:

> Nous avons dit, qu'il y a un plan qui doit avoir quelque rapport avec les Orbites des Planetes, dans lequel ces Orbites tâchent de se réünir; que ce plan est situé au milieu des Orbites, et enfin qu'il est, selon toutes les apparences, le même que celui de Equateur solaire . . .

3. Nevertheless (says BERNOULLI in his § 22, p. 116):

> Mais comme la position de l'équateur solaire est fort incertaine; de telle manière que, selon quelques-uns, son inclinaison avec l'écliptique ne surpasse pas deux degrés, on pourroit peut-être sans absurdité, feindre une telle position, que son inclinaison moyenne avec toutes les orbites planetaires, fût la moindre, a laquelle condition l'on peut satisfaire en essayant un grand nombre de positions: ainsi, par exemple, dans la précedente hypothese l'inclinaison moyenne des orbites avec l'équateur solaire, est de 5°11′: mais si l'on supposoit que cet équateur fît avec l'écliptique un angle de 3°22′ et que son Pole Boreal répondît au 20°des Poissons, alors l'équateur solaire seroit coupé par l'orbite de Saturne, sous un angle de 1° 51′ (the five other planets including the earth are listed their angles taken to be positive quantities) et l'inclinaison moyenne des orbites . . . ne seroit plus de 2° 23′. Je ne scais si on ne pourroit pas préferer cette position de l'équateur solaire, quoiqu'appuyée sur une pure conjecture, et trouvée à posteriori, aux autres positions, fondées sur les taches du Soleil, en attendant que les Astronomes nous donnent une methode Astronomique plus exacte.

In other words, BERNOULLI's method is tantamount to constructing a plane so that the sum of the angular deviations of the given planes (orbits) from it would be minimal, the deviations considered to be positive. This means that BERNOULLI's additional condition was

$$| \Delta l_1 | + | \Delta l_2 | + \ldots + | \Delta l_n | = \min .$$

On the other hand, BERNOULLI, as he himself stated, had been unable to give any formulae corresponding to this principle. Nevertheless, his reasoning should not be forgotten. A few decades later BOSCOVICH recommended a similar method for adjusting arc measurements.

His method[34a], sufficiently well known[34b], consisted in assuming two conditions

---

34a. C. MAIRE, R. J. BOSCOVICH, *Voyage astronomique et géographique dans l'Etat de l'Eglise enterpris par l'ordre et sous les auspices du Pape Benoît xiv, pour mesurer deux dégrés du méridien et corriger la Carte de l'Etat ecclesiastique.* Paris, 1770.

$$\begin{cases} \Delta l_1 + \Delta l_2 + \ldots + \Delta l_n = 0 \\ |\Delta l_1| + |\Delta l_2| + \ldots + |\Delta l_n| = \min \end{cases}$$

for which he explicitly gave a qualitative probabilistic explanation. Quotations from BERNOULLI (see above) prove that he also had such considerations in mind.

Not elaborating on the further history of BOSCOVICH's method I shall make only two short comments. Firstly, BERNOULLI, having assumed all his Δl_i positive, naturally did not use the first of BOSCOVICH's conditions. Secondly, BOSCOVICH's method became one of the main ones for adjusting indirect observations at least until the advent of the method of least squares.

5.2. Adjustment of direct observations

I have discussed BERNOULLI's subsequent work[35] elsewhere[36]. Not repeating this discussion, I shall only notice that instead of the usual arithmetic mean

$$\frac{\sum\limits_{i=1}^{n} x_i}{n}$$

BERNOULLI introduced an estimator

$$\frac{\sum\limits_{i=1}^{n} p_i x_i}{\sum\limits_{i=1}^{n} p_i}$$

34b. I. TODHUNTER, *History of the mathematical theories of attraction and the figure of the earth*, vol. 1–2. London, 1873. Reprint: New York, 1962, in one volume. See §§ 511 and 514. On the adjustment of arc measurements by **Boscovich** see also R. WOLF, *Handbuch der Astronomie, ihrer Geschichte und Literatur*, Bd. 2, Dritter Halbband. Zürich, 1892. See § 425; C. EISENHART, *Boscovich and the combination of observations*. In: R. J. BOSCOVICH. *Studies of his life and work*. Editor, L. L. WHYTE. London, 1961, pp. 200–213. A shorter version of this appeared in *Actes du symposium international Boscovich* 1961. Beograd, 1962, pp. 19–25.

35. D. BERNOULLI, *Dijudicatio maxime probabilis plurium observationum discrepantium atque verisimillima inductio inde formanda*, 1777 (1778). Published together with EULER's commentary (ENESTRÖM 488). See L. EULER, *Opera omnia*, ser. 1, t. 7, 1923, pp. 262–279 (BERNOULLI's memoir) and pp. 280–290 (EULER's commentary). An English translation of the memoir and commentary is in *Biometrika*, vol. 48, No. 1–2, 1961, pp. 3–18, preceeded by a short commentary by M. G. KENDALL. The translations are reprinted in the book mentioned in note 27, pp. 155–172.

36. *Mathematical treatment of observations by L. Euler. Archive for history of exact sciences*, vol. 9, No. 1, 1972, pp. 45–56.

of the unknown quantity ξ which is interesting in that the extreme observations influenced it more than the middle ones (the $p's$ increased to the tails of the observational series), a fact which should have been considered rather strange; only recently a confirmation of it was obtained by E. H. LLOYD (LLOYD's best linear estimators).

BERNOULLI did not explicitly state this fact (although it is plainly seen in his examples) and for this reason EULER, in his commentary, misunderstood him. Being rather old, BERNOULLI did not bother himself to correct EULER (which, as seen in § 2, was not the case in his earlier life).

BERNOULLI's memoir is also interesting in that it contains the second use of the principle of maximum likelihood (due to Lambert[37]).

BERNOULLI's last memoir on probability appeared in 1780[38]. Containing interesting ideas and methods, it nevertheless had been completely forgotten, possibly being classified, owing to its title, as pertaining to practical mechanics.

BERNOULLI does consider the influence of various errors in pendulum observations (extensively used in those times and still in use now for the determination of the figure of the earth). However, main points of interest are connected with probability.

Discussing various kinds of errors BERNOULLI concludes his memoir (§ 22, p. 128) thus: Essentially, there exist two *species* of errors in pendulum observations: *chronicarum* and *momentanearum*; the influence of the first ones are often almost proportional to the corresponding time interval, but the second ones may act proportionally to the square root of this interval.

This reasoning being extremely important I give now the complete quotation:

> § 22. Summa pertracti argumenti nostri in hoc consistit, ut intelligatur duas inesse horologiis aberrationum species, chronicarum quae diutius subsistere possunt; et momentanearum, quae continue hinc et illinc evagantur: priores saepenumero recte statui poterunt temporibus ellapsis propemodum proportionales, posteriores vero sequentur potius horum temporum radices quadratas, haeque aliquando post breve temporis intervallum, si recte coniecto, sensibiles aliquantillum esse poterunt, praesertim in horologiis non ultima accuratione elaboratis.

---

37. O. B. SHEYNIN, *J. H. Lambert's work on probability, Archive for history of exact sciences* 7, No. 3, 1971, pp. 244–256.

38. D. BERNOULLI, *Specimen philosophicum de compensationibus horologicis, et veriori mensura temporis. Acta academiae scientiarum imperialis Petropolitanae,* for 1777, pars posterior (1780), pp. 109–128.

BERNOULLI came to this conclusion after employing the normal law, thus (see §§ 13–17 of his memoir):

let the number of pendulum oscillations per day be $2N$, $(N + \mu)$ being retarded and $(N - \mu)$ accelerated. As pendulums were usually regulated so that the period of oscillation equaled one second, $2N \approx 86,400$ and the actual periods are respectively $(1 + \alpha)$ and $(1 - \alpha)$ the retardation being supposed equal to the acceleration.

The total error during 24 hours would then be equal to

$$\delta = 2N - [(N + \mu)(1 + \alpha) + (N - \mu)(1 - \alpha)] = -2\mu\alpha$$

and, if $\mu = 100$ and $\alpha = 0.01$, $\delta = -2^s$. BERNOULLI then refers to the first part of his *Mensura sortis* (see my § 4.3) and notices that if $|\mu| \leqslant 100$ there is an equal probability of $|\delta| < 2^s$ and $|\delta| > 2^s$ (in more modern language: $|\mu| \leqslant 100$ corresponds to a probable error of 2^s) and that the corresponding annual and hour errors are respectively

$$\delta\sqrt{365} \qquad \text{and} \qquad \delta/\sqrt{24} .$$

In his *Mensura sortis* BERNOULLI gave the normal law essentially as

$$p(\mu) = \frac{1}{\sqrt{\pi N}} \, e^{-\mu^2/N}$$

with

$$\frac{1}{\sqrt{\pi N}} \int_{-\mu}^{\mu} e^{-x^2/N} \, dx = 1/2$$

for $N = 10,000$ if $\mu = 47 \ 1/4$. Now, for $N = 43,200$, BERNOULLI naturally arrives at

$$\mu = 47 \ 1/4 \ \sqrt{\frac{43,200}{10,000}} \approx 100$$

and notices, as stated above, that this μ (and the corresponding δ) varies as the square root of N. This, of course, is a simple corollary of the normal law being used but actually, as is well known since the works of GAUSS, the same property (but expressed in terms of the mean square error) holds regardless of the law of distribution.

123

It is not difficult to offer criticisms. (i) BERNOULLI considered only the simplest law of distribution of errors (retardation numerically equal to acceleration). He did not consider the more general triangular distribution due to T. SIMPSON (1756) possibly because the central limit theorem being of course unknown he would have been unable to use his normal law. Moreover, he tacitly supposed oscillations to be mutually independent, which is hardly true.

(ii) Considering only pendulums, BERNOULLI did not generalize his account on observations as such which would have been necessary for a (general) theory of errors.

(iii) Being undoubtedly in a position to offer general rules for using laws of distribution (or at least the normal law) BERNOULLI did not do so.

(iv) A special remark: no connection exists between this memoir and the *Dijudicatio*.

With all this in mind, BERNOULLI's memoir is extremely important. At the age of 80 BERNOULLI gave

(i) the first use of the normal law in the theory of errors. In the nineteenth century this law happened to be almost the only one used in the theory of errors, not only in the limiting case as in BERNOULLI, but also as the law of distribution of actual errors.

(ii) the first explicit bifurcation of observational errors in two species, viz., systematic (*chronicarum*) and accidental, random (*momentanearum*) with an explanation of their different laws of propagation.

(iii) the first use of a modern measure of precision in the theory of errors.

6. Urn problems

We have seen (§ 4.2) that urn problems had been used by BERNOULLI for direct applications to political arithmetic. Another, theoretical application was to illustrate the use of different approaches in probability.

In *De usu*[25] BERNOULLI considered an urn with $2n$ strips of paper, n of them white and n black. Strips of each colour are numbered from one through n so that each two strips different in colour but having the same number constitute a pair. The problem consists of finding out the number x of paired strips remaining in the urn after $(2n - r)$ random extractions without replacement (after the urn contains r strips only out of the original number, $2n$).

At first BERNOULLI uses a combinatorial approach: after extracting one more strip the number of paired strips diminishes by $2x/r$. Actually BERNOULLI (as already in the formulation of his problem) is dealing with the mathematical expectation of this number:

$$Ex \;=\; x \,\frac{r-2x}{r} \;+\; (x-1)\,\frac{2x}{r} \;=\; x - \frac{2x}{r}$$

which is $2x/r$ less than before this extraction; $(r-2x)$ is the number of single strips in the urn at the moment of extraction.

Using similar calculations for the second, third etc. extractions, BERNOULLI arrived at a general formula: after $(2n-r)$ extractions

$$x \;=\; \frac{r\,(r-1)}{4n-2}$$

where, again, Ex should be read instead of x , and, for $r \to \infty$ and $n \to \infty$.

$$x \;=\; \frac{r^2}{4n} \tag{3}$$

He independently arrives at this last formula by using a differential approach: if r diminishes by dr , the corresponding dx either equals zero $(r-2x$ cases) or $dx = dr$ $(2x$ cases) so that

$$dx \;=\; \frac{(r-2x)\,0 + 2x\,dr}{r} \;,\quad x \;=\; \frac{r^2}{4n}$$

the initial condition being $r = 2n$ when $x = n$. But suppose the white strips are for some reason extracted more often than the black ones, then

$$x \;=\; \frac{st}{n} \tag{4}$$

where s and t are the numbers of black and white strips left in the urn respectively.

BERNOULLI arrived at (4) by the same differential approach and generalized the problem by considering strips of several different colours. For $s = t = r/2$ (4) of course coincides with (3).

In the second part of his memoir (from § 11) BERNOULLI solves the same problem for strips of two colours taking

125

$$\frac{ds}{s} = \varphi \, \frac{dt}{t}$$

where φ is the constant or variable ,law' of the relative likelihood of the extraction of strips, and gives special consideration for the case of $\varphi = 2$. He concludes his memoir with a numerical comparison of results in this last case $(\varphi = 2)$ with results obtained in the previous case $(\varphi = 1$, formula (3)). In fact he is posing a problem of distinguishing between two hypotheses, see also § 4.3.

BERNOULLI returned to urn problems in his *Disquisitiones analyticae*[39]. His first problem is: two urns are given, the first one with n white balls, the second one with n black balls. To find x, the number of white balls in the first urn, after r extractions had been made out of each urn with balls extracted from the first urn being each time put into the second urn and vice versa.

Employing the combinatorial approach BERNOULLI arrives at

$$x = \frac{1}{2} \, n \left[1 + \left(\frac{n-2}{n} \right)^r \right] \approx \frac{1}{2} \, n \, (1 + e^{-2r/n}). \qquad (5)$$

The differential approach leads him to

$$dx = -\frac{x}{n} \, dr + \frac{n-x}{n} \, dr \qquad (6)$$

from which the last approximation (in (5)) is readily derived.

In § 6 BERNOULLI, quite appropriate for the author of *Hydrodynamica*[39a] that he was, compares the differential approach with a continuous inter-mixing of liquids in vessels. In § 8 BERNOULLI generalizes his problem considering three urns and balls of three different colours being cyclically moved from one urn into the next one.

---

39. *Disquisitiones analyticae de nouo problemate coniecturale. Novi commentarii academiae scientiarum imperialis Petropolitanae* t. 14, pars 1, 1769 (1770), pp. 3–25, Year 1759 is erroneously given on title page.

39a. In part 10 of his *Hydrodynamica* BERNOULLI gives what is essentially a kinetic theory of gases and liquids. It would have been extremely interesting to find any connections between his kinetic theory and his ideas in probability. Such a connection may well have been implicity used by BERNOULLI but I could not find it in his book. On p. 504 of the Russian translation of *Hydrodynamica* (1959) one of its editors, K. K. BAUMGARDT, notices that an almost complete Latin manuscript, being the original version of the *Hydrodynamica* had been discovered at the Archive of the USSR Academy of Sciences.
The published parts of this manuscript (only in Russian translation) contains no explicit relation to probability either.

Combinatorially BERNOULLI derives formulae for the number of white balls to be found in each urn after r such moves and notices (§ 9) that these formulae could be arrived at by a process of summation of a certain infinite series: consider the series $[(n-1)+1]^r$. The sum of the first, fourth, seventh, . . . terms divided by n^{r-1} equals the number of white balls in the first urn; for the other urns the sums of the second, fifth, eighth, etc, and the third, sixth, ninth, etc. terms correspondingly are taken.

Noticing also that a similar process is possible for the first, more simple problem, he then calculates the sums of the three series. For the first urn, with large n and r, he arrives at

$$A = \frac{1}{n^{r-1}} \left[\binom{r}{0}(n-1)^r + \binom{r}{3}(n-1)^{r-3} + \binom{r}{6}(n-1)^{r-6} + \dots \right] \approx$$

$$\approx \frac{(n-1)^r}{n^{r-1}} \left[1 + \frac{r^3}{3! \, n^3} + \frac{r^6}{6! \, n^6} + \dots \right] \approx n e^{-r/n} S \qquad (7)$$

where S, the whole expression in the square brackets, satisfies the differential equation

$$S \, \frac{dr^3}{n^3} = d^3 S$$

and is therefore equal to

$$S = \alpha e^{r/n} + \beta e^{-r/2n} \sin \frac{r\sqrt{3}}{2n} + \gamma e^{-r/2n} \cos \frac{r\sqrt{3}}{2n}, \qquad (8)$$

where, because of the corresponding initial conditions, $\alpha = 1/3$, $\beta = 0$ and $\gamma = 2/3$.

BERNOULLI deduces similar formulae for the number of white balls in the other urns, for which he again employs the differential method. Denoting x, y, and $[n-(x+y)]$ the numbers of white balls in the urns, he proceeds from

$$dx = -\frac{x}{n} dr + \frac{n-x-y}{n} dr, \quad dy = -\frac{y}{n} dr + \frac{x}{n} dr \qquad (9)$$

and, after some hard work, achieves complete coincidence with the previous formulae.

127

The end of the memoir is devoted to the investigation of all these formulae. BERNOULLI calculates the number of extractions for which the number of white balls in the first urn is maximal and notices the existence of a limiting case, i. e. the case of an equal number of balls of each colour in each urn.

Considering the symmetry of the problem, the easiest method for proving this fact, or, rather, the corresponding fact about the expectation of the number of balls, and, furthermore, for generalizing the proof to the case of any finite number of urns, is by referring to the theorem on the existence of a limit transition matrix in homogeneous MARKOV chains.

A physicist could have noticed here a prototype of the once popular probabilistic model of the thermal death of a (finite) universe.

TODHUNTER (§§ 417—420) solved BERNOULLI's second problem by using equations in finite differences and, also, by operational calculus. He also noticed that the differential equations (9) could be more conveniently written as

$$dx = \frac{dr}{n} (z - x) \ , \quad dy = \frac{dr}{n} (x - y) \ , \quad dz = \frac{dr}{n} (y - z) \ .$$

Lastly, TODHUNTER noticed that the sum S is equal to

$$S = \frac{1}{3} \left[e^{ar/n} + e^{\beta r/n} + e^{\gamma r/n} \right]$$

where α, β and γ are the three different values of $\sqrt[3]{1}$. The x in the first problem (formula (5)) and the S and A (formulae (8), (7)) depend upon a (discrete) parameter, 'time' r/n. For mathematical expectations to possess such a dependence is characteristic for non-stationary random processes. A similar dependence exists for s/ξ, see formula (2).

Es seien N Kugeln (z. B. 100) fortlaufend numeriert über zwei Urnen verteilt Die Urne A enthalte P_0 (z. B. 90), die Urne B also $Q_0 = (N - P_0)$ Kugeln . . . . Überdies befinden sich in einem Sack N Lotteriezettel mit den Nummern 1—N. Nach je einer Zeiteinheit wird ein Zettel gezogen und zurückgelegt. — Jedesmal, wenn eine Nummer gezogen wird, hüpft die Kugel mit dieser Nummer aus der Urne, in der sie gerade liegt, in die andere Urne und bleibt dort so lange liegen, bis gelegentlich wieder ihre Nummer gezogen wird.

This is the celebrated EHRENFESTS' model[40] recently subjected to a computerized test[41]

---

40. P. and T. EHRENFEST, *Über zwei bekannte Einwände gegen das Boltzmannsche H-Theorem* (1907) in: P. EHRENFEST, *Collected scientific papers.* Amsterdam, 1959, pp. 146—149.
41. M. KAC, *Probability. Scientific American,* vol. 211, No. 3, 1964, pp. 92—108.

Played on a computer, an Ehrenfest game with 16,384 hypothetical balls and 200,000 drawings took just two minutes. Starting with all the balls in container A , the number of balls in A . . . declined exponentially until equilibrium was reached with 8,192 balls . . . in each container. After that fluctuations were not great.

KAC does not fail to add that, according to POINCARÉ, a dynamical system would eventually return arbitrarily close to its original state.

As the EHRENFESTS' model may be described by BERNOULLI's differential equations (6) we would say now that this model more properly should be called the D. BERNOULLI-EHRENFESTS' model.

On the urn problems in BERNOULLI see also URBAN[42].

7. Addendum

7.1. D'Alembert

I referred to D'ALEMBERT and his criticisms of DANIEL BERNOULLI in § 2.

D'ALEMBERT published quite a number of memoirs and shorter articles on the theory of probability[43]. In these writings D'ALEMBERT expresses his doubts as to the correctness of the basic principles of the theory of probability and, besides this, criticizes other scholars. Being sometimes completely erroneous, his criticisms hardly influenced contemporaries and subsequent scholars; but on the other hand the criticisms were indicative of an insufficient accurateness in the formulation of some propositions of the theory of probability and, especially, in the principles of its applications.

In parlicular[44], he just did not regard this calculus as

un calcul exact et précis également net dans ses principes et dans ses résultats.

On the other hand, D'ALEMBERT held a sober opinion on the practical impossibility of rare events occurring at the very first instance so that the mathematical expectation of gains in games of chance seemed to him useless. He also demanded that a qualitative distinction be made between *absolute certitude* and the *largest probability*.

The first reasoning is due to BUFFON, and D'ALEMBERT refers to him. The second one, as it seems, contradicts the first. However, both of them

---

42. F. M. URBAN, *Das Mischungsproblem des D. Bernoulli.* Atti del Congr. intern. matematici, t. 6. Bologna, 1928. Congress Bologna, 1932, pp. 21—25.
43. His main writings on probability are in his *Opuscules mathématiques*, t. 2, 1761; tt. 4 and 5, 1768; tt. 7 and 8, 1780 and also in his *Mélanges de littérature, d'histoire et de philosophie*, t. 5, 1786.
44. TODHUNTER, § 515.

taken together indicate D'ALEMBERT's prudent approach to applications of probability: it is impossible to expect a rare event to occur in a single instance, and it is also impossible not to expect such an event in a long run. A more thorough probabilistic analysis of this problem (the strict law of the large numbers) occurred only in the XXth century.

D'ALEMBERT pays attention to problems of political arithmetic. In this field he does not manage without erroneous opinions either. For some reason he is confused by the existence of two different durations of life, the probable and the mean ones, a problem perfectly well understood by HUYGENS, and supposes this fact to be an additional evidence against the theory of probability at large. He makes mistakes in his commentaries of D. BERNOULLI's *Essai*, but in this case he also expresses sensible considerations.

D'ALEMBERT notices that BERNOULLI's assumptions concerning epidemics of smallpox are oversimplified and that detailed statistical data is necessary. He then justifiably states that a final inference concerning the benefits of inoculation should not be made only on the grounds of the prolongation of the mean duration of life: not everyone would agree to be inoculated and, consequently, be exposed to a risk, even if to a small one, of an immediate death in exchange for a remote prospect of living a few years more. Besides this, there exists a moral side of the problem, e. g. concerning inoculation of children, so that a comprehensive mathematical analysis is impossible. Nevertheless, D'ALEMBERT supports inoculation suggesting compensation or special medals for families of persons perishing of inoculation.

D'ALEMBERT's own method of comparing risks from smallpox and inoculation is cumbersome and, it seems, not reasoned out but it is based, as it is possible to say now, on an idea of a loss function, i. e. actually the idea of D. BERNOULLI.

D'ALEMBERT several times mentions BERNOULLI's probabilistic reasoning in astronomy (see our § 2). He supposes this reasoning to be worthless, as an argument *ad hominem*[45].

We have elsewhere noticed objections of other scholars to similar reasonings[46] and it is our opinion that here, as in many other instances, D'ALEMBERT's criticisms make sense.

---

45. *Opuscules mathématiques*, t. 4, p. 292.
46. See O. B. SHEYNIN, note 7, § 2.2.1.

7.2. Buffon

The history of probability and especially of the moral expectation (see my § 3) should include the name of the great naturalist, BUFFON (1707–1788), to whom probabilistic reasonings were not foreign. In particular, his is the celebrated problem involving throws of a needle and geometric probabilities[47]. His *Histoire naturelle*[48] contains interesting features:

1. A letter to CRAMER dated Oct. 3, 1730, which is prior to the publication of BERNOULLI's *Specimen* (§ 6, pp. 75–77 of the *Essai d'arithmétique*) where we find (p. 76) an assertion that a given person should differentiate between a certain amount of money and its subjective value. This is actually a qualitative assertion of which the quantitative approach of BERNOULLI was a further development.

2. In § 8 (p. 57) BUFFON publishes his letter to BERNOULLI (1762) where he introduces the concept of *l'homme moyen*, later to become the central idea of A. QUETELET:

> l'homme moyen, c'est-à-dire, les hommes en général, bien portans ou malades, sains ou infirmes, vigoreux ou foibles . . .

3. In § 15 BUFFON discusses the Petersburg problem, notices the contradiction between *bon sens* and *calcul* (§ 16) and makes several proposals, one of which, to consider small probabilities strictly equal to zero, he mentioned in his letter to BERNOULLI. This proposal had been accepted by D'ALEMBERT (§ 7.1).

Conclusion

In my opinion DANIEL BERNOULLI was one of the main precursors of LAPLACE in that he introduced most important mathematical methods into probability and opened new fields of application for probability thus helping to establish the theory of probability as a mathematical tool for natural sciences at large.

---

47. N. T. GRIDGEMAN, *Geometric probability and the number* π *Scripta math.*, vol. 25, No. 3, 1960, pp. 183–195.
48. *Supplement*, t. 4. Paris, 1777. This contains BUFFON's *Essai d'arithmétique morale* (pp. 46–148).

Zusammenfassung

Der vorliegende Artikel beschreibt D. BERNOULLI's Arbeiten zur Anwendung der Wahrscheinlichkeitstheorie, die wichtige Beiträge zur Bevölkerungsstatistik, Astronomie und Theorie der Beobachtungsfehler enthalten. Gleichzeitig förderte D. BERNOULLI die Wahrscheinlichkeitstheorie als solche wesentlich. Nach Ansicht des Autors war sein Einfluß auf LAPLACE vergleichbar mit dem DE MOIVRE's. Ergänzend werden relevante Beiträge von Zeitgenossen diskutiert.

CONTRIBUTIONS TO THE HISTORY OF STATISTICS

H. WESTERGAARD
(1932)

CHAPTER VI

PROGRESS IN THE MIDDLE OF THE EIGHTEENTH CENTURY

22. AFTER the stagnation in the first part of the eighteenth century a new era opened in the history of statistics. *Sweden* deserves first mention, as this country was the first to give political arithmetic a solid basis by a system of official statistics. Direct observations were also made in *France* and *Holland* (annuitants, members of tontines, monks and nuns). And though the Bills of Mortality in *England* made little progress, the observations at hand were at least treated more critically than before, as shown by Th. Simpson's mortality-table for London.

In *Sweden*, as in most other countries, it was felt as a great drawback that the population was too sparse to undertake all the economic tasks arising within the kingdom in agriculture and industry, not to speak of the problem of raising the military forces judged necessary for the protection of the country. Anders Berch's remarks on the population problem were mentioned above (Art. 19). In 1744 Salander wrote on the same problem.[1] If France had 20 millions, Sweden had probably only 3 millions. But he maintains that the soil could produce sufficient food for 20 millions. If this appears doubtful, he is willing to reduce the number to 10, and even to 5 millions, but at all events he had no doubt whatever that agriculture in Sweden could produce sufficient food for these 5 millions.

It is evident that loose estimates of this kind could

[1] Försök, om Sverige kan ved egen Växt föda sina Invänere" (*Kongl. Wetenskaps Academiens Handlingar*, Vol. V, 1744).

not in the long run prove satisfactory to the discussion of the population problem, and after long deliberations a bill for making tabular records of the population was approved by the King and became law on February 3rd, 1748. For many years the Swedish clergy had had the duty of keeping parish registers, containing lists of the members of the congregation, of marriages, births and deaths, and of persons entering or leaving the parish. Incomplete as these lists may have been, they quite naturally became the foundation-stone of the Swedish system of official vital statistics. Occasionally they were used for local statistical investigations, as in a paper by Wassenius on the vital statistics of a parish.[1] But a deeper investigation was made by the mathematician Pehr Elvius (1710–49), Secretary of the Swedish Academy of Science, who had undertaken to compile lists of births and deaths for the whole kingdom in order to determine the probable number of inhabitants. The result was a report (1746) which the Academy sent to Parliament and which probably played a significant part in prevailing on the Parliament to pass the above-named Act.[2] As Elvius, like most political-arithmeticians of those days, treated his observations freely, it is not possible to reconstruct his calculations with absolute accuracy, though his method is on the whole clear enough. He regards 70,000 as the normal yearly number of deaths, and he distributes

[1] "Wassenda Församlings Forökelse genom födda och vigda, så ock aftagande genom döda . . ." (*Kgl. Wet. Acad.*, Vol. VIII, for 1747).

[2] As to the History of Swedish official statistics, see E. Arosenius, *Bidrag till det Svenska Tabelverkets Historia*, Stockholm, 1928, and the same author's *The History and Organization of Swedish Official Statistics* (The History of Statistics collected and edited by John Koren, New York, 1918).

A. Hjelt, *De första officiela relationerna om svenska tabellverket, åren* 1749–57, Helsingfors, 1899.

A. Hjelt, *Det svenska tabellverkets uppkomst, organisation och tidigare verksamhet*, Helsingfors, 1900.

Lundell, *Den politiska arithmetikens uppkomst och utveckling*, Helsingfors, 1911.

G. Sundbärg, *Bevölkerungsstatistik Schwedens*, 1750–1900, Stockholm, 1907.

Further, *The Official Vital Statistics of the Scandinavian Countries and the Baltic Republics* (League of Nations, Health Organisation, Geneva, 1926).

them according to age after observations made in certain parts of the kingdom. Following in the main Halley's method for finding the number of inhabitants in Breslau, on the supposition that the population was stationary, he would, for instance, proceed thus: Out of the 70,000 persons who died yearly, about one-third were under 3 years, and about 29,300 under 10 years of age. Supposing, in order to arrive at the same numerical result as Elvius, that 8,000 died between 3 and 10 years of age, and 21,300 under 3 years, we shall find that 70,000 newly-born altogether have passed about 178,000 years before 3 years of age, viz. 1·5 (70,000 + 48,700), and about 313,000 between 3 and 10, viz. 3·5 (48,700 + 40,700). He thus calculates 491,000 years of life, which, on the hypothesis of the stationary population, is the same as the number of persons living under 10 years of age. For the following ages he uses decennial averages. The total number of inhabitants of all ages he finds to be 2,097,000, which is not far from the number which some years later was found by direct observation.

Elvius did not regard this result as more than a preliminary attempt, and was himself an advocate of a regular system of official vital statistics, such as was established in 1748. According to this system, somewhat complicated schedules had to be filled out every year for each parish. Thus there were particulars, for each calendar month, of *baptisms* of legitimate and illegitimate children, with distinction as to sex, further, *weddings* and number of marriages dissolved by death, the number of *deaths*, separately for each sex, for children under 10 years of age, unmarried and married people, with additional remarks on still-births, plural births, etc. Another table gave a classification of *deaths according to age, sex and cause of death*. The children were classified according to age between 0 and 1 year, 1 and 3 years and 3 and 5 years respectively, then follow 17 quinquennial age-groups and, finally, deaths above 90. This detailed combination of age and cause of death was indeed a remarkable step forward in vital statistics.[1]   The nomenclature of the causes of death, 33 in number, would not of course satisfy a modern medical statistician, but it gives much interesting information about that period, notably with regard to violent deaths and zymotic diseases.   The first report, for 1749, shows 12 per cent. of the deaths as due to small-pox and measles, 6 per cent. as due to scarlet fever, and 5 per cent. to whooping-cough (all of which causes, from 1911 to 1920 inclusive, only counted 1 to 3 per cent. of the total deaths); 14 per cent. died from consumption and lung disease.   Several secondary causes of deaths were noted, such as dropsy and jaundice.

Finally, there were details with regard to the *population*. The number of persons of each sex in the same age-groups as the deaths was required.   Further, the population was distributed as to sex and conjugal condition (children under 15 years, single, married, widowers and widows), and according to occupation and rank (sometimes with particulars as to children under 15 years, and young persons).   In addition, the number of households, of inns and public-houses, etc., was given.   It is interesting to note the average number of persons belonging to each household, which, compared with modern times, was high (six to seven members against three to four nowadays).

To a certain extent the magistrates in the towns took part in preparing the lists of the population, but the main work devolved upon the clergy, the study of a pastor being in fact a small statistical bureau.   The clergy complained bitterly and obtained relief to the extent that the lists of population were only required to be prepared every three years.   It could not of course be expected that the material which was sent by the clergy should be faultless, in fact there were complaints that the lists were often very imperfect.   Perhaps not the least trouble, especially in the towns, was caused

[1] Sundbärg has calculated rates of mortality according to age and sex, from *tuberculosis*, 1776–1800, l.c., p. 148, Table 60.

by the particulars which had to be entered on removal to and from a parish.

A serious drawback was the decentralisation which characterised the whole institution. From the parish the lists were sent to the deaneries, where they were summarised. These summary tables were again forwarded to the consistories which had the duty of sending their condensed reports to the provincial governors for that part of the diocese which belonged to his province. Finally, the provincial government sent a summary for the province to the "Kanslikollegium." [1]

23. Pehr Elvius died in 1749. "Kanslisecretary" E. Carleson was given the task of preparing a general summary, with the assistance of some of the members of the Academy of Science. The most active of these was the astronomer Per Wargentin (1717–83), the successor of Elvius as the secretary of the academy. By a royal rescript of 1756 this committee was made a permanent institution, the Tabular Commission ("Tabell-kommissionen").

Before the latter event Wargentin had published some papers in which he treated the death-lists in 1749. [2]

So far there is nothing new in these papers as Wargentin only uses well-known methods. But it was very important that here for the first time detailed observations embracing a whole country were at hand, and, moreover, Wargentin gives the programme of future investigations. He maintains that it would be possible to draw safe conclusions, if not only the numbers of births, marriages and deaths were known but also the numbers of the living. He adds that Halley has shown a short-cut, and the following investigations are based on his method. He does, not, however, describe it very clearly, so that it may be doubtful whether at

---

[1] Arosenius, *The History and Organization of Swedish Official Statistics*, l.c., p. 541.

[2] "Om nyttan af Förteckningar på födda och döda" (*Kgl. Vet. Acad.*, Vol. XV, for 1754, and Vol. XVI, for 1755).

that time he had thoroughly understood it, but in a later investigation (1767) he gives a quite clear statement.

Wargentin gives the distribution per mille of deaths 1749 according to age (still-born included) in the whole Kingdom and in six provinces separately ("Höfdinge dömen"), where the frequency of epidemic diseases had been relatively small. But of these two tables combined with certain other observations (Halley, Kersseboom, Deparcieux) he forms a new table. This is interesting inasmuch as it explains how Süssmilch some years later calculated the mortality-table, which carries his name and which in spite of its evident defects enjoyed a high reputation. It will be mentioned below.

Wargentin looks upon Halley's table as trustworthy even for Sweden, as regards mortality, and by comparing the figures he draws various conclusions with regard to the growth of the population in Sweden. His remarks are, however, not very clear.

The Tabular Commission took up its task with great zeal; two reports from 1751 and 1761 gave various results of the work, with more general remarks on the population problem, particularly burning in Sweden which, as may be surmised, had suffered severely from the long wars at the beginning of the century. The first report complained very strongly of the many difficulties which must be felt in so thinly populated a country. Many persons died yearly whose lives might have been saved; it was urged that there ought to be public access to medicinal drugs, that inoculation against small-pox ought to be tried, that too many children died in infancy through overlaying, that there were too few physicians. The report also suggests that students of divinity should acquire some knowledge of medicine before being appointed as clergymen. Further, that there ought to be more inducement to marry. It was a pity that so many thousands were forced to live as beggars (according to the lists there were 29,000 "paupers not in hospital"). The second report considered in detail the problem of emigration. In the course of three years 24,000 persons

138

had emigrated; the report complained of the emigration to Copenhagen, to Norway and to Russia. Later, however, Wargentin acknowledged that the supposedly high emigration-figures were largely due to inaccuracy of the registers.

The report also drew certain conclusions with regard to the causes of *fertility*. Opulence and luxury were a hindrance to fertility, the number of conceptions being relatively small in the months when the peasant had the most plentiful supply of provisions.

These reports were not published.[1] As in so many other countries statistical information, particularly on population, was regarded as a State secret. In Sweden, particularly, it was feared to allow the serious complaints as to deficiency of population to be known in the neighbouring States. Gradually, however, this silence was abandoned, and several statistical observations were printed in the proceedings of the Academy of Science, so intimately connected with the Commission. Various papers were quoted above. In 1762 the Commission finally got permission to publish a yearly report in the Acts of the Academy, and various important results were published in the following years; in 1764–5 the Secretary of the Commission, E. F. Runeberg, gave the number of inhabitants in Sweden and Finland in 1760 (2,360,000), and in 1766 Wargentin published his famous mortality-tables for the nine years 1755–63.[2] Here, for the first time, mortality-tables for a whole country are to be found, based on the observations of the living population as well as on deaths. Wargentin compared, for instance, the average number of deaths in the triennium, 1755–7, in one of the twenty-one classes of age, with the corresponding number of inhabitants registered in 1757, asking among how many persons one death would occur in each year. Having these enumerations at his

---

[1] They are reproduced by Hjelt in his above-quoted treatise of 1899.

[2] "Mortalitäten i Sverige i Anledning af Tabell-Verket" (*Kgl. Vet. Acad.* Vol. XXVII). Republished by "Lifförsäkrings-Aktiebolaget Thule," Stockholm, 1930.

disposal, he might have used an interpolation in order to find the average population in the triennium, but the population was growing so slowly, that a correction of this kind would have been insignificant, and as a first attempt at finding the mortality of a whole population by direct observation, these calculations may justly be looked upon as sufficiently reliable. It must not be forgotten, that, as mentioned above, there were considerable inaccuracies which no interpolation could do away with. In his later years Wargentin made an investigation on this subject and found that whereas some of the enumerations were tolerably exact, as, for instance, those of 1751 and 1763,-others showed obvious defects.

Wargentin did not go further in his investigation concerning mortality; he did not, for instance, calculate the expectation of life or the numbers surviving of a certain generation at various ages. On the whole he did not enter much into theoretical questions. But at all events he deserves praise for his contributions to vital statistics. As one important result we can refer to the fact that for the first time he proved that in a general population the rate of mortality of the female sex was smaller than that of males.

It may be added that *Iceland* already had a general census in the year 1703,[1] which was carried out on the proposal of an Icelandic Commission, appointed to inquire into the economic condition of the country. The name, occupation and age of each individual was recorded; paupers were registered separately, as also persons who were temporarily present in the place concerned but had their home elsewhere. Unfortunately there was at that time no organisation to work out the results. The census is going to be published, and it will form a most interesting basis for the study of the structure of the population in this remote country, two centuries ago.

Other Scandinavian countries also had occasional enumerations. Thus in *Norway* as early as in 1662 a

[1] Th. Thorsteinsson, "Den islandske Statistiks Omfang og Vilkaar," *Nordisk Statistisk Tidskrift*, 1922.

census of males above 12 years of age had been taken, for military purposes; in Denmark partial enumerations were taken in 1645 and 1660 for the purposes of taxation. But the results of enumerations of this type remained mostly unpublished and they could therefore have no influence on the evolution of political arithmetic. This was also the case with several local or general numerations in various countries outside of Scandinavia in the course of the seventeenth and eighteenth centuries.

24. In the same year in which Pehr Elvius published his investigation a highly interesting work on mortality appeared in *France*, by Deparcieux (1703–68).[1] Here old and new methods come in use. Thus in order to calculate mortality-tables for monks and nuns he distributes the deaths between 1685 and 1745 in several cloisters, according to age, under the supposition that the number of inmates in these institutions were on the whole constant. But for the Benedictine monks separately he calculates a table for the period 1607–1745 on a thoroughly correct principle. The monks concerned entered in 1607–69, aged 17 to 25, and all of them died before 1745. As none of them left during the period of observation it was easy to calculate the numbers of persons exposed to death at each age, and consequently to find the proper rates of mortality. The next step would be to take the fluctuations of the numbers into consideration in cases where members are leaving during the period of observation, as also to find the numbers of years, where several members are still alive at the close of the period. This would have been necessary if the period 1607–1745 had been divided into various periods in order to observe changes in mortality, a problem which, however, quite naturally did not occur to political arithmeticians in this epoch. But Deparcieux also mastered this last step: he calculates mortality-tables for the members of two tontines, of 1689 and 1696, embracing 5,911 and 3,345 persons respectively, of whom, in 1745, 711 and 616 respectively were still alive.

[1] *Essai sur les probabilités de la durée de la vie humaine*, Paris, 1746.

This is done quite correctly, the numbers exposed to risk being calculated from the third year of age.

In order to calculate the mean duration of life, Deparcieux recommends an approximate method. In a stationary population with a regular balance of births and deaths the expectation of life can be formed by dividing the number of inhabitants by the number of births. But if there is a surplus of births he takes it for granted that the expectation of life is higher, whereas the result would be too high, if he used the number of deaths as denominator. He therefore suggests taking the mean of births and deaths as denominator. This method or similar formulas were frequently recommended in the following 100 years. The calculation cannot of course give very exact results. Thus we find that in Denmark 1840–9 Deparcieux's method would give 39 years compared to 42·2 years correctly calculated. 1921–5 would give 60 and 61·1 years respectively.

After having published his chief work Deparcieux did not contribute much to political arithmetic. He had a discussion with an author (Thomas) who had criticised his results, and in 1760 he published a supplement to his book, with various observations, for instance, Swedish death-lists 1754–6. It appears from some recently published letters to Wargentin (1760–7)[1] that he was planning a second revised edition. This is also mentioned by a nephew of his who in 1781 published a treatise on annuities.[2] With the assistance of the bishops Deparcieux had obtained death-lists from 162 priests in various parts of France for sixteen years. Evidently he wished in this way to get a broader foundation for mortality statistics, thus meeting objections concerning the applicability of his observations from the narrow sphere of tontines and religious institutions, even though he did so at the expense of correctness of method. But death prevented him from realising his plan, and attempts

---

[1] Arosenius, *Bidrag till det Svenska Tabelverkets Historia*, 1928, pp. *38–*45.
[2] Deparcieux, *Traité des annuités, accompagné de plusieurs tables très utiles*, Paris, 1781.

made after his death at treating the material received from the priests proved unsuccessful.

Other important contributions to political arithmetic were published in *Holland*. Thus Struyck (1687–1769) published an interesting work, chiefly containing contributions to astronomy and geography, but with several observations on vital statistics.[1] Besides detailed observations he naturally tries to give various estimates. The world population is supposed to be 500 millions, and he estimates the number of deaths in every hour to be 2,000 (35 per mille yearly). But as to other questions his basis is more solid. He discusses ably the difference of mortality of males and females, partly based on German observations. This he further elucidates by observations on annuitants, following these quite correctly, according to age, from quinquennium to quinquennium, thus finding the number exposed to risk for each age and sex.

In a later publication[2] Struyck collected several observations on vital statistics in Holland and other countries. There are several other interesting observations, on mortality in childbed with distinctions as to time passed between the birth and the death, further on twins and multiple births, on mortality of sailors on the long journey from Holland to the Cape (which was extraordinary in comparison with modern observations). He is interested in the problem of climacteric years, and here as in other directions he shows much critical sense.

In statistical literature, Kersseboom (1691–1771) is perhaps more frequently quoted than Struyck, although his contributions to vital statistics appeared in a less regular form, being partly of a polemic nature. These contributions were published 1737–48.[3]

[1] *Inleiding tot de Algemeene Geographie benevens eenige Sterrekundige en andere Verhandelingen*, Amsterdam, 1740.

[2] *Vervolg van de Beschryving der Staartsterren, en nader Ontdekkingen omtrent den Staat van't Menschelyk Geslagt*, Amsterdam, 1753.

[3] See, for instance, three *Verhandelingen over de probable Meenigte des Volks in de Provintie van Hollandt en Westfrieslandt* (1738–42) (later republished under the title : *Proeven van politique rekenkunde*, 1748). A good account of his contributions to political arithmetic was given by G. F. Knapp, *Theorie des Bevölkerungs-Wechsels*, Braunschweig, 1874.

Kersseboom has not the mathematical training of which Struyck was possessed, though his conclusions are generally clear.　Owing to the form of his publications a complete system cannot be expected, he was too much engaged in attack and defence, in criticism and anticriticism.　It may even occur that he uses in one paper a method of which he thoroughly disapproves in another one, as, for instance, when he criticises van der Burch for having calculated a mortality-table from the London Bills, though the number of inhabitants could not be found in that way, and two years later uses the very same observations to calculate a mortality-table and the population.[1]

Van der Burch (1673–1758) did not in fact deserve this bitter criticism; he seems to have written clearly on the problems he took up, as when he treated annuities from a financial point of view, maintaining that the price of an annuity ought to depend on the age of the annuitant.[2]

Even though this desultory production prevented Kersseboom from a systematic treatment of the problems concerned, his thoughts are always circling around the same ideas; he is constantly returning to the question how to find the population from the number of births and the mean duration of life.　He is particularly interested in estimating the number of inhabitants in the provinces of Holland and Westfriesland.　In order to reach this goal he examines a rather large amount of material available concerning annuitants.　Like Struyck, he finds quite correctly the number of persons who have been exposed to risk at each age, thus arriving at a life-table for well-to-do people in his country in the seventeenth and the earlier part of the eighteenth century.　He is not to be blamed for having generalised these results, so that he considered that his table would hold good for a whole population, and not for annuitants

---

[1] *Observatien waarin voornamentlyk getoont word wat is gelijktijdigheid,* 1740; *Derde Verhandeling,* etc., 1742.

[2] *Bouwstoffen,* pp. 155 *sq.*

only. At all events, his mortality-table for annuitants deserves the reputation it enjoyed, at least for all ages above one year. As to infant mortality, he calculated on numbers for the whole population, but these were only directly observed in some cases and depended partly on estimates, though he took much trouble to prove the reliability of his figures. The supposed annual number of births in the two provinces was 28,000, out of which 5,500 were supposed to die in the first year of life. His life-table begins with 1,400, or a twentieth of the annual number of births given, and from this table he finds the expectation of life to be 35. He seems to have found this number by adding up the numbers of living at the beginning of each year of age, without taking the decrease in the course of the year into consideration.

Having found the mean duration of life, he calculates the birth-rate at 29 per mille, 35 being the number of inhabitants with one birth yearly. 29 per mille is rather a low birth-rate for that period, and Kersseboom did not escape criticism on account of this point. Probably the rates of mortality in the general population were higher than among annuitants, and if he had had a suitable mortality-table at his disposal he would have found an expectation of life lower than 35, even if he were right in supposing the population to be stationary.

As usual with political arithmeticians, he seems to have treated the observations freely, it being therefore impossible to reconstruct quite exactly his calculation as to the age distribution of the population. As regards the first five years of age, the number alive according to his supposition seems to be somewhat too high. Comparing his age distribution with the Swedish numbers in the middle of the eighteenth century, we find a pretty good correspondence. As we should quite naturally expect, the numbers above 80 seem to be relatively high judged by Kersseboom's table.

As with other political arithmeticians of his time, he is inclined to transfer results from one country to

145

another. Thus he uses King's estimates of the distribution of the population according to matrimonial classes as reliable also for Holland.

Sometimes his conclusions are a little clumsy. In one of his papers [1] he deals with observations (for each sex separately) on annuitants who entered at the same time and at the same age, and of whom none was alive at the time when the investigation was made. Summing up the ages which all the annuitants had reached, he found the mean duration of life. But he failed to use his material to calculate the rates of mortality.

In spite of all objections, however, Kersseboom justly deserves his reputation as one of the most prominent statisticians of the eighteenth century.

25. While great progress was made in this epoch on the Continent with regard to statistical observations, political arithmetic in *England* was chiefly engaged in revising and improving the material at hand. Though annuities on life were well known, no attempt was made at treating the material of the life-offices, nor did Parliament follow the example of Sweden regarding a system of official statistics. An important reform was made with regard to the Bills of Mortality, in so far as the age distribution of the deaths was known from the year 1728. The material was, however, rather defective, many deaths—particularly of dissenters—were not registered. Interesting comments on these defects were made in an anonymous introduction to a work which was published in London, 1759: *A Collection of the Yearly Bills of Mortality from 1657 to 1758 inclusive.* In a treatise included in this work Corbyn Morris recommends various important improvements.[2] He proposes, for instance, to divide the deaths of infants into three groups, viz. under 1 month, 1–3 months and 3–12 months. Further, the lists ought to give the distribution according to *birth-year* as well as to *age.* He complains

[1] *Derde Verhandeling . . ., 1742.*
[2] *Observations on the past Growth and present State of the City of London,* 1751, 2nd ed., 1757.

very much of the disproportion between the number of christenings registered and of births.

Even if improvements of this kind could be obtained the observations had the obvious defect that *migrations* had an influence on the age distribution of the deaths which made it impossible to apply Halley's method directly, especially in a place like London, where immigration was so conspicuous. First of all, the mathematician Th. Simpson tried to find a solution in a work on annuities.[1] He tries to correct the observations by combining Smart's table for London, 1728–37, with Halley's for Breslau. He considers the immigration after the age of 25 as insignificant; above this age therefore Smart's table can be used without amendment, but for the younger years a correction in respect of this afflux from outside will be necessary. He does not explain his method clearly, the main result being that 140 per mille of all the deaths in London occur among immigrants. He asserts that he has compared the number of christenings and burials and observed by help of Halley's table "the proportion which there is between the degrees of mortality at *London* and *Breslaw* in the other parts of life, where the ages are greater than 25."

Simpson seems to have simplified the problem by supposing that immigration into London was concentrated around the age of 25. If 140 out of 1,000 deaths belonged to immigrants. and 860 to natives, the deaths could be divided into two sections, one pertaining to the age under 25, altogether 574, natives solely, and the other to 426 above 25, out of whom 286 were natives and 140 immigrants.

In order to combine these two sections into one correct table of mortality he could either reduce all the numbers above 25 in the proportion of 286 to 426, or he could raise all numbers below that age in the proportion 426 to 286. Choosing the latter alternative, he will have to increase the 860 native new-born by 49 per cent., the table thus beginning with about 1,280.

[1] *The Doctrine of Annuities and Reversions*, London, 1742.

A similar method was used several years after by
Richard Price in his Observations on Reversionary
Payments.[1] He wished to calculate a life-table for
London on the basis of the numbers of deaths 1759–68,
and supposed that about half of the persons dying in
London above 18–20 were immigrants who settled at
that age.   On similar principles tables were constructed
for *Northampton* and *Norwich*.   These two tables were
much appreciated, although the material was rather
limited.   In Norwich the distribution according to age
below 10 was unknown.   Price therefore uses a proportion
similar to that in Northampton.

While Simpson and Price tried to take *migrations* into
consideration, another attempt at improving the death lists
was made on the Continent by the famous mathematician,
Euler (1707–83), under the supposition that migrations
were trifling, whereas there was a regular *surplus of births*
over deaths.[2]   If, for instance, the yearly surplus is 1 per
cent. of the population, and if the rates of mortality can be
considered as constant, then persons aged 70 will belong
to a generation about half as numerous as the generation
born at the present moment.   It will thus be easy to
find exact rates of mortality from the numbers of deaths,
if the yearly rate of surplus is known.

These are the first attempts at taking the movement
of the population into consideration by the construction
of mortality-tables.   Generally, political arithmeticians
took it more easy.   This is the case with the French
naturalist, Buffon (1707–88), who in his *Histoire Naturelle*
(*De l'homme*, Tome II, 1749) deals with a material which
Dupré de Saint-Maur had placed at his disposal, containing lists of deaths in some parishes in Paris and
outside of it.   He takes no notice of the influence of

[1] First edition, 1771, here quoted after the third edition, London, 1773.
Confer Sutton, "On the Method used by Dr. Price in the Construction of
the Northampton Mortality-Table" (*Journal Institute of Actuaries*, XVIII,
1875).
[2] "Recherches générales sur la mortalité et la multiplication du genre
humain," Berlin, 1767 (*Histoire de l'académie royale des sciences et belles-lettres*,
année 1760).

the movements of the population. He is chiefly interested in the probable duration of life—the age at which just half of the generation in question will be alive ("les probabilités de la durée de la vie"). In other respects his treatise is not uninteresting, with useful remarks on the concentration of the numbers on the round years of age, and his observations on the numbers of foundlings throw light on the social history of that epoch.

Other authors of this period may be read with interest. In Holland N. Duyn, who died 1745, wrote upon the influence of seasons on mortality.[1] Th. Short, in a work from 1750,[2] takes up a number of problems, though unfortunately in a rather superficial way. Thus he tries to show the influence of the geologic situation and of the soil on mortality. If circumstances are favourable in this respect, the number of miscarriages will be relatively small, and there will be many male births. Astronomic phenomena have influence on mortality, particularly eclipses. Children born in the cold months have a greater mortality than others. Like Wargentin, he maintains that "the Months of the greatest Ease, Repletion, Insolence, and the smallest Discharge, are most improper for Procreation" (l.c., p. 143). The months of October, November and December, with relatively few births, have proportionally many male births, "for strong Labour and Exercise has strung the Nerves and purified the Blood."

---

[1] *Bouwstoffen*, pp. 167 *sq.*
[2] *New Observations, Natural, Moral, Civil, Political and Medical, on City, Town, and Country Bills of Mortality . . . with an Appendix on the Weather and Meteors*, London, 1750.

CHAPTER VII

SÜSSMILCH AND HIS CONTEMPORARIES

26. MANY of the conclusions drawn by the political arithmeticians of those days were ill-founded and premature and their methods often very imperfect. Still, to some extent these results justified the high opinion with regard to the new science which frequently found expression in literature. It will therefore be worth while to get a bird's-eye view of what the statisticians of that epoch knew or believed they knew.

First we must mention the *regularity* of certain phenomena in vital statistics, which deeply impressed political arithmeticians of those days. These phenomena were frequently looked upon as proofs of a Divine Order. This was particularly the case with the question of the balance between the two sexes. Thus Thomas Short in his above-mentioned *New Observations* finds a remarkable providence in the small surplus of male births; as greater dangers menace boys than girls, a balance between the sexes would be the result, and he is led to the conclusion that "Polygamy is a most ridiculous, monstrous Custom." Wargentin writes in the same vein.

Two clergymen who contributed to political arithmetic maintained with particular force that the regularity is the result of Divine Providence: Will Derham (1657–1735) in England, and J. P. Süssmilch (1707–67) in Germany. The former published a frequently quoted series of sermons under the title: *Physico-Theology.*[1] He

---

[1] *Physico-Theology or a Demonstration of Being and Attributes of God from the Works of Creation*, 1713. Here quoted after the eighth edition, London, 1732. Cf. James Bonar, *Theories of Population from Raleigh to Arthur Young*, London, 1931, pp. 136 *sq.*

did not, however, collect many statistics, as did Süssmilch, who, in 1741, published a work on the *Divine Order* which appeared in a very enlarged edition in 1761.[1]

Süssmilch collected material with great care wherever it was accessible. His correspondence with Wargentin[2] testifies to his energy, and his work is a fairly complete compendium of all the statistical literature up to his time.

He is not a very original author. Having no mathematical training, he was naturally led to accept without much criticism the results which other authors, such as Deparcieux, had found. He is, however, by no means lacking in critical judgment, and many of his remarks show a sound common sense.

Süssmilch belongs to the army of authors in the eighteenth century who considered the population question the most important object of politics. His work contains in fact a discussion of all the causes which might influence the growth of population. His starting-point is chosen from the Bible (Gen. i. 28): "And God blessed them, and God said unto them: Be fruitful and multiply and replenish the earth, and subdue it" (Introduction, § 1). In order to reach this goal a Divine Order is established, regulating the proportion between births and deaths so as to give a normal surplus as the result. He maintains further that if there is a law of mortality, there must also be an order in diseases. It is true that epidemic diseases may cause some disturbance if certain years only are considered, but these irregularities disappear if twenty years or more are combined. The balance of sexes is a chief point in his discussion of the Divine Order: if, for instance, one male were born for every three females, then polygamy would be the natural consequence, but the wise Creator gave a proportion which led naturally to monogamy.

[1] *Die göttliche Ordnung in den Veränderungen des menschlichen Geschlechts aus der Geburt, dem Tode und der Fortpflanzung desselben erwiesen.* After his death the work was republished by his son-in-law Baumann, also a clergyman, who added a Supplementary Volume (Berlin, 1775–6). This fourth edition is here quoted.

[2] Reproduced by Arosenius, l.c., pp. *63–5.

151

In vital statistics Süssmilch finds (§ 12) a constant, general, great, perfect and beautiful order. He maintains that the duration of life is constant (§ 13). As it was 3,000 years ago, so it is in Europe nowadays: in Finland, Sweden, England, Holland and France. On account of this supposed permanence everywhere he feels justified in trying to calculate normal rates of mortality on the basis of various observations (§ 35), viz. in the country in "good years" 1 : 42 to 1 : 43, in "mixed years" 1 : 38, and as a general mean 1 : 40. Smaller towns have a mortality rate of 1 : 32, larger places, such as Berlin, 1 : 28, and still larger places, Rome and London, 1 : 24 to 1 : 25. For a whole province he considers 1 : 36 as normal, and he adds with great deference (§ 42) that every age and sex, all professions and all diseases, must contribute their share in order that the rate of mortality —one out of 36—may be the result. Dropsy, as well as convulsions and fevers, all have their part in the great tribute to the grave.

It is easy enough in our days to criticise these assertions. We find the regularity less striking than Süssmilch and his contemporaries did. In explaining his life-table he repeats his remarks on the *Divine Order*. He was struck by the harmony between the observations on monks and nuns in Paris and peasants in Brandenburg. But in fact the variations were not inconsiderable. The average for the second year of life in the country, viz. 78, was, for instance, calculated from the following four numbers: 49 and 59 respectively, 100 and 104. And as to the alleged constancy of duration of life Süssmilch had no reliable observations at all on mortality in ancient times. Modern life-tables show an enormous increase of the expectation of life, of which no eighteenth-century writer could dream.

But even though quite naturally we are induced to take notice of deviations from the average, and to ask how these irregularities can be explained, it is just as natural that our predecessors first of all were struck by the regularity and cared less for the deviations.

Süssmilch devotes a couple of chapters to marriage— and birth-statistics. He is particularly interested in the question of how long a time a doubling of the population will require. Here he was assisted by Euler, who furnished him with a table which showed the doubling period under varying circumstances. Of chief importance to Süssmilch was the probable increase of the inhabitants on the earth before the Flood. In another chapter he discusses various obstacles for increase: the plague, wars, famines, etc.

The succeeding chapters are uninteresting to the history of statistics, in so far as Süssmilch here chiefly enters into a discussion of the economies of population from the well-known mercantilistic standpoint. He brings arguments against polygamy, discusses proposals as to supporting married couples with numerous children, as to hygienic measures, luxury, etc.

After this long digression the author resumes his statistical investigations. One chapter is devoted to the population on the earth, according to various authors and to his own estimates. He considers 14,000 millions as being within the range of possibility, but the estimated population is only 1,080 millions and consequently far behind that limit. In another chapter he treats birth-statistics, particularly with regard to sex-proportion. He finds that out of 100 births 51 will be males; this holds good for Europe, and a rather limited material from Trankebar in India points in the same direction.

27. By far the most important chapter deals with *mortality statistics*. Even though his life-table is very primitively constructed, and in spite of evident defects, it enjoyed a great reputation. In his calculations he follows Wargentin closely in the above-quoted report of 1755 on the mortality in Sweden in 1749. Süssmilch, like Wargentin, evidently looks upon a mortality-table from one section of the earth's region as holding good for the whole region. Having two such tables at disposal, even though based on material of very different proportions, they felt justified in taking the mean of the

rates of mortality as an approximation to the truth, so much the more that problems of the limits of the deviations from the mean as the size of the material varies, had not yet claimed the general attention of statisticians. Wargentin takes the mean of the deaths, distributed per mille, according to Halley's and Kersseboom's tables, the Swedish experience for the whole kingdom and for the provinces where epidemic diseases had been relatively rare in 1749. A similar method is used by Süssmilch (§ 441), the only difference being that he replaces Halley's and Kersseboom's tables by two tables calculated from observations on deaths in some parishes in Brandenburg. He misunderstood Wargentin in so far as he considers the table for Sweden as only pertaining to the provinces where epidemic diseases were relatively prominent, whereas it embraced the whole kingdom, the one table thus overlapping the other. But if all the four tables can be considered as equally good observations on the normal mortality, this objection is irrelevant. According to Süssmilch's opinion, the resulting table gives a good picture of mortality in a rural population.

Süssmilch is well aware that the supposition of a constant population does not hold good (§ 463), but he looks upon his tables as approximately correct. In fact one of the Brandenburg tables is based on 1,072 deaths (stillborn included), whereas the numbers of christenings was 1,437.

Having calculated a table for the rural population, he again follows Wargentin by adding three tables for select classes, which he supposes to enjoy the same health as the peasant population. Wargentin borrowed these tables from Deparcieux, viz. one for members of tontines, one for Benedictine monks, and one for nuns, but whereas Wargentin only gives them for comparison, Süssmilch calculates one table out of all the seven tables, as applying to select classes ("Landsleute und ausgesuchte Personen"). Deparcieux's table for tontines begins with 1,000 at the age of 3, 814 of these being alive at 20. This age again is the starting-point for the tables for monks and nuns, all the tables concerned beginning with 814. Wargentin reduces the numbers in order to compare them with the other tables; this he has done somewhat freely. Each of the tables for monks and nuns begins (at the age of 20) with 523, whereas Süssmilch's table for the rural population has 521. The corresponding number for the tontines is 537. Süssmilch, however, takes no notice of these differences, and they disappear in the mean of the seven tables. This mean consequently consists of two sections, one for the first twenty years, identical with the table for rural districts, and another for the ages above 20, slightly different from that table.

Süssmilch then proceeds to calculate tables for provincial towns. The material is much restricted, the observations pertaining to two small towns in Brandenburg, altogether with a few thousand inhabitants. The average is calculated in just the same way as above.

For the city population he takes the mean of five tables (two for Berlin, in different periods, one for Paris, Vienna and Braunschweig respectively). Finally, Süssmilch joins the three tables, for the country, the towns and the cities into one (§ 461). As the original tables mostly are based on quinquennial classes of age, he undertakes an elementary adjustment so as to distribute the deaths for each year of age.

As we have seen, Süssmilch was well aware that there were considerable differences in the health conditions among the rural and the urban populations. When joining the three tables into one he might have taken the actual numbers of inhabitants into consideration. In that period the rural population was as a rule much larger than in the towns, whereas here its weight is only one-third. Here again the idea of a uniform law of mortality seems to be predominant, in spite of the evident fact that the chances of death are different.

Having constructed his mortality-table, Süssmilch calculated the numbers of the surviving at each age, and from these numbers he draws various conclusions just as Halley or Graunt did. He does not, however, make himself quite clear with regard to the ages, for strictly speaking his numbers should show how many were alive on their birthday, not how many on an average were living between two subsequent birthdays. But after the infant age this objection is of relatively small consequence. A curious estimate of the number of married couples may be mentioned here (§ 466). According to his table there are 8,794 between 20 and 40. Altogether he supposes the number of married people to be somewhat more, about 9,000. The number of deaths being 1,000, and the population according to the supposition being stationary, the 4,500 couples will give 1,000 births annually, or in other words, one marriage will on an average give a birth in $4\frac{1}{2}$ years. Another curious estimate concerns the number of inhabitants in the old Nineveh (§ 470). He interprets what is said in Jonah iv. 11: "Nineveh, that great city wherein there are more than six score thousand persons that cannot discern between their right hand and their left hand." These words, according to Süssmilch, should be taken as a statement of the number of children under 4 years. His table shows that one-ninth of the population was under that age; consequently Nineveh had 1,080,000 inhabitants.

Süssmilch calculates the probable duration of life (§§ 473 *sq.*), asking, after how many years will half of a certain generation still be alive. He speaks as if this is just the same as the expectation of life, though he explains (following Deparcieux) the proper way of finding the mean duration of life, with the addition however that the two methods will often give the same result. His own table shows that this is a mistake; thus he finds as the probable duration of life at birth 19 years, whereas the mean duration of life according to his own calculations is 29 years.

He compares his results with various tables, frequently quoting statistical phenomena which he finds remarkable, as for instance Kersseboom's comparison of the mortality of boys and girls.

In his first edition on the Divine Order Süssmilch discussed popular ideas concerning climacteric years according to which the age of 63 (7 × 9 years) had a particular risk. Instead of this he found an increase of mortality at the round years, 50, 60, etc. Later, however, he acknowledged that these irregularities might be explained as arising from errors in the lists (§ 490).

After a digression on tontines and life-annuities he resumes mortality statistics and devotes a chapter to the causes of death, chiefly with reference to English statistics. The Divine Order is also to be found here in spite of the above-mentioned irregularities caused by epidemics. He discusses (§ 528) the burning question of inoculation against small-pox. Further, he gives observations on the influence of seasons on mortality.

A succeeding chapter contains estimates of the number of inhabitants in a large number of towns and cities based on the yearly number of deaths. Finally, he devotes some paragraphs to King's and Davenant's contributions to political arithmetic.

This review can, of course, only give the main contents of Süssmilch's work. Several interesting remarks are scattered about in the text, for instance, on the source of errors arising from the fact that several persons die away from the place where their home is (§ 51), or on the two censuses in Berlin, 1747, with a week's interval, which differed but little from each other (§ 143), or the curious discussion at the royal table (§ 469) on the probable number of persons above 80: the question being whether fifty persons of this age were living in Berlin. In great haste inquiries were made, and more than 400 were found. Süssmilch himself expected 1,147, according to his life-table.

28. C. J. Baumann, the son-in-law and colleague of Süssmilch, added a supplementary volume to the fourth edition of the *Göttliche Ordnung*. In many respects we meet quite the same views. On the whole, Baumann

accepts Süssmilch's method of calculating a life-table, and the statistical observations which he has collected —chiefly on births and deaths—do not deviate in the main from the material in the original work. Nor does Baumann's long treatise on a favourite question of the time, that of funds for widows, give weighty contributions to the discussion. And we meet the same mercantilistic ideas as in Süssmilch's work; Baumann speaks (p. 187) of money as the blood in a body, an idea which we so frequently meet in the economic literature of that period.

Still, there is evidently some progress. He discerns clearly between the probable and the mean duration of life (pp. 409 *sq.*, pp. 523, 536). His critical sense is awake to observations on macrobiots, and he gives a curious example from his own experience (p. 419). He is well aware, by the calculation of life-tables, of the disturbing influence of the surplus of the births over deaths. Therefore Süssmilch's table will not tell us the number of persons among whom 1,000 will die yearly, but, on the other hand, according to his opinion, the table shows how 1,000 persons born at the same time will gradually die out.

He corrects various numerical errors in Süssmilch's life-table (pp. 361 *sq.*), and he leaves out monks and nuns as well as tontine members from the table for rural districts. And whereas Süssmilch aims at finding normal values of the rates of mortality, Baumann asserts that there is a great difference between one country and another. In consequence he does not look upon the tables for rural districts, towns and cities as essentially alike, and the average should therefore be calculated by taking the numbers of inhabitants (pp. 367 *sq.*) into consideration. In order to get a mortality-table for the Churmark Brandenburg he would, for instance, let the table for the rural districts have the weight of $\frac{2}{3}$, whereas the other two tables were to have $\frac{1}{6}$ each.

CHAPTER VIII

ESTIMATES AND ENUMERATIONS OF POPULATION

29. SEVERAL French authors in the second half of the eighteenth century took up the problem of approximate estimates of the population in France, particularly l'Abbé d'Expilly, Messance and Moheau.

D'Expilly (1719–93) was a very fertile author. He had planned a comprehensive geographical dictionary and succeeded in completing five volumes (ending with Q).[1] For several places he gives the number of souls or of hearths. In a long article on population in which he quotes various authors, such as Wargentin (with whom he corresponded),[2] he gives the results of original investigations regarding the population of France, to a great extent achieved at his own cost. In some cases there were direct enumerations at his disposal, and in Provence, as an exception, with distinction of males and females, boys and girls under 12, male and female servants, etc. In other cases he calculates the population by means of births, deaths and marriages. As a general result he found 22 millions which he distributes according to age based on the experience for Sweden.

D'Expilly quotes approvingly the work of Messance,[3] who in turn supports d'Expilly. This author was Receveur des Tailles de l'Élection de Saint-Etienne. According to him the number of inhabitants can be found

[1] *Dictionnaire géographique historique et politique des Gaules et de la France*, 1762–8.

[2] Arosenius, l.c., *45–7.

[3] *Recherches sur la population des Généralités d'Auvergne, de Lyon, de Rouen et de quelques provinces et villes du royaume avec des réflexions sur la valeur du bled tant en France qu'en Angleterre depuis 1674 jusqu'en 1764*, Paris, 1766.

by means of the births, the marriages or the number of households; among these methods he prefers the first. His calculations are a little curious. Thus, having on an average 1,020 births yearly and a population of 25,025, each birth will correspond to 24 $\frac{1}{2}$ $\frac{1}{40}$ $\frac{1}{80}$ persons, viz. 24 $+ \frac{1}{2} + \frac{1}{40} + \frac{1}{80} =$ 24·5375. By the usual division we find 24·5373. He adjusts this quantity to 25, and finds that the province of Auvergne with its 24,604 births has a population of 615,100. In order to find the population in the whole kingdom he finds the number of parishes in a select population with 59,894 yearly births, or a population, with a birth-rate $\frac{1}{25}$ of nearly 1$\frac{1}{2}$ millions. The 2,152 parishes have on an average 696 inhabitants each; he reduces the number to 600 and finds that 39,849 parishes have altogether about 23·9 millions.

Several years after Messance, Moheau published his contributions.[1] He enumerates several methods of finding the population. Thus we can base the calculations on the consumption, on tax-lists, households or houses, and on births, marriages and deaths. Out of these he chooses births as the safest starting-point. For the whole kingdom there will be one birth out of about 25$\frac{1}{2}$ inhabitants, whereas there will be about 121 or 122 for each marriage. Altogether there were in five years on an average 928,918 births yearly, the population thus being 23,687,409. The number of marriages being 192,180, he finds—with less certainty—at least 23 millions, whereas a death-rate of $\frac{1}{30}$ and 793,931 deaths will give 23,817,930 inhabitants. He thus arrives at about the same result as Messance. By means of some special observations he tries to distribute the population according to age, 7 per cent. being above 60 years, 40 per cent. below 10.

He is less original concerning mortality, the mean duration of life being according to him identical with the mean age at death. Imperfect as his material is, he

---

[1] *Recherches et considérations sur la population de la France*, Paris, 1778. The book was probably finished in 1774.

draws comparatively correct conclusions as to the proportion between the sexes, the climacteric years (defined
as the ages at which females reach puberty and cease
being fertile), and his investigations as to the growth of
population are not without interest.   In opposition to
the current opinion that population was regularly decreasing, he maintains that enumerations in certain parts
of Auvergne give an increase of $\frac{2}{43}$ in 15 years, and he
concludes that if this holds good for the whole country
the population will double in less than 250 years (the
exact number is 229 years).   According to his opinion
an increase of that extent would not be improbable.

Necker also was interested in the problem.[1]   He
chooses as birth-rate one to 25·75, as death-rate one to
29·6, and as marriage-rate one to 113$\frac{1}{3}$.   In the four
years 1777–80 there was a yearly average of 940,935
births, the population consequently being 24·2 millions,
the marriage- and death-statistics leading to about the
same result.   By taking the births in the quinquennium
1776–80 he gets a somewhat higher number.   Necker
does not, however, enter much into the material, nor
does he explain how he found the various rates, so that
it is impossible to judge of his results properly.   The
reader naturally grows a little suspicious on seeing that
births, deaths and marriages all lead to the same result.

In the above-quoted (art. 21) work by le Chevalier des
Pommelles the author chooses the same rates as Necker,
but he uses an average of 10 years (966,240 births).
Making a correction for Paris, he arrives at 25 millions.
In order to find the proportion between males and females,
the distribution as to marital condition, etc., he bases
himself on the rates in selected districts where it proved
possible to find the number of inhabitants.   That he is
sometimes tempted to rather unfounded assertions is
natural, as when he tries to show that marriages are most
fertile in elevated places where the air is warm and dry,
the reverse being the case where the air is heavy and the
soil marshy and low-lying.

[1] *De l'Administration des Finances,* 1784.

30. Looking through the various attempts at finding the population of France, we get the impression that those estimates may not have been far from the truth. But it cannot be denied that the authors have possibly not been quite unbiassed, and indeed political arithmeticians were at a loss for methods to control their results. Here the famous mathematician Laplace (1749–1827) made a long step forward. In a paper of 1786 on births, marriages and deaths in Paris he proposed to find the population of France by means of the birth-rates in certain parts of the kingdom, and he makes an investigation as to the probable limits of the deviation from the real number. After the fall of *l'ancien régime* he obtained the support of the Government for a practical experiment.[1] Thirty departments in France were chosen, and in each of these a number of places, where the *maire* was sufficiently intelligent and zealous. Here the inhabitants, on September 22nd, 1802 (the republican New Year's Day), were enumerated, altogether about 2 millions, and the number of births in the three preceding years, September 22nd, 1799, to September 22nd, 1802, was counted. It was found that there was one birth yearly per 28·352845 inhabitants. If now the yearly number of births in the French Empire within its boundaries at the moment when Laplace's *Theory of Probabilities* appeared was $1\frac{1}{2}$ millions, the total population would be 42,529,267. He further finds that there will be 1,161 to one that the error will not reach half a million.

Looking at this experiment with the experience of the following three generations in mind, we have no difficulty in finding objections. Curiously enough, in the third edition (1820) the whole argument is repeated verbally, with the only exception that Laplace, according to the intermediate regulation of the frontiers of France, reduces the yearly number of births to 1 million and the population consequently to about 28 millions. But he does not recalculate the limits, and in the following he therefore uses 42 millions as the supposed number of inhabitants.

[1] Laplace, *Théorie analytique des Probabilités*, 1812, pp. 391 *sq.*

Here, evidently, the readers' doubts will set in. How could Laplace know that the normal number of births in Greater France was $1\frac{1}{2}$ millions, and in the reduced Kingdom 1 million? In fact, few years in the course of the nineteenth century show over a million births in France; as a rule, the number was below this limit. By studying this passage in Laplace's work a suspicious reader will naturally fix his attention on these two numbers: 1·5 and 1 million of births respectively.

Again, a modern statistician would wish to see the actual numbers in the various places which the investigation embraced. The limits are calculated in accordance with the results of the calculus of probabilities, but it will be necessary to know whether the various birth-rates are grouped according to the binomial law, or whether there are several centres in the country around which the numbers are grouped. This will, of course, all have influence on the limits within which the total population is to be found.

A modern statistician might also object that the number of inhabitants ought to have been taken in the middle of the triennium from which the births had come. This objection is, however, rather irrelevant, as movements of population in those days were generally much slower than later on.

But in spite of all these objections it must be acknowledged that Laplace by his solution gave a very important impulse to statistical knowledge, even though very few persons understood the range of this experiment. The modern "representative" statistics which are based on the study of selected sections of the material ("sampling"), from which the conclusions are extended to the whole material, are in reality greatly indebted to Laplace. Curiously enough, there was a general census in France in 1801, but it seems to have been anything but a success, so that Laplace's experiment was by no means superfluous. But naturally representative statistics were laid aside for a while as in the course of the nineteenth century general enumerations gradually won the confidence of the statisticians.

31. *In England*, the birthplace of political arithmetic, it seemed more difficult than in France to get reliable estimates of population.   In March, 1753, a Bill was introduced in Parliament to provide for an annual counting of the people.   But this Bill met with such opposition and the proposal was viewed with such alarm that the Bill was rejected by the House of Lords.[1]   Curiously enough, a very interesting private experiment was at the same time made in Scotland.   The Rev. Alexander Webster succeeded in getting into communication with all the ministers in Scotland and to get an account of the whole population in the year 1755.   But unfortunately this report was never published.[2]   In England nothing was done in this direction except local censuses here and there.   And for many years it was under discussion whether the population in England was increasing or decreasing.   It was the underlying supposition in Malthus's *Essay on Population* (1798) that there was an increase, but other authors maintained the reverse; for instance, Richard Price.   In vain Arthur Young suggested a census.[3]   He remarks that some assert that the population declines, "that we have lost a million and a half since the Revolution; and that the decrease now continues strong; others are of a direct contrary opinion." He rejects altogether the number of houses as the basis of an estimate, "for by what rate is the number of souls per house to be determined?   How is the medium to be found between the palace and the cot?"   He therefore proposed a regular census every five years "unless a change of national circumstances called for variations." But the public turned a deaf ear to it.   Still, in the year 1800 Sir Frederick Morton Eden had to make an

[1] *Official Vital Statistics of England and Wales* (League of Nations Health Organisation, Geneva, 1925), p. 17.

[2] "Note on An Account of the Number of People in Scotland in the Year 1755," by Alexander Webster, one of the Ministers of Edinburgh. *Journal of the Royal Statistical Society*, LXXXV, 1922.

[3] *Proposals to the Legislature for Numbering the People*, 1771.

estimate of the population.[1]   Knowing the number of houses, he tried by sampling to find the average number of inhabitants in each house, as well as the number of births.   The result was a total of 322,000 births and a birth-rate of one out of $27\frac{3}{4}$.   This would give a total population of 9 millions; various reasons led him to expect that the number was even higher.   Price had in 1779 maintained that the population was only 5 millions. By that time, however, the opposition against a general census was declining.   A Bill for ascertaining the population of Great Britain was under consideration by the Legislature, and in 1801 the first decennial enumeration took place, giving 9 millions as result.

In addition, the clergy of each parish were required to prepare a statement of baptisms and burials for each decennial period from 1700 to the end of the year 1780 with distinction as to sex, and of marriages, 1754–80. Thus a foundation was laid for English official statistics.

Among the authors who took part in the efforts to solve the problem of the number of inhabitants, Will Black can be mentioned.   Although he considers a census the best way to ascertain the population, he recommends as an alternative to use the number of houses as a basis, allowing 5 or $4\frac{1}{2}$ to each house.[2]

32. In other countries it was also felt desirable to find the number of inhabitants by direct enumeration.   The constitution of the *United States of America* provided for a regular census of the population with a view to determine representation in the Congress, which had to take place every tenth year, and which was taken for the first time in 1790.   But already before this time enumerations were well known, even in the Colonial period, and had been employed in order to obtain information of value

[1] *An Estimate of the Number of the Inhabitants in Great Britain and Ireland,* London, 1800.
[2] *An Arithmetical and Medical Analysis of the Diseases and Mortality of the Human Species,* London, 1789. On the discussion in the last decennium of the eighteenth century about the population in England, see Edw. C. K. Gonner, "The Population of England in the Eighteenth Century," *Journal of the Royal Stat. Soc.,* LXXVI, 1913.

in the administration of the Colonies. Thus in 1756 a census embracing all the inhabitants of *Connecticut* was taken, in 1764 in *Massachusetts*. The articles of Confederation as originally reported in 1776 provided for a triennial enumeration of the population as a basis for apportioning the charges of war, and even though this was altered, several of the States made an enumeration.

The census of 1790 was rather summary. It returned the number of free white males above and under 16, the free white females, and, further, the slaves without distinction of sex or age, and it has been remarked that the institution of the Census seems to have been "a political incident, little regarded at the time except as a practical means of apportioning representatives and taxes." But it was at all events a remarkable step forward.[1]

Also in *Norway* and *Denmark* interesting progress can be recorded. Both countries had a census taken in 1769, and in 1787 a new census was taken in Denmark. There was no committee appointed to prepare the census reports, the work being left to private persons. The tables concerning the enumeration of 1787 were finished in the course of three years, but unfortunately no report was published; only a few results were made known to the public. Finally, in 1797 a statistical institution was established in Norway and Denmark, a "Tabelkontor," but its influence was very small, its chief duty being to collect and revise accounts concerning taxes and other public revenues. Vital statistics had only a secondary place in its activities.

In *Sweden*, the stronghold of official vital statistics, conditions were of course much more favourable. Still, no great progress can be recorded in the latter part of the eighteenth century. For several years the Tabular Commission worked with great zeal, but at last it grew tired. Wargentin died in 1783 after a protracted illness, and

---

[1] John Cummings, "Statistical work of the Federal Government of the United States," *The History of Statistics*, 1918, p. 670; see further, Charles F. Gettemy, "The Work of the Several States of the United States in the Field of Statistics," the same volume, pp. 711–12.

the commission seemed rather inactive till it was reconstructed in 1790 and the astronomer Nicander was
appointed secretary. At the close of the century a more
liberal grant was allotted to the commission, so that it
could more easily discharge its duties. The work had
been simplified in 1773 in so far as the lists of population
were hereafter only to be sent in every five years and
only for each diocese, not for the provinces. In 1792
an important reform was made, lists for the deaneries
had now to be sent direct to the commission, thus an
important step towards centralisation had been taken.
After 1775 the records contain particulars concerning the
age of *mothers bearing children* (in five-years age groups),
a most remarkable step forward.

33. The preceding list of political arithmeticians is not
complete. In several countries we meet authors whose
names deserve mention. In Italy, for instance, Marco
Lastri can be quoted.[1] He gives several census results
for Florence, also numbers of births per month, for each
calendar year, each sex separately. The schedules are,
curiously enough, arranged so that the empty spaces can
be filled up by the author's successors up to the year 1850.
Many of the data throw light on the evolution of Florence,
though it is impossible to bring complete harmony into
the whole material.

Another highly interesting treatise was published by
a *Swiss* clergyman, Muret,[2] on the vital statistics of the
country of Vaud. There were 100 christenings for
$79\frac{1}{3}$ deaths, and the rate of mortality was one out of
$45\frac{1}{9}$. He concluded that the surplus of births will be
one out of 173 yearly and the population will double in
120 years. Then follow some curious calculations concerning the doubling period. 375 mothers have altogether borne 2,093 children, out of whom 494 males
and 562 females are supposed to survive at 30. As
one-eighth of the males remain unmarried, 434 males

---

[1] *Ricerche sull' antica e moderna popolazione della città di Firenze per mezzo di registri di Battisterio di S. Giovanni dal 1451 al 1774,* Firenze, 1775.

[2] *Mémoire sur l'état de la population dans le pays de Vaud,* Yverdon, 1766.

are left, a number which again corresponds to 478 marriages, as about 10 per cent. marry twice. Twenty of these marriages are supposed to be sterile—a very low proportion according to modern experience, though Muret maintains that it is probably too high. There being 458 marriages with issue and the number of children being on an average 5.58 for each mother (375 mothers having as above mentioned 2,093 children), the new generation will count 2,556. He finds (not quite correctly) an increase of 23 per cent. compared to the previous generation born thirty years ago. Euler's calculations show a doubling in 110–12 years, not far from the above-mentioned 120 years.

As to life-tables, Muret does not go beyond the distribution per mille of the observed deaths as to age. In discussing the material he shows, however, much common sense, for instance, concerning the connection between the mean age at death and the birth-rate.

Though *America*, as we have seen, presented fairly good conditions for an evolution of vital statistics, contributions to any research were somewhat scanty. A rather interesting attempt at measuring mortality in an increasing population was made by Wigglesworth[1] on the supposition that the number of deaths was half the number of births. Unfortunately he does not explain his method sufficiently, so that it is hardly possible to reconstruct his calculations.

In *England*, as in America, observations on vital statistics continued to be made in this period. Richard Price's contributions were mentioned above. Two medical men may be cited, viz. Haygarth and Heysham. The former took particular interest in the problem of inoculation against small-pox; moreover, he issued a report on the mortality in Chester.[2] Heysham, who

---

[1] "A Table showing the Probability of the Duration, the Decrement and the Expectation of Life, in the States of Massachusetts and New Hampshire formed from sixty-two Bills of mortality . . . in the year 1789" (*Mem. American Acad.*, II, Part I, Boston, 1793).

[2] Westergaard, *Die Lehre von der Mortalität und Morbilität*, 2nd Ed., 1901, p. 58.

seems to have been a very original and energetic man, was also interested in the problem of small-pox, but his name is particularly connected with the famous *Carlisle-table*.[1] For several years Heysham prepared lists of marriages, births, diseases, etc., in Carlisle, at that time rather an insignificant town with about 200 deaths yearly. A local census was taken in 1780, and again in 1787. Heysham was, however, not satisfied with the latter and undertook a private census himself. He found the official census had omitted about 6 per cent. of the population. Many years after, his material came into the hands of J. Milne, who constructed a life table on the census results and the lists of deaths for 1779–87, after a very careful revision of the observations, based on a long correspondence with Heysham.[2] This clearly shows how difficult it was in those days to get reliable statistical material.

On the *Continent* several investigations regarding *widows' funds* may be quoted, a subject which in the eighteenth century claimed much attention. But it must be admitted that the actual results were on the whole trifling. One of these authors was Kritter, who shows little originality with regard to statistical investigation.[3] Like the Danish mathematician, Tetens, he tried to find the difference between the mortality of males and females. As will be mentioned below, the latter deserved credit for his work in connection with life-insurance and theoretical statistics, but did not contribute much to mortality statistics. In a great work, published 1785–6,[4] he discusses the mortality experience at hand. He accepts Süssmilch's table, with Baumann's

---

[1] Henry Lonsdale, *The Life of John Heysham, M.D., and his Correspondence with Mr. Joshua Milne, relative to the Carlisle Bills of Mortality*, London, 1870.

[2] J. Milne, *A Treatise on the Valuation of Annuities and Assurances on Lives and Survivorship*, London, 1815.

[3] E.g. "Sammlung wichtiger Erfahrungen bey den zu Grunde gegangenen Wittwen- und Waisen-Cassen, Leipzig, 1780. Untersuchung des Unterscheides der Sterblichkeit der Männer und der Frauen von gleichem Alter," *Göttingisches Magazin der Wissenschaften u. Litteratur*, B. II, 1781; confer Westergaard, l.c., pp. 56–7 and 290.

[4] *Einleitung zur Berechnung der Leibrenten und Anwartschaften*, I–II, Leipzig, 1785–6.

corrections, as sufficiently exact for Northern Europe, even though he finds that further corrections are possible. He is well aware that migrations may have influence, but on the other hand he warns against too much altering of the original observations (l.c., 1, p. 77). If the lists are correct, then the numbers in the mortality table can be considered as facts; if we alter these facts we must give reasons for the supposition we have made. He recommends, however, Price's Northampton table, which he finds classic.

34. In many respects statistical observations in those days were treated with critical sense, but in other fields much had yet to be learnt, the data often being taken as correct in spite of evident defects of the material. This may particularly be observed in the chapter of statistics which deals with *longevity* ("Macrobiotics"). The authors writing on this subject were often amateurs with regard to statistics, not having availed themselves of the latest progress.

Very naturally the question arose as to the utmost limits of human life. Instead of trying to find how a certain generation would gradually die out, the question was what age a human being could reach under the most favourable circumstances. This problem required less technical insight than the calculation of a life-table, but on the other hand observations had to be sifted with the utmost care if the efforts to solve the problem were not to be useless. Several authors touched on this problem, for instance Süssmilch. The famous scientist A. v. Haller may also be quoted here; he discusses longevity in his physiology.[1] In England, James Easton published a long list of "macrobiots",[2] and in America Will Barton discussed the problem.[3] One of the most

[1] A. v. Haller, *Elementa Physiologiæ Corporis Humani*, Tome VIII, 2nd ed., Lausanne, 1778.

[2] J. Easton, *Health and Longevity as exemplified in the Lives of Six hundred and Twenty-three Persons deceased in various Parts of the Globe remarkable for having passed the Age of a Century*, Salisbury, 1799 (new ed., 1823).

[3] W. Barton, "Observations on the probabilities of the Duration of Human Life and the progress of Population in the United States of America" (*Transact. of the American Phil. Soc.*, III, 1793).

characteristic contributions of this kind was C. W. Hufeland's book on the art of living long.[1] This book enjoyed a very great reputation, which may have been well deserved as far as the medical advice which it contained is concerned, whereas the statistical value is very small. The observations on "macrobiots" are uncritical, and his calculations as to the frequency of deaths from various causes are extremely vague (l.c., pp. 366–7). He also gives a mortality table (p. 214), which he describes as being based on experience, though without quoting its origin. It seems to be calculated freely from Graunt's table. According to this table, out of 100 newly-born, 50 will be alive at 10 years of age, 14 at 40 years, and 6 at 60 years.

[1] Hufeland, *Die Kunst das menschliche Leben zu verlängern*, 1st ed., Jena, 1797.

CHAPTER IX

PROGRESS OF THEORY AT THE CLOSE
OF THE EIGHTEENTH CENTURY

35. AMONG the famous mathematicians of the Bernoulli family Daniel Bernoulli (1700–82) has just claims to be remembered in a history of statistics. The problem of inoculation against small-pox has been mentioned above. Dan Bernoulli added a weighty contribution to the discussion.[1] The material for his investigations was by no means faultless. As the basis for his calculations he used, for instance, Halley's table, and here he even misunderstood the numbers, supposing that the initial number 1,000 corresponded to the age 1 year. In order to get back to 0 year he adds, rather arbitrarily, 300 to the number of births, thus beginning with 1,300. He wanted to know the frequency of attacks of small-pox as well as their lethality (the rate of mortality among persons who have the disease). He also supposed the lethality as well as the frequency of attacks to be constant throughout life; both these qualities are estimated at one-eighth, one out of eight thus being yearly attacked, and one-eighth of these succumbing. All this may be doubted, but the merit of his paper is the masterly way in which he treated the theoretical problem. At first sight it is very complicated. Taking, for instance, the year as unity, we may ask how many persons of a given age contracted the disease in that year, how many recovered, and how many again of these died from other

[1] D. Bernoulli, "Essai d'une nouvelle analyse de la mortalité causée par la petite vérole et les avantages de l'inoculation pour la prévenir" (*Histoire de l'Acad. royale des sciences, année* 1760, *avec des mémoires de mathématique et de physique pour la même année*, Paris, 1766).

causes in the course of the year concerned. This Bernoulli simplifies by using *infinitely small units* of time. Dealing with a single moment he has only to find the probability of a person dying in that moment, without being embarrassed by future events. Using the force or intensity of mortality as the stepping-stone for further calculations, he obtains elegant formulæ, the problem being reduced to the solution of a simple differential equation. Having solved this question, he can easily find the relation between the number of persons—surviving according to the life-table which has been chosen —who have never had small-pox, and how many have had an attack but have recovered.

The paper concludes with some tables with numerical results. According to his supposition, out of 100 persons surviving at the age of 24 only 5–6 would never have had the disease. This may be wrong but, in regard to method, his investigations represent a great progress, and only correct observations will be required in order to reach reliable results. By his invention of the *continuous method* he opened up a wide field for statistical investigations. Unfortunately his method was not generally understood; it was not until late in the nineteenth century that it was thoroughly appreciated.

Soon after sending his paper to the academy he was vigorously attacked by d'Alembert,[1] who was, however, not very just in his criticism. D'Alembert is right in maintaining that the individual member of a population will necessarily look upon inoculation with other feelings than the State (Society). It may be of profit to the latter if inoculation adds some years to the expectation of life, but the private citizen when considering profit and gain will ask whether there is a risk of dying as a result of the inoculation. Daniel Bernoulli seems to have seen this clearly himself, and in other respects d'Alembert's objections carry little weight.

A single mathematician may be mentioned who

---

[1] d'Alembert, "Sur l'application du Calcul des Probabilités à l'inoculation de la petite Vérole" (*Opuscules Mathématiques*, II, 1761).

understood the method thoroughly, viz. Duvillard
(1755–1832). He had in 1787 published a book
on interest.[1] At the beginning of the nineteenth century he was in the public service (population statistics).
The newly introduced vaccination against small-pox gave
him quite naturally an impulse to take up the problem
which D. Bernoulli had treated; the result was a remarkable volume which unfortunately became very little
known.[2] He speaks with veneration of D. Bernoulli's
work and follows his methods, but, having more complete
observations at hand, he is enabled to go more into
detail. It may be objected that his mortality-table is not
essentially better than the table which D. Bernoulli used.
It was constructed on about 101,000 deaths on the
supposition of a stationary population. He was himself
of the opinion that the table was a good expression of the
state of health in France before the Revolution. But it
was not his principal aim to calculate a life-table but
only to obtain means for a comparison of the effects of
vaccination, and in this respect it would probably serve
fairly well. Unfortunately, several authors who afterwards criticised Duvillard have only had this table in
view, without thinking of the problem under discussion.
As stated, his book is, on the whole, very little known,
only very few copies seem to be preserved.

As to the occurrence of, and death-rate from, small-
pox, various observations with distribution according to
age were at his disposal (Geneva, the Hague, Berlin),
and of these he took an average which he used for his
calculations.

His book contains a considerable number of tables.
The population corresponding to the life-table is divided
into various classes: persons who it is assumed will never
be attacked by small-pox, who will recover from an
attack, who will die from the disease, etc., on the whole

[1] *Recherches sur les rentes, les emprunts et les remboursemens* . . . Par M.
Du Villard, Genève, 1787.
[2] E. E. Duvillard (du Léman), *Analyse et tableaux de l'influence de la petite
vérole, et de celle qu'un préservatif tel que la vaccine peut avoir sur la population et la longévité*, Paris, 1806.

following the line of Bernoulli. It may be difficult for readers in our days to judge the assumptions with absolute clearness. He found that out of 100 persons aged 30 years, only 3 had not had small-pox, and that two-thirds of all newly-born will be attacked. The mortality of persons suffering from small-pox was, according to his calculations, no less than one-third at the age of one year, 3 per cent. at 10 years. If small-pox disappeared, a new-born child would on an average gain $3\frac{1}{2}$ years in expectation of life.

Even if Duvillard's work had been generally known it is doubtful whether his results would have been accepted. The leading economists of those days did not believe in great changes in mortality and expectation of life, such as Duvillard did. If economic conditions did not improve they would expect that somehow or other the gain would be out-balanced by losses, so that the result would be reduced to zero.

Strictly speaking, Duvillard's book should have been treated in a following chapter, but it was natural to mention it here on account of its close relation to D. Bernoulli's treatise.

36. Although the continuous method was very little noticed, another important progress may be mentioned though only indirectly connected with the history of statistics. J. N. Tetens has, in his above-quoted work, used a method of great actuarial usefulness, what in modern technical terms may be called the method of Dx and Nx columns. We have seen that the mathematicians of the eighteenth century, such as de Moivre and Simpson, had thoroughly grasped the problem of finding the present value of sums payable with a certain probability at given dates. But if an assurance embraced many payments at various dates, as, for instance, an annuity on life, the calculation might require much time. De Moivre tried, as mentioned above, to find a short cut by his hypothesis on the law of mortality, but his solution was not general enough. The problem is analogous to that of finding the value with compound interest at a

given date of several instalments in a savings-bank. The greater part of the work can be done in advance, by calculating the value of each instalment at a previously fixed date, for instance, the beginning of the current calendar year, the problem thus being reduced to finding the value of the aggregate of these sums at another date. Just the same can be done in life insurance, choosing, for instance, the age o as starting-point.[1] This invention was in fact a real Columbus-egg. Tetens' method was, however, not much known. English actuaries, to whom Tetens' work was unknown, even invented the method many years after.

Credit is also due to Tetens for making an energetic attempt at finding the risk incurred by a life-insurance society. He shows that the total risk of a widow-fund increased as the square-root of the number of members. He refers to de Moivre, but he reached his result by an original process.

Finally, Tetens was one of the first to use the method of *expected deaths* in order to compare the mortality in a widow-fund with that of a certain mortality-table—in this case Süssmilch's table (l.c., II, pp. 1 *sq.*). His method is, however, a little unpractical though perfectly clear. The period of observations is 1767–83, and he includes the members who enter in the same half-year in a separate group, asking how many of these would be expected to survive in 1783 according to the table, and comparing the result with that actually experienced. The method does not enable him to find the mortality in each age of life. For females, the main result was 743 actual and 1,048 expected deaths.

37. If it was difficult to gauge the population, there were still greater obstacles for getting correct figures concerning *economic statistics*. Here and there progress was made. Thus the United States had official statistics

---

[1] l_x being the number of survivors at the age x and r the rate of interest, we have $D_x = l_x(1 + r)^{-x}$ and $N_x = \Sigma_x . D_x$ and N_x being calculated beforehand, the calculation of annuities and other insurances can be simplified very much.

of foreign commerce compiled each year beginning with 1789. Agriculture was not much dealt with statistically although occasionally details may occur in the writings, for instance, figures showing the number of cattle and horses in Alsace in Boulainvilliers' above-quoted work, but generally very little was known even much later in the century, in spite of the great interest aroused by the physiocrats. At the close of the century, however, a remarkable little book appeared which made notable attempts in this direction. The sixty-four pages contain contributions from four authors, first of all the famous chemist Lavoisier (born 1743, decapitated 1794) and the mathematician Lagrange (1736–1813).[1]

Lavoisier had already in 1784 begun his investigations in this field. He does not consider his report as complete, but the Assemblée Nationale ordered its publication. He refers to Moheau and de la Michaudière (Messance) regarding the number of inhabitants, and without going into detail as to how he has arrived at the result he gives an estimate of the total population at about 25 millions, which he again distributes according to various classes, supposing 8 millions to live in towns, whereas $2\frac{1}{2}$ millions are getting their living from vineyards, etc. His principal assumption is that the consumption per head of certain necessities of life is nearly constant within each class, and after having found plausible support from statistics on consumption in Paris, he finds that the yearly consumption of wheat, rye and barley, including seed, was 14,000 millions of "livres pesant," and neglecting export and import, he considers this quantity as the yearly average produce. He now pro-

---

[1] *Collection de divers ouvrages d'arithmétique politique par Lavoisier, Delagrange et autres*, Paris an IV. The four papers are the following: Lavoisier, "Résultats extraits d'un ouvrage intitulé"; "De la richesse territoriale du Royaume de France, 1791"; "Réflexions d'un citoyen propriétaire sur l'étendue de la contribution foncière et sa proportion avec le produit net territorial converti en argent" (author unknown), 1792.

Le citoyen Delagrange, *Essai d'arithmétique politique sur les premiers besoins de l'intérieur de la République.*

Antoine Diannyère, *Preuves arithmétiques de la nécessité d'encourager l'agriculture.*

ceeds to question how many ploughs and how much cultivated area will be sufficient to produce this quantity. Various investigations combined with his own practical experience led him to the result that a plough drawn by horses will suffice for 27,500 livres pesant, whereas a plough drawn by oxen cannot give more than 10,000. One plough will again correspond to 90 and 60 arpents respectively. Knowing, further, in which part of the country it is customary to use horses, and where oxen are prevalent, he can find the number of animals occupied in this labour, but various groups will have to be added, thus for horses, the number used in agriculture where oxen are prevalent, also the numbers in the towns, and in the army. Here he is obliged to use various estimates. He admits himself that this part of his work is very hypothetical,[1] but he hopes by further observations to get more correct results. The total number of horses will, on his estimate, be 1,781,000, whereas there are supposed to be 7,089,000 cattle, 20 million sheep and 4 million pigs.[2]

As the main result with regard to the area, Lavoisier finds that more than one-third of the total area was left uncultivated, and he seems himself a little surprised. He proposed a great institution—a statistical department, where all particulars concerning agriculture, commerce and population, etc., might be collected. He maintained that there was only one country where an institution of this kind could succeed, viz. France; it depended only on the will of the Assemblée Nationale.[3] Here the whole economic result of the French agriculture —in goods as well as in net income, should be calculated. He makes a characteristic remark reminding all of the

[1] Cette partie de mon travail est, comme l'on voit, fort hypothétique.

[2] Comparing with modern statistics we find (1926) 7 horses per 100 inhabitants; the same result has Lavoisier, cattle 35 against 28 according to his estimate, sheep 26 and 80 respectively, swine 14 and 16. Taking into consideration the great changes in agriculture which have taken place, Lavoisier's results seem to be in rather good harmony with modern figures.

[3] "l'Assemblée nationale n'a que le désirer et le vouloir"; cf. above, art. 21, Fénélon.

physiocrats (among whom he quotes Quesnay as well as Turgot) that a report of this kind might contain in a few pages the whole science of political economy, or rather this science would cease to exist, the problems being so easily solved that no disagreement could be possible.[1]

A following paper, by an anonymous author, continues Lavoisier's estimates of the produce of the French agriculture and tries to give a more optimistic impression, whereas Lagrange in the next paper deals with the consumption in the whole empire, particularly investigating the consumption in the army. The last article, by Diannyère contains an attempt at proving the correlation between prices of corn and mortality. On the whole, all these articles are worthy supplements to the weighty treatise by Lavoisier, though they cannot by any means compare with it, but the small volume gives an insight into the difficulties with which political arithmeticians still had to contend at the close of the eighteenth century, at the same time showing the ingenuity which enabled Lavoisier to draw outlines of economic statistics.

---

[1] "Un travail de cette nature contiendroit, en un petit nombre de pages toute la science de l'économie politique, ou plutôt cette science cesseroit d'en être une; car les résultats en seroient si clairs, si palpables; les différentes questions qu'on pourroit faire, seroient si faciles à resoudre, qu'il ne pourroit plus y avoir de diversité d'opinion."

LEADING BRITISH STATISTICIANS OF THE
NINETEENTH CENTURY
Paul J. FitzPatrick
(1960)
Catholic University of America

This paper explores statistical contributions of eleven outstanding British statisticians of the nineteenth, century. They are Playfair, Porter, Babbage, Farr, Guy, Newmarch, Jevons, Rawson, Galton, Giffen, and Edgeworth. This treatment represents one aspect of British statistical thought not previously developed.

INTRODUCTION

THIS paper aims to present leading British statisticians whose main statistical contributions were made in the nineteenth century. So far, very little history has been written about British statistical thought. The eleven individuals who are considered here stand out among the British statisticians of that century as having made the best contributions to the field of statistics by means of original statistical ideas and techniques and by their direction of outstanding statistical organizations. Four different kinds of statistical work are distinguished, namely, (1) techniques of presentation; (2) bodies of material compiled; (3) substantial investigations employing statistical techniques; and (4) contributions to statistical theory. On the occasion of the search for statistical contributions of leading American statisticians of the nineteenth century, statistical activities of eleven British statisticians came to light. It was considered desirable to develop this aspect of the history of statistics so that American students of statistics might become familiar with their work.

These British statisticians are William Playfair, the founder of graphic methods of statistics; George R. Porter, head of the statistical department of the Board of Trade who directed so well the development of this newly-created organization; Charles Babbage, the founder both of Section F—Statistics—in the British Association for the Advancement of Science in 1833 and of the Statistical Society of London in 1834, as well as the inventor of calculating machines; Dr. William Farr, the founder of British vital statistics and well known statistician of the Annual Reports of the Office of the Registrar General; Dr. William A. Guy, another leading authority in the field of British vital statistics, editor of the *Journal of the Statistical Society of London*, and honored by the establishment of the famous Guy medal; William Newmarch, leading authority on monetary and banking statistics, editor of the *Journal of the Statistical Society of London*, and one of the few British statisticians of his time to perceive the need for utilizing a greater measure of mathematics in describing and analyzing economic and social problems; W. Stanley Jevons, more famous as an economist and logician, who made a number of important statistical contributions in the form of the ratio chart, the geometric mean, and measures for analyzing secular trend, seasonal variation and cyclical fluctuations; Sir Rawson W. Rawson, an authority on international statistics, editor of the *Journal of the Statistical Society of London*, and first president of the International Institute of Statistics (1885–98); Sir Francis Galton, an eminent scientist, who developed the idea of correlation and other statistical measures, including the quartile deviation, the median, and the index of correlation; Sir Robert Giffen, well-known head of the statistical department of the Board of Trade, and editor of the *Journal of the Statistical Society of London;* and Francis Ysidro Edgeworth, the philosopher of statistics, probably the outstanding statistician in the nineteenth century because of his work in probability, correlation and index numbers, and the distinguished editor of the *Economic Journal.*

1. WILLIAM PLAYFAIR (1759–1823)

William Playfair, economist, journalist, inventor, and statistician, is regarded as the founder of graphic methods in statistics. He wrote several works containing excellent charts between 1786 and 1821 [49, p. 190; 50, p. 101; 56], but his contemporaries paid little or no attention to these volumes. Playfair had earned much ill-will because of previous caustic and unfriendly writings, and no English economist or statistician took any notice of his charts until 1879 when the famous English economist and statistician W. Stanley Jevons remarked at the June 17, 1879 meeting of the Statistical Society of London that "Englishmen lost sight of the fact that William Playfair, who had never been heard of in this generation, produced statistical atlases and statistical curves" [32, Vol. 42, p. 657].

Indeed, we find many statistical tables in English economic and statistical works and journals in the first half of the nineteenth century, but very few charts. Funkhouser's investigation reveals that graphs first appeared in the *Journal of the Statistical Society of London* in 1841. The first fifty volumes of this *Journal* (1837–1887), contain about fourteen charts. As far as the United States is concerned, little or no evidence of the use of graphs appears before 1843, when George Tucker's work, *Progress of the United States in Population and Wealth in Fifty Years* appeared with three charts, two being line charts and one a bar chart. Much the same condition prevailed in western Europe, notwithstanding the favorable reception of Playfair's works in France. Moreover, some continental scholars, such as Jacques Peuchet (1805) and P. A. Dafau (1840) in France, and Carl Knies in Germany, were strongly opposed to the use of graphs.

Playfair's first volume (1786) containing charts bears the title *The Commercial and Political Atlas.* Its long sub-title reads "representing by means of stained copper-plate charts, the exports, imports and general trade of England . . . with observations. . . . To which are added, charts of the revenue and debts of Ireland." This volume contains forty-four charts, all but one being time series, the other a bar graph. Funkhouser describes them in these words:

> They are well executed copper-plate engravings colored by hand in three and four colors. Twenty of these represents the trade of England with other countries from the year 1700. The line of imports is stained yellow, that of exports, red; the space between is colored blue when the balance is favorable to England and pink when the balance is unfavorable [16].

The work was again published in 1787 and in 1801. In the introduction to the third edition of this work (1801), pages ix–xii, Playfair explains the use of his "lineal arithmetic" as follows:

The advantage proposed, by this method, is not that of giving a more accurate statement than by figures, but it is to give a more simple and permanent idea of the gradual progress and comparative amounts, at different periods, by presenting to the eye a figure, the proportions of which correspond with the amount of the sums intended to be expressed.

Suppose the money received by a man in trade were all in guineas, and that every evening he made a single pile of all the guineas received during the day, each pile would represent a day, and its height would be proportioned to the receipts of that day; so that by this plain operation, *time, proportion,* and *amount,* would be all physically combined.

Lineal arithmetic then, it may be averred, is nothing more than those piles of guineas represented on paper, and on a small scale, in which an inch (suppose) represents the thickness of five millions of guineas . . . as much information *may be obtained in five minutes as would require whole days to imprint on the memory* . . . by a table of figures.[1]

A French edition had appeared in 1789, published by H. Jansen of Paris, bearing the title *Tableaux d'arithmétique linèaire du commerce, des finances, et de la dette nationale de l'Angleterre.* This work also carries a long sub-title. This translation was very well received in France. In 1801, Playfair published in London his volume, *The Statistical Breviary,* with the long sub-title: *shewing, on a principle entirely new, the resources of every state and kingdom in Europe; illustrated with stained copper-plate charts, representing the physical powers of each distinct nation with ease and perspicuity.* This volume contained four plates, three being circle charts of different sizes, proportional to the nature of the data presented. They employ the colors, green, pale red, red, and yellow. A French edition appeared in 1802.[2]

Funkhouser, who has made a detailed study of Playfair's works, points out that Playfair:

> published his many excellent examples of the line graph, circle graph, bar graph and pie diagram and accompanied them with pointed expositions of the advantages of the new method for the discovery and analysis of economic trends [16].

Playfair obtained his ideas about charts from several sources.[3] Funkhouser and Walker quote this statement by him:

> At a very early period of my life, my brother, who, in a most exemplary manner, maintained and educated the family his father left, made me keep a register of a thermometer, expressing the variations by lines on a divided scale. He taught me to know that whatever can be expressed in numbers may be represented by lines [17].

Later on, Playfair worked for James Watt, the inventor of the steam engine, who had developed a gauge on his engine for registering the steam pressure.

---

[1] Italics are Playfair's.

[2] In 1796 another work appeared with the title *A Real Statement of the Finances and Resources of Great Britain; illustrated by two copper-plate charts.* In 1798 Playfair published *Lineal Arithmetic,* bearing the long sub-title: *applied to show the progress of the Commerce and Revenue of England during the present century, which is represented and illustrated by thrity-three copper-plate charts.* In 1805 and again in 1807 he published *An Inquiry into the Permanent Causes of the Decline and Fall of Powerful and Wealthy Nations, illustrated by four engraved charts. Designed to shew how the prosperity of the British Empire may be prolonged.* In 1805 Playfair translated a small pamphlet entitled *A Statistical Account of the United States of America,* written by D. F. Donnant, a Frenchman, and published it as a supplement to *The Statistical Breviary.* In this pamphlet, Playfair refers to Thomas Jefferson's "Statistical Account of Virginia." (The correct title is: *Notes on the State of Virginia* (1787), London: John Stockdale.) In 1821 he wrote *A Letter on our Agricultural Distresses, Their Causes and Remedies, accompanied with tables and copper-plate charts, shewing and comparing the prices of wheat, bread, and labour, from 1565 to 1821.* Addressed to the Lords and Commons. It contained bar charts.

[3] As this paper goes to press, an article "A Note on the History of the Graphical Presentation of Data" by Erica Royston comes to my attention. It was published in *Biometrika,* 43 (1956).

Moreover, Playfair, having been a draftsman, was familiar with the Cartesian system of plotting lines.

As Funkhouser and Walker point out:

> from the standpoint of the history of statistics his graphs are astonishing in that they were made at a time when large collections of reliable quantitative data were not yet available, when the passion for weighing, measuring, counting, and tabu.ating was not yet consonant with the spirit of the age, and when the development of statistical method still waited upon the collection of large bodies of mass data [17].

Playfair, the fourth son of the Rev. James Playfair, born at Benvie near Dundee, in Scotland, received little formal education. His older brother, John Playfair (1748–1819), well-known mathematician and geologist, was at one time professor of mathematics and philosophy at the University of Edinburgh. This brother, elected a Fellow of the Royal Society in 1807, was one of the original members of the Royal Society of Edinburgh. Their father died when William was thirteen years old, and John undertook to care for the family. William was sent to serve as an apprentice to the millwright Andrew Meikle of Prestonkirk, the inventor of the threshing machine. John Rennie, later to become famous as engineer of the Waterloo bridge, was a fellow-apprentice. In 1780, at the age of twenty-one, Playfair served as a draftsman for the firm of Boulton and Watt at Birmingham. Watt was the James Watt, famous as the inventor of the steam engine. Possessing an inventive talent, Playfair secured a number of patents in the field of mechanics. He left his Birmingham employment to open a shop in London in order to sell various items which he had invented and made, but he was unsuccessful in this venture. In 1787 Playfair went to Paris, where he became interested in the promotion of the Scioto (Ohio) land company. About 1792 he attacked the 1789 French Constitution in his writings, and for several years he became involved in French politics. He thought it best to leave France and about 1793 he returned to London, after visiting Frankfort. He opened a so-called security bank to handle small loans, but this venture was also unsuccessful [1; 10, Vol. 15, pp. 1300–1; 42, Vol. 3, pp. 116–7].

About 1795 he engaged in various writings, one attacking the French Revolution, and another advocating an issue of forged assignats. About 1795 he established a "critical and satirical newspaper (called) the *Tomahawk*," which he edited and owned. In 1808 he founded a weekly paper, *Anticipation*, which published about twenty-five issues during its short life. Later he went to Paris again, and edited "*Galignani's Messenger* newspaper for a short time." He published in 1806 his annotated edition of Adam Smith's *Wealth of Nations* with some uncomplimentary remarks, which earned him the ill-will of the influential *Edinburgh Review* and of others. Playfair was a prolific writer. It is estimated that he wrote over a hundred items, mostly articles. *The Gentleman's Magazine*, an English periodical, listed forty of his writings in its June 1823 issue.

2. GEORGE R. PORTER (1792–1852)

George Richardson Porter, economist, executive and statistician, is well known for his contribution to British statistics by directing so well the development of the newly-created statistical department of the Board of Trade. Porter was a pioneer in England in advocating the use of statistics to place economics on a sound scientific basis. His famous work, *The Progress of the Nation*, contains numerous statistical tables.

Porter's chief task as a civil servant was to digest and arrange for the Board of Trade the mass of information appearing in Parliamentary reports and papers, and this position furnished him an excellent opportunity to show his statistical talent. Sir Jervoise Athelstane Baines, C.S.I., President of the Royal Statistical Society (1909–1910), reports:

> The Statistical Branch . . . was started by Porter, by whom the incoherent mass of periodical tables (official government returns) then prepared was for the first time reduced to orderly and comprehensive returns, accompanied by lucid explanations of the meaning and limitations of the figures. Moreover, he took advantage of the wide scope afforded by his commission to collect returns from other sources, adding them to his review, and giving it a comparative character by including the figures for a series of years [5].

Baines also reveals that "this Board was the only government department in which official statistics were dealt as a special subject, and to this day, it stands out as the premier representative of the scientific interpretation of publications" [5].

Porter played a strong part in the formation of the Statistical Society of London. When this Society was formed in 1834, Porter was one of its active founders, and he served as a member of its first council. He was a member of the Publication Committee of the *Journal* of the Society when it was first published in May 1838 under the editorship of Rawson W. Rawson, and he was also a contributor to this periodical. Porter served, moreover, as treasurer of the Society from 1841 until his death in 1852 [51, pp. 15–16, 57, 298]. He was regarded as one of its "most esteemed members." After his death, the council of the Society ordered his contributions to its *Journal* to be bound in a separate volume "partly as a permanent tribute to his memory, and partly for convenience of those who may wish to peruse his valuable papers in a collected form" [47]. Furthermore, Porter was one of the active members of the British Association for the Advancement of Science from the time of its founding, and contributed several papers to Section F—Statistics. F. W. Hirst, editor of the *London Economist*, regarded Porter as "a thoroughly painstaking statistician."

Porter is best known as the author of *The Progress of the Nation in its Various Social and Commercial Relations from the Beginning of the Nineteenth Century to the Present-Day*, published in three small volumes, London, 1836–43. A revised one-volume edition appeared in 1847 and 1851. This work "was a statistical and descriptive study of the social, economic, commercial and fiscal changes which took place in the United Kingdom during the first half of the nineteenth century." In 1912, it was republished with additional material under the direction of F. W. Hirst. Professor Hewins, Director of the London School of Economics and Political Science, termed this work as "an invaluable record of the first half of the nineteenth century. It is remarkable for the accuracy and variety of its information, and for the skill with which the results of statistical inquiry are presented." A review in the *Journal of the Statistical Society of London* called it "a compendious and valuable library of British Statistics."

Porter was a pioneer in the use of index numbers, as Westergaard reveals:

> He treated the prices of 1833–7 in the same way as Shuckburgh Evelyn, but his material was much more complete. For each month in these five years he gives the average of index-numbers for fifty articles comprising the "principal kinds of goods that enter into foreign commerce." It is his aim in this way to find "the mean variation in the aggregate of prices from month to month" [57, p. 203]. `

Porter contributed Section Fifteen, entitled "Statistics," to Sir John F. W. Herschel's *Manual of Scientific Inquiry* (1849), which was prepared for the use of Her Majesty's Navy. His twenty-page treatment of statistics deals with the taking of a census and suggests subjects such as population, manufactures, agriculture, mining, education, domestic and foreign trade, etc. He did not include any discussion of statistical methods. Incidentally, the fourth edition of Porter's contribution was slightly corrected by William Newmarch in 1871.

Porter was born in London, the son of a London merchant, and was educated at the Merchant Taylors' School. His father intended him to manage the family's sugar-broker business, but Porter failed to make a success of it. He prepared a paper on Life Assurance for Charles Knight's *Companion to the Almanac* in 1831 which attracted attention. He married Sarah Ricardo, a well-known writer on educational subjects and a sister of the famous economist David Ricardo [10, Vol. 16, p. 178; 13, Vol. 4, p. 946; 42, Vol. 3, p. 170]. In 1832 Porter was appointed to the Board of Trade by Lord Auckland upon the recommendation of Charles Knight, after the latter had refused the position. He served as head of the newly-established statistical department for many years.[4] In 1840 he was made senior member of the railway department of the Board of Trade, and in 1841 he was appointed joint secretary of the Board. Porter proved to be an able official [57, p. 137]. He was a Fellow of the Royal Society and a member of a number of learned societies.

Porter probably influenced Professor George Tucker of the University of Virginia to take a deep interest in statistics. When Tucker spent the summer of 1839 in England, he lodged part of the time with Porter. Tucker mentioned that he enjoyed meeting Thomas Tooke, the author of the well-known work, *The History of Prices*, and he also met Professor George Long, the editor of the *Penny Cyclopaedia*. Tucker later published his famous American statistical work, *Progress of the United States in Population and Wealth in Fifty Years*, (1843) which originally appeared in installments in Hunt's *Merchant's Magazine* from July 1842 to December 1943 (Vols. 7, 8, and 9). Walter Willcox of Cornell University considered this work as "the most important American book on statistics to appear in the first half of the nineteenth century." He added that Tucker "displayed remarkable insight in utilizing scanty census material" [15, p. 690].

In 1845, Porter collaborated with Professor George Young and Professor George Tucker in publishing a volume *America and the West Indies*. Besides his other writings, he contributed several papers for Lardner's *Cabinet Cyclopaedia*, including "A Treatise on the Origin, Progressive Improvement, and Present State of the Silk Manufacture," and "A Treatise on the Origin, Progressive

---

4 As this article goes to press, a book, *The Board of Trade and the Free-Trade Movement, 1830–42*, New York: Oxford University Press, 1958, by Lucy Brown, comes to my attention.

Improvement and Present State of the Manufacture of Porcelain and Glass" [49, p. 191–2; 50, p. 102].

3. CHARLES BABBAGE (1792–1871)

Charles Babbage, economist, inventor, mathematician, scientific mechanician, and statistician, played a very prominent part in the founding in 1833 of Section F—Statistics—in the British Association for the Advancement of Science of which he was the first president. He played a similar part in the founding of the Statistical Society of London in 1834. Dr. William Farr, in his presidential address before the Statistical Society of London on November 21, 1871, made the statement that Babbage "was, in reality, more than any other man its founder" [32, Vol. 34, p. 411]. Babbage was also one of the founders of the British Association for the Advancement of Science in 1831, an organization that was formed in part as a result of his attack on the Royal Society in a work *Reflections on the Decline of Science in England* (1830).

In establishing Section F, Babbage was assisted by the famous economist, Rev. Thomas R. Malthus, Rev. Richard Jones, Professor of Political Economy, King's College, London, and Adolphe Quetelet, eminent astronomer, mathematician and statistician, director of the Royal Observatory at Brussels who was at that time in England attending the meeting. Babbage reports: "The Section was formed for the purpose of promoting statistical inquiries which were of considerable importance. They had been assisted by a distinguished foreigner, Quetelet, possessing a budget of most valuable information." Although the British Association for the Advancement of Science insisted that Section F should adhere "to facts, relating to communities of men, which are capable of being expressed by numbers, and which promise when sufficiently multiplied to indicate general laws," Babbage was urged by Quetelet to form a statistical society in London. A meeting of the Committee of the Section F—Statistics—was held on February 21, 1834, at Babbage's home. Those present were Charles Babbage (President), William Empson, an economist, Rev. Richard Jones, Rev. Thomas R. Malthus, William Ogilby, Lieut.-Col. Sykes, G. W. Wook, M.P., and John Drinkwater (Secretary). Edward Strutt, M.P. and W. W. Whitmore, M.P. were also present. On a motion made by Malthus, seconded by Jones, it was unanimously voted to establish a statistical society in London. At a public meeting, March 15, 1834, held at the rooms of the Horticultural Society, with the Marquis of Lansdowne, a descendent of Sir William Petty, the well-known author of *Political Arithmetick*, presiding, Babbage moved: "That a Society be established in the name of the Statistical Society of London, the object of which shall be the collection and classification of all facts illustrative of the conditions and prospects of Society, especially as it exists in the British Dominions." Rev. Richard Jones seconded the motion, which was unanimously carried. Quetelet was chosen the first foreign member of the society. It was then moved, seconded and unanimously voted that "Charles Babbage, Esq., M.P., Rev. Richard Jones, M.A., Henry Hallam, Esq., and John Elliot Drinkwater, Esq., M.P., be appointed a Provisional Committee to prepare Regulations for the conduct of the Society." Their report with some changes was later accepted by the Society [51, pp. 4–11; 22–28; 57, p. 174; 33, p. 15].

About 1812 Babbage conceived the idea of inventing a calculating machine, the forerunner of our current calculating machines, and this model was completed around 1822. He described the design of this machine and its workings in a brief paper before the Astronomical Society in June 1822 where it was favorably received. The Society awarded him its first gold medal on June 13, 1823. Babbage then enlisted the support of the Royal Society in the construction of a larger-scale calculating machine. He also contacted Mr. Robinson, Chancellor of the Exchequer, for a government grant, and he was awarded the sum of 1500 pounds. Later on, Babbage received additional grants. In 1828, upon his return from France where he had gone to improve his health, he attempted to secure additional government funds, but he was unsuccessful. Babbage later decided to construct a calculating machine on an entirely different principle. However, only a smaller machine was built and exhibited at the 1862 International Exhibition. One of his machines is now in the South Kensington Museum. Thus Babbage spent much time and money between 1822 and 1843 developing and perfecting his calculating machines [4].

Babbage wrote several statistical papers: "Sur les constantes de la nature," (1853–55), "Notice statistique sur les Phares," (1853–55), and "On the Antecedents of International Statistical Congresses," (1860–61), all three appearing in the *Congres International de Statistique*. One paper appeared in the *Journal of the Statistical Society of London* (Vol. 19, 1856) bearing the title "Analysis of the Statistics of the Clearing House during the year, 1839." This pioneering study of seasonal variations contained nineteen tables and several charts "too large for engraving." Another paper, "An Examination of Some Questions Connected with Games of Chance," was read March 21, 1820, and appeared in Volume 9 (1823), of *Transactions of the Royal Society of Edinburgh*.

Babbage was born near Teignmouth in Devonshire and received his early education at private schools at Alphington and Enfield. His father was a member of the banking firm of Praed, Mackworth and Babbage. In 1811 Babbage enrolled at Trinity College, Cambridge, and graduated in 1814. He received the M.A. degree from this institution in 1817. In 1816 he was elected a Fellow of the Royal Society. He was one of the founders of the Analytical Society in 1812 and of the Astronomical Society in 1820, holding several offices in the latter. From 1828 to 1839 he held the Lucasian Chair of Mathematics at Trinity College, Cambridge [10, Vol. 1, pp. 776–8; 12, Vol. 2, p. 374; 42, Vol. 1, pp. 75–7].

He wrote on a variety of subjects, including infant mortality, geology, life insurance, light-houses, mathematics, taxation, and others. His chief work, *On the Economy of Machinery and Manufactures* (1832; third edition, 1833, fourth edition, 1835), is an excellent description of the subdivision of labor and economic function of machines. It contains a wide range of practical illustrations of the factory system. This work was translated into four foreign languages and republished in the United States. Babbage is the author of altogether some eighty writings, many being brief papers or sketches or pamphlets. The titles are listed in the appendix of his book *Passages from the Life of a Philosopher*, his autobiography. His *Table of the Logarithms of the Natural Numbers from 1–108,000* (1827) was well regarded and reprinted in several foreign countries. Other works are *A Comparative View of Various Institutions for the Assurance of Lives* (1826), *Thoughts on the Principles of Taxation* (1848; second edition, 1851, third edition, 1853), and *The Exposition of 1851* (1851) [49, pp. 15–16; 50, p. 5].

4. DR. WILLIAM FARR (1807–1883)

Dr. William Farr, physician and founder of British vital statistics, is widely known as the statistician of the *Annual Reports* of the British Office of the Registrar General. He joined this newly-established organization in 1838 as Compiler of Abstracts in the statistical department. Later he was made superintendent of the department. He gave up medical practice and remained with this organization as Deputy Registrar-General until his retirement in 1879. Farr first served under T. H. Lister, who held office until 1842 when he was succeeded by Major George Graham. It should be remembered that the registration of all deaths and causes of death was first started in 1837. Farr foresaw the urgent need of placing English public health on a scientific basis, and was a pioneer in the use of statistical data and techniques to achieve this objective. He played a great part in developing the nomenclature of causes of death. Farr succeeded so well that one noted authority, Sir Arthur Newsholme, claims:

> Farr is rightly regarded as the founder of the English national system of vital statistics. For over forty years he supervised the actual compilation of English vital statistics, introduced methods of tabulation which have stood the test of time and a classification of causes of death which has been the basis of all subsequent methods. On the basis of national statistics he compiled life tables still used in actuarial calculations [12, Vol. 6, p. 133].

Another statistician, Simeon North, President of the American Statistical Association (1910), points out:

> The world acknowledges with undying gratitude the inspired genius with which Dr. William Farr, of England, organized this work of registration. . . . Under his hands, the great problems to which vital statistics are the key and clew, were converted into scientific truths, and the general principles established which determine the relationship of density of population and hygienic conditions, to disease and death. Dr. Farr was the pioneer in the protection of the people against a thousand insidious sources of infection. He first showed, by statistical method, the relation of cause and effect. He organized the British "Annual Reports of the Registrar General of Births, Deaths, and Marriages,"—a splendid and unrivalled series of demographic statistics . . . [41, pp. 30–1].

The first census in England was taken in 1801. The census of 1851, organized under Dr. Farr, is said to have been the first fairly complete one. In the 1851 and 1861 censuses he was an assistant commissioner, in the 1871 census, commissioner [40, pp. 25–26; 22, pp. 65–70].

As superintendent of the statistical department of the Office of the Registrar General, Farr was responsible for more than forty volumes of reports on births, marriages, and deaths. After Farr's death, Sir Robert Giffen, in his presidential address before the Statistical Society of London, spoke highly of Farr by indicating:

> At least two remarkable monuments of his later labours, the special report to the Registrar General on the mortality of the 1861–1871 decade, which was completed only seven or eight years ago, and his paper on the mode of estimating the value of stocks having a deferred dividend. What he has left is a noble monument of industry and ingenuity, full of example to all of us who have devoted time and strength to statistics [51, pp. 115–116].

Farr compiled three English Life tables (1843, 1853, and 1864), the third being the most elaborate. They are based on English censuses of 1841 and 1851 and records of deaths, 1838–54 [57, pp. 137, 144, 147, 161, 219].

Farr was a very active member of the Statistical Society of London, and read many papers relating to vital statistics before this Society which were published in the Society's *Journal*. He was a liberal donor to its library. He served on the council from 1840–1882 with the exception of 1847, and he held the office of treasurer from 1855 to 1867, and that of president from 1871 to 1875 [51, pp. 60–61, 95–96]. He proposed and seconded no less than 216 persons as members of the Society. In 1864 he was president of Section F—Economic Science and Statistics—of the British Association for the Advancement of Science. He was also president of the Public Health Section of the Social Science Association in 1866. Moreover, being an official delegate of the British Government, he took an active part and manifested a deep interest in the work of the nine International Statistical Congresses which were held in Brussels (1853), Paris (1855), Vienna (1857), London (1860), Berlin (1863), Florence (1867), The Hague (1869), St. Petersburg (1872), and Budapest (1876). He and Dr. D'Espine played a prominent part at the First International Statistical Congress in 1853 in attempting to bring about the adoption of an international list of the causes of death. Their recommendations were adopted at the Second International Statistical Congress in 1855 [40, pp. 173, 177; 14].

Farr was born at Kenley in Shropshire, and, as his parents were in humble circumstances, was adopted at an early age by Mr. Joseph Pryce, squire of Dorrington, near Shrewsbury. Both Mr. Pryce and the Reverend J. J. Beynon directed his early education. Farr assisted Mr. Pryce in his various affairs. From 1826–28 he studied medicine under Dr. J. Webster, a promising young physician, and assisted Mr. T. Sutton, a surgeon at the Shrewsbury Infirmary. Mr. Pryce at his death in 1828 left Farr a legacy of 500 pounds for his future education, and Dr. Webster at his death in 1837 left Farr a similar amount of money along with his library. In May 1829 Farr went to Paris where he enrolled at the University of Paris to study medicine for two years, and it was while he was in that city that he first became interested in medical statistics [10, Vol. 6, pp. 1090–1; 11, Vol. 10, p. 187; 13, Vol. 6, pp. 993–4]. One of his teachers was the famous French physician, P. Ch. A. Louis, who is generally regarded as the father of medical statistics. In 1831 he returned to London to study at University College, and in March 1832 he became a licentiate (L.A.S.) of the Apothecaries' Society. In the same year he started to practice medicine. He then edited the *Medical Annual*, wrote for medical journals, and in 1837 with the assistance of his close friend, Dr. R. Dundas Thompson, he edited the *British Annals of*

189

Medicine. He wrote an article on "Vital Statistics" for McCulloch's *Statistical Account of the British Empire*, Vol. 2 (1837) [42, Vol, 2, pp. 33–35].

Farr was a prolific writer who contributed not only to the *Journal of the Statistical Society of London* and the *Congres International de Statistique*, but also to the *Lancet*, the *Reports* of the British Association for the Advancement of Science and the Social Science Association [49, p. 76; 50, p. 42]. Many of his important views may be found in a memorial volume entitled *Vital Statistics* (1885), edited by Noel A. Humphreys for the Sanitary Institute of Great Britain [27].

Farr was honored in many ways. The Royal Society elected him a Fellow in 1855. The University of New York gave him the honorary degree of M.D. in 1847, Oxford University bestowed upon him the honorary D.C.L. in 1857, the Royal Medical and Chirurgical Society elected him an Honorary Fellow in 1857, and the King and Queen's College of Physicians in Dublin also elected him an Honorary Fellow in 1867.

5. DR. WILLIAM A. GUY (1810–1885)

Dr. William Augustus Guy, editor, physician and statistician, while well known for his writings in public health, is better known for his many activities on behalf of the Statistical Society of London. Guy, like Farr, was strongly of the opinion that statistics was seriously needed for the study of medical problems. At King's College Hospital, he collected data on out-patients which resulted in three papers relating to the "Influence of Employments Upon Health," read before the Statistical Society of London and published in its *Journal* [57, p. 157]. Some other medico-statistical papers read before the Society and published in its *Journal* were: "On the Health of Nightmen, Scavengers and Dustmen," "Temperature and Its Relation to Mortality," "Mortality of London Hospitals, and Deaths in the Prisons and Public Institutions of the Metropolis," and "Annual Fluctuation in the Number of Deaths from Various Diseases." Guy's contribution to statistics rests primarily on the compilation of bodies of material relating to public health, and on his informative papers dealing with the history of statistics.

Guy was a very active member of the Statistical Society of London, serving for many years as its honorary secretary, 1843–1869, editor of its *Journal*, 1851–1856, and as its president, 1873–75. He was also for many years treasurer of its informal group known as the Statistical Dining Club. Incidentally there are two references to this club. One account reports:

> Outside the work of the Society as such, but still closely connected with it, is the Statistical Dining Club. The only detailed reference to it in the Minutes of the Council is to be found under date 11 January 1839, where it is entered that Mr. Porter reported that a statistical Club had been formed "and had appointed to dine together on the days of the Ordinary Monthly Meetings; that the terms were an annual subscription of one guinea, and 10 s. each on dining with the Club." The Club is limited to forty members and "clubability" is an indispensable prerequisite for election; at each gathering the lecturer of the evening is received as a guest and treated hospitably. It has few rules, no minutes, no records, and only one officer, the Treasurer. The Club is a select body [51, p. 69].

The second account reveals, at the time of its one hundredth anniversary, that:

At a meeting of the Council of the Statistical Society on January 11, 1839, Mr. George R. Porter reported that a Statistical Club has been formed and "had appointed to dine together on the days of the Ordinary Monthly meetings." That Club has now completed the first hundred years of its life, and the members, in order to mark the occasion, decided that a special Centenary dinner should be held and that Mr. Macrosty should compile a record of the Club for circulation among the members. That account has now been printed, with "recollections" by prominent members, and though no records exist for the first fifty years, it has been found possible to recover the names of 163 past and present members. The membership is limited to forty and there were in February three vacancies.

The Centenary Dinner was held at the Trocadero after the Ordinary Meeting on February 21st, 1939, under the Chairmanship of the President of the Society, Professor A. L. Bowley. Fifty-one persons took part, namely twenty eight members, six Club guests, and seventeen private guests [54].

This club is still going strong under the same rules. It is reported that at one time it had one of the best cellars in the city, but this was destroyed by the bombing of London.

Guy contributed many papers which were read before the Society and published in its *Journal*. One statement records "that in twenty years, 1844–63, Dr. Guy read 15 papers" [57, pp. 60–2, 96, 103]. Noting his death in December 1885, another statement records:

The Council minuted: "Dr. Guy was a constant and liberal donor to the Library and the numerous papers which he read before the Society, exceeding in number and variety of subjects those of any other Fellow were of exceptional value. He further testified to his constant interest in the prosperity of the Society by the large number of Fellows whom he introduced to it, and finally by bequeathing to it a legacy of £250 and a reversionary interest of considerable value" [51, p. 152].

Some of Guy's papers relating to statistics, which were read before the Society and published in its *Journal*, were: "On the Relative Value of Averages Derived from Different Numbers of Observations," (Vol. 13, 1850), "On the Original and Acquired Meaning of the term 'Statistics,' and on the Proper Functions of a Statistical Society," (Vol. 28, 1865), "John Howard as Statist," (Vol. 36, 1873), "John Howard's True Place in History," (Vol. 38, 1875), and "On Tabular Analysis" (Vol. 42, 1879) [51, pp. 150–1, 153, 156–7]. In the *Jubilee Volume* of the Society (1885), he has a paper, "Statistical Development, with Special Reference to Statistics as a Science." The 1861 issue of *Congres Internationale de Statistique* contains his "Statistical Methods and Signs" [33, pp. 72–86, 363-4; 49, pp. 105–6; 50, pp. 53–4].

Guy was held in high esteem as a statistician. Sir Rawson W. Rawson, K.C.M.G., C.B., President of the Society (1884–86), called attention to "his industry, and high capacity, his professional knowledge, and statistical insight." Sir Arthur Newsholme, K.C.B., M.D., F.R.C.P., an outstanding authority in the field of vital statistics, regarded Guy as "one of the ablest early English statisticians [40, p. 314].

Because of Guy's many activities on behalf of the Society, the Royal Statistical Society voted in 1891 to establish in his honor, the "Guy Medal." One account reveals:

> At the Council Meeting on 21 May, 1891, a motion was put forward by Sir R. W. Rawson and T. H. (afterwards Sir Thomas) Elliott that in memory of Dr. Guy a gold medal should be awarded "at the discretion of the Council in recognition of the original statistical work placed at the disposal of the Society." This was approved at the Annual Meeting in June. . . . The definition was further expanded by the Council. "The Guy Medal of the Royal Statistical Society—founded in honour of the distinguished statistician whose name it bears—is intended to encourage the cultivation of statistics, in their strictly scientific aspects, as well as to promote the application of members to the solution of the important problems in all the relation of life to which the numerical method can be applied, with a view, as far as possible, to determine by its methods the laws which regulate them." Then the scheme was expanded: a Gold Medal for "work of high character founded upon original research performed specially for the Society"; a Silver Medal for "work founded on existing materials" [51, pp. 160–2].

As to the Guy Gold Medal, two persons were awarded this honor in the nineteenth century, The Rt. Hon. Charles Booth, F.R.S., in 1892 and Sir Robert Giffen, F.R.S., in 1894. From 1900 to 1930 inclusive, six additional persons won this honor, namely, Sir Jervoise Athelstane Baines, C.S.I., in 1900, Professor Francis·Y. Edgeworth, F.B.A., in 1907, Major P. G. Craigie, C.B., in 1908; Professor George Udny Yule in 1911, Dr. T. H. C. Stevenson, C.B.E., in 1920, and Sir Alfred W. Flux, C.B. in 1930. As to the Guy Silver Medal, five persons won it during the nineteenth century, and twenty persons between 1900 and 1930 [51, pp. 301–2].

Guy was born in Chichester, "where his male ancestors for three generations had been medical men." He studied at both Christ's Hospital and Guy's Hospital. In 1831 he was awarded the Fothergillian medal of the Medical Society of London for the best paper on asthma. He enrolled at Pembroke College, Cambridge, receiving the M.B. degree in 1837. His college career in England had been interrupted by two years at Heidelberg and Paris, where he studied under leading medical men. In 1838 Guy was appointed to the chair of forensic medicine at King's College, and in 1842 he was made physician to King's College Hospital, having the care of outpatients. From 1846 to 1858 he served as dean of the medical faculty, and in 1869 he was also appointed professor of hygiene. In 1844 he was admitted a Fellow of the Royal College of Physicians, and he served as censor in 1855, 1856 and 1866, and as examiner in 1861–63. At the Royal College he was also Croonian (1861), Lumleian (1868) and Harveian (1875) lecturer. In 1862 he was examiner in forensic medicine at the University of London. He was Swiney prizeman in 1869. Because of his intense interest in vital statistics, he retired from medical practice. He served on a number of commissions [10, Vol. 8, pp. 835–6; 24].

Dr. Guy was an outstanding physician and was "often consulted in medico-legal cases." He wrote several medical works, as *Principles of Forensic Medicine* (1884), which was frequently reedited, and *The Factors of the Unsound Mind* (1881). Another work was *Public Health; A Popular Introduction to Sanitary Science*, part I (1870) and part II (1874).

192

6. WILLIAM NEWMARCH (1820–1882)

William Newmarch, banker, economist, editor and statistician, an authority on monetary and banking statistics, editor of the *Journal of the Statistical Society of London*, was one of the few British statisticians of his time to perceive the need of utilizing mathematics in describing and analyzing economic and social problems.

Newmarch's contributions to Tooke's *History of Prices* is regarded as a masterly statistical review of the economic history of Great Britain, and this work contains many elaborate statistical tables. He was an early user of index numbers. He was honorary secretary of the Society from 1854 to 1862, its president for two years, 1869–71, succeeding the Rt. Hon. W. E. Gladstone, M.P., and served as editor of its *Journal* for five years, 1855–61, making a number of important changes in this periodical for which he was praised. The Society's publication also records that "the Council expressed 'their approbation' of Mr. Newmarch's 'valuable services' and their knowledge of 'the practical and scientific character of the *Journal* under his editorship.' It is for others to say how far his successors have lived up to his standard" [51, p. 88].

Newmarch, in his presidential address, "Progress and Present Conditions of Statistical Inquiry," before the Statistical Society of London in 1869, reveals greater insight and foresight than most of his contemporaries when he said, among other things:

> Let me now state what appears to me to be the fields of statistical research which in this country most require early attention.

Then he goes on to enumerate eighteen fields, the last one being:

> 18. Investigations of the mathematics and logic of Statistical Evidence; that is to say, the true construction and use of Averages, the deductions of probabilities, the exclusion of superflous integers, and the discovery of the laws of such social phenomena as can only be exhibited by a numerical notation.

Later on in this address, he emphasizes:

> The last subject (division eighteen) in the list, relates to the mathematics and logic of Statistics, and therefore, as many will think, to the most fundamental enquiry with which we can be occupied. . . . It is certain that by means of averages, and variations of increase and decrease, presented by large masses of figures representing social phenomena which occur within longer or shorter intervals of time and within defined limits, it is possible to arrive at conclusions which so far resemble the law of several cases that they justify the enunciation of probabilities and predictions [32, Vol. 32, pp. 365–6, 373].

Newmarch enjoyed two distinctions as president of the Statistical Society of London. First, until he became president of the Society in 1869, all previous presidents had been either high government officials or members of the royalty. Secondly, he instituted the custom "of a regular series of Presidential Addresses." The inaugural addresses are given at the commencement of each presidential term. Previously they had been made at the close of the term—a practice established in 1851 by the Rt. Hon. Earl of Harrowly, K.G., D.C.L., at the end of the second term of office. As the *Annals* records: "Since that time each succeeding President . . . has enriched the records with an address, and in their mass their addresses form a contribution to the history and development of statistics which is unrivalled elsewhere."

Newmarch was an outstanding member of the Statistical Society of London. He was regarded as "one of its most eminent members." One publication of the Society records: He "had for more than thirty years been identified with its work, having contributed many papers on the leading economical questions of the day, and taken a prominent and guiding part in its discussion" [51, p. 115]. Robert Giffen, K.C.B., F.R.S., a distinguished economist and statistician, claims that Newmarch "was remarkable not merely as a statistician but as a man of business and as an economist" [51, p. 115].

Newmarch was an early user of index numbers. Westergaard reports that:

> In 1859 W. Newmarch published index-numbers for nineteen articles with the New Year, 1851, as a starting point, and in the following two years he treated a similar material, extending his investigations to twenty-two articles, with the years 1845–50 as a basis [57, p. 204].

Newmarch published several articles for the *Journal of the Statistical Society of London*, after having read them before the Society. Some are: "Progress and Present Condition of Statistical Inquiry" Vol. 32, p. 359; "Electoral Statistics of Counties and Boroughs in England and Wales from the Reform Act of 1832 to the Present Times" Vol. 20, p. 169; "Electoral Statistics of England and Wales, 1856–58," Vol. 22, p. 101; "Attempts to Ascertain the Magnitude and Fluctuations of the Amount of Bills of Exchange in Circulation at one time in Great Britain, England, Scotland, Lancashire, and Cheshire, respectively, and of Bills drawn on Foreign Countries during each Year, 1828–47," Vol. 14, p. 143. An article "On the Statistical Society of London" appeared in the 1860–61 issue of the *Congres International de Statistique*.

Newmarch collaborated with Thomas Tooke in completing the two concluding volumes, 5 and 6, of the well-known work, the *History of Prices and the State of the Circulation, From 1793 to 1857* (London, 1857), the six volumes covering the period from 1793 to 1856. Newmarch's two concluding volumes, covering the years from 1836 to 1856, are not only a masterful statistical review containing many elaborate statistical tables but also a careful monetary and banking analysis. He entertained great hopes of bringing the *History of Prices* up to date, but pressure of many duties prevented it. These two volumes attracted much attention and were translated into German and used in several German universities [42, Vol. 3, pp. 17–8]. When Mr. Tooke passed away, Newmarch played a leading part in securing funds to establish the "Tooke Professorship of Economic Science and Statistics" at King's College [48, Vol. 34, pp. xvii–xix].

In 1863 he inaugurated an annual section, "Commercial History of the Year," in the London *Economist* which continued up to 1882. During his connection with the *Economist*, he served under the editorship of Wilson, Bagehot and Palgrave.

Newmarch was an authority on monetary and banking problems and appeared several times before a number of Parliamentary committees. He also appeared in 1857 before the Select Committee of the House of Commons investigating the Bank Act, being a leading critic of the famous Peel's Bank Act passed in 1844 [23]. After 1846 he was a frequent contributor to the *Morning Chronicle* [48, Vol. 34, pp. xvii–xix]. Some of his articles dealing with the supply of gold attracted much attention and were later published in 1853 with additions as a book entitled *The New Supplies of Gold*. Part of this work had been read as a paper before the Statistical Society of London in 1851. Besides his contributions to the *Morning Chronicle*, he wrote for the *Economist*, *Fortnightly Review*, *Pall Mall Gazette*, the *Statist*, and the *Times*. Some of his writings were anonymous [33, pp. 367–8; 49, p. 176; 50, p. 93].

Newmarch was born at Thirsk, Yorkshire, and educated in the schools at York. He held several positions as a clerk in his hometown, first under a stamp distributor and then with the Yorkshire Fire and Life Office. Newmarch moved to Wakefield in 1843 to serve as one of the cashiers in the banking house of Leathem, Tew, and Co., and remained with this firm until 1846 when he joined the managerial staff of the Agra Bank of London. This change furnished him the opportunity to become acquainted with many leading persons, some being Thomas Tooke, Alderman Thompson, M.P., and Lord Wolverton [39]. In 1851 he resigned to become actuary and secretary of the Globe Insurance Company and distinguished himself by carrying out several important financial transactions. In 1862 he resigned to be appointed manager of the banking house of Glyn, Mills and Company and he remained with this firm nineteen years until his retirement in 1881 when he was striken with paralysis. He was also a director of several business enterprises [10, Vol. 14, pp. 352–4; 12, Vol. 11, pp. 368–9]. He was elected a Fellow of the Royal Society in 1861. He was, moreover, for many years secretary of the Political Economy Club, and also an active member of both the Adam Smith and the Cobden Clubs. Besides serving one time as secretary, Newmarch was in 1861 also president of Section F of the British Association for the Advancement of Science [9 (1861), pp. 201–203]. Incidentally, Arthur Bowley in his presidential address before Section F (1906) states that "from 1835 to 1855 Section F of the British Association was devoted to 'Statistics,' and it is only from 1856 onwards that it has received the curious name, 'Economic Science and Statistics' " [32, Vol. 69, p. 540].

His colleagues' high esteem of Newmarch is partly reflected by three memorials: After his death the Statistical Society of London "subscribed twenty guineas to the Newmarch Memorial Fund" [51, p. 115]. Mr. H. D. Pochin, a member of the Council of the Society placed at its disposal 100 pounds for a Newmarch Memorial Essay [33, p. 35]. Finally, 1420 pounds and 14 shillings were subscribed toward "the foundation of the Newmarch Professorship of Economic Science and Statistics at the University College, London."

7. W. STANLEY JEVONS (1835–1882)

William Stanley Jevons, economist, logician and statistician, is well known for his influential writings in the fields of logic and economics. He should be equally well known for his important statistical contributions. Jevons was a pioneer in statistical methods in several ways: First, in emphasizing the superior value of the geometric mean over the arithmetic mean; second, in strongly recommending the use of the chart, now known as the ratio chart, as a graphic means of showing per cent of change; third, in calling attention to the several problems involved in the proper construction of an index number; and finally, in improvising statistical means of measuring time series in the form of secular trend, seasonal variation and cyclical fluctuations.

Jevons' early interest in statistics dates at least from October 1860, when at the age of twenty five years "he began to form diagrams to exhibit some statistics" he had been collecting in the British Museum Library for the purpose of developing a *Statistical Atlas*. Later in a letter dated April 7, 1861, he wrote to his brother, Herbert:

> I am very busy at present with an apparently dry and laborious piece of work, namely, compiling quantities of statistics concerning Great Britain, which are to be exhibited in the form of curves, and if possible, published as a *Statistical Atlas*. . . . Almost the whole of the statistics go back to 1780 or 1800, a large part extend to 1700 or 1720, and some—for instance, the price of corn—as far back as 1400. The quantity of statistics which I shall exhibit in about thirty plates will, I think, rather astonish people.

He then goes on to enumerate the various items dealing with population, foreign and domestic commerce, money and banking, agriculture, government debt, etc, which he intends to cover. Then in this same letter, he says:

> Most of the statistics, of course, are generally known, but have never been so fully combined or exhibited *graphically*. . . . The mode of exhibiting numbers of curves and lines has, of course, been practiced more or less any time on this side of the Deluge. At the end of the last century, indeed, I find that a book of *Charts of Trade* [correct title of Playfair's work was the *Commercial and Political Atlas*, 1786] was published, exactly resembling mine in principle; but in statistics, the method, never much used, has fallen almost entirely into disuse. It ought, I consider, to be almost as much used as *maps* are used in geography [30, p. 157–8].

In a letter dated December 3, 1861, to his brother Tom, Jevons remarks: "My statistical matters proceed slowly, and the mere drawing of diagrams takes up an incredible deal of time."

Jevons wrote in his journal on December 8, 1861:

> About October 1860, having then recently commenced reading at the Museum Library, and met some statistics, I began to form some diagram to exhibit them. . . . After doing two or three diagrams the results appeared so interesting that I contemplated forming a series for my own information. Then it occurred to me that publication might be possible, and I finally undertook to form a statistical atlas of say thirty plates, exhibiting all the chief materials of *historical statistics*. For the last year this atlas has been my chief employment. . . . Towards the end of last October I had some twenty-eight diagrams more or less finished in the first copy, and thought it time to arrange for publication [30, p. 161].

However, Jevons was unsuccessful in his efforts with several publishing houses to have the atlas published. They were of the opinion that the work would involve too great an expense in view of the limited market, he records in his journal on December 8, 1861. He also records how Mr. Newmarch "looked at my diagrams without *interest*, and almost without a word, so that I soon left him" [30, p. 162]. It appears that the academic and the business worlds were not ready to appreciate these statistical tools and the knowledge they could impart about the immediate future.

In his letter of December 28, 1862, Jevons wrote his brother Tom:

> I am at present going on with my old work of diagrams. I am now thinking of a small atlas with plates about 6 by 8 inches, from 1844–62, comprising monthly quotations

of prices, exports, imports, etc., all fully reduced, analyzed, etc., so as to make quite a small gem of work. . . . It is somewhat the same idea with which I just began nearly two years ago but I have learnt so much by experience that my first diagrams were quite laughable besides the little gems I now produce. . . . The atlas would contain perhaps twelve plates [30, p. 173].

However, this atlas met the same unfortunate fate.

In September 1862 Jevons sent two papers to be read at the annual meeting of Section F of the British Association for the Advancement of Science. One paper entitled "On the Study of Periodic Commercial Fluctuations" was read and approved, while the second paper, "Notice of a General Mathematical Theory of Political Economy," was merely read [9, Vol. 32, pp. 157–9]. This little-noticed second paper was to be further developed later and published as a book, *The Theory of Political Economy* (1871). This mathematical description of economic principles was to earn Jevons a world-wide reputation as an economist. He states in the preface of the second edition (1879) that: "I do not write for mathematicians, nor as a mathematician, but as an economist wishing to convince other economists that their science can only be satisfactorily treated on an explicitly mathematical basis." This work thus reveals Jevons to be a pioneer in the field of mathematical economics, and, as will be seen, he was a trailblazer also in the field of econometrics.

In the former paper, "On the Study of Periodic Commercial Fluctuations," Jevons studied the nature of seasonal variations by computing monthly as well as quarterly averages. He found that "it is interesting to observe that the monthly and quarterly variations are of precisely the same character." He employed four diagrams revealing "Average Rate of Discount, 1845–61 and 1825–61," "Total Number of Bankruptcies, 1806–60," "Average Price of Consols, 1845–60," and "Gazette Average Price of Wheat, 1846–61." This investigation enabled Jevons to discover the nature of secular trend movement of prices [29, pp. 2–11].

Analyzing this study, Keynes points out that Jevons:

was not the first to plot economic statistics in diagrams; some of his diagrams bear, indeed, a close resemblance to Playfair's with whose work he seems to have been acquainted. But Jevons compiled and arranged economic statistics for a new purpose and pondered them in a new way. . . .

Jevons was the first theoretical economist to survey his material with the prying eyes and fertile, controlled imagination of the natural scientist. He would spend hours arranging his charts, plotting them, sifting them, tinting them nearly with delicate pale colours like the slides of the anatomist, and all the time pouring over them and brooding over them to discover their secret. It is remarkable, looking back, how few followers and imitators he had in the black arts of inductive economics in the fifty years after 1862 [35, pp. 523–4; 53].

Next year he brought out his pamphlet of seventy-three pages, *A Serious Fall in the Value of Gold Ascertained, and Its Social Effects Set Forth With Two Diagrams* [28], a very important statistical treatment. For one thing, Jevons, in this pioneering study explains on page 7 the value of using the geometric mean as an average in place of the arithmetic mean to combine wholesale monthly prices near the middle of the month, the prices being those of 39 chief commodities for the years 1845 to 1862 [57, pp. 203–4]. He proposed the use of the geometric mean by calculating the arithmetic mean of the logarithms instead of using the original numbers. Secondly, he demonstrates the use of diagrams with logarithmic vertical scale for observing "proportional variation of prices" with the horizontal scale showing arithmetic progression. This diagram, the forerunner of our current ratio chart, appears to be a variant of Playfair's charting technique. Thirdly, he offers an excellent demonstration of the problems involved in constructing index numbers by examining various aspects such as the question of weighting, the choice of an average, the number and kinds of commodities to include, etc. Seventy-nine minor items were also employed as a check on his results. He even applied the theory of probability to his work. Thus Jevons is regarded by some as "the father of index numbers."

Keynes again appraises Jevons:

> For unceasing fertility and originality of mind applied, with a sure touch and unfailing control of the material, to a mass of statistics, involving immense labours for an unaided individual ploughing his way through with no precedents and labour-saving devices to relieve his task, this pamphlet stands unrivalled in the history of our subject. The numerous diagrams and charts which accompany are also of high interest in the history of statistical description [35, 525–6; 53].

It is, indeed, unfortunate that after Jevons' pioneering efforts the theory and use of index numbers were to mark slow progress until 1887 when Professor Francis Y. Edgeworth commenced his excellent studies in this area [9, (1887), pp. 247–301; (1888), pp. 181–232; (1889), pp. 133–64; (1890), pp. 485–8].

The last page of Jevons' pamphlet contains an advertisement reporting that he had "in preparation" *The Merchant's Atlas and Handbook of Commercial Fluctuations*. This original effort indicates that Jevons was again a pioneer in planning to sell business men information about the current status of business conditions. No reason, however, can be found for the failure to publish this *Merchant's Atlas*. It is quite likely that the poor sales for his pamphlet, *A Serious Fall in the Value of Gold*, may be the answer. It sold only 74 copies.

In 1865 he brought out another statistical paper "On the Variation of Prices, and the Value of the Currency Since 1872" which he read before the Statistical Society of London in May 1865 and published in its *Journal* in June 1865 (Vol. 28). This paper represented a further development of the theory of index numbers, and contained data going back to the eighteenth century. He continued to use the geometric mean and the ratio chart. In 1865 his famous book, *The Coal Question*, appeared, which attracted considerable attention. In Chapter 9, entitled "Of the Natural Law of Social Growth," he pointed out that many economic and social phenomena experience a law of geometric growth, some at a greater rate than others. He goes on to apply this idea to the growth of Great Britain. Jevons advanced the thesis that future prosperity of Great Britain would increase the demand for coal in the form of geometric progression leading to a possible exhaustion of coal. This book resulted in the appointment of a royal commission to examine the available coal reserves. Thus Jevons can be considered as a pioneer in the field of secular trend measurement. In 1866, he brought out another statistical paper, "On the Frequent Autumnal Pressure in the Money Market, and the Action of the Bank of England," which was read before the Society in April 1866, and published in its *Journal* in June 1866 (Vol. 29).

Jevons was a student of business cycles, then known as commercial crises, pointing out that there is a strong relation between the solar period and the price of corn. His first paper, "The Solar Period and the Price of Corn," was read in 1875 before Section F—Economic Science and Statistics—of the British Association for the Advancement of Science. Corrections were made in subsequent papers, entitled "Commercial Crises and Sun Spots," which were published in two articles in *Nature:* Part I appeared in the November 14, 1878 issue and Part II in that of April 24, 1879. Another paper, "The Periodicity of Commercial Crises and its Physical Explanation," was read in August 19, 1878 before Section F of the British Association for the Advancement of Science and published in Volume 7 of the *Journal of the Statistical and Social Inquiry Society of Ireland.* Wesley C. Mitchell, in his famous book *Business Cycles* (1927), pays tribute to Jevons by stating: "It was left for W. Stanley Jevons to give the first powerful impetus to statistical work in economic theory."

Jevons was a very active member of the statistical Society of London and "made numerous donations to its Library." He served as its honorary secretary from 1877 to 1880 [51, p. 115], and read several papers before the Society which were published in its *Journal.* Some are: "On the Variation of Prices, and the Value of the Currency since 1782" (Vol. 28, 1865), "On the Frequent Autumnal Pressure in the Money Market, and the Action of the Bank of England" (Vol. 29, 1866), "Condition of the Metallic Currency of the United Kingdom, with Reference to the Question of International Coinage" (Vol. 31, 1868), and "Statistical Use of the Arithomometer" (Vol. 41, 1878). The latter article referred to the use of a French calculating machine. The *Annals* reports that "no other economist so distinguished was so closely connected with the Society." In 1870 he served as president of Section F of the British Association for the Advancement of Science.

While living in Manchester Jevons was also an active member of the Manchester Statistical Society. He served as its vice president in 1868–69, and as its president, 1869–71 [2]. He read several papers before this Society, for example, "The Work of the Manchester Society in connection with the Question of the Day" (1869–70), and "The Progress of the Mathematical Theory of Political Economy, with an Explanation of the Principles of the Theory" (1874–75). Furthermore, he wrote a paper, "The Periodicity of Commercial Crises and its Physical Explanation," which was published in the *Journal of the Statistical and Social Inquiry Society of Ireland* (Vol. 7, 1876–1879). This paper examined the nature of the relationship between commercial crises and sun spots.[5]

Jevons, born the ninth of eleven children, at Liverpool, was the son of an iron merchant who was a writer on economic and legal matters. He received his early education at the hands of a private tutor, and at the Mechanics Institute High School. But he remained only a short time at this institution and then

---

[5] Jevons was the author of a number of other works [49, pp. 134–5; 50, p. 66]. In economics, *Primer of Political Economy* (1878) which was translated into French and German, *Money and the Mechanism of Exchange* (1875), *The State in Relation to Labour* (1882), and *Methods of Social Reform* (1883). In the field of logic, *Elementary Lessons in Logic* (1870), *Primer of Logic* (1876), and *Pure Logic and Other Minor Works* (1890), *Studies in Deductive Logci,* (1880). Another work was *The Principles of Science: A Treatise on Logic and Scientific Method,* 2 volumes (1874), and one volume (1877). He was a contributor to Australian newspapers in earlier years. Later he contributed to the *Spectator, London Quarterly Review, Contemporary Review, National Review, Times, Macmillan Magazine.*

enrolled at Beckwith's private school. In 1850, at the age of fifteen, he attended the University College School at London, and in 1851 he enrolled at the University College, where he studied mathematics, chemistry, biology, and metallurgy. By the midsummer of 1851 he had won five prizes—three being first prizes and two second prizes. Because of financial circumstances (his father having become a bankrupt in January 1848), Jevons was forced to leave college when halfway through his studies. This business failure of Jevons and Sons probably resulted from the depression of 1847. Late in 1853 at the age of eighteen he left England to accept the position of assayer in the newly-established Royal Mint at Sydney, Australia. Gold had recently been discovered in Australia. En route he stopped off at Paris where he spent two months at the Paris Mint studying assaying. He arrived at Sydney in October 1854 [48, Vol. 35, pp. i–xii]. At first he was interested in meteorology, but his interest later shifted to Adam Smith's *Wealth of Nations* and John Stuart Mill's *Logic*. His residence of about five years in Australia is said to have given him much opportunity to reflect on various problems and subjects, a development which is well indicated in his writings after his return to England. Early in 1859 he resigned his position and returned to London by way of South America and the United States, where he visited a number of cities. In October 1859 he enrolled at the University College to study mathematics, political economy and logic, receiving the A.B. degree in 1860, and the Ricardo scholarship and the M.A. degree in June 1863. He won the gold medal in philosophy and political economy [31; 12, Vol. 8, pp. 389–91]. In 1863 at the age of twenty eight he became a tutor at Owens College, a young institution in Manchester, where in 1866 he was appointed professor of logic and mental and moral philosophy and in 1867 also Cobden Lecturer in Political Economy [10, Vol. 10, pp. 811–5; 11, (1957), Vol. 13, pp. 30–1]. In 1864 he joined the Statistical Society of London and remained an active member for the remainder of his life. While living in Manchester he became an active member of the Manchester Statistical Society. In 1868 Jevons was appointed an Examiner in Political Economy at the University of London. In 1874 and 1875 he served as an Examiner in the Moral Science Tripos at the University of Cambridge. In 1876 he was Examiner in Logic and Moral Philosophy at the University of London. In 1872 he was elected a Fellow of the Royal Society, the first economist so honored since Sir William Petty, famous author of *Political Arithmetick*. In 1876 he resigned his teaching position at Owens College in order to accept the chair of Political Economy at the University College at London. Because of ill health, because of his dislike for lecturing, and because of his intense desire to devote all his time to his writing projects, he resigned from teaching in 1880. In 1875 he received the honorary degree of LL.D. from the University of Edinburgh [42, Vol. 2, pp. 474–9; 38, p. 1202].

8. SIR RAWSON W. RAWSON, C.B., K.C.M.G., (1812–1899)

Sir Rawson William Rawson, administrator, editor and statistician, an authority on international statistics, is remembered as the first editor of the *Journal of the Statistical Society of London* (1838–40), and is well known as the first president of the International Institute of Statistics during its formative years (1885–98).

Rawson became a member of the Statistical Society of London in March 1835, was elected to the council in 1836, served as honorary secretary from 1836 to 1842, and as editor of its *Journal* beginning with the first issue of May 1838 [45]. This scholarly periodical continued as a monthly until April 1839 when it became a quarterly. It has been published continuously since that date and is probably the most outstanding statistical journal in the world. As editor, Rawson was assisted by "a Publication Committee (Mr. Porter, Dr. Lister, Mr. Heywood, Mr. Romilly, and Mr. Boileau)" [51, p. 57]. In 1840 Rawson was succeeded by Joseph Fletcher as editor. The *Annals* reports that during the first decade of the Society, "Mr. R. W. Rawson contributed 6 of the papers." It also mentioned "13 contributions on as many separate subjects by R. W. Rawson" [51, pp. 60–2]. His articles were of a critical nature because it was "the custom to comment on parliamentary papers and other state documents." *Proceedings of the Statistical Society, 1834–1837*, a volume published by the Society, contains papers prepared and read during the years before the publication of its *Journal*. In this volume Rawson has four papers, one being "On the Collection of Statistics."

"On his return from a distinguished colonial career," he again took an active part in the Statistical Society of London [51, p. 155]. In 1876 he was reelected to the council and remained a member of it until his death. Five times he was elected vice president during the period from 1876 to 1884, and he served as president in 1884–86. In 1885 the official title of the Society was changed to the Royal Statistical Society. In his address at the Golden Jubilee meeting of the Society, he mentioned that "my public career in the colonies afterwards separated me from active participation in the work of the Society for the third of a century." This Golden Jubilee, postponed one year on account of the death of the Duke of Albany, was held in London, June 22, 23 and 24, 1885. In planning this Golden Jubilee, the Society had set up a Committee "to consider in what manner the Jubilee of the Statistical Society may be utilized for the advancement of Statistical Science and the extension of the Statistical Society." The objectives of the Jubilee became: "1. To review the work of the Statistical Society during the past fifty years. 2. To consider what has been achieved by the International Statistical Congresses, or by other means, in the direction of the uniformity of statistics, and by what means that object may be further promoted. 3. To consider the possibility of establishing an International Statistical Association" [51, pp. 139–40].

The Golden Jubilee meeting was an outstanding one with distinguished guests in attendance from many countries. General Francis A. Walker, President of the American Statistical Association, was the sole representative from the United States. An excellent set of statistical papers, read by distinguished economists and statisticians, such as Edgeworth, Levasseur, Galton, Guy, Marshall, Mouat, Giffen, Korosi, von Neumann-Spallart, and others, are published in the Society's *Golden Jubilee* volume. Sir Rawson, presiding over these sessions, delivered the opening address [33, pp. 2–12]. He was elected the first president of the International Institute of Statistics, founded at this meeting, which held its first meeting in Rome in 1887 [57, pp. 246, 260].

Rawson was regarded as "the Nestor of British statisticians." He suggested in one of his papers "the use of varying price of an *average ton* of exports or imports as a sort of an index number for measuring the change in the value of money" [46]. Rawson was chairman of the Library Committee of the Society, and he left the statistical portion of his library to the Society. One obituary records that "the Society has been deprived of both its senior Fellow and of one who has done more than perhaps anyone else toward placing it in its present satisfactory condition" [45].[6]

Rawson, the eldest son of Sir William Rawson, K.B., was born in London and educated at Eton. In 1830 he was appointed private secretary to Mr. Poulett Thompson, vice president of the Board of Trade, and served in the same capacity to Mr. Alexander Baring who succeeded Mr. Thompson in 1834. He again served in the same capacity to William Gladstone in 1841 who was then vice president. In 1842 he was appointed civil secretary to the Governor General of Canada, and in 1844 treasurer to Mauritius. In 1854 he became colonial secretary at Cape of Good Hope, during which period he was honored with the C.B. In 1864 he was appointed governor of the Bahamas, and in 1869 he succeeded to the governorship of Windward Islands. In 1875 he retired from public service, and in the same year he was honored with K.C.M.G. [59, Vol. 1, pp. 588–9].

9. SIR FRANCIS GALTON (1822–1911)

Sir Francis Galton, eugenist, explorer, psychologist and statistician, the father of correlation analysis, created a number of statistical tools. As early as 1869, Galton, in his work *Hereditary Genius*, began to develop statistical techniques, stating on page 26: "The method I shall employ . . . is . . . theoretical law of 'deviation from an average'," which he acknowledges has been used by the famous Belgian statistician Quetelet. Galton claims that he is the "first to treat the subject in a statistical manner." In the preface of this work, Galton says:

> The theory of hereditary genius, though usually scouted, has been advocated by a few writers in the past as well as in modern times. But I may claim to be the first to treat the subject in a statistical manner, to arrive at numerical results, and to introduce the "law of deviation from an average" into discussion on heredity.

As to the idea of correlation, Galton describes the incident which gave rise to its development as follows:

> As these lines are being written, the circumstances under which I first clearly grasped the important generalisation that the laws of Heredity were solely concerned with deviations expressed in statistical units, are vividly recalled to my memory. It was in the grounds of Naworth Castle, where an invitation had been given to ramble freely. A temporary shower drove me to seek refuge in a reddish recess in the rock by the side of the pathway. There the idea flashed across me, and I forgot everything else for a moment in my great delight [18, p. 300].

---

[6] Besides contributing many papers to its *Journal*, Rawson also was the author of a number of other works, including Reports on Mauritius Census of 1851, *Immigration of Coolies and Valuation of the Rupee in Mauritius, 1845–54, Statistical Description of the Bahamas, and an Account of the Hurricane of 1866 in those Islands, Reports on Barbados Census of 1871 and Rainfall of Barbados 1873–74, British and Foreign Colonies (1884), International Vital Statistics (1855), Synopsis of the Tariffs and Trade of the British Empire, 1884–85, Our Commercial Barometer (1890–91), Ocean Highways or Approaches to the United Kingdom (1894)* [49, pp. 200–1; 50, p. 107].

The date of this famous idea is 1888. Galton also relates another incident, which pertains to the regression line:

> I had given much time and thought to Tables of Correlations, to display the frequency of cases in which the various deviations say in stature, of an adult person, measured along the top, are associated with the various deviations of stature in his mid-parent, measured along the side. I had long used the convenient word "mid-parent" to express the average of the two parents, after the stature or other character of the mother had been changed into its male equivalent. But I could not see my way to express the results of the complete table in a single formula. At length, one morning, while waiting at a roadside station near Ramsgate for a train, and poring over the diagram in my notebook, it struck me that the lines of equal frequency ran in concentric ellipses. The cases were too few for certainty, but my eye, being accustomed to such things, satisfied me that I was approaching the solution. More careful drawing strongly corroborated the first impression [18, 302].

Galton first used the term "correlation" on December 20, 1888, when his paper, "Co-relations and Their Measurement" was read before the Royal Society. It was not until the publication of his *Natural Inheritance* in 1889 that the terms "correlation" and "regression" became known [57, pp. 226–7, 268, 270–2].

Thus 1869 marked the beginning of Galton's many contributions to the theory of statistics. He devised the ogive curve (1875), the symbol of the coefficient of correlation, *r*, (which first meant reversion, but later changed to regression), (1877), the quartile deviation (1879), the median (1883)—although Fechner had the same idea independently—the percentile system (1885), the index of correlation (1888), and the use of the normal curve for grading children. Galton was responsible for the introduction of graphical methods in mapping the weather, and he was the originator of the use of statistical methods in the field of biology. Galton's statistical contributions are so well described by two well-known authorities, Karl Pearson and Helen M. Walker, that readers are urged to consult them [43, 55].

Galton joined the Statistical Society of London in 1860, "but his association with the Society was not close" [51, pp. 179, 225]. He served on the council from 1869 to 1879, and was vice president in 1875. He read three papers before the Society.

Galton, born in Duddeston, in Warwickshire, was the youngest of seven children. His father, a member of the Society of Friends, was a banker, and his mother was related to a number of prominent persons. His two grandfathers were both Fellows of the Royal Society, and his cousin was the famous Charles Robert Darwin (1809–1882). After being educated at several private schools, Galton attended King Edward VI's grammar school, then studied at the Birmingham Hospital, and completed his medical education at King's College. His parents wished him to be a physician, but he later changed his mind about medicine and enrolled in 1840 at Trinity College, Cambridge, graduating in 1843. His father passed away in 1844, leaving him a considerable financial inheritance, and so he was able to spend much time in foreign travel for the next few years. In 1850 he, along with Dr. Charles J. Andersson, explored certain unknown areas in Africa. An account of his experiences resulted in a book, *The Narrative of an Explorer in Tropical South Africa* (1853), which went through several editions. This exploration earned him in that year the gold medal of the Royal Geographical Society and in 1854 the silver medal of the French Geographical Society [19]. In 1855 Galton wrote another work, the *Art of Travel*, which also went through several editions. In 1856 he was elected a Fellow of the Royal Society. During the years 1860–63 he was the editor of an annual volume, *Vacation Tourists and Notes of Travels*. Galton now became interested in meteorology and in 1863 published his work *Meteorographica*. This was followed by other writings in meteorology so that he became a member of the meteorological committee and its successor, the meteorological council, as well as being connected with the Kew Observatory. He was thus associated with the meteorological committee for about forty years, ever since its beginning [10, Supp. Vol. 2, pp. 70–3; 11, Vol. 11 (1910) pp. 427–8; 12, Vol. 6, pp. 553–4]. The publication of the *Origin of Species* by Charles Darwin in 1859 encouraged Galton to make a study of heredity which resulted in several works, *Hereditary Genius* (1869), which went through several editions, *English Men of Science* (1874), *Inquiries Into Human Faculty and Its Development* (1883), and *Natural Inheritance* (1889). At the 1884 International Health Exhibition in London he set up the first Anthropometric Laboratory, which measured over 9,000 persons. At the close of this exhibition, the laboratory was established at the Science Museum, South Kensington. This later became the foundation of the well-known biometric laboratory at the University College, London [38, pp. 1167–72; 44]. Because of the possible anthropological significance, Galton now turned his talents to the study of fingerprints, and several works appeared, namely, *Finger Prints* (1892), *Blurred Finger Prints* (1893), and *Finger Print Directories* (1895). This system is now employed all over the world.

Galton was a most prolific writer [49, p. 89; 50, p. 47]. His *Memories* in 1908 lists 182 writings. Pearson has listed over 220 papers and fifteen books. He was, moreover, a member of many scientific and learned societies at home and abroad. He was the recipient of honorary degrees from Oxford, which conferred the D.C.L. on him in 1894, and from Cambridge, which honored him with the D. Sc. in 1895. The Royal Society bestowed upon him three medals: The Royal Gold Medal in 1886, the Darwin Medal in 1902, and the Copley Medal in 1910. In 1908 the Linnaean Society gave him a medal in honor of the Darwin-Wallace Celebration. In 1901 the Anthropological Institute awarded him the Huxley Medal. He was knighted in 1909 [58, pp. 562–3; 59, pp. 265–6].

Galton was general secretary of the British Association for the Advancement of Science from 1863 to 1868, and twice he declined the presidency. Four times he was president of its Sections; twice of its Geographical Section in 1862 and 1872, and twice of its Anthropological Section in 1877 and 1885. He was president of the Anthropological Institute from 1885 to 1888. Galton served for several years on the Council of both the Royal Geographical Society and Statistical Society of London, and was vice president of the latter in 1875. He was also Chairman, Committee of Management, Kew Observatory of the Royal Society 1889–1900. In his will he left a sum of 45,000 pounds for the establishment of a Chair of Eugenics to be occupied by his close friend, Karl Pearson.

10. SIR ROBERT GIFFEN (1837–1910)

Sir Robert Giffen, economist, economic journalist, editor and statistician, is well known as the head of the statistical department of the Board of Trade, as the editor of the *Journal of the Royal Statistical Society* (1876–1891), and as a prolific writer on economic statistics. In 1867 Giffen became a member of the Statistical Society of London, and he first served on the council in 1871. He was elected its honorary secretary for 1873–74 and 1876–82, and he was editor of its *Journal* from 1876 to 1891. He was vice president of this Society, 1880–81, and its president for two years 1882–84 [51, pp. 227–8]. He read eleven papers before the Society and three before Section F of the British Association for the Advancement of Science. He was twice president of Section F in 1887 and 1901. He assisted in the founding of the International Statistical Institute in London in 1885. He was an outstanding member of the Political Economy Club from 1877 to 1910. He was also one of the founders of the *Statist*, to which he contributed a number of articles, and of the British Economic Association in 1890, now known as the Royal Economic Society, in which he held office as vice president at one time. He contributed articles, known as *City Notes*, "received from R.G.," to the *Economic Journal* from its first volume in 1891 up to his death in 1910. The Royal Statistical Society honored him with its Guy Gold Medal in 1894.

As a statistician Giffen was highly regarded. He was chief statistical adviser to the British Government, for which he prepared various reports, and he was frequently called upon to present his views before royal commissions and committees. One obituary notice observed that he "was the most popular, if not the ablest, statistician of modern times. . . . He was singularly painstaking and careful in weighing statistical data, and his power of imagination was of immense use in suggesting to him the key to many an economic problem" [20, p. 529]. Another notice contains this quotation:

> I think that one of the features of Sir Robert Giffen's work which impressed me most was its extraordinary rapidity and certainty, whether he was piercing to the heart of a complicated mass of statistics and extracting their real significance, or whether he was composing the luminous and original memoranda, which he tossed off at lightning speed with little apparent effort. He has an almost unique power of carrying his statistics in his head; they were always at his command, and he was never overwhelmed by them. In the most complicated mazes of figures he never lost his grip on the realities for which the figures stood, and he never seemed to lose his bearings or his fine sense of proportion. . . .
>
> With an acute perception of the things that were not measured or unmeasurable, he first surrounded the official statistics with an atmosphere of caution, and then cleared away the mist by the use of bold estimates. For these estimates he had an arithmetical sense almost amounting to genius, a feeling for the probable error of the factors used, and a courageous rejection of measurement where the inaccuracy was too great. He had an intuitive feeling for the relative importance of numbers. He used to express his conclusion as to the adequacy of the data by saying he could, or could not, "give a figure." He appears to have had little or no knowledge of the modern mathematical theory of statistics, but arithmetical sense was so strong that he was able to proceed safely and with knowledge through calculations whose validity could only be established mathematically [21, pp. 319–21].

This obituary closes with the last sentence reading: "Giffen deserves to be honored with the Masters of Statistical Science."

One writer states:

> Giffen, a prolific writer on economic, financial, and statistical subjects, possessed a luminous and penetrating mind, great store of information, an intimate acquaintance with business matters and methods, and shrewd judgment. His instructive handling of statistics and his keen eye for pitfalls contributed greatly to raise the reputation and encourage the study of statistics in this country, though he did not develop its technique by the higher mathematical treatment [10, Supp. Vol. 2, pp. 104–5].

Another writer declares: "Giffen's numerous statistical studies are models of clear exposition and of legitimate statistical inference. He paid little attention to the mathematical analysis of statistical data and was acutely aware of the limitations commonly inherent in quantitative material" [12, Vol. 6, pp. 656–7].

Some of his papers are regarded as classics. One is his presidential address before the Statistical Society of London in 1883, entitled "The Progress of the Working Classes in the Last Half Century," which was followed in 1886 with "Further Notes on Progress of Working Classes During Last Half Century." Another is "Recent Accumulations of Capital in the United Kingdom," (1878) followed by "Accumulations of Capital in the United Kingdom, 1875–85," (1890). Still another is "The Use of Import and Export Statistics," (1882). His book, *The Growth of Capital* (1889), is one of the early estimates of national wealth. His papers and books comprise a remarkable record for a top official who wrote most of them outside his regular departmental duties. They reveal his wide acquaintance with many problems and his "unusual power of accurate generalization from voluminous and complex evidence." Richmond Mayo-Smith, a distinguished American statistician, who termed Giffen as "the greatest living statistician in England," was critical at times of Giffen's handling of some statistics, although at other times he praised Giffen's statistical writings [37].

During his twenty-one years with the board he was mainly responsible for the noteworthy improvements in official economic statistics, and he rendered much valuable assistance to royal commissions and committees. He directed the first national census of wages in 1886.[7]

Giffen was born in the small Lanarkshire town of Strathaven, and received his early education in the village school. His father was a small merchant and an elder of the Presbyterian Church. At the age of thirteen, Giffen moved to Glasgow, where he spent several years (1850–1855) as a clerk in a solicitor's office. He attended part of the time classes at the University of Glasgow, but did not graduate. In later years (1844) this University honored him with the degree of doctor of laws. In 1860 he became a reporter and sub-editor of the *Stirling Journal*. In 1862 he moved to London where he worked for the *Globe*

---

[7] He was the author of many books, some being *American Railways as an Investment* (1872); *Stock Exchange Securities; An Essay on the General Causes of Fluctuations in Their Price* (1877); *Essays in Finance, First Series* (1880; fifth edition 1890); *Essays in Finance, Second Series* (1886; third edition 1890); *The Growth of Capital* (1889); *The Case Against Bimetallism* (1892; second edition in the same year); and *Economic Enquiries and Studies*, two volumes (1904) which contains miscellaneous writings and addresses. He left a manuscript which was published in 1913 as a book, *Statistics*, edited by Henry Higgs with the assistance of G. Udny Yule. This volume contains nothing on statistical methods [49, p. 99; 50, p. 49].

as sub-editor and served until 1866. He then joined the *Fortnightly Review* under John Morley (later Viscount Morley), and in 1868 he became associated with the *London Economist* under the famous Walter Bagehot, as an assistant editor for the years 1868–76. During part of this period of 1873 to 1876, he contributed articles to the *Times* and the *Spectator*, and also served as city editor of the *Daily News* [11, Vol. 12 (1910), p. 4]. He joined the Board of Trade in 1876 as chief of its statistical department. In 1882 he was made assistant secretary of the board, and was also placed in charge of the commercial department. This had previously been entrusted to the foreign office, but was now made a part of the statistical department. In 1892, a third department, labour, was merged with the statistical department and Giffen was then appointed Controller-General of the Commercial, Labour and Statistical Department [57, p. 184]. Giffen was responsible for the considerable improvement of official economic statistics. In 1897, at the age of sixty, he retired from this position. He was honored in 1895 with the rank of K.C.B., after being made a C.B. in 1891 [58, p. 582].

11. FRANCIS YSIDRO EDGEWORTH (1845–1926)

Francis Ysidro Edgeworth, originally named Ysidro Francis Edgeworth, economist, editor, and statistician, is highly regarded by many as the philosopher of statistics. Some claim he is the oustanding statistician of the nineteenth century. His writings, scattered in many English and foreign periodicals, cover a wide variety of subjects, including probability, law of error, law of change, correlation, index numbers, types of averages, method of least squares, banking, prices, rates of births, deaths and marriages [57, p. 261; 8].

The decade of the 1880's marks Edgeworth's initial and strong interest in the theory of statistics. At that time six papers on the theory of probability appeared (1883–84), the first one bearing the title of "The Law of Error" and appearing in the *Philosophical Magazine* (1883). This paper was the foundation of a later paper of the same title appearing in the *Cambridge Philosophical Transactions* (1905). At the beginning of this same period in 1880, Edgeworth was appointed Lecturer in Logic at King's College. In 1885, at the Golden Jubilee of the Royal Statistical Society, he delivered a remarkable paper, "Methods of Statistics," employing ideas of leading thinkers such as Laplace, Lexis, and Venn, wherein he advanced the scientific basis for the theory of statistics [33, pp. 181–217; 57, pp. 230–1]. This epoch-making paper served to bring the calculus of probability into practical use and demonstrated that "in the apparatus for eliminating chance the most important piece of mechanism is the *law of error* or *probability* curve" [51, p. 179–80]. During this decade Edgeworth was greatly influenced by three works: Lexis' *Zur Theorie der Massener-scheinungen*, Todhunter's *History of Probability* and Venn's *Logic of Chance*. In the 1880's Edgeworth was a lonely pioneer in the somewhat unknown world of statistical theory, and it was not until around 1895 that Bowley, Pearson and Yule became his statistical companions [51, p. 180]. From 1887–90, Edgeworth was busily engaged in the study of index numbers, holding also office as secretary of the committee of Section F of the British Association for the Advancement of Science. In this scholarly work he not only examined carefully such aspects as the relative value of averages, the appropriate weights to employ, the application of the law of error, etc., but he also showed considerable interest in the several economic angles of this problem [9, 1888–1891 volumes]. In 1890 he was appointed Tooke professor of economic science and statistics at King's College, a chair distinguished for its outstanding professors, some being Reverend J. E. Thorold Rogers, Reverend William Cunningham, E. J. Urwick, and Friedrich A. von Hayek. In 1891 he resigned this chair to succeed Thorold Rogers as Drummond Professor of Political Economy at Oxford University, and served until he retired as Emeritus Professor in 1922 [10, Vol. 1922–30, pp. 284–5]. In 1892 his first paper on correlation bearing the title, "On Correlated Averages," appeared in the *Philosophical Magazine*. As to his statistical writings, Bowley points out that:

> The numerous statistical studies published between 1893 and 1926 are to a very large extent the working out of ideas expressed or latent in the papers already named, with numerous applications to a great variety of problems and with critical and explanatory references to the work of other writers. Throughout the two score papers listed for these years run the thread of the importance of sound fundamental ideas on probability in all mathematical statistics as opposed to purely empirical work [52, p. 622; 7, p. 118].

Edgeworth, regarded as "one of its most admired and trusted leaders," joined the Royal Statistical Society in 1883, served on its council, with short intervals, from 1886 to 1912, and was its president in 1912–14. The *Annals* records:

> His work for the Society lay mainly, either through papers or through contributions to Miscellanea, in the application of mathematics to the study of social and other problems. No subject was too great or too small for the use of his analysis—the theory of banking, the flow of wasps through a cycle of operations, variations in the rates of births, marriages, and deaths, chance in examinations, the rationale of exchange, psychical research were only a part of the material to which he vigorously applied his tools. . . . He was the greatest academic figure in the inner circles of the Society in the last fifty years and the most charming of friends to all those who were honoured by his acquaintance [51, pp. 237–9].

Since Bowley has so well classified and described Edgeworth's statistical writings, the reader is referred to this valuable memorial work. In the introduction Bowley states:

> In the arrangement of subjects I have endeavoured to follow the logical sequence that was always present to his mind. . . . First comes "Probability" and "Credibility," in which the philosophic basis of the whole is laid. Secondly, "The Law of Error" and the "The Method of Translation," in which the implications of the postulate of plural causation are worked out in the light of the theory of pure probability. Thirdly "Applications to Special Problems," where, cases being taken to which the theory is definitely applicable, the use of the method in measuring variations, and in distinguishing the accidental (or random) from results of direct causation ("the elimination of chance") is illustrated in many practical statistical problems. Edgeworth held very definitely the opinion that it was not sufficient to measure the variation of a statistical result simply by the statement of a standard deviation, but that it was necessary to connect this standard deviation with a law of error, to assign the probability that it (or a multiple of it) would be exceeded, and to relate this probability to credibility by the inverse method, The modulus in this use always performs this complete function. Fourthly, a short section on "The Best Mean" is mainly devoted to an explanation of Edgeworth's championship of the median, which depends on an understanding of parts of the previous sections. There follows an account of his

early contributions to the theory of correlation, so that it may be determined how far a claim for priority in its development is valid, and finally a note on his concept,on of the relations between the theories and methods of probability and of political economy [6, pp. 4–5].

Edgeworth was an active member of the British Association for the Advancement of Science, holding office as president of Section F in 1889. He was elected a Fellow of the British Academy in 1903. He was a very active member of the Royal Economic Society, serving a term as vice president. He distinguished himself, moreover, by serving first as editor, then as chairman of the editorial board, and finally as joint editor with John Maynard Keynes, of the *Economic Journal* since its first issue in March 1891. He served up to the day before his death, February 13, 1926. Under his inspiring influence, this scholarly journal achieved international prominence.

Edgeworth, born at Edgeworthstown, County Longford, Ireland, was the youngest of five sons. He was educated at home under tutors until the age of seventeen, and in 1862 attended Trinity College, Dublin, but apparently did not graduate. In 1867 he went to Oxford, and in 1868 to Balliol College, where he was awarded first class honors in Literis Humanioribus, the great school of Philosophy, the following year, although he did not receive the A.B. degree until 1873 [58 (1926), pp. 875–6]. In 1877 he obtained the M.A. degree and in the same year he was admitted to the bar, but he never practiced. He preferred to spend his time writing and lecturing [12, Vol. 5, pp. 397–8]. In 1877 he published a paper-covered volume of 92 pages, *New and Old Methods of Ethics*, being a commentary on Henry Sidgwick's *Methods of Ethics* (1874). In 1881 he published a slender volume of about 150 pages, *Mathematical Psychics: An Essay on the Application of Mathematics to the Moral Sciences*. In 1883 he wrote his first paper for the *Journal of the Statistical Society of London* (now the Royal Statistical Society), bearing the title "The Method of Ascertaining a Change in the Value of Gold." In 1884, his paper "The Philosophy of Chance" appeared in *Mind*, this being a critique of Venn's *Logic of Chance* (1883). In 1887 he published his third and last work, *Metretike, or the Method of Measuring Probability and Utility*. These three works are the only books Edgeworth produced in his life. He was apparently more inclined to write numerous articles and many book reviews, as well as to edit the *Economic Journal*, the official quarterly of the Royal Economic Society. These activities took the greater part of his time during the last thirty-five years of his life [34, p. 234; 36, p. 151]. His principal writings in economics were selected, edited and revised by Edgeworth, and published in three volumes, *Papers Relating to Political Economy*, by the Royal Economic Society in 1925. They contain thirty-four papers and seventy-five reviews. His contributions to statistical theory, embracing seventy-four papers and nine reviews are collected and arranged by Bowley in 139-page volume entitled *F. Y. Edgeworth's Contributions to Mathematical Statistics*, which the Royal Statistical Society published in 1928.

CONCLUSIONS

Many interesting features stand out in the lives and accomplishments of the men whose work has been reviewed in the preceding pages. Of the eleven leading British statisticians, six, Babbage, Edgeworth, Farr, Galton, Guy and Jevons, were college graduates, while five, Giffen, Newmarch, Playfair, Porter, and Rawson, were not. Only Edgeworth had any legal training. Two, Farr and Guy, were physicians, intensely interested in vital statistics, for one thing, to find some means of reducing the death rate. Only four, Babbage, Edgeworth, Galton, and Jevons, had post-graduate training. Not one taught a course in statistics. Only one, Farr, was connected with the national census, namely that of 1851, 1861 and 1871. Eight, Babbage, Farr, Galton, Giffen, Guy, Jevons, Newmarch, and Porter, were Fellows of the Royal Society, while three, Edgeworth, Playfair and Guy, were not. Four, Giffen, Guy, Newmarch and Rawson, were editors of the *Journal of the Statistical Society of London*, now the *Journal of the Royal Statistical Society*. Six, Edgeworth, Farr, Giffen, Guy, Newmarch and Rawson, were presidents of the Statistical Society of London, known since 1885 as the Royal Statistical Society, while five, Giffen, Guy, Jevons, Newmarch and Rawson, served as honorary secretaries. Three, Galton, Giffen and Rawson, were knighted by their government. Giffen won the Guy Gold Medal of the Royal Statistical Society, while none received the Guy Silver Medal. Rawson was for many years president of the International Institute of Statistics, founded in 1885. Finally six, Babbage, Edgeworth, Farr, Giffen, Jevons and Newmarch, were at one time president of Section F—Statistics—of the British Association for the Advancement of Science.

Three outstanding contributors to the theory of statistics were Edgeworth, Galton and Jevons; Edgeworth in the areas of probability, correlation and index numbers, Galton in the field of correlation, and Jevons in the fields of averages, index numbers, ratio chart, seasonal variation, secular trend, and crises, now known as business cycles. Edgeworth distinguished himself as editor of the *Economic Journal*. Newmarch made a significant contribution when he suggested the use of variations from the average, now known as dispersion, and particularly in his 1869 presidential address when he stated that statistics should be placed more on a mathematical basis. He thus seems to have possessed more statistical insight and foresight than most of his contemporaries. Babbage will be remembered as the founder of two important statistical organizations, Section F—Statistics—of the British Association for the Advancement of Science in 1833, and the Statistical Society of London in 1834—warmly aided by his friend Quetelet, the famous Belgian statistician. Playfair will be remembered as the founder of graphic methods in statistics, and Farr for his outstanding work in developing British vital statistics; Porter and Giffen for establishing and developing the well-known statistical department of the Board of Trade.

Finally, it appears that, even in the nineteenth century, British statisticians were making newer and more significant contributions to the theory of statistics than American statisticians, probably because of the fact that British statisticians in general had a better mathematical training.

REFERENCES

[1] *Annual Biography and Obituary*, Vol. 8, London: Longman, Hurst, Rees, Orme, and Brown, 1824.
[2] Ashton, T. S., "The Work of Stanley Jevons and Others," *The Manchester Statistical Society, 1833–1933*. London: P. S. King and Son, Ltd., 1934.

[3] Babbage, Charles, *Passages From the Life of a Philosopher*, London: Longman, Green, Longman, Roberts and Green, 1864.

[4] Babbage, Henry Prevost, *Babbage's Calculating Engines*, London: E. and F. N. Spon, 1889.

[5] Baines, Sir Jervoise Athelstane, C. S. I., "The History and Development of Statistics in Great Britain and Ireland," *The History of Statistics*, John Koren, editor, NewYork: The Macmillan Company, 1918.

[6] Bowley, A. L., *F. Y. Edgeworth's Contributions to Mathematical Statistics*, London: Royal Statistical Society, 1928.

[7] Bowley, Arthur L., "Francis Ysidro Edgeworth," *Econometrica*, 2 (1934), 113–24.

[8] Bowley, A. L., "Miscellaneous Notes; F. Y. Edgeworth," *Journal of the American Statistical Association*, 21 (1926) 224.

[9] British Association of the Advancement of Science, *Reports of*, London: various volumes.

[10] *Dictionary of National Biography*, London: Oxford University Press, various volumes.

[11] *Encyclopaedia Britannica*, Cambridge: Cambridge University Press, 1910, various volumes.

[12] *Encyclopaedia of the Social Sciences*, New York: The Macmillan Company, various volumes.

[13] *English Cyclopaedia, Division III, Biography*, London: Bradbury and Evans, various volumes.

[14] "Dr. William Farr, C.B., D.C.L., etc.," *Journal of the Statistical Society of London*, 46 (1882), 350–1.

[15] Fitzpatrick, Paul I., "Leading American Statisticians of the Nineteenth Century—II," *Journal of the American Statistical Association*, 53 (1959).

[16] Funkhouser, H. Gray, "Historical Development of the Graphical Representation of Statistical Data," *Osiris*, 3 (1937), 280–90. Also contains several photographs of Playfair's charts.

[17] Funkhouser, H. G. and Walker, Helen M., "Playfair and His Charts," *Economic History*, 3 (1935), 103–9. Also contains several photographs of Playfair's charts.

[18] Galton, Francis, *Memories of My Life*. London: Methuen and Co., 1908.

[19] "Obituary—Sir Francis Galton, D.C.L., D.Sc., F.R.S.," *Journal of the Royal Statistical Society*, 74 (1911), 314–20.

[20] "Obituary—Sir Robert Giffen, K.C.B., F.E.S., LL.D.," *Journal of the Royal Statistical Society*, 73 (1910), 529–33.

[21] "Sir Robert Giffen," *Economic Journal*, 20 (1910), 318–21.

[22] Greenwood, Major, *Medical Statistics from Graunt to Farr*, Cambridge: Cambridge University Press, 65, 69–71, 1948.

[23] Gregory, T. E., *Select Statutes, Documents and Reports Relating to British Banking, 1832–1928*, Vol. 2, Oxford: Oxford University Press, 1929.

[24] "Dr. William A. Guy, F.P.C.P., F.R.S.," *Journal of the Statistical Society of London*, 48 (1885), 650–1.

[25] Hayek, F. A., "The London School of Economics, 1895–1945," *Economica*, 13 (1946), 6–8.

[26] Hotelling, Harold, "British Statistics and Statisticians Today," *Journal of The American Statistical Association*, 25 (1930), 186–90.

[27] Humphreys, Noel A., *Vital Statistics*, London: Edward Stanford, 1885.

[28] Jevons, W. Stanley, *A Serious Fall in the Value of Gold Ascertained, and Its Social Effects Set Forth in Two Diagrams*, London: Edward Stanford, 1863.

[29] Jevons, W. Stanley, *Investigations in Currency and Finance*, H. S. Foxwell, ed., London: Macmillan and Co., 1884.

[30] Jevons, H. A., *Letters and Journal of W. Stanley Jevons*, London: Macmillan and Co., 1886.

[31] "Professor William Stanley Jevons, F.R.S.," *Journal of the Statistical Society of London*, 45 (1882), 484–7.

[32] *Journal of the Statistical Society of London*, various volumes.

[33] *Jubilee Volume*, Statistical Society of London, London: Edward Stanford, 1885.

[34] Keynes, John Maynard, *Essays in Biography*, London: Rupert Hart-Davis, 1951.

[35] Keynes, J. M., "William Stanley Jevons, 1835–1882," *Journal of the Royal Statistical Society*, 99 (1936), 523–4.

[36] Keynes, J. M., "Francis Ysidro Edgeworth," *Economic Journal*, 36 (1926), 151.

[37] Mayo Smith, Richmond, book review of Giffens' *Essay on Finance* (Second Series), *Political Science Quarterly* (1886), 337–8.

[38] Newman, James R., *The World of Mathematics*, Vol. 2, New York: Simon and Schuster, 1956.

[39] "The Death of Mr. William Newmarch, F.R.S.," *Journal of the Statistical Society of London*, 45 (1882), 115–21.

[40] Newsholme, Sir Arthur, *The Elements of Vital Statistics*, rev. ed., London: Allen and Unwin, 1923.

[41] North, S. N. D., "Seventy-Five Years of Progress in Statistics," *The History of Statistics*, John Koren, Ed. New York: The Macmillan Company, 1918.

[42] Palgrave's *Dictionary of Political Economy*, London: Macmillan and Co., 1926, various volumes.

[43] Pearson, Karl, *Life, Letters and Labours of Francis Galton*, Vol. 3A, Cambridge: Cambridge University Press, 1924 and 1930.

[44] Pearson, Karl, *Francis Galton, 1822–1922—A Centenary Appraisal*, Cambridge: Cambridge University Press, 1922.

[45] "Sir Rawson W. Rawson," *Journal of the Royal Statistical Society.* 62 (1899), 677–9.

[46] "Sir Rawson Rawson, K.C.M.G.," *Economic Journal*, 9 (1899), 665–6.

[47] "Report of the Council," *Journal of the Statistical Society of London*, 16 (1853), 97–8.

[48] Royal Society of London, *Proceedings of*, London: various volumes.

[49] Royal Statistical Society, *Catalogue of the Library of the Royal Statistical Society*, London: The Royal Statistical Society, 1908.

[50] Royal Statistical Society, *Catalogue of the Library of the Royal Statistical Society*, London: The Royal Statistical Society, 1921.

[51] Royal Statistical Society, *Annals of the Royal Statistical Society, 1834–1934*, London: The Royal Statistical Society, 1934.

[52] Spiegel, Henry W., "Bowley on Edgeworth," *The Development of Economic Thought.* New York: Wiley, 1952.

[53] Spiegel, Henry W., "Keynes on Jevons," *The Development of Economic Thought*, New York: Wiley, 1952.

[54] "The Statistical Dinner Club, 1839–1939," *Journal of the Royal Statistical Society*, 102 (1939), 292–4.

[55] Walker, Helen M., *Studies in the History of Statistical Method*, Baltimore: Williams and Wilkins, 1931.

[56] Wenzlick, Roy, *William Playfair and His Charts*, privately printed, 1950.

[57] Westergaard, Harald, *Contributions to the History of Statistics*, London: P. S. King and Son, Ltd., 1932.

[58] *Who's Who (1904)*, London: Adam and Charles Black, Ltd., various volumes.

[59] *Who Was Who (1897–1916)*, Vol. I, London: Adam and Charles Black.

Notes on the History of Quantification in Sociology— Trends, Sources and Problems

*By Paul F. Lazarsfeld**
(1961)

INTRODUCTION

THE three major nouns in the title of this paper are necessarily vague. Quantification in the social sciences includes mere counting, the development of classificatory dimensions and the systematic use of "social symptoms" as well as mathematical models and an axiomatic theory of measurement. The notion of history is ambiguous because some of these techniques evolved several hundred years ago while others were developed within the last few decades. Finally, there is no precise line between sociology and other social sciences; with the economist, the sociologist shares family budgets, and with the psychologist he makes the study of attitudes a joint concern.

The task of sketching out the history of quantification in sociology is made more difficult by the fact that it rarely has been seriously attempted. Both the history and the philosophy of science have been concerned almost exclusively with the natural sciences. Their discoveries have been linked step by step with their antecedents; their relation to the political, social and religious events of the time has been spelled out; even their effect on belles lettres has been traced. The few comparable studies in the social sciences have usually been concerned with broad, semi-philosophical systems. There has been hardly any work on the history of techniques for social science investigation. In following some of these procedures back to their origins, it was often necessary to draw attention to historical situations or to men with whom the American reader is not likely to be familiar, and to report something about the broader political and ideological contexts in which the pioneers of sociological quantification worked.

The need for such details required a severe restriction in the scope of the paper. Actually, it deals with only three major episodes. They were selected because they carried the seeds of many subsequent developments and foreshadowed discussions which continue today. To give the three major sections of this paper a proper frame, a few words are needed about how a future history of quantification in sociology might look. It would begin with a preparatory phase lasting approximately from the middle of the 17th to the beginning

* Columbia University

of the 19th century. These first 150 years were dominated by the sheer difficulty of obtaining numerical information on social topics. Many historians
of statistics and demography have described this period, and I shall not try
to re-trace the ground which they have covered. I shall, instead, suggest points
at which sociological ideas entered into the work of some of the more famous
writers on society of the period. But my main attention will be focused on
the life, work and followers of one man—Hermann Conring. As I slowly
pieced together whatever information I could find about him, I became increasingly impressed with his importance. He saw the same problem faced by
his British contemporaries whom we remember today as the founders of political arithmetic. But his efforts took a very different turn. The first section
of my paper sketches his work, tries to explain it in the context of the times,
and traces its consequences.

A second period in this history begins with the work of the Belgian, Quetelet, and the Frenchman, LePlay. Both men started out as natural scientists, acquiring their interest in the social sciences during the period of social
unrest which culminated in the French-Belgian Revolution of 1830. Quetelet
was an astronomer who wished to uncover for the social world eternal laws
similar to those he dealt with in his main field of investigation. LePlay was a
mining engineer and metallurgist who believed that the minute attention to
concrete details which made him a success in his main occupation could also
provide the foundations for a true social science. The spirit in which these
two men worked and the role their ideas played in subsequent developments
are correspondingly different.

Quetelet concerned himself almost exclusively with the interpretation of
large-scale statistics which became available, at the beginning of the 19th
century, as a by-product of the rapidly expanded census activities undertaken
by various government agencies. He anticipated with varying degrees of
precision many basic concepts of quantification, and his writings led to sophisticated controversies which continued into the 20th century. It seemed to me
useful therefore to single out some of these ideas and to show how they were
slowly clarified. Section II thus will report the Quetelet story in reverse. My
implicit starting point will be some modern ideas on quantification, and I shall
trace these back to the writings of Quetelet, his opponents and his commentators.

It is more difficult to fit LePlay into my narrative because, in spite of his
many assertions, he never was nor really meant to be a detached scientist.
During his lifetime he created a number of ideological movements, and these
both attracted and distracted his followers. Those of his disciples who intended
primarily to develop his ideas on social research could never free themselves
entirely from the political position of the founding father. This was so in a
two-fold sense. They continued the organizational activities which LePlay
initiated, and had an interestingly ambivalent attitude toward the methods
which he had developed. They succeeded in making considerable improvements in these methods, but they experienced such achievements as impious
disloyalty to their great master. The history of the LePlay school following his
death is a curious example of what happens when a research tradition assumes

a sectarian form. So far as I know, this story has never been traced, and in Section III I shall give it somewhat more attention than strict adherence to the topic of quantification would require.

My paper thus concentrates on the development of some of the basic notions and broader ideas which introduced quantification into the study of social affairs. Any history of science must include at least three elements: the intrinsic intellectual nature of the ideas, their historical social context and the peculiarities of the men who made the major contributions. It will become obvious why I give most space to the intellectual element in Section II, the historical in Section I, and the biographical in Section III.

Throughout the paper, I have had to discuss repeatedly the historiography of the field itself. Because professional historians of science have paid so little attention to the social sciences, their history was often written by amateurs— specialists in social research who only occasionally looked into its past. As a result, quite a number of legends have been passed on from one to the next. While I am an amateur myself, I have for the major parts of this paper gone back to the original sources, including the commentators who previously have given them careful attention. At points where I felt that a further pursuit would exceed the time or the material available to me, I have brought specific unsolved questions to the attention of the reader. In a postscriptum to this paper, I shall indicate the topics I have not dealt with, my reasons for their exclusion and the places where one can find pertinent information.†

THE PREPARATORY PERIOD

The Political Arithmeticians

The idea that social topics could be subjected to quantitative analysis acquired prominence in the first part of the 17th century. There are conventional explanations for this emergence: the rational spirit of rising capitalism; the intellectual climate of the Baconian era; the desire to imitate the first major success of the natural sciences; the increasing size of different countries which necessitated a more impersonal and abstract basis for public administration. More specifically, one can point to concrete concerns: the rise of insurance systems which required a firmer numerical foundation, and the prevailing belief of the mercantilists that size of population was a crucial factor in the power and wealth of the state.[1]

Problems of demographic enumeration were the first topics to be discussed systematically. No reliable data were available, and no modern census machinery was in sight. Two obstacles are mentioned by the authors who have dealt with the work of this period: the unwillingness of the population to give information, because of their fear of increased taxes; and the tendency of governments, whenever statistical information was available, to treat it as highly classified, because of its possible military value.[2] Thus, the ingenuity

---

† In the bibliographical footnotes, foreign titles are given in their original form; in the main text these titles are usually translated and abbreviated so as to support the narrative.

[1] For an instructive survey and a new look see Trevor-Roper, "The General Crisis of the 17th Century" in *Past and Present*, 1959, *16*.

[2] Secrecy regarding statistical information collected by government agencies was maintained by some countries well into the 19th

of early scholars was directed mainly toward obtaining estimates of population size and age and sex distributions from meager and indirect evidence. Multiplying the number of chimneys by an assumed average family size or inferring the age structure of the population from registered information regarding age at the time of death were typical procedures in what was then called political arithmetic.

Today it is hard for us to imagine the lack of descriptive information available in the middle of the 17th century. The ravages of periodic outbreaks of the plague, for instance, made it impossible for anyone to know whether the population of England was increasing or decreasing. As a matter of fact, the first mortality tables, published in 1662 by Graunt, who is considered the originator of modern demography, were based partly on public listings of burials; they had acquired news value for the average citizen—somewhat comparable to the list of victims which nowadays is published after an airplane accident.

But soon the supply of facts increased, the analytical techniques were improved, and by about 1680 the art of "political arithmetic" was well established under English leadership. I have the impression that something like a community of *aficionados* developed: all over Western Europe, empirical data were traded for mathematical advice. Thus, for instance, in 1693, the English astronomer Halley published a paper on mortality based on registration figures of births and funerals in the city of Breslau. How had Halley obtained these figures? German historians discovered that Leibnitz was an intermediary. He had learned about the material through a friend, a clergyman in Breslau who, together with a local physician, was an ardent and capable amateur demographer. Leibnitz brought the data to the attention of the Royal Society, which asked Halley to express an opinion.[3]

A century later natural scientists still considered descriptive social statistics appropriate topics to be worked on. In 1791, Lavoisier, the chemist who was to be guillotined three years later, published a treatise for the National Assembly dealing with the population and economic condition of France; in this he expounded the idea that a revolutionary government had the opportunity and the duty to establish a central statistical bureau. At the beginning of the 19th century, the mathematicians, LaPlace and Fourier, dealt with population statistics; and, as we shall see, their work played an important role in Quetelet's life.[4]

I do not intend to pursue the development of political arithmetic in this paper. But I want to suggest that the sociological implications of some of these early writings be reexamined. To indicate the kind of analysis I have in mind,

---

century. The parallel to contemporary secrecy about atomic physics is obvious. Several revolutionary governments made it a point that statistical data should be made available to the public. I wonder whether the explicit mentioning of the decennial census in the United States Constitution had partly such ideological implications.

[3] This episode is interesting, incidentally, because it shows the efforts which even minor political arithmeticians made to put their work to practical use. The Breslau group wanted to counter the contention of astrologists that certain years in a man's life are especially dangerous. The pertinent historical papers are reviewed by Victor John, *Geschichte der Statistik* (Stuttgart: Ferdinand Enke, 1884).

[4] The facts mentioned up to this point can be found in any of the histories of statistics mentioned in the footnotes.

I shall briefly sketch out the work of two men. One of these is William Petty (1623-1687) who worked with Graunt and created the term "political arithmetic." After the Restoration, he decided to use his experience in Ireland to formulate a general theory of government based on concrete knowledge. He was convinced that to this end "one had to express oneself in terms of number, weight, and measure." He argued that Ireland was a good case study, not only because he knew it so well, but because it was a "political animal who is scarce twenty years old," and a place, therefore, where the relation between the social structure of the country and the chances of good government could be studied more closely. Thus originated his *Political Anatomy of Ireland* (1672).

A sensitive biography by E. Strauss[5] describes the political and social settings of Petty's work: for this reason alone it makes very worthwhile reading. Two chapters provide a more detailed guide through those parts of Petty's writings which are relevant to my paper. He anticipates ideas which only recently have been considered noteworthy intellectual discoveries. A few years ago, the Harvard economist Dusenberry argued that sociological factors must be taken into account in the economics of saving. He pointed out that whites save less than Negroes on the same income level, because white people have a broader range of social contacts and therefore must spend more money on conspicuous consumption. Compare this with the following passage from Petty:

> When England shall be thicker peopled, in the manner before described, *the very same people shall then spend more, than when they lived* more sordidly and inurbanely, and further asunder, and *more out of the sight, observation, and emulation of each other;* every man desiring to put on better apparel when he appears in company, than when he has no occasion to be seen.[6] (emphasis mine)

During the depression of the 1930's, a number of studies appearing in this country and abroad made it clear that, for psychological reasons, work relief is preferable to a straight dole. Petty also argued for unemployment benefits, and, while he phrases his beliefs in less humanitarian words than we would use today, the psychological foundations of his argument that even "boondoggling" is preferable to dole are certainly very modern:

> tis no matter if it be *employed to build a useless pyramid upon Salisbury Plain, bring the stones at Stonehenge to Tower Hill, or the like;* for at worst this would keep their minds to discipline and obedience, and their bodies to patience of more profitable labours when need shall require it.[7] (emphasis mine)

The second exhibit in my appeal for a sociological reconsideration of some of the early Political Arithmeticians is the German J. P. Suessmilch (1707-1767), who too first studied medicine. But he then turned to theology, and spent most of his adult life as a pastor, first with a Prussian regiment and later at the court of Frederick the Great. In 1741, a year after Frederick II ascended the throne, Suessmilch published his book on the "Divine Order as

---

[5] E. Strauss, *Sir William Petty* (Glencoe, Ill.: The Free Press, 1954).

[6] Strauss, *op. cit.*, p. 203.
[7] Strauss, *op. cit.*, p. 221.

proven by birth, death and fertility of the human species (Geschlecht)."[8] In his work, Suessmilch collected all the data published by his predecessors, and his book is considered the most complete compendium of the time. In addition, historians of statistics credit him with having been the first to focus attention on fertility (in addition to birth and death rates). But all of these reviews omit any reference to Suessmilch's broad-gauged interpretation of his findings. For instance, Westergaard[9] says at one point when describing Suessmilch's major work, "the succeeding chapters are uninteresting to the history of statistics insofar as Suessmilch here briefly presents arguments against polygamy, discusses proposals as to supporting married couples with numbers of children, as to hygienic matters, luxury, etc. After this long digression, the author resumes his statistical investigations."

Even a cursory look at this part of Suessmilch's text shows that much would be gained by a careful examination. For example, when he finds that the number of marriages has declined in a certain part of Prussia, he offers a variety of explanations: an increase in the number of students attending universities, of people called into military service, a shift to industrial work, increases in food prices, and so on. All in all, the *Goettliche Ordnung* is filled with social analysis. Suessmilch considers a growing population of crucial political importance; he therefore tries to uncover the political and social conditions which make for such growth, so that he can advise the king effectively.

It is true that Suessmilch frequently turns to theological arguments. He finds that slightly more boys than girls are born; he attributes this to the wisdom of the Creator, because young boys, who grow up under less sheltered conditions, have a somewhat higher mortality rate than do girls. At the time of marriage, the two sexes are in balance. At the time of widowhood, however, there are more women than men; but widowers have a greater chance of remarrying repeatedly (he created the term "successive polygamy"); in other words, even in the later phases of life, the sex ratio is functionally useful. Altogether, one could probably find surprising parallels between modern functionalism and Suessmilch's efforts.[10]

Having done no special research myself on the political arithmeticians, I feel somewhat hesitant about making one more suggestion before I leave the

[8] Suessmilch had become interested in demography through reading the work of William Derham, an English cleric, whose book, "Psycho-Theology, or a Demonstration of the Being and Attributes of God from His Works of Creation," had already gone through several editions by the beginning of the 18th century.

[9] Harald Westergaard, *Contributions to the History of Statistics* (London: P. S. King & Son, Ltd., 1932).

[10] It has often been noted that, in spite of the fact that Suessmilch's was the first serious discussion of the relation between standard of living and population growth, he had no direct intellectual effects or followers. Malthus, whose work did not begin until fifty years later (and who, incidentally, used many of Suessmilch's computations), received all the acclaim. In a later context, I shall try to ex-

plain this neglect of the Prussian pastor by his academic contemporaries. I have found only one English summary of Suessmilch's work which goes beyond the conventional histories of statistics, a dissertation by F. S. Crumm, "The Statistical Work of Suessmilch," in *Quart. Publ. Amer. Statist. Ass.,* 1901. It is a rather dry, but very specific and therefore useful, guide through Suessmilch's main writings. In the Festschrift for Toennies' 80th birthday, one contribution by a Georg Jahn is entitled, "Suessmilch and the Social Sciences of the 18th Century." It is very disappointing. The author gives a brief description of Suessmilch's work and expresses the hope that sociologists will one day pay more attention to it. Jahn's own contributions are some remarks on how Suessmilch fitted in with the rational theology of his period.

topic. Given the fact that political arithmetic, especially in its early phases, was equivalent to obtaining a quantitative foundation for broad social problems, it is surprising that it seems to have had so little relation with another stream in English intellectual history—the Scottish moral philosophers. Some of them, like Adam Ferguson, are cited as the precursors of modern empiricism, mainly because they wanted to substitute concrete anthropological observations for mere speculation about the origins of society.[11] But they were also much concerned with human nature. And, for these concerns, they could have derived much information from the work of the political arithmeticians, which was well developed by the middle of the 18th century. Yet, I have not been able to find a study of the points at which these two traditions merged, and, if they did not, an explanation of what accounts for this separation.

The Story of the Two Roots

In 1886, August Meitzen, a professor at the University of Berlin, published a book on statistics.[12] The first part of the volume dealt with the history of the field. It contains no detailed analysis of specific writings, nor does it pretend to be a history of ideas. Meitzen's main aims were to record the times and circumstances under which the early statistical organizations were founded, to list and describe the early publications of statistical data, and to provide brief sketches of the major writers who made notable contributions. The book apparently seemed important at the time, for in 1891 the American Academy of Political and Social Science published an English translation in two supplements to their regular series. (The second part contains sound advice on the collection, tabulation, and interpretation of demographic and social data, which might explain why the translation seemed desirable.)

In describing the main historical trends, Meitzen put forth the idea that the statistics of his time developed from two different roots. One was represented by the political arithmeticians whom I have just described. He correctly places their origins in the middle of the 17th century. The other root was an intensive interest in characteristic features of the state, from which the term "statistics" was derived. This brand of statistics considered anything which seemed noteworthy about a country, and was in no way restricted to the topics now covered by the term. As a matter of fact, numerical data played only a small role in this tradition.

Sometime toward the end of the 18th century, Meitzen writes, the English root of political arithmetic and the German root of university statistics (as it came to be called) became involved in a controversy about which of the two was more scientific and more useful. The battle was won, in Germany as well as elsewhere, by the political arithmeticians. From the beginning of the 19th century onwards, they also monopolized the title of statisticians. Whatever was left of the former activities of university statisticians was thereafter considered a part of political science.

[11] Herta Jogland, *Ursprünge und Grundlagen der Soziologie bei Adam Ferguson* (Berlin, Duncker & Humblot, 1959). The author informs me that the first use of statistical data she could trace with a moral philosopher is in a book by a George Combe, *Moral Philosophy or the Duties of Man* (New York: 1841).

[12] August Meitzen, *History, Theory, and Technique of Statistics* (Phila.: Am. Academy of Political and Social Science, 1891).

Meitzen designated a Göttingen professor, Gottfried Achenwall (1719-1772), as the founder of the German non-quantitative root. In 1749, Achenwall had published a book which included in its title the phrase, "The Science of Today's Main European Realms and Republics." He is, says Meitzen, *"therefore called the father of statistical science."* This paternity has been accepted by many contemporary writers. Thus, George Lundberg follows Meitzen's example.[13] While his sympathies, of course, are on the side of the victorious party, he acknowledges that Achenwall's book "gained such general recognition when it was translated into all languages that Achenwall was long hailed as the father of statistical science." George Sarton[14] and Nathan Glazer[15] write in the same vein.

There is something strange in this story, however. What happened in Germany during the ninety years between Graunt and Achenwall? Why did the political arithmeticians have so little influence in the German universities? How could the Göttingen professor create a "second root" so quickly that, within a few decades, it acquired equal standing with the English tradition to the point that, as we shall see, the final battle was fought in all the countries of Western Europe?

Part of the answer was given in another book which actually appeared two years before Meitzen's but which did not come to the attention of American social scientists for a long time. In 1884, Victor John, docent at the University of Bern in Switzerland, also published a history of statistics, but of a very different kind. He properly calls his work a source book,[16] and it is indeed a volume of remarkable scholarship in which he either summarizes or quotes sources that today are quite inaccessible.[17] He had noticed the queer hiatus between 1660 and 1750, left unexplained by traditional stereotypes. He was able to fill in this gap by focusing attention on the work of Hermann Conring (1606-1682). Conring was one of the great polyhistors of his time, holding three professorships at the Brunswick University of Helmstaedt: first in philosophy, then in medicine, and finally in politics. In 1660, at practically the same time that Graunt and Petty started their work, he began a series of systematic lectures under the title, "Notitia Rerum Publicarum." These lectures were first published as notes taken by his students, but later appeared, at the beginning of the 18th century, as part of a multi-volume collection of Conring's writings and correspondence. He had a large number of important students, some of whom went into public service and some of whom taught at other German universities. Many compendia of his system were in use throughout the Empire. And as John proves convincingly, the book which Achenwall published in 1749 was essentially the first systematic presentation in the *German* language of the Conring tradition which until then was available

[13] George A. Lundberg, "Statistics in Modern Social Thought," *Contemporary Social Theory*, Barnes and Becker (eds.), (New York: D. Appleton-Century, 1940).

[14] George Sarton, "Preface to Vol. XXIII of *Isis* (Quetelet)," *Isis*, 1935, *65*: 6-24.

[15] Nathan Glazer, "The Rise of Social Research in Europe" in Daniel Lerner (ed.) *The Human Meaning of the Social Sciences* (New York: Meridian Books, 1959).

[16] Victor John, *op. cit.*

[17] John's work grew out of Knapp's broad interest in the history of statistics, to which I shall return in the section on Quetelet. Meitzen himself, who started the original confusion, corrected himself later. He wrote the entry on Conring in the influential German Encyclopedia of the Social Sciences.

almost exclusively in Latin. John demonstrates that all of Achenwall's basic ideas had been explicitly developed by Conring. As a matter of fact, Achenwall always conceded this: he was a pupil once removed of Conring's, and wrote his dissertation about him, still in Latin, incidentally.

Thus, John clarifies at least one point. There was no hiatus. The English root of political arithmetic and the German root of university statistics developed at the same time. But this fact only raises a number of new and more interesting questions. For one, why did the two countries develop such different answers to what was essentially the same intellectual challenge? Compare the programmatic statements of the two authors:

Petty	Conring
Sir Francis Bacon, in his "Advancement of Learning," has made a judicious parallel in many particulars, between the *Body Natural and Body Politic, and between the arts of preserving both in health and strength:* and as its anatomy is the best foundation of one, so also of the other: and that to practice upon the politic, *without knowing the symmetry, fabric, and proportion of it, is as casual as the practice of old women and empirics.* (Emphasis supplied)[18]	Just as it is impossible for the doctor to give advice for the *recovery or preservation of health when he does not have some salient knowledge of the body,* so it is also impossible for anyone who does not have *knowledge and awareness of the facts of public life to cure it* either in its totality or in some of its parts.[19] (Emphasis supplied)

Here were two men, equally concerned with problems of government, both trained in medicine and intermittently acting and thinking as physicians, both bent on finding empirical foundations for their ideas. And yet they took two completely different roads. The Englishman, citizen of an empire, looked for causal relations between quantitative variables. The German, subject of one of 300 small principalities and, as we shall see, involved in the petty policies of many of them, tried to derive systematically the best set of categories by which a state could be characterized. As I sketch out a tentative explanation of this difference, a collateral problem will arise. Once the German tradition was established, why was, it so impermeable and, in the end, even so hostile to the quantitative tradition of the political arithmeticians? This second question is highlighted by the fact that 100 years later Suessmilch, writing in German at the same time as Achenwall, made no dent on the work of the German professors. And, finally, what was it in the European intellectual scene which kept Conring from having any international influence as a statistician when, at the time of the great battle 150 years later, German university statistics and English political arithmetic were pretenders of equal strength and prestige throughout Western Europe?[20]

[18] Strauss, *op. cit.,* p. 196.

[19] John, *op. cit.,* p. 58, here translated from the Latin quotation.

[20] Anyone who wishes to trace the theory of Conring in English studies will encounter difficulties. Both Frank Hankins and Westergaard agree with John's views that Conring is to be considered the founder of the German tradition, yet both men devote a mere paragraph to Conring and several pages to Achenwall. They might have had difficulties with John's review which always quotes Conring in Latin, and without translation, a tradition incidentally which continues in the more recent

Hermann Conring and German Universitätsstatistik

It is difficult to visualize intellectual life in Germany in the decades following the Thirty Years' War. The educated layman knows the connection between Locke and the Glorious Revolution; he is aware of the brilliance of the French theater when Louis XIV established his hegemony in Europe through a series of wars. But who remembers that at about the same time (1683) the imperial city of Vienna was besieged by the Turks and saved in the nick of time by a Polish army—except, perhaps, if he has heard that this brought coffee to Europe and thus greatly affected the intellectual life of London?

The physical devastation of large parts of Germany at the time of the Westphalian Peace (1648) and the drastic decline in the population would lead one to expect a complete blackout of the mind. And yet, recovery had to come; and indeed, it did come, but it gave German intellectual activities a peculiar complexion. First of all, there was abject poverty. It is true that both Petty and Conring were involved in shady financial transactions, but the differing social contexts affected the scale of their misdeeds. Petty dealt in huge fortunes, while Conring's correspondence resounds with begging and waiting for a few gold pieces promised to him for some political service. In the two-score German universities, professors remained largely unpaid and the life of the students was at an all-time low. No middle class existed, no intellectual center, no national aristocracy which might have supported the work of artists and scientists. Whatever help there was had to come from the three hundred princes who ruled their ruined little countries with absolute power.

These princes were, by and large, all concerned with the same problems. One was to maintain their independence against the Emperor whose power had been greatly weakened by the great war. At the same time, there was enough national feeling to make a possible invasion by France a common concern, especially in the western part of the realm. The relation between Catholics and Protestants, although formally settled by the peace treaty, was still very much in flux. And finally, the jealousies and battles for prestige between the various courts kept everyone on the move. This competition is especially relevant for my narrative. As one could expect in as many as three hundred principalities, there was a typical distribution of ability and culture among the rulers; for the better kinds, the advancement of knowledge and the improvement of government was a serious concern or at least an important competitive tool. It may sound somewhat strange to compare this situation with what happened in the Italian city states during the Renaissance. There was, of course, little money and probably less taste to build palaces, to paint murals, or to collect sculptures. But people were cheap. Therefore, if a man acquired some intellectual renown, several courts might bid for his services. The natural habitat for an intellectual was still the university corporation; so a relatively large number of small universities were created (and disappeared) according to the whims of some of the more enlightened rulers.

publications. The original writings of Conring were not available to me. But the German sources give many and often overlapping quotations; thus frequent internal checks were possible.

But again, what was expected from the learned doctors was colored by the peculiar situation. All in all, the critical German problem of the time was civic reconstruction. Problems of law and of administration had high priority. The competition between the principalities pressed in the same direction: the struggle over a little piece of territory; the question of which prince should have which function in imperial administration; questions of marriage and succession among ruling houses were discussed and settled in the light of precedent, and by the exegesis of historical records. International law started a few miles from everyone's house or place of business. No wonder then that it was a spirit of systematically cataloguing what existed, rather than the making of new discoveries that made for academic prestige. This, in turn, prolonged the life of the Scholastic and Aristotelian traditions which had dominated the medieval universities and by then had withered away in most other parts of Western Europe. In the second half of the 17th century, when English and French intellectuals wrote and taught in their own language, the form of communication among German academicians was still exclusively Latin, even in the numerous "position papers" they were asked to publish by the princes whom they served.

The life and work of Hermann Corning must be examined against this background. He was born in 1606 in Friesland, son of a Protestant pastor. His gifts were soon recognized and he was taken into the house of a professor at the University of Helmstaedt. This town belonged at the time to the Duchy of Brunswick which in turn was part of the general sphere of influence of the Hanoverian Duchy.[21] In 1625, Conring went to the University of Leyden, probably because his family still had many connections with the Netherlands from which they had originally come. (Four years before, Grotius, who was also a student at Leyden, had fled his native country because he had been involved on the liberal side of religious and political controversy.) The best source on Conring's study period in Holland is Moeller.[22] Conring stayed in Leyden for six years and was greatly attracted by the breadth of its intellectual life. He tried very hard to get a Dutch professorship, but did not succeed. So in 1631, he returned to Helmstaedt as professor of natural philosophy and remained there for the rest of his life.

From what we know about him today, we can conclude that he would undoubtedly have become a European figure like Grotius who was twenty years his elder and Leibnitz who was forty years his junior, if like them, he had spent part of his mature life outside Germany. I have no space here to document in detail what is known about Conring as a person. Instead, I shall briefly compare him with these two men with whom we are so much more familiar. Conring never met Grotius, but he came back to Helmstaedt imbued with his ideas. Whenever it was not too dangerous for him, he stood up for religious tolerance and if possible, for reunification of all Christian churches. In many respects the two men seem to have been similar, as may be seen, for

---

[21] It should be remembered that not long after the death of Conring in 1681, Hannover and Brandenburg were competing for top prestige in Germany. When the Brandenburg Elector acquired the title of Prussian King by a legal trick, the Hannoverian tried to balance this success by accepting somewhat reluctantly, the crown of England.

[22] Ernst V. Moeller, *Hermann Conring* (Hannover: Ernst Geibel, 1915).

example, from a character sketch of Grotius by Huizinga.[23] Huizinga described how Grotius was "permeated in every fiber with the essence of antiquity": that in his writings on public affairs he mixed contemporary cases and "examples from antiquity in order to give advice to his own day"; that his knowledge in all spheres of learning was so great that "the capacity and the alertness of the humanist memory has become almost inconceivable." Practically the same terms are used of Conring by his contemporaries and by the historians who tried to reconstruct his image from his very extensive correspondence. The only difference is one of morality. All authorities on Conring agree on his veniality and servility, although some point out that this was characteristic of German academicians of the time.

Leibnitz and Conring had repeated personal contact. A brief summary of how this came about will bring the description of Conring's life one step further. From early in his Helmstaedt career, Conring lectured on politics although officially he was made professor of this subject only in 1650. One of his students was the young Baron J. C. Boineburg (1622-1672) who defended the dissertation he had written under Conring in 1643. A few years later, after having changed his religion, Boineburg entered the service of the Archbishop Elector of Mainz, who was one of the leading rulers of Germany in the period after the Westphalian Peace; and Boineburg became possibly the most prominent German statesman of the time. The crucial problem of the principalities in the Rhine area was how to contain the power of Louis XIV. The Mainz Elector changed his position, being first in favor of appeasement and then in favor of organizing defensive alliances with Protestant countries. Boineburg remained an appeaser throughout his political career which brought him repeatedly into conflict with his own prince, but at the same time made him an important bridge between Germany and the West. Whenever he had to make a political decision, he turned to his old teacher for advice. Many times he asked Conring to write pamphlets on current issues to support a specific position. Boineburg was also very much concerned with the possibility of reunifying the Christian churches which, incidentally, was also Grotius' main interest towards the end of his life (1645). At one time, Boineburg suggested an exchange of statements by relatively conciliatory representatives of the Catholic and Protestant positions. Again, he called upon Conring who wrote a number of monographs from the Protestant point of view; the whole affair came to nothing. These "expert opinions," always based on extensive historical and legal research, are one of the sources of knowledge about Conring. An instructive and well organized inventory can be found in Felberg.[24]

At the same time that Boineburg turned to his teacher either for advice or for backing of his various political schemes, he was on the outlook for other intellectuals he could use for similar purposes. His attention was drawn to a young Saxonian who, at the age of 24, had written a treatise on "A New Method of Teaching Jurisprudence" (Methodus Nova), a work in which he wanted to apply to legal studies and research the same ideas which Bacon had

---

[23] Johann Huizinga, *Men and Ideas* (New York: Meridian Books, 1959).

[24] Paul Felberg, *Hermann Conrings Anteil* *am Politischen Leben Seiner Zeit* (Trier: Paulinus-Druckerei, 1931).

sketched in his Novum Organum for the natural sciences. Boineburg prevailed upon Leibnitz to enter the service of the Elector of Mainz; he accepted, and remained in this position for ten years—from 1666 to 1676. Boineburg wanted to be sure of his judgment and sent Leibnitz' drafts to Conring.[25] The latter was not overly impressed, but Boineburg retained Leibnitz nonetheless. As a result, Leibnitz and Conring came into continuous contact with each other. Both were Protestants, and the one was directly, the other indirectly, attached to a Catholic court. Boineburg called on Conring for informal or formal expression of expert opinion. Leibnitz he used more as a personal representative. For four years, beginning in 1672, Leibnitz was Boineburg's representative in Paris, a stay which was interrupted only by a brief visit to London. This was the period in which Leibnitz made contacts with French and British academicians and laid the foundations for his international fame. The only return Conring got out of it was a small French pension for which he was expected to contribute to the fame of Louis XIV in his public writings.

Leibnitz never completely escaped Conring's shadow. When he moved to Hannover in 1676 to become court librarian, a position he held until the end of his life (1716), his administrative superior was a man who again had been a student of Conring.[26] Between Conring and Leibnitz an atmosphere of mutual respect and ambivalence prevailed. The former was probably jealous of the rising fame of this new, and last, German polyhistor. He may also have had the feeling that he had been born a few decades too early. When he was thirty, Germany was still in the middle of the great war, and the main problem of a German professor was to keep out of the hands of various occupying armies and to clothe and feed his family. When Leibnitz was thirty, the agent of a German statesman could make trips all over Europe and take active part in public affairs. By then, Conring was old; and while he was famous, his participation in the recovery period was restricted to the written word.

Conring's isolation during a crucial intellectual period may also explain a final disagreement between Conring and Leibnitz. By the middle of the 1670's, Leibnitz was already deeply involved in mathematical studies and his reputation in this field, however controversial, was at least as great as his reputation as a social scientist. Conring thoroughly disapproved of all such mathematical ideas and advised Leibnitz not to waste his time on them.[27] This blind spot in Conring cannot, however, completely be attributed to Conring's age. All in all, he seems to have had only a limited understanding of mathematical thinking. A careful and systematic compilation of everything which Conring wrote on population problems[28] shows that he had a static view of the problems

---

[25] It is possible to trace all this in some detail in Guhrauer's biography of Leibnitz. (G. C. Guhrauer, *Gottfried Wilhelm Freiherr V. Leibnitz* [Breslau: Ferdinand Hirt's Verlag, 1846].) This two-volume book puts great emphasis on Leibnitz' personal contacts and either quotes directly from correspondence or gives at least references as to where further information can be found. It is, of course, written with Leibnitz as the central figure and therefore the many allusions to Conring are

often brief. It should be worthwhile to follow up Guhrauer's references and to piece the picture together with Conring in mind. Among the monographs on Conring, I have not found one with this emphasis.

[26] Guhrauer, *op. cit.*, p. 212.

[27] Guhrauer, *op. cit.*, pp. 213f.

[28] Reinold Zehrfeld, "Hermann Conring's Staatenkunde" in *Sozialwissenschaftliche Forschungen*, 1925, 5: 79ff.

involved. While interested in studying size and social structure of populations in relation to public policy, he had no conception of or interest in birth and death rates or any of the other dynamic ideas so characteristic of contemporary British political arithmeticians.[29] Whether further research could explain these blind spots, I cannot tell. One rather obvious root is Conring's continuous concern with Aristotle's writings. They were the topic of his dissertation; later, he published many commentaries on and new editions of Aristotle's political texts; his own work, which created the tradition of German university statistics, was deeply influenced by Aristotelian ideas. I feel it was necessary to sketch this general background before I could turn to this part of Conring's efforts.

Conring wants to bring order into the available knowledge about various countries. His purpose is explicitly threefold: he looks for a system which should make facts easier to remember, easier to teach, and easier to be used by men in the government. To this end it is necessary to have categories of description which are not accidental, but are deduced, step by step, from basic principles. His "model," as we would say today, is *the state as an acting unit*. The dominant categories are the four Aristotelian causae. His system is consequently organized under four aspects.

The state as an acting body has a goal or *causa finalis*. The second aspect is a *causa materialis* under which Conring subsumes the knowledge of people and of economic goods. The *causa formalis* is the constitution and the laws of a country. The *causa efficiens* is its concrete administration and the activities of its elite. Under each of these main categories, Conring systematically makes further subdivisions. The causa efficiens, for example, describes the concrete ways by which the state is governed. They are either principales or instrumentales. The former are the statesmen themselves; the latter are again subdivided into animatae (staff) and inanimatae. At the point where he has arrived at a "causa efficiens instrumentalis inanimata," his main example is money. And under this rubric he then develops elaborate monetary ideas which, I gather, were quite advanced for his time.

One should notice, behind this forbidding terminology, a number of very modern topics. Thus, contemporary social theory is much concerned with the goals (causa finalis) and subgoals of organizations, their possible conflict and the duty of the "peak coordinator" to attempt their integration. The distinction between causa formalis and efficiens corresponds almost textually to the distinction between formal and informal relations, which is fundamental for all modern organizational analysis. The many examples which Conring attaches to his definition can be gleaned from Zehrfeld.[30] And Conring does not stop at a merely descriptive presentation of his categories. He often adds what we would call today speculative "cross-tabulations." For example, con-

[29] As far as I know, political arithmetic is about the only topic of contemporary knowledge on which Leibnitz himself did not write. It is therefore well possible that Conring did not even know of the work of his English contemporaries. He was, however, informed on English developments in at least one other field: among his medical writings, I found mention of a treatise on Harvey's discovery of the circulation of blood (in addition to texts on skorbut, fractured skulls and iatrochemistry).

[30] Zehrfeld, *op. cit.,* pp. 15ff.

QUANTIFICATION IN SOCIOLOGY

sider his conjoining the causa formalis and the causa materialis: in a democracy, all people should be studied; in an aristocracy, knowledge of the elite is more relevant.

These ideas are first developed as a general system and then applied consistently to one country after another. Stress is laid on interstate comparisons. Conring's richest material pertains to Spain, which he still considered the leading European power. It would not be too difficult to reconstruct from his very extensive comments an "Anatomy of Spain." By comparing it with Petty's work on Ireland, one should get a better picture of the difference in style of thinking which distinguishes these two authors. Conring himself, incidentally, was very explicit about his method. He classifies the type of sources available to him and gives detailed criteria as to how to judge their reliability; he tries to separate his own work from that of the historian, the geographer, the lawyer, etc.; and again in the frame of Aristotelian logic, he discusses elaborately the kind of inferences which can be drawn from descriptive facts of rules of conduct for the responsible statesman.[31]

The first publication of Conring's on the "notitia rerum publicarum" were unauthorized notes by one of his students; the original manuscripts were published only after Conring's death. Soon, his students began to give the same course at other universities. Various compendia appeared, usually under a title such as "Collegium Political-statisticum." (It is controversial when and how the term "statistics" was introduced for this tradition.) By the beginning of the 18th century, the Conring system was taught all over Germany.[32] It had the advantage of being eminently teachable even by minor men, and gave an academic frame of reference to the training of civil servants, which remained a common problem to all the little German states up to the end of their existence in the Napoleonic era. Conring's political activities helped in the diffusion of his main idea. He spent some time in about ten other German principalities in his capacity at temporary advisor; it can be taken for granted and could probably be proven from a perusal of his correspondence that on such occasions he established academic contacts.[33]

In any case, when Achenwall was a student at various German universities in the years around 1740, he met a well-established tradition. He began to collect statistical information in the Conring sense and when in 1748 he received a call to Göttingen, he was prepared to make it the base of his main course. As a matter of fact, his inaugural lecture was a defense of the whole system against representatives of related disciplines who feared the competi-

[31] Zehrfeld, op. cit., pp. 46ff.

[32] One day the history of universities should be rewritten from a social scientist's point of view. Stephen D'Irsay (Histoire des Universités [Paris: Auguste Picard, 1933]) is altogether too superficial, although his footnotes on sources are very valuable. Paulsen's "History of Academic Instruction in Germany" is a fine piece of analytical writing and contains much information on the period after the Thirty Years' War (Friedrich Paulsen, Geschichte des Gelehrten Unterrichts [Leipzig: Veit & Comp., 1919]). His main interest, however, is with the ups and downs of the classical studies and he pays less attention to the men and institutions relevant to my narrative.

[33] Several princes gave him the title of "Personal Physician and States Counselor" which shows how the many facets of his reputation were fused. I do not discuss here the similar positions he held for a while in Denmark and in Sweden. There is evidence that in 1651 he hoped to move to Sweden permanently; but again nothing came of this effort to escape the small town atmosphere of Germany.

tion with their own prerogatives. I have already mentioned that Achenwall's writing in German helped to focus attention upon his work. But the image of a "Göttingen school of statistics," which got abroad rather quickly, was much strengthened by an institutional factor.

The Göttingen School

The University of Göttingen opened in 1737. To it, historians of higher education trace back some of the main ideas of 19th century German university life. Professors should do original research, not only transmit available knowledge; sons of upper-class families, who traditionally were educated in private schools, should be attracted; and they should not only listen and recite, but be active in their studies. (The prestige of Göttingen was greatly enhanced by the creation of a library which was considered the outstanding one in Europe.) It is not surprising that this university became the center for the development of statistics understood as knowledge of the state. It was interesting to young men who came from a more literate background; it gave opportunity to do research in terms of the accumulation of more and more information about more and more countries; the material was useful for future civil servants.[34] In addition, however, it provided for several generations of professors a methodological continuity; and nothing is more conducive to institutional fame and stability.[35]

From Achenwall on, we are in often described territory. He and his successors refined Conring's categorical system, raised new methodological questions (e.g., what is *relevant* for a necessary and sufficient description of a society), and added a great deal of substantive material. The main figures in the Göttingen tradition are Schloezer, Achenwall's successor, who in 1804 created the slogan that statistics is history at a given point in time while history is statistics in flux; and Nieman, who, in 1807, published the most elaborate classificatory system which he, incidentally called a "theory."[36] By then the confrontation with the political arithmeticians was in full swing. It developed by a somewhat roundabout route. Among the offshoots of the Göttingen group were men who were especially concerned with the *presentation* of comparative information on countries. This led to the idea of two-dimensional schemata: on the horizontal dimension, the countries to be compared and on the vertical dimension, the categories of comparison. Originally the entries in those "matrices" were verbal descriptions or references to sources. But this schematization was naturally conducive to the use of figures wherever they were available, if for no other reason than that they took less space. This in turn favored topics which lent themselves to numerical presentation (more and more such information became available due to the increasing number of

---

[34] Achenwall presided over a weekly seminar where transient explorers and diplomats reported on their experiences abroad; it was the task of the student to subsume this information under the proper categories of "the System."

[35] Denifle, in his book on the origin of medieval universities (*Die Entstehung der Mittelalterlichen Universität* [Berlin, 1885]) raises the question as to why the University of Paris remained intact after the death of Abelard, while preceding foundations disappeared after the death of the charismatic master. He thinks that it is the *methodological* idea of the disputation, the finding of truth by a staged dialectic controversy, which provides the explanation.

[36] John, *op. cit.,* p. 98 f.

government agencies collecting census materials). The guardians of the Göttingen tradition were afraid that this would give a "materialistic" flavor to the comparative study of states and deflect from the educational, social, and spiritual significance of their teaching. They invented the derogatory term, "table statisticians" which today would be considered a pleonasm, but was very apt in the context of the time. A distinction was made between "refined and distinguished" in contradistinction to "vulgar" statistics. John gives a number of quotations from the *Göttinger gelehrte Anzeigen* to indicate the vehemence of feelings at the beginning of the 19th century. The vulgar statisticians have depraved the great art to "brainless busy work." "These stupid fellows disseminate the insane idea that one can understand the power of a state if one just knows its size, its population, its national income, and the number of dumb beasts grazing around." "The machinations of those criminal politician-statisticians in trying to tell everything by figures . . . is despicable and ridiculous beyond words."[37] The only thing missing here is the apt expression of quantophrenia which Sorokin has recently created in a similar mood.[38]

In spite of John's very detailed documentation of this fight,[39] I do not feel satisfied that it has yet been clarified. Thus, for example, it should not be forgotten that the period of the Napoleonic wars led to a temporary collapse of Göttingen and the meteoric rise of the University of Berlin after its foundation in 1810. It is also likely that the helplessness of the German states in the face of the Napoleonic onslaught discredited the claim of the Conring tradition that it provided the factual foundation for successful statecraft. And the beginning of a broader German nationalism probably made the academic world more receptive to the kind of general causal relations which the British political arithmeticians looked for under similar political circumstances one hundred fifty years earlier. There is here still a piece of history which deserves much more careful investigation than is available in the literature so far.[40]

But the story sketched in the preceding pages helps to explain why German academia was for so long a period impervious to the political arithmeticians. What Conring started served the most immediate needs of his country. It could not afford the intellectual investment which in England led to a better rate of growth in general knowledge; one can be concrete in surmising that Conring, even if he had thought of them, could not have afforded the computations with which Graunt started. Later the system served well the educational needs of the many little German universities, staffed by mediocre people. When Achenwall gave it additional prestige in Göttingen, something like vested interest began to play a role. Schloezer worked for a while in Berlin and knew of Suessmilch—yet why should a professor at the leading university pay attention to a military pastor?[41] But their king, Frederick the Great, did.

[37] John, *op. cit.*, pp. 129f.
[38] Pitirim Sorokin, *Fads and Foibles in Modern Sociology* (Chicago: Henry Regnery, 1956).
[39] John, *op. cit.*, pp. 128-139.
[40] One might record the victory of the table statisticians with an ironical smile. Today, they themselves are beginning to be looked upon as mere social bookkeepers. Many a mathematically trained statistician would not deal with the problems the political arithmeticians inaugurated; he would reserve the term "statistics" to an abstract theory of inference.

[41] Actually Schloezer promised in one of his books to one day present Suessmilch's ideas, but "never got around to it."

Upon Suessmilch's advice he ordered a number of statistical surveys and therewith contributed to the downfall of the university tradition.

Meantime, the Germans had acquired an international audience. Things had greatly changed since Conring's time. In 1774, twenty-five years after Achenwall's first publication, Goethe published Werther. When Schloezer's book came out, Fichte's philosophy was well developed and the whole romantic school was in full flourish. I have no knowledge of an early contact between the Weimar-Jena humanists and Göttingen which was the center of the German social sciences in the broadest sense. (At the creation of the University of Berlin in 1810, the convergence is well documented.) But as the rest of Europe paid very serious attention to the German humanistic flowering, it should not be too surprising that even the more specialized efforts of the university statisticians were considered abroad as a major intellectual development which deserved as much attention as the contemporary work of the political arithmeticians in England. The fact that by now the former are practically forgotten while the latter are considered the foundation of the modern notion of statistics should not keep us from realizing that the intellectual balance might have looked very different at the end of the 18th century. It would not be too difficult to trace how influences from these two centers radiated to other countries and where and how they overlapped. From a cursory inspection, I have the impression that early Italian writers were especially influenced by the Germans, while in 18th century France combinations began to emerge.

The question can be raised why I consider the development of classificatory systems a legitimate part of the history of quantification in the social sciences. I want to postpone my answer until I have described another effect of this kind by the LePlay school. First, I turn to the writer who marks the beginning of what one should consider modern efforts at sociological quantification. I agree with John's judgment that a major turning point came about when Quetelet's first publication appeared. "He is the focal point, at which all the rays of the first great period of statistical history converge, to be redirected through moral statistics and similar developments into many new fields."[42] While this simile too will need some correction, it is certainly a deserved tribute to the man who is the central figure of my next section.

QUETELET AND HIS "STATISTIQUE MORALE"

Life and Writings

Quetelet was born in 1796 in the Belgian city of Ghent. He grew up in modest circumstances, but his abilities were recognized early. He originally wanted to become a painter and sculptor, and for a while he was an apprentice in a painter's studio. He also had strong literary interests: he published poetry at an early age and had a play performed in a local theatre. In order to make a living, he took a job at the age of eighteen as a teacher of mathematics in a lycée. When he was about twenty, he came under the influence of a mathematician at the University of Ghent who prevailed upon him to extend his mathematical studies. In 1819 Quetelet received his doctorate. His thesis, on

---

[42] John, *op. cit.*, p. 9.

a problem in analytical geometry, was considered a major contribution. However, he retained his humanistic interests, and up to 1823, when he was twenty-seven years old, he continued to publish essays on literary criticism and to translate Greek poetry into French. Many of his friends at this time were writers and philosophers.

Quetelet's thesis brought him to the attention of the Ministry of Education and in 1819 he was called to Brussels to teach mathematics at one of the several institutions of higher learning in the capital of the Belgian part of the Dutch kingdom.[43] Part of his teaching involved what we would today call adult education, and this seems to have fitted in well with his literary inclinations; repeatedly thereafter he wrote popular monographs on a variety of scientific topics. I am stressing Quetelet's many-sided and humanistic background, because the quality of his later work cannot be understood without knowledge of it.

Quite soon after coming to Brussels, Quetelet became deeply involved in a plan the origins of which are unknown: he wanted to start an astronomical observatory. For almost a decade until the plan became a fact, it had priority among all his activities. It also brought about the main turn in his intellectual life, although this occurred in an unexpected way. In 1823 he was sent to Paris to study astronomical activities and to find out what kinds of instruments would be needed for the Brussels observatory. While in Paris he became acquainted with the great French mathematicians whose headquarters were at the Ecole Polytechnique. He was especially impressed by Fourier and LaPlace and by their work on probability of which he evidently had known nothing before. Both of these men, as was mentioned earlier, had analyzed statistical social data. This combination of abstract mathematics and social reality obviously provided the ideal convergence of the two lines along which Quetelet's mind had developed. He quickly realized that a whole new field of activities, embracing all of his interests, could be opened up. In later reminiscences he said that after he had become acquainted with the statistical ideas of his French masters, he immediately thought of applying them to the measurement of the human body, a topic he had become curious about when he was a painter and sculptor. With his return from Paris in 1824 at the age of twenty-eight, the basic direction of his intellectual career was essentially set.[44] Soon the Dutch government gave him an opportunity to do something about it.

In 1826 Quetelet was asked to help the work of the Royal Statistics Com-

[43] The University of Brussels was not created until a decade later.

[44] In connection with his work for the observatory, he made a number of other trips, including one to Germany in 1829 when he stayed in Weimar for a week and had several meetings with Goethe, then 80 years old. Goethe was quite fascinated by his young visitor. While Quetelet meant to stay only a few hours, Goethe kept him for a week and remained with him in correspondence for the remaining years of his life. There exist several contemporary reports on this encounter and Quetelet himself wrote at the age of 70, a reminiscence. The material has been collected by Victor John "Quetelet bei Goethe," *Festgabe für Johannes Conrad*, H. Paesche, ed. (Jena: Gustav Fischer, 1898). Goethe discussed with Quetelet repeatedly his favorite notion of "types." John surmises that this might have given Quetelet the idea for his terminology of the "Homme Moyen." Some of John's material has been briefly summarized in a contribution by Auguste Collard to *Isis* ("Goethe et Quetelet," *Isis*, 1933, 1934, *20*).

mission, and he participated in the preparation of plans for a Belgian population census. His first publications covered quantitative information about Belgium which could be used for practical purposes including mortality tables with special reference to insurance problems. In 1827 he analyzed crime statistics but still with an eye to improving the administration of justice. In 1828 he edited a general statistical handbook on Belgium which included a great deal of comparative material obtained from contacts he had formed during his stays in England and France. The unrest leading to the Belgian insurrection in 1830 intensified Quetelet's interest in social topics.[45]

Quetelet's two basic memoranda appeared in 1831. By then he had decided that from the general pool of statistical data, he wanted to carve out a special sector dealing with human beings. He first published a memorandum entitled "The Growth of Man" which utilized a large number of measurements of people's size. A few months later he published statistics on crime under the title "Criminal Tendencies at Different Ages." While the emphasis in these publications is on what we today would call the life cycle, both of them included many multi-variate tabulations, such as differences in the age-specific crime rates for men and women separately, for various countries, and for different social groups.[46] In 1832 a third publication, giving developmental data on weight, made its appearance. By this time the idea of a social physics had formed in his mind, and in 1835 he combined his earlier memoranda into one entitled "On Man and the Development of his Faculties," with the subtitle, "Physique Sociale." With this publication all of Quetelet's basic ideas were available to a broader public.

Quetelet's literary background and the fact that his humanist friends remained an important reference group for him help to explain the manner in which he published his works. When he had new data or had developed a new technique, he first published brief notes about them, usually in the reports of the Belgian Royal Society, and sometimes in French or English journals. Once such notes had appeared, he would elaborate the same material into longer articles and give his data social and philosophical interpretations. He finally combined these articles into books which he hoped would have a general appeal. He obviously felt very strongly that empirical findings should be interpreted as much as possible and made interesting to readers with broad social and humanistic concerns.

The statistical data which he published in the period just reviewed were

[45] Two years before Quetelet's birth, Belgium was still one of the provinces under the domination of the Austrians. During the Napoleonic Wars, it became a part of France, and after the settlement of 1815 it was made a part of the Netherland Kingdom which combined Belgium and the old Dutch provinces to serve as a buffer state against possible future aggression on the part of the French. The French Revolution of 1830 spread into this new kingdom, however, and led to the separation of Belgium and Holland. Since that time, Belgium has existed in the form in which we know it today. The unrest leading to the Belgian insurrection in 1830 intensified Quetelet's interest in social affairs. So far as I can tell, Quetelet was never active in politics in the narrow sense, and this may account for the fact that his emphasis was always on social science in general, not on the role of the government which was of such great concern to the Political Arithmeticians.

[46] Quetelet talks often about the "social system" and its "equilibrium." What he means is the fact that the social strata contribute each their own rates, which are constant over time but different from each other.

always averages and rates related to age and other demographic characteristics. Around 1840, however, he became interested in the *distribution* of these characteristics. It occurred to him that the distribution of the heights and weights of human beings, when put in graphic form, looked very much like the distribution of errors of observations which had been studied since the turn of the century. This led him to the conviction that the distribution of physical characteristics could be looked at as if they were binomial and normal distributions. They were currently discussed in terms of two models. One was the idea of trying to hit the bull's eye of a target with a rifle. Measuring the distance radially from the center, one would find that most of the hits were near the center and the less frequent the greater their distance from the center. This could be represented mathematically by assuming that a very large number of incidental causes, such as movements of the air, involuntary movements of the triggering finger, and so on, affected the tendency of the marksman to hit the center of the target. To obtain the proper derivation through the application of probabilistic mathematics, one had to assume that these incidental factors were independent of each other.

The other model used the time-honored urn of balls. Suppose that in such an urn one has, say, one hundred balls, thirty of them black and seventy of them white. By drawing many samples of ten balls from this urn (always replacing the balls after each draw), one would find that most of the samples contained three black balls. The frequency of samples with very few or very many black balls would decline the more the number of black balls deviated from the "true number" in the urn, here three out of ten. The distribution of black balls in many samples could again be computed mathematically.

It was Quetelet's idea that the distribution of people's size originated in the same way. Nature, like the marksman, kept trying to hit a perfect size, but a large number of accidental and independent causes made for deviations in the same way that the rifle shots deviated from the bull's eye. Or using the urn scheme: the "true size" corresponding to the number of black balls in the urn was due to basic biological factors. People were samples in which varying numbers of black balls were drawn. This was just another way to account through probability theory for the actual distribution of size which Quetelet saw in his data.

In his earlier publications Quetelet had been primarily interested in the fact that the *averages* of physical characteristics and the *rates* of crime and marriage showed a surprisingly stable relation over time and between countries with age and other demographic variables. It was these relations which he had pointed to as the "laws" of the social world. By now, however, he was concerned with the distributions about the averages. He was convinced that if he could make enough observations, his distributions would always have the normal or binomial form. The notion of "law" was now extended: the distributions themselves and their mathematical derivations, as well as their constancy over time and place, became laws.

In the middle of 1840 Quetelet published a number of articles along these lines. One dealt with "probability as applied to moral and political sciences," another with the "social system and the laws which govern it." He now developed in addition another terminological distinction. Originally he had talked about "physique sociale" when he had dealt with people rather than with general statistical information about commerce, armament, and so on. He now distinguished within this area between physical and other characteristics of men and women. The quantitative study of non-physical characteristics was called "statistique morale." The same data and the same ideas were again used and re-used in a large variety of publications, including popular treatises on probability theory itself.[47]

In 1855 Quetelet suffered a stroke from which he recovered only slowly. After a while he resumed work, but never again developed any new ideas. His publications continued, however, and in them he reported new data and re-iterated his basic propositions. In 1869 he published the two volumes on which his international fame is based, entitled Physique Sociale or an Essay on the Development of Man's Faculties,[48] thus reversing the order of the title used in 1835. The work was introduced by a translation of an article of the English astronomer Herschel who, writing in the *Edinburgh Review* in 1850, had drawn the attention of the English public to Quetelet's applications of probability theory to the social data. The first volume, divided into two parts, dealt in the main with population statistics, such as birth, death, fertility, and so on, and with the application of these statistics to public administration and medicine. The first part of the second volume dealt with physical characteristics such as size and weight, and added some data on physiological investigations. It is the second part of this volume, Book 4, which contained the material relevant to the social sciences. Intellectual abilities and insanity were treated in the first chapter. A second chapter deals with the development of "moral qualities." The emphasis was on crime statistics, but data on suicide and on all sorts of behavioral manifestations such as drunkenness, duels, were also included. These two chapters will serve as the basis for our specific discussion of Quetelet.[49]

Anyone who draws his information from this book, generally considered the standard source for Quetelet's writings, should be warned that it is a very confusing compilation. Large parts of it are verbatim reprints of previous publications including the most important ones from the early 1830's. Sometimes footnotes are added to enlarge on an idea; in other places, Quetelet adds chapters containing more recent data and offers an interpretation which overlaps almost word for word with interpretations of similar data in other chapters. Furthermore, because his thinking on the social and philosophical implications of his work varied over the thirty years of his productive career, some statements on quite basic ideas contradict others made in a different part of the "Physique Sociale." Even in his lifetime the confusing nature of Quetelet's prodigious literary output had created concern. The German economist G.

---

[47] While his early publications on geometrical problems are considered important mathematical contributions, Quetelet did not do any original work in probability theory. He only extended its applications to social phenomena.

[48] Adolphe Quetelet, *Physique Sociale,* Vol. I, C. Muquardt (Bruxelles, 1869).

[49] A short book 5 is an extension of book 4. But its introductory pages give a good picture of Quetelet's ideas on the relation of his work with literature and art; he develops a notion which today we would label "national character."

F. Knapp published in 1871 an analytical catalogue of Quetelet's publications in order to trace where various parts of his texts appeared for the first time and where he subsequently repeated them more or less verbatim in other publications. Knapp also wrote extensively on Quetelet's substantive ideas and created great interest in him among his own students and contemporaries.[50] A very helpful book on Quetelet was published in 1912 by a Belgian, Joseph Lottin.[51] This is an outstanding example of intrinsic analysis. Organized by topics, it combines and confronts extensive quotations in order to bring out what Quetelet most likely intended to convey. Lottin's book probably contains all available biographical material on Quetelet. It provides a well-organized chronology of his publications and, most valuable of all, it lists almost all the German, English and French literature on the various debates which started during Quetelet's lifetime and continued well beyond his death.[52] English-speaking readers will find Frank Hankins' dissertation[53] the best source of information about Quetelet. This author was well aware that there is no English equivalent for the French term "statistique morale" or the German "Moralstatistik." He decided to use the phrase "moral statistics," and I shall follow his example.

Throughout his life Quetelet was active in many enterprises, such as the organization of international statistical congresses, the improvement of census work in his own and in other countries, and so on. Because of the specialized focus of my paper, I shall not give attention to these aspects of Quetelet's life and works, but rather try to elucidate some of his ideas on quantification.

The Distribution of Non-Physical Human Characteristics

I mentioned previously Quetelet's work on the distribution of size and weight in large-scale human populations. He was convinced that similar distribution curves could be found for intellectual and "moral" characteristics if appropriate measures were used. He made explicit his belief that such measure-

---

[50] Knapp became famous mainly because of his monetary theory. The article about him in The Encyclopedia of the Social Sciences is informative, but the appended bibliography does not include any of his numerous publications on Quetelet's moral statistics, nor his editions of statistical classics. (A series of pertinent papers by G. F. Knapp appeared in the 1871 volumes of Bruno Hildebrand, *Jahrbücher für Nationalökonomie und Statistik* (Jena: Friedrich Mauke).

[51] *Quetelet, Statisticien et Sociologue* (Louvain: Bibl. de L'Inst. Sup. de Philosophie, 1912).

[52] Lottin's book also contains valuable material on Quetelet's relation to his contemporaries. Comte, as is well known, complained that Quetelet had stolen the term "social physics" from him, and that he had therefore had to invent the word, sociology, to identify his approach. In his review (*op. cit.*, pp. 357-368), Lottin argues that the Belgian probably did not know of the Frenchman's terminological

invention. And he adds interesting comments on the very different images of the social sciences held by the two men: Comte was trying to derive from history broad developmental trends which could be projected into the future, while Quetelet was bent on finding precise regularities which would help to explain the contemporary social scene. Quetelet rarely engaged in controversy. The most serious one was with the Frenchman, Guerry, who invented the term, "moral statistics," and who was analyzing crime statistics at the same time that Quetelet was. Lottin provides a detailed analysis of the relevant drafts and publications (*op. cit.*, pp. 128-138). The question of priority is a subtle one, because, although their interest in crime statistics developed independently, they were informed of each other's work through a common friend. According to Lottin, Quetelet wins in a photo finish.

[53] *Adolphe Quetelet as Statistician* (New York: Columbia Univ. Press, 1908).

ments were possible in principle, and that the lack of data was due only to technical difficulties. Much of his general interpretive writing on moral statistics hinges on the kind of quantitative data he used as substitutes for the variables he felt would be really desirable. Debates about Quetelet's work usually focus on his idea of an "average man" or his shifting pseudo-psychological comments on his data. A careful reading would show, however, that his underlying theory of measurement, partly brilliant and partly a confused foreshadowing of later developments, gives the clue to much of his work.*

On pp. 148 ff. Quetelet makes a statement not on his findings themselves nor their interpretation, but on the *formal* nature of the variables which can be used to *"measure the qualities of people which can only be assessed by their effects."* The passage which follows is a somewhat condensed summary of his presentation which I will reanalyze in the light of contemporary thinking about measurement in the social sciences.

One can use numbers without absurdity in the following cases:

(1) When the effects can be assessed by a *direct* measure which shows their *degree of energy:* effects, for instance, which can be seen by strength, by speed, or by work efforts made on material of the same kind.[54]

(2) When the *qualities* are such that their *effects* are about the *same,* and are *related to the frequency of these effects,* as, for instance, the fertility of women, drunkenness, etc. If two men placed in the same circumstances got drunk, the one once per week and the other twice, then one could say that their propensity to drunkenness has a 1:2 ratio.[55]

(3) Finally, one can use numbers if the causes are such that one has to take into account the *frequency of their effects as well as their energy* ... this is especially true when it comes to moral and intellectual qualities such as courage, prudence, imagination, and so on.[56]

When the effects vary as to their energy but appear in approximately the same proportions, the matter is greatly simplified. One can then disregard the element of energy, and work only with frequencies. Thus, if one wants to compare the propensity to theft of 25 year old and 45 year old men, one could, without too much error, just consider the frequency of thefts committed at these ages. Variations in the degree of serious-

---

* All page numbers without an additional reference are from the second volume of Quetelet's *Physique Sociale*. Direct quotations from the original are my translations; the underscoring of sentences is done to emphasize points in the present discussion and does not appear in the original.

[54] Here Quetelet inserts an elaborate footnote that the measurement of memory might be of the same type if records were made of the length of time different people remembered a text they had learned. He emphasizes that such measures can be used to study variations by age, just as one can study developmental differences in eyesight or acuity of hearing.

[55] The most frequently quoted example is the measurement of courage by counting the number of courageous deeds. It is followed by

the suggestion of a courage-poll (p. 147) : "Assume now that society, *in a more perfect state,* keeps track of and evaluates all acts of courage and virtue, *just as one does today in regard to crime;* would this not make it possible to measure the relative propensity to courage and virtue in different age groups?"

[56] Quetelet had previously mentioned (p. 143) that in order to measure the productivity of writers one would have both to count the number of works and weigh each item according to its literary merits; having elaborated on this and similar examples at some length, he obviously felt that he could deal with the general idea briefly. He therefore proceeded immediately to the question of feasible approximations.

ness of these infractions can be assumed to be about the same in both age groups.

I think that *all the qualities of people which can only be assessed by their effects can be classified in the three categories which I have just established;* I believe also that the reader will feel that the temporary impossibility of using numbers in such assessments is due to the *unavailability of data rather than to shortcomings in the methodological idea.*

For the moment, let us disregard the references to cause and effect, a topic to which I shall return presently. The three types of variables described here are then quite familiar. The first is a continuous variable which we would exemplify today in the "moral" area by the amount of time an individual watches television, the distance a child walks on an errand without being distracted by something which makes him forget his original goal, and so on. Whether a variable is continuous or not is a question of empirical fact, which has nothing to do with the substantive content of the subject matter to which it is applied.

The same is true for the second of Quetelet's types. These we would call discontinuous variables (or rather, variates, if we wanted to be very precise). Examples would be the number of times per year that an individual buys a weekly magazine, the number of times he goes to the movies, and so on. The natural sciences also deal with such discontinuous variables and do so for a broad range of phenomena, whether it be the number of meteorites in a certain sector of the sky or the number of atoms per section of a bombarded screen.

Finally the third, or mixed, type is well known today. A price index combines price changes of a variety of products; a scholastic aptitude test counts the number of mistakes made, and might give different weights to different items. To simplify matters, we shall forget about the mixed type for the time being, and just distinguish between continuous and discontinuous variables.

Why were these distinctions important to Quetelet? The answer is simple. The type of discontinuous variables which Quetelet had in mind plays an important role in the social sciences. But the observations needed to establish them usually require a considerable amount of time. They have nevertheless one great advantage: if we only wish to compare subgroups in a population, we can often *substitute a one-time observation of many people* for *repeated observations of one person.* This is especially true if we are only concerned with averages, which in this case take the form of rates. Suppose we want to know whether men or women are more faithful readers of a particular weekly magazine. The logically correct procedure would be to select a sample of both sexes, find out from each person the number of weeks in a year he had bought the magazine, and then compute the average. This would be the number of weeks per year the average person of both sexes performed this act. Dividing this average by the number of weeks in a year gives us the probability that the average man (or woman) will buy the magazine. Instead of proceeding in this way, under many conditions we can approach the problem differently. We can interview a sample of men and women *just once* and ask them whether they have bought the magazine *this week.* The proportion of positive replies will be the same as the probability obtained through the more precise procedures.

It is this substitution of observations over people for observations over time which is one of Quetelet's central ideas. And as long as one wishes only to compare averages, this is the standard procedure in social research today. But when it comes to measuring *distributions* of these discontinuous variables, we are of course in a very different situation. In these cases we would need the full variable over time for each person and therefore observations over time. In many of the major writings of Quetelet and his commentators, the existence of such distributions is implied. A digression on two of these commentators will be enlightening at this point.

Lottin states repeatedly that Quetelet never wanted to make measurements of such discontinuous variables on *single* individuals. This is one of the few points on which Lottin is wrong. As the preceding quotation shows, and as can be corroborated in many other citations, Quetelet considered such measurements desirable and, in principle, feasible. One would misunderstand much of his writing if one overlooked this point, for Quetelet was convinced that, if such measurements were available, they would show distribution curves just like the ones he published for size and weight.[57] He gave these hypothetical distributions exactly the same interpretation that he gave his biological data, substituting society for nature. Society aims at a certain average which it obtains in a majority of cases but from which it deviates up and down according to the laws of probability. Today we deal with many empirical distributions of the sort which Quetelet had more or less clearly in mind. These include scores on intelligence tests, frequencies of sociometric choices addressed to each of a group of individuals, the number of times an individual did not vote in a series of elections, and so on. Within the range of precision which Quetelet thought of, the distributions of such measures do indeed look like distributions of physical characteristics.

Yet there is an important difference between the first two types of measurements in Quetelet's list; this leads to an interesting episode in the criticism of Quetelet. Durkheim's disciple Halbwachs wrote a monograph on Quetelet in 1912.[58] He was not clear about the important difference between looking at averages of discontinuous non-physical variables and studying their distribution, just as Quetelet himself was never really explicit about this distinction. But one can deduce from Halbwachs' text that he had these distributions in mind when he launched his main attack on Quetelet. The distribution of physical characteristics, he argued, might be accounted for in the same way as are shots or astronomical observations, namely as the effect of a large number of minor *independent* factors which make for deviations from the average. But clearly this could not be true for data in the area of moral statistics, because here the interdependence of social action becomes crucial. Halbwachs stated and restated the notion that individuals influence each other, that the various sectors of society are dependent upon each other, that contemporary society is affected by past ideas and experiences. From such con-

---

[57] It is true that Quetelet repeatedly says that he does not deal with individual measures. This occurs, however, only when he wants to argue for the validity of his rates. As can be seen from our main quotation and from many other passages, he always insisted that, in principle, quantitative information on individuals could and should be obtained.

[58] Maurice Halbwachs, *La Theorie de l'Homme Moyen* (Paris: Felix Alcan, 1912).

siderations he developed an argument against Quetelet's basic idea of probabilistic analysis which rests on a serious misunderstanding. Because of the dominant position of the Durkheim School in French and in modern sociology, it is worthwhile to consider this point in some detail. We are confronted here with the following situation. An early author, Quetelet, has a correct idea which he develops, however, only within certain limits. A subsequent author, Halbwachs, instead of adding what was missing, misjudges the partial contribution which his predecessor had made. As a result, instead of steady progression, we find temporary discontinuity in intellectual developments. Only many decades later when the idea of stochastic processes developed was the trend resumed. And it is fairly safe to say that the modern version was created by men who knew nothing of Quetelet's work.

Typical of Halbwachs' arguments are the following:

> All these circumstances *preclude the idea that the laws of chance play a role in the social sciences.* Here the combination of causes (represented by individuals) are connected with and dependent upon each other, because, as one of them comes about, similar ones are reinforced and tend to occur more often; therefore, we are not in the same situation here as we are in games of chance, where the players as well as the dice are not supposed to acquire habits, to imitate each other, or to have a tendency to repeat themselves. (*op. cit.*, p. 146)
>
> Society and the moral acts of its members are probably, of all phenomena, the area in which it is least possible to consider an individual and his acts in isolation from the behavior of all others; this would mean leaving out what is really essential. *This means, at the same time, that this is an area in which the calculus of probabilities is least applicable.* (*op. cit.,* p. 174)

The crucial sentences are those I have underscored. Halbwachs sets up an antithesis between social interaction and the application of probability mathematics. This has been proved to be completely wrong in more recent applications of mathematical models in the social sciences. Let us consider an example. Suppose that we are observers at a ball attended by men and women who do not know each other and suppose moreover that more women than men are present. As the music starts, each man chooses a partner at random by drawing a woman's name out of a hat, and this is repeated for ten dances. At the end of the ten dances, we can classify the women according to the number of times they had partners. We will find a normal (more exactly, a binomial) distribution: the women will have had some average number of dances, the lucky ones who exceed this average being balanced by the number who were more or less out of luck. This would indeed be a situation corresponding to an error curve of the kind Quetelet had in mind.

Now let us change the situation slightly. At the time of the first dance each woman has the same probability of being chosen by chance. But suppose that the men watch the situation and believe that the women chosen the first time are the more desirable partners; a first choice thus increases their chances of being selected a second time. One might assume, to translate this situation into probability terms, that at the second dance the names of the women chosen the first time are put into the proverbial hat twice. Then even if the partners for the second dance are again chosen at random (by drawing slips from the hat), obviously the women who were chosen the first time will be more likely to be chosen again. Suppose that we allow this process to repeat itself for ten dances, and, after the tenth dance, we make a statistical count of how often each woman was chosen. The number of women with very many and with very few partners will now be clearly larger than in the previous "model," although the average number of "successes" will be the same.[59]

Here then is a situation in which, contrary to Halbwachs' opinion, social interaction is taken into account and probability considerations are still applicable. In the same way there is no difficulty in developing so-called stochastic processes in which probabilities of individual choices at time $(t + 1)$ depend upon the total probability distribution at time (t). Modern mathematical sociology has proven the Durkheim School wrong and Quetelet right. It is true, however, that Quetelet himself was unaware of the fact that his sociological thinking exceeded the specific mathematical model upon which he drew. In his discursive writings, Quetelet often talks about social interaction and the effects of the past. As far as I am aware, however, he never went beyond the classical normal distribution which, indeed, makes no provision for these more complex processes.

Let us return now to the way Quetelet worded his distinction between the three types of measurement. He talked of "qualities which can only be assessed by their effects." Today authors would talk interchangeably of hypothetical constructs, disposition concepts, underlying characteristics, or mathematical latent structures. But his three types of measurements classify manifest, observable data, not the underlying qualities themselves. Quetelet thought that certain things, such as size and weight, were human qualities that could be measured directly. Other qualities, such as a tendency to suicide or marriage or, most of all, to crime, could be measured only indirectly. Now here he obviously confused two completely different problems. Certain variables are quite conventional: an individual's height, the prices of commodities, or the amount of income. As one turns to the new area of social investigation, other variables have to be developed: the number of times a man gets drunk or the amount of money a person contributes to charity (another example which Quetelet himself gives). These variables may somehow sound different, but as far as the required observations go, all are on the same level of reality. A difference arises only *if one wants to make different uses of such information*. One might be interested in the size of people; then that is all there is to it. Or one might be interested in the charitable tendencies of people; then the amount of money they give to organized charity is an *indicator*, perhaps only one of several that might be used. This fact of course, is equally true for physical characteristics. Suppose one were interested in a child's "propensity to physical growth." Then actual size would be a reasonable indicator,

---

[59] Early work on these more complicated distributions was carried out by the German economist and mathematician, Wilhelm Lexis. He was a charter member of the Verein fur Sozialpolitik in 1872, and taught at Strassburg where, somewhat later, Knapp formed a center for work on Quetelet and the history of statistics in general. Lexis had broad interests, co-edited the German Encyclopedia of the Social Sciences, and wrote himself on moral statistics. Whether the Lexis "distribution with contagion" was developed with reference to Quetelet, I do not know, but the question suggests a good topic for further research.

although physical anthropologists could certainly give us many others. The relation of propensities (or tendencies or however else one wants to translate Quetelet's favorite term, "penchant") to manifest data is always the same, regardless of subject matter. It is true that in the social sciences the inferential use of manifest data is more frequent than in the natural sciences. But medicine is an interesting borderline case: diagnosticians use manifest physiological data to make inferences regarding unobservable physiological propensities. It is this general problem and its partial misconception which we now have to trace in Quetelet's writings and the subsequent literature they inspired.

Propensities and Their Measurement

The matter is best explained in connection with Quetelet's writings on criminal tendencies ("penchant au crime") which fill the largest part of Book 4 of his "Physique Sociale" (pp. 249-363). What he had before him were crime rates carefully computed for a large number of subsets in the population. However, he was not satisfied with what we today would call descriptive correlations. He wanted to fit his findings into a much broader picture. In order to do so, he considered his rates as measures ("échelles") of underlying tendencies. From his point of view, this had two advantages. He could talk about his findings in more dramatic language, and he could speculate on interpretations which seemed to him of great human interest. If murder was more frequent among younger than among older people, that gave occasion to write about the violent nature of youth. The higher crime rate among men allowed him to speculate about the restrained nature of the female personality.

Quetelet constantly refined his arguments. Often he stresses the comparative uses of his data. It is true, he would say, that crimes are committed more often than criminals are caught. But, as police vigilance is a rather constant phenomenon, this plays no role if we just want to *compare* the criminal tendencies of various age or other population groups. It is true that the same criminal tendency can lead to manifest crime under one set of conditions and not under another. So we measure only the "penchant apparent" and not the "penchant réel." But for comparative purposes, either serves equally well. On p. 343, he demonstrates this interchangeability by analyzing the age distributions of indictments, convictions, and acquittals.

At other points, Quetelet met a more general statistical difficulty. He looked at crime rates as probabilities. In order to compute this probability for a specific group, he had to know both the absolute size of the group and the number of crimes committed by its members. But often he only had data on the criminals themselves, without the corresponding population figures. Thus, for instance, he could distinguish his criminals according to whether they were illiterate, could read and write a little, or had a considerable amount of education (L'Influence des Lumières). But no educational statistics were available for the country as a whole. Again, relative rates were the solution. He gives a table cross-classifying French crime reports by sex and level of education (and, incidentally, for two different years). And then follows a typical interpretation (p. 297): "I think that one could explain these findings by saying that in uneducated strata, the habits of women are similar to those of men; in more educated strata, women have a more retiring style of life and consequently have less opportunity to commit a crime, all other things being equal."

But none of this touches on the central issue: What is implied when one makes inferences from a committed crime to a criminal tendency? Does Quetelet do more than substitute for the observed crime rate the *word* criminal tendency? It is not difficult to gather his own position. For him, the crime is an *effect* of the tendency which he thinks of as a *cause*. Today, we would talk of a propensity and its indicator rather than of a cause and its effect. This may be only a difference in linguistic fashions; I will follow the modern usage without attaching any importance to the difference.[60]

Quetelet assumes a *deterministic* relation between the hypothetical construct and its manifestations. This obviously derives from his training as a natural scientist. In the natural sciences, for example, the acceleration of an object is related to a force in a deterministic way. But in the social sciences, indicators or symptoms have a *probabilistic* relation to the underlying propensity. It is curious that Quetelet never hit on this idea in view of the fact that he was so imbued with probabilistic thinking in other contexts. Much of what I quoted before should have led him to this notion. Crimes are not always detected; whether or not they are committed depends on opportunity—all this would be best formulated by saying that the individual criminal tendency should be measured by the probability of committing a crime. This would have clearly raised the question how such probabilities assigned to individuals could be ascertained.

But Quetelet overlooks this question because of another idea which he repeats many times in his writings. He continually stresses that the only assumption he introduces—we would call it an axiom today—is that "the effects are proportional to the causes." To put this in modern terminology again, he assumes a *linear* relationship between the latent continuum he tries to measure and the probability of the manifest indicator he can observe. Even in the deterministic world of classical physics, this is undoubtedly wrong: the angle by which the needle of an instrument deviates is by no means always linearly related to the strength of an electric current which is to be measured. Measurement theory, as developed in modern social science, has made it quite obvious that such linearity does not prevail. The matter is important enough to justify two examples.

Suppose we want to develop a conventional test to measure conformity towards social regulations. One item in the test might be the question, "Are there situations in which white lies are justified?" The probability of a positive answer will be great all along the intended continuum: only at the extreme, let us say, right end, where people have very strict feelings about always doing the right thing, will this probability be small. A graph relating these probabilities to the underlying axis—authors call it variously a traceline or an operating characteristic—will look convex upwards. Take then the ques-

[60] In Book 3, Quetelet discusses in great detail the measurement of human strength with the help of dynamometers. These sections give leads to the imagery by which he was probably guided, when he tried to explain his ideas on indicators of moral qualities.

tion, "Are people entitled to break rules if it is clearly to their advantage?" The probability of a positive reply will be high only at the extreme left where people have quite loose moral standards; it will then rapidly decrease and remain low along the rest of the continuum. The corresponding graph will look concave upward. In other words, the relation between the effect (the answer to the question) and the cause (the underlying attitudes) can take various forms.[61] The situation becomes even more interesting when we look for an example which comes closest to Quetelet's use of rates. He takes it for granted that a high aggregate rate of criminality indicates a high average criminal tendency in a given population. He would certainly also take it for granted that a high proportion of male births expresses a high average tendency to have male children mediated "obviously" by the tendency of families to go on having children until at least one of them is a boy. But Leo Goodman has recently shown that the relation can be the reverse (personal communication). If all families pursue this practice, the couples who biologically are likely to have more girls would produce more children. The "effect," the rate of male births, would go in the opposite direction than the "cause," the penchant to have male children. Quetelet thus makes a much over-simplified assumption on the relation between a tendency and its manifestation.

The matter becomes even more complex if more than one indicator is at stake. On page 224 Quetelet talks about the possibility of measuring foresight (prévoyance) especially in its economic implications. He argues that the propensity to foresight existing in a certain country can be measured and he gives a series of examples on what data could be used: the amount of savings people have, the amount of insurance they purchase, how often they use pawn shops (because they do not properly balance their income and their necessary expenses in advance), the frequency of gambling, the number of bankruptcies; even visits to nightclubs are suggested as measures.[62] At this point, where Quetelet is only speculating, he sees that the "cause" he wants to measure, lack of foresight, can have a variety of effects or indicators, and that the inference as to the underlying tendency would be the safer the more of these indicators were used. But how would they be combined? Quetelet is dimly aware of this problem and this is obviously why he introduces his third type of variable as described in the main quotation above. He would probably correctly say that crude measurements of "lack of foresight" would be an additive counting of all these instances just listed. But he does not see the implication this has for the criminal tendencies in which he is interested.

Measuring the productivity of writers or the lack of foresight of householders is an *intrinsic procedure*. A measurement has to be constructed either

[61] Readers unfamiliar with recent developments in theory of socio-psychological measurement might find useful Paul F. Lazarsfeld, "Latent Structure Analysis," *Psychology: A Study of a Science*, Sigmund Koch (ed.), (New York: McGraw-Hill, 1959), pp. 477-491.

[62] This idea has come up all through the period covered by my paper. Thus Petty made this suggestion as "the measure of vice and sin in the nation": "the quantity spent on inebriating liquors, the number of unmarried persons between fifteen and fifty-five years old, the number of corporal sufferings and persons imprisoned for crimes" (Strauss, *op. cit.*, 196). And 250 years later, William James exemplified the pragmatists' notion of a "prudent" man: "that means that he takes out insurance, hedges in betting, looks before he leaps." Lazarsfeld, *op. cit.*, p. 480.

by a mathematical model or by some cruder kind of reasoning; but there is no *outside* criterion by which it could be validated. No single one of the eligible indicators is a priori more relevant than any of the others.[63] The same idea would have to be accepted for the measurement of criminal tendencies. Starting out with a conceptual analysis, one would have to look for indicators of aggression, contempt for law, and so on. One *might* include the actual performance of a criminal act as one of the indicators. But this would spoil the main use to which such a measure could be put. What we would like to know is the *empirical* relation between the criminal tendencies, measured independently, and the frequency of criminal acts in various populations and under various social circumstances. Actually Quetelet at one point adopts (page 250) a conceptual distinction which really implies what has just been said here. He states that the criminal act is the resultant of three factors: the *general* criminal *tendency,* the *skill* to perform a certain *class* of crimes, and the *opportunity* to go ahead and do it on a *specific* occasion. But Quetelet never proceeds from this formal distinction to a clear awareness that therefore *independent measures of propensities have to be developed.* He thought that in his comparative rates he had partialled out skill, opportunity (and chance of detection) and that they measured therefore dispositions, tendencies only. Subsequent research has shown that these assumptions are wrong. It is necessary to develop measures of criminal tendencies independent of the criminal act itself—provided that the notion of "penchant" is to be maintained.

But should it be maintained at all? This was the question raised by the German philosopher and mathematician, M. W. Drobisch, one of the first who brought Quetelet to the attention of his German colleagues by a review published in 1849. In 1867 he wrote a monograph,[64] the first part of which brings the review of Quetelet's work up to date and focuses on the question, whether criminal tendencies could be imputed to all people or should be considered restricted to criminals only. He argues strongly for the latter position: "the regularities demonstrated by moral statisticians have bearing only on certain classes of arbitrary human action and refer always to small proportions of a country's people who are especially disposed to these actions." (Translated from p. 52.) Such a statement is trivial if it means that the crime rate pertains to criminals only. But if it implies something about criminal dispositions then it has to be tested empirically. An independent instrument of measurement to support Drobisch would have to show a *bimodal* distribution of these tendencies; on the contrary if the tendencies of criminals —whatever their origin—are only extreme forms of everyone's experiences, then the distribution should be unimodal. As long as only *rates,* or to put it more generally, the *averages* of propensity distributions, are known, the issue cannot be decided. A uni-modal and bi-modal distribution might well have the same average. It is surprising how many prominent writers participated in a "discussion bizantine" as Lottin calls it in a review of the pertinent literature (*op. cit.,* pp. 550ff).[65]

---

[63] In the parlance of modern measurement theory, this is called construct-validity.

[64] Moritz Wilhelm Drobisch, *Die moralische Statistik und die menschliche Willens-*

freiheit (Leipzig: Leopold Voss, 1867).

[65] Drobisch himself deserves a more detailed study. His exposition of Quetelet is extensive and clear. He was a logician interested in all

Halbwachs drew another conclusion from Quetelet's analysis without noticing that he here accepted probabilistic thinking which he disapproved of so strongly in other contexts. Combining his functionalism with Quetelet's idea of normally distributed criminal tendencies (and elaborating on a remark made by Durkheim in his Rules of Sociological Method), he makes the following statement:

> In order to supress all crimes, it would be necessary to instil in all people a deep collective aversion against the qualities leading to such acts, such as ruthless ingenuity, a spirit of intrigue and manipulation. But this maybe is not desirable; in fact, *as society permits the regular occurrence of a stable proportion of crimes, it undoubtedly feels that it would do more harm than good if this number were further reduced.* (*op. cit.,* p. 151, emphasis supplied.)

In quantitative terms, what Halbwachs says is this: One cannot just cut off the extreme tail of a distribution of attitudes or traits: if one wants to curtail certain *excessive* degrees of vitality, one would have to move the *whole* distribution of the relevant propensities towards a lower level of intensity. While substantively Halbwachs' take-off from Quetelet seems less realistic than the one Drobisch made, it certainly is logically more in the spirit of what Quetelet was driving at.

It took a hundred years after Quetelet before his effort to quantify penchants was taken up again in Thurstone's well-known paper, "Attitudes Can Be Measured."[66] I am confident that Thurstone did not know about Quetelet; even Allport, who traces the conceptual antecedents of the notion of attitude measurement, finds the first seeds in Spencer.[67] I have the impression that this discontinuity is greater in the social than in the natural sciences. But I have no evidence on this; nor do I have a good explanation, even if I should be right.[68]

After Quetelet

From time to time Quetelet published speculations on the philosophical implications of his work. Did the apparent constancy of crime, marriage, suicide and other rates over time imply that human beings have no free will? He never could make up his mind, but he certainly engendered a large literature on the topic. The debate led to the formation of a "German School"

sorts of statistical applications. I have not had access to a paper of his which seems to be one of the first examples of content-analysis: "Statistischer Versuch Über die Formen des Lateinischen Hexameters," published by the Saxonian Scientific Society in 1865.

[66] L. L. Thurstone, "Attitudes Can Be Measured," in *J. Sociology,* 1928.

[67] Gordon W. Allport, "Attitudes" *Handbook of Social Psychology,* Murchison, Carl (ed.), (Worcester, Mass.: Clark University Press, 1935).

[68] Stephan remarks on the same discontinuity in his field: "The foregoing examples suggest that *modern sampling procedure might have developed at least a century sooner than it did if it had received more attention from the scientists of the day.* Only the officials of statistical bureaus...were preoccupied with ...the important problems of trade, finance, industry, agriculture, public health, etc., for which statistical data were needed. Hence they favored complete censuses or the closest approach to them that was feasible." See footnote 112. Quetelet himself is a good example for Stephan's argument. At one point he says that midwives would be a bad source for anyone who wanted to establish the sex ratio at birth; their observations are based on small numbers. He does not argue for a *representative* sample but for a *complete* enumeration.

of moral statisticians which fought a rather imaginary "French School" with undertones quite explicitly related to the Franco-Prussian War of 1870. One gets a good idea of this rather curious discussion from a survey paper by Knapp *(op. cit.)* who also refers to the main literature up to 1871. The topic would deserve a new review in the light of modern ideas on the role of mathematical models as mediating between statistical findings and theories on human behavior; in the confines of my present paper I cannot do more than draw attention to the matter.[69]

Moral statistics as a topic of empirical research expanded rapidly during the 19th century. Also, more and more areas of social life were made the object of enumerations: literacy, the circulation of newspapers, voting, and so on. Correspondingly, new substantive fields of statistics were created parallel to or as subdivisions of moral statistics: thus came into being educational statistics, political statistics, social statistics, and so on. The Germans were particularly apt in thinking of appropriate classifications. The most comprehensive effort is undoubtedly the voluminous textbook by George V. Mayer. Meantime, sociology as a discipline had penetrated the universities and its relation to the data collected by descriptive statisticians became a topic of discussion. Quite a literature on "Sociology and Statistics" emerged written by representatives of both disciplines; in a way it can be considered the beginning of the modern debate on the role of quantification among sociologists. An especially perceptive contribution, containing also interesting historical material was a short monograph by the Austrian, Franz Zizek.[70] The most creative effort to give structure to the ever-increasing mass of data was made by the Italian, Alfredo Niceforo. From the end of the 19th century on, he had been interested in what he called "the measurement of life."[71] He finally exemplified his main ideas in a small book on the measurement of civilization and progress, published in French.[72]

The title is characteristic: instead of just classifying available data, Niceforo starts out with the problem of how one would characterize quantitatively a civilization in "space and time." In Chapter 2, he defines what he means by civilizations, and in Chapter 3, he proposes what today would be called the dimensions of this concept. For each dimension he then proposes a number of crucial (signalétiques) indicators. In his first chapter there is an especially clear discussion of this three-level relation between concepts, dimensions and indicators. He calls his whole effort "social symptomatology," and the logical clarity of his ideas is remarkable. Niceforo incidentally is the earliest sociolo-

---

[69] Since the appearance of Buckle's 'History of English Civilization" (1861), writers mention a Quetelet-Buckle position. A study of Buckle's book shows that he makes only some very inferential references to Quetelet, never seriously uses his material, nor adds any new statistical data. But Buckle mentions Quetelet very early in his long book, and there uses him as a witness for his own social determinism. It is curious that such a completely external coupling has led to a continuously repeated but substantively quite unjustified stereotype which I was able to trace even in books appearing in 1960.

[70] *Soziologie und Statistik* (Leipzig: Verlag von Duncker & Humblot, 1912).

[71] Alfredo Niceforo, *Les Indices Numeriques de la Civilization et du Progrès* (Paris: Ernest Flammarion, 1921) and Alfredo Niceforo, *La Misura Della Vita* (Turin: Fratelli Bocca, 1919).

[72] *Les Indices numeriques de la Civilization et du Progres* (Paris: Ernest Flammarion, 1921).

gist I found who used correlation coefficients explicitly, and who competently demonstrated their place in such a study.[73]

One cannot talk today about a Quetelet school; his thinking, his way of analyzing data have become an integral part of empirical social research.[74] In one respect, however, he did not transcend the intellectual climate in which he worked. While he repeatedly mentioned the idea that special data could be collected to form the empirical basis of a new concept, he never set a concrete example; he only reinterpreted material collected by the social bookkeeping procedures of contemporary society. The tradition of starting with an idea and collecting observations under its guidance must be credited to the man to whom our last section is devoted.

LEPLAY AND HIS "METHODE D'OBSERVATION"

LePlay's Life and Writings

LePlay was born in 1806, and thus was ten years younger than Quetelet. He grew up in a small Norman fishing village, and received his early education at various regional schools. In 1824, at the age of eighteen, he moved to Paris where he soon thereafter graduated as a mining engineer. In his early jobs, he had to inspect mines and pass judgment on the possible commercial values of mineralogical deposits. In 1829, he made a trip through Germany in the company of a friend from his student times with whom he had hotly debated social issues over the years. They stayed for a while with the family of a miner and wrote down detailed observations on its way of life. This became the first of several hundred family monographs which formed the body of LePlay's empirical social research.

In 1830, during the July Revolution, LePlay was hospitalized because of an accident in a laboratory experiment. The turbulent events of the day reinforced his decision to devote himself systematically to the study of social conditions. For eighteen years, he was equally active in both of his chosen fields. He became a professor of metallurgy, head of the official committee on mining statistics, but at the same time, continued his travels, collecting his family monographs and pondering their use for the sake of social reform. In 1848, the second revolution occurred in his lifetime. He decided to give up his regular profession and to devote himself completely to the cause of social reform as he understood it. In 1855, he published a selected number of his monographs under the title of "The European Workers." A year later, he founded an international society for social economics which organized the

[73] Recently Raymond Cattell has used correlation analysis to develop dimensions relevant for the description of regions and countries. He is obviously not aware of Niceforo's work, another example of the discontinuities referred to previously. Niceforo gives in the last two chapters of a historical review of early quantifications of cultural phenomena. As one who has done work of this kind, I was embarrassed to find how many techniques considered contemporary inventions have been used many decades ago.

[74] It would be worthwhile to follow more in detail the ways in which Quetelet's ideas penetrated into specific subject areas. An interesting example, accidentally noticed, is the close personal contact between him and Florence Nightingale, described in E. W. Kopf, "Florence Nightingale as a Statistician," *Journal of the American Statistical Association*, December 1916.

collection of family monographs all over the world, and published them under the title, "The Workers of Two Worlds" in a series which continued until after the First World War.

After 1848, LePlay's source of income came partly from the Academy of Science, and partly from public offices which were entrusted to him by Napoleon III. Most conspicious were his activities as French representative for a series of international exhibitions. The reports on these exhibitions stress his organizational ability and describe him as a master of classificatory devices which facilitated the exhibition of products and the orientation of visitors among the vast variety of activities going on at these international affairs. In 1864, he was appointed Senator by Napoleon III. By that time, he was a well-known public figure, and had published a large number of books and pamphlets on current affairs.

In 1871, the Paris Commune was the third upheaval which LePlay witnessed at close range. From then on, he devoted himself completely to the setting up of various reform organizations, supervised the writing and distribution of numerous pamphlets, and finally founded a periodical, "La Réforme Sociale." His travels brought him repeatedly to England, which he admired because of its social stability; among his many publications is one on the British Constitution. Beginning in 1877, he published the six volumes which form the basis for our subsequent discussion. The book is known as the second edition of "The European Workers."[75] The first volume was newly written and presents in 650 pages an autobiography, a detailed account of the methods he used in his social investigation, and numerous prescriptions about how the results should be used for public policy. The other five volumes reprint the family monographs as they were presented in the first publication of 1855, including the earlier introductions. In addition each volume contains a newly written appendix which comments on the changes which had taken place between 1855 and 1877 in the regions he had studied. The family monographs themselves contain a great many policy considerations which are either summarized or extended in the new appendices. The cases are classified according to what LePlay considers their degree of social disorganization. In the second volume he starts with families from Eastern Europe, where he sees the highest degree of social stability; he then moves to the Scandinavian and English countries, where he still finds a considerable amount of stability (Volume III); Volumes IV and V then divide the Western continent into two groups of increasing difficulties; the sixth volume is devoted to family monographs mainly collected in the Mediterranean region, and they all are examples of disintegration.

In scrutinizing the literature about LePlay, I have found that many authors refer to the first volume only; it is indeed an impressive document. But as to his research techniques it tells what LePlay thought he did, not his actual procedure; this can be gathered only from a reading of the monographs themselves. I shall, when I discuss his methodology, also draw on Volumes II to VI. It is necessary to describe the rigid external form in which the fifty-eight family monographs are presented. At the center of each, as Sections

[75] Frederic LePlay, *Les Ouvriers europeens*, Vol. I (Alfred Mamé et Fils, 1879).

14, 15 and 16, one finds the famous budget, divided by accounts of incomes, expenses and supplementary details. The budget is preceded by thirteen sections which again are divided into four groups under the following headings:

A. Description of the locality, the occupational organization, and of the family itself (Sections 1 - 5)

B. Sources of subsistence (Sections 6 - 8)

C. Style of life (Sections 9 - 11)

D. History of the family (13)

Each of the first eleven sections is about a page long and consists in the main of vivid descriptive detail, the purpose of which is to help interpret the main budget account. Beginning with Section 17, each monograph is followed by about four to eight additional sections headed by the general title: "Various Elements of the Social Structure (Constitution Sociale)." The impact of the monographs derives mainly from these miscellaneous sections which range in length from two to ten pages. Some of them are based on direct personal observations to which I shall come back presently. Others are summaries of what today we would call organizational and institutional arrangements: descriptions of the steps by which a young worker in a certain region advances from apprenticeship to the status of master; classification of the kind of contracts which existed between workers and entrepreneurs; detailed information on the economic and technical aspects of the industry in which the family worked; wherever possible, LePlay draws attention to the inheritance laws or customs to which he attaches great theoretical importance. The sources of these descriptions are sometimes personal interviews with informants who know the area, and sometimes quotations from books. Another group of these accessory paragraphs consists of interesting but uncorroborated statements on what we would today call empirical correlations. He classifies workers into certain categories and claims that they vary according to the rate of illegitimate births; he divides the population of an area according to their ethnic characteristics and states that they also vary according to their colonizing ability—the topic of emigration is of great interest to him. Still another type of comment deals with one of the central topics of all his writings: What is it in the personal habits of a family or the conditions of their work which facilitates or inhibits their rise in the social scale?

The purpose of these sections of the family monographs is quite easily discernible. LePlay is not concerned with the families for their own sake. He is convinced that his case studies are the best means of understanding the working of the whole social system. He therefore wants to link the individual facts he has established up to Section 16 with whatever information he can gather about broader social structures. His contemporaries understandably admired him equally for his naturalist approach to the individual family and for his ability to abstract, during a relatively short trip, an enormous amount of information from crucially located informants and from literature only locally available. Even now, readers will be captivated by the vividness of his descriptions and the plausibility with which he argued connections between phenomena observed on very different societal levels. It is no wonder, then, that it took quite a while, as we shall see, before anyone asked whether his stories were correct and his interpretations sound. I am confident that much of his contemporary fame was due to the fact that he never just reported something the way a traveler would tell about a "cute" observation. Everything was part of a description of the functioning or malfunctioning of a coherent social system.[76]

LePlay tried to make each volume of "The European Workers" self-contained. As a result there is much repetition in explaining the purpose of his work and emphasizing the conclusions he wants drawn. Here is a brief summary of his main position.*

1. LePlay was convinced that he was creating an objective social science similar to the kind of mineralogy so familiar to him.

> In order to find the secrets of the governments which provide mankind with happiness based on peace, I have applied to the observation of human societies *rules analogous* to those which had directed my own mind in the *study of minerals and plants*. I *construct a scientific mechanism*. (Vol. I, first page of the Foreword)

He stated repeatedly that "the conclusions of 'The European Workers' are logically derived." (Vol. I, pp. 432, 436)

2. He did not see any contradiction between his claim to objectivity and the many statements that his personal history made him discover so many "truths" (les verités) overlooked by other students.

> My first impressions, which I mention here because they are so vivid, *devel-*

---

[76] Another element which accounts for the persuasiveness of the whole work is the skill with which he uses classificatory devices. I have already indicated the way each monograph in the six volumes of "The European Workers" is explicitly organized. In turn, the various introductions and appendices contain a great many cross references. In the first volume, which has a very complex organization, LePlay repeatedly stops to explain to the reader at what point he finds himself at the moment and where the following sections are supposed to lead him. This classificatory urge has obvious relations to his work as a metallurgist and reminds one also of his contributions to the international expositions mentioned above. It is undoubtedly also related to his grave doubts about whether writing books is an appropriate means of communication. In a long passage (Volume 1, p. 549), he explains that he has created so many reform organizations because his main reliance is on personal influence. At the same time, he knows that printed empirical descriptions are very important; so he continuously looks for ways of overcoming the obstacle of reader inertia. Charac-

teristically, each volume contains an alphabetic glossary explaining his main terms as well as pointing out where his factual material can be located in the volume.

I have not drawn upon the long series of monographs in "The Workers of the Two Worlds." Having read a fair sample of them, I agree with the judgment of several French writers that they are inferior. While LePlay was, for the first years, the general Secretary of the International Society which organized the collection, he seemed to have had relatively little contact with most of the individual contributions. It is incidentally worth translating the full title of the international collection: "Studies on the work, the domestic life and the moral conditions of the workers of various countries and the relations which connect them with other classes."

* If in the following pages references are given only by volume and page, they refer to *Les Ouvriers europeens;* the quotations are my translations from the second edition; I had no access to the first one; the underscoring is 'mine.

oped over the salutary influence of religion, national catastrophe and poverty. (Vol. I, p. 400)

He always disliked Paris. It is easy to explain why some people disagree with him.

> They were educated in urban agglomerations, and in educational institutions where in France all forms of error are accumulated; *they have acquired preconceived ideas,* the sinister influence of which I have just explained. (Similar references—Vol. I, pp. 41, 432, Vol. VI, p. 32)

Both statements are in the tradition of good sociology of knowledge. It is just not clear why the environment in which LePlay grew up led to truth while the upbringing of his adversaries bred error.

3. LePlay was quite explicit about the use he wanted made of his monographs. In his opinion, they bring out, by comparative analysis, the conditions under which people are happy or unhappy. This knowledge was to be conveyed to the elite of a country (les classes dirigeantes). They, in turn, were supposed to take the necessary measures so that favorable conditions prevailed. This is the main sense in which he uses the term "reform." The word he often uses to describe the groups responsible for the well-being of a society is "patronage." Patron is the French word for the small entrepreneur who still has quite close contact with his workers; he uses the term also to include the resident landlord. Occasionally, however, he personalizes this social function of a specific group:

> I clearly understood that the greatest interest of my fellow citizens is to *escape the errors which push them to disaster;* and I have even clearly seen that for someone who is in a favorable situation and has the qualities necessary for our times, it would have been easy to lead our people (race) back on the road to salvation. *I have not stopped looking for such a saviour.*

I cite this quotation, not only because of its characteristic wording, but because it appears in a chapter (Vol. I, Chap. 13) which for more than forty pages details his relations with various public figures of the July Monarchy and the Second Empire. His reports are undoubtedly biased; but an historian acquainted with the period will find in these pages important leads to study the relation between public policy and the beginning of social research in that period.

4. The large majority of the working people should contribute to the stability of society by conducting their family life according to two major principles: observance of the Decalogue and strengthening of paternal authority. These two terms are repeated well over a hundred times through the six volumes of "The European Workers." They are reiterated in the introductions and appendices to each volume; they are likely to appear in any of the sections of the cases themselves; and they form almost the theme song of the first volume. The word, "Decalogue," is not used in any metaphorical sense; LePlay is convinced that the Ten Commandments do indeed contain the essence of social wisdom.

5. Actually LePlay is concerned with the life of people at a relatively early period of industrialization in Western Europe. But he describes it in exclusively moral terms. People have vices and are corrupt; evil prevails in all strata of society; the population is degraded and offers no resistance to the "invasion du mal"; the "pratique du bien" disappeared; we live in an "époque d'erreur et de discord"; only the "reform" which is derived from LePlay's Methode d'Observation can bring help. Even the most descriptive parts of his monographs are so permeated by this terminology that it is quite pointless to pick out specific page references. The reader cannot glance at any three pages without meeting this vocabulary; in addition, each volume registers them in the glossary.

6. None of these evils is explained in terms of social and economic dynamics or as due to changes in technology. They are the result of "false teachings" and "fundamental errors" which have been disseminated. There are essentially two sources of these social heresies which have infected all strata of society. One is the writings of Rousseau, and especially his idea that man is born good; the child is a "dangerous barbarian" and it is the main task of a strict paternal regime to teach him order, obedience and submission to authority. The second source of all false ideas is Thomas Jefferson with his insistence on equal rights. He speaks of Jefferson's "deplorable presidency" (Vol. III, p. 426; for other examples of attacks on Jefferson, see, for instance, Vol. I, p. 626, Vol. VI, p. 28 and p. 546; I am sure that I have by no means caught all of them.) LePlay considered the United States the most vicious social system of all he knew, and predicted for it an early and complete disintegration. This was his position in the 1855 edition of "The European Workers" and it is reiterated with even more emphasis in the second edition (1877).[77]

7. According to LePlay, the success of his reforms will depend on the outcome of an "eternal battle between the good and the evil": the good is defined by the "hommes de tradition"; they stand for the "éternelles verités." The enemy are the "hommes de nouveauté." Most of them are intellectuals—the "lettrés." One repeatedly finds the opposition of "les lettrés" versus "les sages," the men who are usually old and without formal education, but imbued with the wisdom of the past. In Chapter 4 of the first volume, LePlay deals with the contributions which various occupational groups make to the welfare of society; it is no surprise to find that the liberal professions have a bad record, indeed. They come about as part of a necessary division of labor. "But earlier or later (they) misuse their authority; they oppress those whom they should protect, and they become dissemination points of corruption." (Vol. I, p. 131) A reference to the glossaries will round out the picture. Under the heading, "écoles," LePlay stresses that professors do harm if they do not consider themselves supplementary agents of parental authority; otherwise, they diffuse "bad principles." The next entry is *Economie Politique,* and it is *defined* as a "regrettable influence exercised by one such principle upon the mutual understanding of employees and workers."

[77] He refers at one point to de Tocqueville's "Democracy in America" as the "most evil book ever written by a man of good will." (Vol. 1, p. 193).

8. LePlay can in no way be seen as a defender of vested interests. He continually stresses the moral duties of the "patronage." He considers Adam Smith a sinister influence (Vol. I, p. 124) because he encourages the entrepreneur to be guided only by his economic interests. He feels that the absentee landowner neglects his moral duties. He is against the repression of free speech; but he argues characteristically that, if there were no free speech, the authorities would be deprived of sources of information which would help them to exercise their functions better.[78] He approves strongly of the incipient British factory legislation, but again, mostly because it sets limits to the freedom of the individual manufacturer. (Vol. III, p. 432ff.)

We are thus faced with a very strange phenomenon. Here is a natural scientist successful as a man of affairs and organizer of great public enterprises. He wants to develop an objective social science, to be based, as we shall see, on quantitative evidence. But the purposes to which he wants to put this knowledge are seen in a system of ideas which is not only conservative in terms of a Burke tradition but which, in modern usage can only be described as "fascistic." Much of the later history of the school which he created (and I shall talk only about those of his followers who continued his empirical investigations) can be understood only if all this is kept in mind.

The Purging of the Saint

From the middle of the 19th century, LePlay attracted waves of men who wanted to learn his research techniques. They shared his political and religious convictions. But in addition, they always looked at the social science which they wanted to practice as a kind of revelation which they were called upon to develop and pass on to subsequent generations. During LePlay's life, they were his apostles, grouped around the organization and the magazine, "Réforme Sociale." In 1896, four years after his death, those who considered themselves scientists as much as reformers started a new journal called "La Science Sociale" which appeared in monthly issues up to 1915. Its approximately 40,000 pages deserve intensive study because of their strange intertwining of a charismatic tradition and empirical research. There does not exist careful analysis of the LePlay tradition similar to Lottin's work on Quetelet. This may be due partly to the fact that LePlay described so extensively his own life and the development of his ideas as he saw it; existing accounts are essentially summaries of what LePlay had written himself. But also the mere labor of going through the thirty years of "Science Sociale" may have appeared too staggering. Nevertheless, I hope that such an effort will be made one day; it would contribute much to the main topic of this paper. The following pages try to indicate its possible value.

The journal, "Science Sociale," was subtitled "according to the Method of LePlay." The articles most interesting in the present context were the elaborations on the method and the periodic reviews of the organized efforts to apply LePlay's ideas to the empirical study of concrete situations.[79] A second group of contributions are new family monographs in the LePlay

---

[78] Frederic LePlay, *Collection des Grands Economistes*, Louis Baudin (ed.), (Paris: Dalloz, 1947).

[79] The methodological emphasis of the group is quite remarkable and a continuous progress towards increased explication and improve-

tradition. A third kind of paper reanalyzed earlier writings: thus, for instance, we find an interesting series of papers on Montesquieu and the kind of evidence he adduced in the "Esprit des Lois"; another scrutinized Necker's writings on public opinion. A fourth group of efforts were devoted to the social interpretation of literary products: What was the social context of the Iliad? What was Balzac's image of the French middle class? The fifth type of essay deals with the historical record of whole cultures like the Assyrian or the ancient Chinese. Finally, there are discussions of contemporary problems, such as the early difficulties in the colonization of Algeria, the reform of the French educational system, the question of anti-semitism, etc.

All these contributions presume that the LePlay school had developed a method by which social subject matters could be treated in an objective and definitive way. Many of these articles contain valuable factual information or lucid insights into other people's writings; all of them reflect a considerable amount of righteousness and self-assurance. The whole style of writing is quite reminiscent of contemporary Christian Science literature. Instead of talking about the Science Sociale, they just refer to it as "La Science"; Le Methode d'Observation becomes "La Methode." And yet, if one scrutinizes the articles carefully, a very significant change in tone becomes noticeable. To make the drama of this play more understandable, some of its actors must first be introduced.

The editor of the journal, up to his death in 1907, was Edmond Demolins. Almost forty years younger than LePlay, he was an historian trained in a provincial Jesuit college, who came to Paris in 1873. He soon fell under the spell of LePlay and began to give public lectures on his methods. During the early years of "Science Sociale," he wrote periodic reviews on the progress of the movement. He was clearly the public relations agent for LePlay; some of the foreigners, who later on disseminated the ideas of the group in their own countries, never met LePlay and refer to Demolins as their main source of inspiration. He was always engaged in a diversity of activities. In the early 1890's, he created an anti-socialist league and tried for a while to use the journal as its communication center.[80]

---

ment of methods can be traced in "Science Sociale" almost from year to year. Nathan Glazer's opinion that LePlay's method was not further developed is obviously based on the fact that he did not have access to the journal (See footnote 15). Special issues summarizing new methodological developments or reassessing the state of affairs will be found in many volumes; see especially November 1912 and October 1913 of "Science Sociale."

[80] At the turn of the century, Demolins created a school for young boys, Ecole des Roches. The school was located in the open country and the students organized their own work and life. Great emphasis was put on observations of nature as well as on field studies in the local communities and farm regions. The school still exists today. At the time, there was a similar movement in Germany, initiated by Lietz; this later on was im-

portant in the history of the German youth movement. Demolins organized exchange visits with other institutions and there is interesting correspondence between him and W. R. Harper, the first president of the University of Chicago. Demolins' own publications, after he became a LePlay convert, were a combination of his early historical interests and LePlay's insistence on contemporary field studies. In recent bibliographical reviews, Demolins is usually classified as a geographical sociologist. The best source on him is the detailed biography which appeared in "Science Sociale" at the time of his death. The journal also published from time to time reports on the activities of the Ecole des Roches. Demolins incidentally published also a variety of more popular pamphlets on the research of the LePlay group which I have not examined.

While Demolins was the activist of the group, the man who was increasingly considered the intellectual heir of LePlay is Henri de Tourville (1843-1903). He was an abbé who lived most of his life in seclusion, trying to make LePlay's work more systematic. His great contribution was a classificatory system which was always reverently called the "Nomenclature" with a capital "N." During the first two years of the journal, de Tourville wrote a series of articles entitled "Is Social Science a Science?" His answer was, of course, affirmative. But he made the point that the genius of LePlay could be utilized by his disciples only if his system was made more explicit. His starting point was the way in which the family monographs were reported. He pointed out that while the first thirteen paragraphs were fairly systematic, the richest insights of LePlay are found in the supplementary paragraphs for which the founder had never provided a systematic guide, thus leaving future workers somewhat up in the air. De Tourville wanted to work out a system of categories which would give a place to every relevant observation LePlay made. I shall come back to this idea presently.

The next major figure in the sequence is Paul de Rousier. His contribution lay in further clarification of the core of LePlay's family monographs—the budget. In issues of "Science Sociale" appearing around 1890, he suggested certain improvements to which we also shall return later. At about the same time, de Rousier also made several trips to the United States, which were to become very important in the further history of the school. His main move at the time was to show that LePlay had been too harsh on the Americans. Here were all these rich people like the Carnegies and the Goulds who had just the sense of responsibility which, according to LePlay, was the duty of the elites. The same men on whom we today look back as economic robber barons Rousier described in "Science Sociale" as the true modern aristocrats. The youngest among the more prominent men in this sequence is Paul Bureau; he joined the group a few years after LePlay's death. By introducing explicitly psychological elements, he was the first to bring about a major change in the Nomenclature. His new category was what he called "representation de la vie" which he proposed as the translation of the German word "Weltanschauung."[81] Bureau was considerably influenced by Gabriel Tarde and, later, wrote an interesting account of the way in which his ideas developed.[82]

Beginning in the late 1890's, a tone of aggression enters into the writings of the group; increasingly a strong ambivalence is apparent. LePlay is still compared with Newton and Galileo and any other great scientific hero who comes to mind. But, in fact, he is being purged. Beginning in 1904, the subtitle of the journal is no longer "according to the method of LePlay" but "Selon la Methode d'Observation." The new series starts with a review of the present state of "La Methode Sociale." The achievements of the last twenty years are no longer described as explications and clarifications of LePlay's own ideas, but as important discoveries made by the younger men. The master left grave

---

[81] Paul Bureau, *Introduction à la Méthode sociologique* (Paris: Blond & Gay, 1923).

[82] Bureau finally broke with the later LePlayistes, accusing them of materialism. He quotes as a witness the French economist, Charles Gide, to the effect that "in reality the new school has not preserved much from the method of their master" (*op. cit.,* p. 116f).

gaps in his monographs; his use of the budget implies a series of errors; his family monographs are monotonous; his family types are badly defined; de Tourville's work is "more scientific and more interesting" (de Rousier). Le-Play made intuitive observations but does not tell on what he based his conclusions; his analysis was oversimplified and incomplete; he was badly deceived and made grievous errors (Demolins). LePlay's family types were quite wrong; he based his classification on the way property was transmitted, but the correct principle of classification would be the kind of educational tradition a family has; his stem-family covers a variety of quite different phenomena. He was especially wrong in his anti-Americanism. Not only are the people in the United States not decadent, their spirit of individualistic enterprise is the best bulwark against the spread of socialism (Pinot). This is the mood in which "La Science Sociale" is written until 1915.[83]

We are thus faced with the following situation. Here is a school created by a charismatic personality who makes an important innovation in social methodology and intertwines it with very strong and activitistic ideological beliefs. One group of his disciples share his beliefs and want at the same time to make methodological progress. Under normal circumstances, a scientific innovator is respected by his students, and it is taken for granted that his successors make continuous advances beyond the teacher. In a charismatic context, this leads to a tension between the scientific and the sectarian element in the tradition. This would be a matter of only secondary interest if it were not for the fact that rather suddenly, the LePlayistes disappeared from the French sociological scene. At least as far as one can see from a distance, the school which was so extensive and articulate up to the First World War has been completely replaced by the Durkheim tradition to which they paid only casual and rare attention in the "Science Sociale." In various reviews of French sociology which Frenchmen have recently written the LePlayistes are not even mentioned.[84] Do we face here a political phenomenon? Did the few relevant university posts all go to the Durkheim group at a time when the French government had an anti-clerical tendency? Did the descriptive fervor of the Le-Playistes exhaust its potentialities, and make it less attractive than the conceptualizing of the Durkheim School to a younger generation? Do we face here the difficulty a charismatic movement has: in spite of their ambivalence to the founder, did the LePlayistes form too much of a sect to be acceptable to the regular academic bureaucracy?

[83] This is the last volume available in Harvard's Widener Library and I suppose that the First World War brought the publication to an end. Unfortunately, a few earlier volumes are missing, among these, the volume of 1903 which precedes the change of name and the review issue I have just described. I do not know, therefore, whether a special explanation was ever given for the change. The volume of 1904 has an imprint "19th year, second period"; I pass over some other technical changes which the editors made. The type of content remains otherwise quite unchanged and Demolins remains editor. The later history can be picked up in the retrospective pages of Bureau

(op. cit., Introduction and pages 115 ff.).

[84] This is strictly true only for Levi-Strauss' contributions to "Twentieth Century Sociology" (New York, Philosophical Library, 1945). Stoetzel mentions LePlay several times in passing when he stresses the empirical tradition of French sociologists in his contribution to "Modern Sociological Theory" (New York: Dryden Press, 1957). He also refers to Bureau and so, incidentally, does Merton in his review of French sociology ("Social Forces," 1934). As far as I can see, Bouglé in his "Bilan de la Sociologie Française Contemporaine" (Paris: Alcan, 1935) never mentions the LePlayistes.

I cannot tell. Certainly the methodological ideas of the school were interesting and susceptible of further development as I shall discuss presently. But first, I must trace briefly the effect which LePlay had abroad, especially in England and in the United States and here, strangely enough, the theme of the purge can be continued. While we find outspoken and clamorous admirers of his in the two Anglo-Saxon countries, they changed his ideas even more than did his French disciples, each in his own way and perhaps without knowing it.

Ramifications Abroad

In 1878, a young Scotch biologist by chance visited a lecture of Demolins. He was deeply impressed, spent the remainder of his Paris study trip in contact with Demolins—he seems never to have met LePlay personally—and in his later writings always described himself as a LePlayiste.[85] There are several biographies of Patrick Geddes, the Edinburgh professor of botany who, in 1902, joined with Branford in creating the Sociological Society in London. All stress Geddes' magnetic personality, his great schemes and his tremendous energy; but they offer little information on his intellectual development. I must therefore depend upon Geddes' own story in "The Coming Polity" which has a special chapter on "LePlay and his method" and many references as to how LePlay's French background can save the social sciences from the evils of "Prussianism" (the book was written during the first World War).[86] I come reluctantly to the conclusion that Geddes never really read LePlay's monographs. In vague terms, he speaks of him as a regionalist and praises him for his fine maps (pp. 183 ff.); Geddes, himself, was famous for his graphical presentation of social facts, but LePlay never published a map, except one indicating the geographical location of the families he studied. Geddes thinks of LePlay as a kind of rural sociologist, and seems unaware of the master's political views which certainly were not congenial to his own position. He and his students[87] kept hammering upon a presumed central formula of LePlay— (place-work-folk)—which does not play any role in his monographs and which can at best be read into some chapter headings in the first volume of the second edition of the "European Workers." Geddes kept in touch with Demolins —he had him as lecturer in summer schools organized by the "Outlook Tower," the famous Edinburgh building, where his farflung activities as city planner were centralized. I guess that the charismatic atmosphere engendered by LePlay appealed greatly to Geddes who himself had the same effect on his disciples. Branford, one of them, was a wealthy businessman who in 1920 donated a house for the work of the British Sociological Society, and called it "LePlay House." Yet from available literature I cannot trace any concrete influence of LePlay's actual research upon the city surveys of the Geddes school. Perhaps British colleagues can provide further clarification.[88]

[85] Victor Branford and Patrick Geddes, *The Making of the Future, The Coming Polity* (London: Williams and Norgate, 1917).

[86] Philip Boardman, *Patric Geddes, Maker of the Future* (Chapel Hill: Univ. of North Carolina Press, 1944).

[87] S. Branford and A. Farquharson, *An In-troduction to Regional Surveys* (Westminster: The LePlay House Press, 1924).

[88] The matter is not clarified by a first scrutiny of the "Sociological Review," the journal created by Branford. As a matter of fact, a curious episode emerges. One of Geddes' followers was the geographer, Herbertson, whose

Another British social scientist presents a more puzzling problem. Charles Booth, the organizer of the great survey of "Life and Labour of the People in London" (the enterprise started around 1880), did in two respects work quite reminiscent of LePlay. He was an avid and skillful observer of family lives, an ability which is well documented by the examples in a biography written anonymously by his wife;[89] and he did at certain points of his work, study budgets. Strangely enough, however, I have in all the writings on Booth not been able to find any evidence that he was even aware of LePlay. The latter is neither mentioned in the chapter of Beatrice Webb's *Autobiography* devoted to her collaboration with Booth nor does he play a role in a recent biography by Simey and Simey.[90] They point out that very little is known about the origin of his ideas; this might be an explanation for this gap in the evidence. But one should also keep in mind that there are vast differences in the basic approach of the two men in spite of the external similarity of their procedure. Booth's thinking was centered around the problem of poverty and he tried to measure it as precisely as possible. LePlay was guided by a vague notion of corruption and never had the idea that some systematic classification of families could ensue from it. Booth is much nearer than LePlay to modern procedures of translating a concept into a well defined system of indices. The difference in technique has a consequence which might exemplify the interplay of ideology and methodology. Both men were started on their inquiries by discussions with friends who were of a radical political persuasion; LePlay and Booth believed that the consequences of industrialization did not require the remedies advocated by their interlocutors. Booth, as a result of his own studies became convinced that he was wrong; LePlay never changed his belief in his own righteousness. The difference is of course mainly due to the general attitudes of the two men. But actually Booth's procedure permitted a check on the amount of existing poverty, while LePlay's observations could neither settle how much "evil" there was, nor to what it was due.

---

wife Dorothy wrote a biography of LePlay around 1900. It was essentially a compilation of statements taken from the first volume of "The European Workers." She sent the manuscript to Branford, who did not know what to do with it. When LePlay House was created, people became curious about the man after whom it was named. Branford remembered the manuscript of Mrs. Herbertson (she had died meantime), edited and published the first four chapters, which summarized the external data of LePlay's life, in "The Sociological Review." He never published the rest. I am satisfied that the reason is as follows: When he came to edit further chapters, Branford noticed that they mainly contained a summary of some of LePlay's outmoded anthropological ideas (Chapters 5 and 6); lengthy descriptions of the categories used in his budget accounts (Chapters 7-9); LePlay's judgment on what various occupational groups contribute to society, seen from his point of

view and based on the situation of fifty years earlier (Chapters 11 and 12); and repetitions of LePlay's statement on his own political mission. It was like taking all the ashes from LePlay's altar and leaving the fire behind. Branford obviously found the manuscript too embarrassing. The present editor of the Review published the whole manuscript in 1950 as a special issue. His short foreword contains a number of interesting historical remarks; he has a slightly different explanation of why Branford discontinued publication. He mentions incidentally that latter-day LePlayistes were active in the Vichy Government during the Second World War. (For addition to this footnote, see p. 5, Section III)

[89] Charles Booth, *A Memoir* (London: Macmillan & Co., 1918).

[90] T. S. Simey and M. B. Simey, *Charles Booth, Social Scientist* (London: Oxford Univ. Press, 1960).

I shall show presently that LePlay and Booth make very different uses of their budget data. Whether their qualitative monographic work is similar or not I cannot tell. The answer would have to come from a very detailed comparison of some of their cases. This has not been done yet but would certainly be worth the effort. The Simeys point out that Booth is badly neglected by contemporary sociologists, and I very much agree with them. When however, they call him "the founding father of the empirical tradition in the social sciences,"[91] they certainly do a great injustice to LePlay.

In the United States, in 1897, the *Journal of Sociology* published a set of instructions for the collection of family monographs.[92] But already, a few years earlier, American social scientists seemed to have become interested in the LePlay school. At the time of one of his visits to the United States, de Rousier was asked to describe its activities; his report was published by the American Academy of Political and Social Sciences.[93] This publication is interesting because it already foreshadows the subsequent ambivalent criticism of the LePlay disciples. To have overlooked this point is the only objection I would raise against the best available English-language presentation of LePlay's own work and that of his followers. Sorokin in his book on contemporary sociological theories devotes more space to them than to Durkheim and Weber taken together.[94] He is mainly concerned with substantive ideas and not with methodological matters; a review of his analysis therefore does not fall within the scope of this paper. Two of his students however made themselves the American exponents of LePlay's family studies and their enterprise requires a more detailed discussion.

In 1935, Zimmerman and Frampton published a book on "Family and Society."[95] The last part of it consists of a 240-page long condensation of the first volume of "The European Workers." The original translation was made by Samuel Dupertuis, and the American version carries the introductory note that the condensation was done "without destroying a single idea." A comparison with the original text shows, however, that somewhere along the line a strenuous effort was made to attenuate LePlay's position so as to make it palatable to an American academic audience. All attacks on America are omitted, the word "Decalogue" never appears—in its place are terms like "moral law," "mores," "universal moral code," etc.; sometimes the phrase "Decalogue and paternal authority" is presented only by the second term. One should also know that LePlay's careful editorial structure, so characteristic of him, is destroyed; in the American version, a major chapter of LePlay's original occasionally begins in the middle of a paragraph. What most emphatically has been eliminated are LePlay's major obsessions without which he cannot be understood at all. Because the Dupertuis version provides American readers with their only access to the French author, I give one example in

[91] *Op. cit.,* p. 190.

[92] Frederic LePlay, "Instruction in the Observation of Social Facts According to the LePlay Method of Monographs on Families," trans. by Chas. A. Ellwood, *The American Journal of Sociology,* 1896-1897, 2.

[93] Paul de Rousier, "La Science sociale," in *Ann. Amer. Acad. Polit. Soc. Sci.,* Jan., 1894.

[94] Pitirim Sorokin, *Contemporary Sociological Theories* (New York: Harper & Bros., 1928).

[95] Carl Zimmerman, *Family and Society* (New York: van Nostrand, 1935).

some detail. Here is a passage from p. 453 in Zimmerman-Frampton, corresponding to about pp. 167-169 in LePlay, Vol. I.

> The sophists of England and Germany inspired by the eloquence of Rousseau, have tried to meet the situation. They conclude that social disorders come especially from the constraints prescribed by the mores (!) and exercised by the family heads, and by the civil, religious and political hierarchies which increase the strength of paternal authority. They seek to abolish these constraints by overthrowing the rulers, if necessary.

Now this statement reads as if it belonged to a distinguished conservative tradition à la Burke (whom LePlay indeed quotes in a footnote). But in the original text the first sentence is preceded by a lengthy passage beginning with: "Around 1750, thinking began to be misled in the literary academies and in the Parisian salons, where the intellectuals, the aristocrats and the financiers met together." The imagery of a conspiracy, the notion that social changes are fostered by "people in the backroom" can never be found in the American version, although it is so characteristic for LePlay; as a matter of fact, a few lines later, the French author repeats that these people "began to pervert the minds of their contemporaries with their sophistries." The first sentence of the Dupertuis translation is correct but he omits about ten lines coming before the second sentence. These Rousseauian opinions are described as "absolutely false" and "contrary to the opinion of all thoughtful men (sages) and to the evidence available daily to the mothers of babies and their nurses." This anti-intellectual element in LePlay's text is avoided throughout the Dupertuis translation. And in the same ten omitted lines, we find still another of LePlay's favorite themes: "Logic applied to Rousseau's fundamental error (leads) to conclusions from which, as a fatal consequence, derives the ruin of any society adopting it." The idea that there are fundamental errors and verities from which logical inferences lead to wrong or correct views about society is the counter-part of LePlay's drive toward empirical observations: a reader cannot really assess the latter if he does not at the same time become aware of LePlay's pseudo-logic which permeates all his arguments. Even the last sentence in our quotation has in its original form an additional implication. The original says: (the sophists think) "that these constraints and hierarchies should be abolished; and if the rulers hesitate to accomplish this task, they should be overthrown." LePlay imputes even to the revolutionaries that they first would try to make the rulers change their evil ways and then only, if this does not work, would they be overthrown.[96]

The French disciples of LePlay were much concerned with the methodology of his work. They were exasperated by the fact that his monographs made fascinating reading while his procedures were loose and often manifestly faulty. The whole history of the "Science Sociale" group can be understood as an effort to capture his spirit and to make it transmittable to others. His Anglo-Saxon admirers probably hardly knew his actual studies which to my knowl-

---

[96] Let me add that in the eighty pages where the authors of "Family and Society" comment on the importance of what they call "LePlay's Theories" none of the actual family monographs is mentioned.

edge are even now accessible only in French. They were fascinated by his programmatic writings and transmitted them in varying degrees of vagueness and distortion.

The first check by an outsider was published in 1913 by Alfred Reuss.[97] His important monograph deals with "LePlay's significance for the development of methods in the social sciences." However, it is not so much concerned with his procedures as with the reliability of his observations and the relation between his data and his conclusions. The most startling part is Reuss' re-analysis of one of the family descriptions that LePlay had made (Volume IV, Case 6). Reuss went back to the German village in which two children of an observed family still lived, and he made use of a great deal of available documentation on local social conditions. He provides a fifty-page translation of the original case study and then confronts it with the material which he himself collected. The discrepancies are of various kinds. Thus, the budget figures make it appear that the man's only recreation was drinking, while actually he was active in a number of civic organizations, interests which are nowhere mentioned by LePlay, which do not show up in the budget, but which in the light of local political habits, explain the large amount of money spent in pubs. (Germans use the term "Bierbank Politiker" to describe a local worthy of this kind.) LePlay's more general considerations made him overlook facts which happened not to fit in with his preconceived notions. Thus, he was greatly concerned with the negative effects of the French inheritance laws which led to the progressive splitting up of farms. In the area on which Reuss checked, LePlay had remarked how the family life of the oldest son had benefitted from his being able to keep his parents' farm intact; but LePlay neglected the bad effects that this had had on the lives of the younger children about whom, according to Reuss, there was ample evidence. As one would expect, the worst misperception occurred on what we today would call labor relations. Reuss documents from contemporary newspapers the occurrence of repeated local riots because of low wages, exploitation by company stores and unsanitary working conditions. Nothing of this is mentioned by LePlay. Reuss's monograph still deserves careful reading; in our context, pages 85-89 on "the rise of figures by LePlay" are specially worthwhile.[98]

One final example combines the charismatic role of LePlay and what happens when "outsiders" enter the scene. In 1886, ten years after Geddes, a young French Canadian spent a few months in Paris and fell under the spell of Demolins. Upon leaving France, he pledged that he would devote his life to the study of Canada in LePlay's spirit. Leon Gerin, who became a distinguished civil servant, collected monographs which were published in "Science Sociale." He also revisited families which had been first reported in "Workers

[97] Alfons Reuss, *Frederic LePlay in Seiner Bedeutung für die Entwicklung der Sozialwissenschaftlichen Methode* (Jena: Gustav Fischer, 1913).

[98] In 1950, Eliot and Hillman, *et al., Norway's Families* (Philadelphia: Univ. of Pennsylvania Press, 1960) studied a Norwegian family to compare it with case 2 in Vol. III of LePlay. Their interest was in analyzing change, not to check on the case LePlay had collected more than 100 years earlier. But from historical records they feel that LePlay painted too rosy a picture of Northern Europe, which is the subject of Volume III. The authors mention in passing that they could not trace the existence of the 1845 family.

of Two Worlds"; he never doubted their authenticity. The official obituary speaks of him as the founder of Canadian sociology. At the age of seventy-five, he published a selection of his cases, several of them introduced by de-voted memories of his initiation by Demolins. Since Gerin's death, however, younger Canadian sociologists have cast grave doubts on the LePlay-Gerin approach. The discussion deals less with research methodology than with con-troversial interpretations of Canadian social history; I therefore mention it only in passing.[99]

LePlay's monographs captivate the reader by his insights, his reckless gen-eralizations, his stream of alleged evidence, his superb style, the clear structure of his writings, and, even if one disagrees with it, the consistency of his philo-sophical position. Outsiders have sided with or against him mainly on emo-tional grounds, since, except for Reuss, none of them has analyzed his empirical work; and Reuss wrote at a time when methodological thinking on social re-search was still in its infancy. The following pages are intended to give an outline of the direction in which a systematic study of LePlay's monographs holds promise.

Quantification and Diagnostics in LePlay's Monographs

Le Play is probably best remembered as the man who introduced the family budget into the tool chest of the empirical social scientist. He is quite out-spoken about its central methodological role. One finds many remarks like the following:

> Every action which contributes to the existence of a working family leads more or less directly to an item of income or expense . . . (Vol. I, p. 225)

> There is nothing in the existence of a worker, no sentiment and no action worth mentioning which would not leave a marked clear trace in the budget. (Vol. I, p. 237)

He repeatedly compares budgetary analysis with the work of the mineralo-gist which he knew so well.

> The surest way an outside observer has to know the spiritual and material life of people is *very similar to the procedure which a chemist uses to under-stand the nature of minerals.* The mineral is known when the analysis has isolated all the elements which enter into its composition, and when one has verified that the weight of all these elements adds up exactly to that of the mineral under analysis. A similar *numerical verification is always available to the student who analyzes systematically the social unit represented by the family.* (Vol. I, p. 224)

LePlay wrote in great detail about the best way to classify and compute the budget items he obtained in periodic talks with his respondent. In this sense,

[99] Philippe Garigue, *Etudes sur le Canada Francais* (Montreal: Universite de Montreal, 1958) and Leon Gerin, *Le Type Economique et Social des Canadiens* (Montreal: l'A.C.-F., 1938). I am indebted to Professor Sigmund Diamond, who, in the course of his own re-search on French Canada noticed the great role of the LePlay tradition among French Canadian sociologists. Some information on Gerin can be found in the 1951 proceedings of the Royal Society of Canada.

he can indeed be considered the fountainhead of an important quantitative technique. But his analysis of the data was quite peculiar and it took others to develop the whole range of possibilities.

In principle, there are three major ways to use budget data; these may be tagged as the analytical, the synthetic and the diagnostic procedure. By *analytical* is meant the study of specific expenses, either in relationship to each other, or to the total income of the family, or to some of its general characteristics, such as occupation, age of children, etc. Already, during LePlay's lifetime, the interest of some of his contemporaries shifted towards the search for such generalizations. In 1857, the German economist, Ernst Engel, published his famous law stating that the proportion of income spent on food increases as the total income of a family decreases. His data were taken in part from LePlay's monographs.[100] Since then, this kind of generalization has been the main objective of an ever-increasing number of budget studies. They represent early forms of multivariate analysis; on the same income level, for example, white collar families spend more on rent than manual workers. I know of no evidence that LePlay was aware of this use of his material. There is very little doubt that he would not have thought well of it. He, as well as his students, found statistical generalization quite pointless.[101]

In the *synthetic* mood of budget analysis one combines all the information to form what in principle are types, although often the information is finally translated into uniform money terms. The best example for our narrative is the way Booth, in his first social survey of life and labor of the people in London, tried to establish his poverty line. For a large number of items of food, clothing, shelter, etc., he listed the minimum supply which, by expert opinion, was needed for the sustenance of a family of given size. If its income did not permit it to supply itself with these items, it was classified as poor. To find the extent and distribution of poverty was the main purpose of this enterprise, as was mentioned before.[102]

What then did LePlay himself do with the pages of budgetary information covered in Sections 14-16 of each monograph? In his methodological introduction, he gives the following example:

> Often a single figure says much more than a long discourse. Thus, for instance, one cannot doubt the degradation of a Paris worker after one has learned from the study of his budget that each year he spends 12% of his income to get drunk, while he does not devote a cent for the moral education of his five children of ages 4-14. (Vol. I, p. 226)

In other words, he selects specific items and uses them for what is best called

---

[100] Engel, himself a mining engineer by training, wanted to combine the LePlay and the Quetelet traditions. He created the term "Quet" for the basic unit in a consumption calculus.

[101] Bureau quotes an extensive diatribe by LePlay against statistics (footnote 82, p. 228) without giving the source. In an issue of "Science Sociale" (Nov. 1912) Descamp compares LePlay's methods with other procedures in social research and again argues against the kind of evidence which today we would call correlation analysis; one of his negative examples is Durkheim's "Suicide."

[102] The Booth survey was repeated later by some of his former assistants. In Vol. III of "The New Survey of London Life and Labour" (London, Orchard House, 1932) one can find a sophisticated discussion of the way Booth tried to translate the notion of poverty into a classificatory instrument. See especially pages 8 ff., 70-77, 97-106.

diagnostic purposes. Space does not permit us to discuss the logical foundations of this kind of social symptomatology. I can list only a few examples here. LePlay uses specific budget items as indicators of broader sentiments or social configurations. A French tinsmith pays high dues to a labor union, which shows how aggressive he feels against upper class people; the family of a London cutlery worker spends much money on food, which allows one to infer that they will have little chance to advance on the social ladder; a German worker's income derives partly from gardening, which accounts for the moral stability still prevailing in his type of family. Most of these observations refer to moral issues: too much money for drink, not enough for religious practices and education, too much for "useless recreations." One has to keep in mind that these budget items are mentioned as part of general observations and discussions which go far beyond the quantitative evidence. One can get the full impact of LePlay's monographs without ever looking at the dreary pages of balance sheets for earnings, occasional incomes, fringe benefits and expenses.[103]

The later LePlayistes increasingly abandoned the budget; as a matter of fact, they became highly critical of it. The nature of this criticism in itself deserves some attention because it signals an interesting trend in the whole history of quantification. When de Rousier in the first issue of the new series of "La Science Sociale" listed all of LePlay's shortcomings, he quoted many examples of the kind which were exemplified above by the summary of Reuss' reanalysis. He thought that LePlay was just following the "habit he had acquired during his professional studies (as a mineralogist)." According to him, LePlay was "seduced by this desire for numerical verification and as a result, left aside the phenomena *which cannot be expressed in numbers and therefore elude such verification*" (emphasis supplied). But if one looks at some of de Rousier's examples, one notices that today they would in no way be considered unquantifiable. Thus, for instance, he says that many families do not spend money, but time, on the education of their children. He obviously did not consider the possibility of a time budget, which today has become quite conventional. He says that expenses for devotional candles are not an appropriate measure of religious devotion, but he does not consider the possibility that records of church attendance or family prayer might at least enlarge the scope of quantitative measurement. In other cases, statistical records of the kind of personal contacts a family has would cover well de Rousier's examples of phenomena which supposedly are accessible only to qualitative comments. LePlay, incidentally, often reports what the members of his families talk about with each other. Even quantified inventories of conversations, although still rare, have cropped up in recent empirical studies.

We face here an episode in the history of quantification which has many parallels. LePlay proposes budget items as a social measure. After a while, the instrument proves deficient and so time budgets or sociometric records are

---

[103] On the other hand, they now form important documents for the economic historian. LePlay repeatedly stressed the relevance of his work for the future historian, who, due to him, would have better information about social conditions during his lifetime than for any other period.

proposed. They cover more ground but after a while, they too, appear to leave out some significant parts of social reality. Bureau, for instance, reproaches the LePlayistes for being materialistic because they fail to consider what today we would call attitudes. Now attitudes are being measured and the objection is that this is an atomistic approach and that it does not take into account "climates of opinion" or "collective norms." These periodic waves of optimism and pessimism are one of the topics which the history of the LePlay school suggests for further investigation.[104]

I now return to another aspect of LePlay's diagnostic procedure. Anyone who has done field work or who has reported to sociologically-minded friends about a personal trip hopes to find incidental observations which throw light on a complex social situation. Professors at the Sorbonne do not list their telephone numbers in the directory; how well does that indicate their exalted status and their social distance from students? In some American towns, families do not lock their doors when they leave the house; to what extent is this an indicator of mutual trust? Anthropologists developed great skill in such observations because of the language barrier between them and the people they visited. In contemporary society, students like Riesman and Margaret Mead have come to symbolize the art of making such incisive diagnostic observations. The logic of the procedure is by no means yet clarified and I will not try to discuss it here.[105] But LePlay's monographs certainly contain pertinent examples. To show the religious indifference of a London cutlery worker, he mentions that the man did not even know how to find a minister when a

[104] In Gabriel Tarde we find a typical remark (condensed translation from his "Lois de l'Imitation" p. 227): "Statistical data are only poor substitutes (for what we really want to know). Only a psychological statistic, reporting on changes in the specific beliefs of individual people—if this were at all possible —would provide the deeper reasons for the ordinary statistical figures." Just as Tarde considers attitude measurements impossible so does Zizek feel about the measurement of occupational prestige, which today has become a research routine. He says (*Soziologie*, p. 25, condensed translation): "In the study of social stratification statistics deal with tangible occupational characteristics, while for the sociologist other aspects, like e.g. social prestige might be of importance. The sociologist will therefore often combine statistical data with non-quantitative information." Toennies (Ferdinand Toennies, *et al., Sektion Soziographie, Siebenter Deutscher Soziologentag* Tübingen: Mohr, 1931]) the great believer in "Sociography" always added that "of course" many things cannot be quantified; but he was diplomatically vague as to where the limits were.

[105] A systematic collection of such global indicators from a variety of community studies has been made by Patricia Kendall (*Qualitative Indicators in Field Work* [mimeo], New

York: Bureau of Applied Social Research, Col. Univ.). The technique is, of course, well known to the historian who often must depend on the interpretation of a single letter or the report on a ceremonial event. Rosenberg has analyzed such procedures in the writings of a number of historians and recreated the transition from the medieval mind to the spirit of the Renaissance. (George Rosenberg, "Without Polls or Surveys," Ph.D. dissertation, 1960, Columbia Univ.)

Among several Geddes biographies the one by Boardman (Philip Boardman, *Patrick Geddes, Maker of the Future* [Chapel Hill: Univ. of North Carolina Press, 1944]) is relatively the most sober one. Even there Geddes' initiation to LePlay through a lecture by Desmolins is described in the following way (p. 42): "In a flash Geddes saw that in LePlay's travels and his actual observation of society, there lay a method of study which both satisfied him as a scientist and inspired him, as one who often puzzled over mankind's ways and institutions, to follow this lead."

In tracing the charismatic nature of the LePlayist world, one will have to keep the age of the actors in mind. LePlay started his role as a "prophet" at the age of fifty. His most immediate French apostles were thirty to fifty years younger than he was.

family member who was ill requested religious consolation. (Vol. III, Case 6)
The high status of women in a nomadic family of the Urals is demonstrated
by a description of how the wife interfered when LePlay interviewed the husband. (Vol. II, Case 1). The social alienation of a tinsmith in French Savoy
is exhibited by the fact that he and his wife collect mischievous gossip about
dignitaries of the town. (Vol. III, Case 4). An impoverished family who won
a special prize spend most of the money buying new clothes which shows that
they hope to regain their former social status. (Vol. VI, Case 7) Living with
the family of a Viennese carpenter, (Vol. VI, Case 1) LePlay tries to teach
the wife an economic lesson: he lends her money so that she can buy sugar
wholesale. The experiment fails because the children keep begging for more
and it is easier for the mother to refuse when she can point to an empty
cupboard. For LePlay this is a sign that the family will never become an
economic success, but at least he appreciates the tender heart of the mother.

What one might call the global indicator game is something which binds
together whole generations of would-be LePlayistes. Lewis Mumford wrote a
short essay on Patrick Geddes and Victor Branford, both of whom he knew
in connection with his interest in city planning.[106] He makes the usual un-
documented and stereotyped references to LePlay's influence on the two men,
but he certainly catches the spirit of the affinity when he characterizes a typical stroll with Branford:

> He would gleefully point out some sinister exhibition of the social process,
> as in the combination of a bank with a meeting hall in the Methodist Center
> in Westminster, or the juxtaposition of the bust of Cecil Rhodes with the new
> examination buildings in Oxford, which sorted out the brains of an imperial
> bureaucracy. (p. 684)

It would clarify the nature of LePlay's work and the logic of this diagnostic
procedure if one were to collect and analyze systematically all pertinent examples in LePlay's monographs. I have the impression, however, that they
are not very numerous. LePlay gives much more space to observations which
go beyond the individual family and link it to broader sociological statements.
Or, to put it more precisely, he soon leaves the specific family and focusses his
attention on broader contexts. Thus, in the case of the London cutlery worker,
he tells us that the poorer workers in his neighborhood go to church in the
evening, while the middle-class people go in the morning; the preacher is aware
of this stratification, and custom tailors his sermons accordingly. The social
ressentiment of the Savoy tinsmith leads to an instructive digression on the
spreading power of the labor unions in this area. One of the most interesting
examples can be found in paragraphs 18 and 19 of Case 7 in Volume 6. Le-
Play obviously selected this family because it got a prize for having produced
a large number of children and bringing them up decently under very restricted economic conditions. In four pages, LePlay makes ten statements as
to the factors which account for high fertility. Formulated as hypotheses,

---

[106] Lewis Mumford, "Patrick Geddes, Victor Branford, and Applied Sociology in England: The Social Survey, Regionalism, and
Urban Planning," *An Introduction to the History of Sociology,* Harry Elmer Barnes (ed.),
(Chicago: Univ. of Chicago Press, 1948).

they would do honor to any modern textbook, and most of them would be controversial even today after a century of empirical research has piled up. LePlay is of course convinced that his opinion is the only one conceivable.

The French disciples of LePlay were very much concerned about the loose connection between the family monographs themselves and the broader social observations which were considered LePlay's most important contributions. This explains the dominant role which de Tourville's "Nomenclature" played in the pages of the "Science Sociale." De Tourville wanted to provide a system of categories which would give a place to every relevant observation and at the same time permit a reorganization of the original work of LePlay. He was also confident that his system would facilitate comparative analysis. The Nomenclature consisted of twenty-five major categories, which approach the families under study, so to say, from two sides. The first nine corresponded approximately to the first thirteen paragraphs of the original monographs: the geographical setting and the type of work done by the family members (a and b), its sources of income, properties, expenses and savings (c - f), the obligations and rights of the family members, the style of life and its history (g - i).

The remaining categories tried to see the family in ever-broadening circles of its social context: the technical, commercial and cultural conditions of the industry in which they worked (j - l); the religious practices, the neighborhood relations and the professional and communal organizations in which the family was embedded (m - q), the broader characteristics of the country in which the family lived, the city, the province and the laws of the whole state inasmuch as they had bearing on the life of the family (r - u); finally, the broad history of the country, its national composition, and its relation (especially emigration or immigration) with other countries (v - z). For a long while, de Tourville's Nomenclature was considered the perfect key to all social analysis. The general argument went about as follows: it guaranteed, so to say, the basic elements needed to describe any social system. After they have been provided, the task of the analyst is fairly easy. He has to find how these elements are related to each other in a specific case and how they vary from one to the next.

Here again is another major wave of categorization, so characteristic of the history of the social sciences. The reader will undoubtedly have anticipated the parallel with Conring and the school of German university statistics. Their aim was certainly the same as the one of the LePlay group, although we can take it for granted that the latter did not know of this earlier effort. It is not too difficult to pinpoint the major difference between the two approaches. The starting point for the Conring school was the state and the administrative tasks of the statesman. In a cameralistic system, he took it for granted that the welfare of the state depended upon the activities of the rulers. Their activities, therefore, were the starting point for the relevant categories: increase of population, defense against potential enemies, improvement of agriculture, monetary policy and so on. Matters like the family would be derivative problems related, for instance, to the number of available conscripts; individual characteristics of diverse population groups would be

worth knowing if the statesman wanted efficient compliance with his administrative measures. The LePlay group took the reverse view. The welfare of
the country depended upon the morality, the industry and the submissiveness
of the citizens at large and upon the sense of responsibility of the elite. These
qualities were formed in the confines of the family. The system of categories,
therefore, had to start out with a description of this primary group; it drew
in the characteristics of the larger context only to the extent that this explained
what happened at the social core. LePlay, so to say, saw society from within
outward. Conring and his school looked at society as a large social system,
the main characteristics of which they wanted to describe; they paid attention
to the primary group only to the extent to which it would affect the actor on
the big scene. Anyone who knows the literature of modern sociology is aware
that the development of nomenclature is still an honorable pursuit. One might
say that recent literature is trying to combine the Conring and the LePlay
traditions.

POSTSCRIPTUM

What other major episode might have belonged in this introductory survey?
I have not described in detail the coming of the British social survey; recently
good summaries have become available, especially Abrams' introductory chapters to his book on social surveys[107] and McGregor's paper on the social
background of the survey movement.[108] The development of quantification in
Germany is a complex topic. Toennies, best known for his conceptual distinction between Gesellschaft and Gemeinschaft, was for years a vigorous pro-
moter of an empirical "Sociography"; but twice, once before the First World
War and again before the rise of Hitler, this development was cut short. Max
Weber is the great symbol of broad-scale historical research. Only rarely is
reference made to his periodic interest in quantitative research and the ambivalence of his efforts. Space limitations have forced me to reserve the German
materials for a future publication.[109] The Italians have an empirical tradition
of their own. Not knowing their language, I had to leave their side of the
story to other students; Niceforo's books provide many leads to historical
sources. Americans came later into the scene of course. Nothing is stranger
than the idea often expressed by European colleagues that quantification is
a U.S. export endangering their tradition. It is true that when this country
took over the European empirical research techniques, it did so on a large
scale. But the steps by which this came about are little known. Here is a
vast area for further inquiry; I have not touched upon it in this paper, because
the tracing of institutional and personal contacts, as well as analysis of the
literature, would be required.

Some time at the end of the 19th century, quantification in sociology takes
on its modern function: to translate ideas into empirical operations and to

[107] Mark Abrams, *Social Surveys and Social Action* (London: Wm. Heinemann, Ltd.,
1951).

[108] O. R. McGregor, "Social Research and
Social Policy in the Nineteenth Century," *The*

British Journal of Sociology, 1957, 8.

[109] Paul F. Lazarsfeld and Anthony Ober-
schall, *History of Quantification in Germany,*
(mimeo) (New York: Bureau of Applied Social Research, Col. Univ.).

look for regular relations between the variates so created.[110] Histories of specific techniques will be needed to clarify this general trend. Helen Walker has done it for correlation analysis[111] and Stephan for sampling.[112] The use of questionnaires has a long past which still waits for its recorder. Mathematical models of social behavior have a curious history. At the end of the 18th century, men like Condorcet worked on them very seriously. For a long while thereafter, the idea was monopolized by the economists. In very recent years, psychologists and sociologists have reentered the scene. The literature increases rapidly; but it is still an object for the book reviewer rather than the historian.

In any case, much work must be done, if we want to match the increasing quantity of sociological quantification by better quality of insight into its history.

[110] For a sketch of this whole trend, see Jahoda, Lazarsfeld and Zeisel, *Marienthal* (Allensbach, Demoskopie, 1961).

[111] *Studies in the History of Statistical Methods* (Baltimore, 1929).

[112] Frederic Stephan, "History of the Uses of Modern Sampling Procedures," *J. Amer. statist. Ass.,* March 1948.

Biometrika (1973), **60**, 3, *p.* 439

Printed in Great Britain

Studies in the History of Probability and Statistics. XXXII

Laplace, Fisher, and the discovery of the concept of sufficiency

By STEPHEN M. STIGLER

University of Wisconsin, Madison

SUMMARY

R. A. Fisher's 1920 discovery of sufficiency while studying competing estimators of standard deviations is discussed. It is shown how Laplace completed a similar investigation a century earlier attempting to improve upon least squares, yet did not abstract the concept.

Some key words: Laplace; Fisher; Sufficiency; History of statistics; History of estimation; Least squares.

1. INTRODUCTION

Because Fisher's concept of sufficiency depends so strongly on the assumed form of the population distribution, its importance to applied statistics has been questioned in recent years. There can be no question, however, as to its importance to theoretical statistics. Sufficiency served as a cornerstone to Fisher's theory of estimation, and has helped illuminate topics as diverse as the optimal design of experiments and Bayesian inference. That the concept did not appear before Fisher's time is not surprising. The very definition of sufficiency requires the notion of the joint distribution of a statistic and the sample, and many statisticians before Fisher were not mathematically able to handle the necessary complicated operations in multiple dimensions, much less abstract the concept. Accordingly, it may come as a surprise to some just how close Laplace came to discovering sufficiency in 1818, following a chain of reasoning remarkably similar to that used by Fisher in 1920.

The aim of this present paper is to examine the relevant works of Fisher and Laplace and show how they set out on parallel routes: Laplace essentially comparing the sample mean and median as estimators of the centre of a symmetrical population, Fisher comparing the sample standard deviation and mean deviation as estimators of the standard deviation of a normal population. Yet we shall see how Fisher continued slightly beyond the point Laplace stopped, and how he was then led to the modern concept of sufficiency.

2. FISHER'S DISCOVERY OF SUFFICIENCY

R. A. Fisher first hit upon the concept of sufficiency in 1920, although he did not use the name 'sufficiency' until the following year. The paper, 'A mathematical examination of the methods of determining the accuracy of an observation by the mean error, and by the mean square error' (Fisher, 1920), was written in answer to a statement of the astronomer A. S. Eddington in his book *Stellar Movements*. Eddington had written:

...in calculating the mean error of a series of observations it is preferable to use the simple mean residual irrespective of sign rather than the mean square residual, to which Eddington added a footnote:

This is contrary to the advice of most text-books; but it can be shown to be true [Eddington, 1914, p. 147].

In his brief paper, Fisher examined Eddington's claim. We shall outline the steps of his argument.

Fisher began by assuming that each of n measurements is normally distributed with mean m and variance σ^2, and that the object was to estimate σ. He considered in particular the estimators σ_1, the mean deviation from the sample mean, suitably scaled, and σ_2, the sample standard deviation. He derived the standard deviations of both σ_1 and σ_2, and observed that, contrary to Eddington's statement, the standard deviation of σ_1 is 14 % greater than that of σ_2 for large n. At this point he permitted Eddington to add a footnote, which read in part:

I think it accords with the general experience of astronomers that, for the errors commonly occurring in practice, the mean error is a safer criterion of accuracy than the mean square error, especially if any doubtful observations have been rejected; but I was wrong in claiming a theoretical advantage for the mean error in the case of a truly Gaussian distribution.

Fisher went on to show that of all estimators based on the sum of the pth powers of the residuals, the estimator with $p = 2$, i.e. σ_2, has the smallest variance for large samples.

Although both Fisher and Eddington were unaware of it at the time, the paper up to this point contained nothing new, save Fisher's elegant geometrical approach to the distribution theory. In fact, Gauss had published essentially the same investigation in 1816! Luckily, Fisher did not stop here. Rather, he went on to a more detailed study of the behaviour of σ_1 and σ_2. He observed that for large n the variances of σ_1 and σ_2 might tell the story, as both are then approximately normally distributed, but that for small n the situation is more complicated. To illustrate their behaviour, he derived the joint distribution of σ_1 and σ_2 for the special case $n = 4$. After calculating the first few moments of σ_1 and σ_2 for this case, a reconsideration of their joint distribution produced the following revelation:

So far the variables have been compared only in respect of the quantitative characters of their frequency distributions. There exists also in the form of the frequency surface a qualitative distinction, which reveals the unique character of σ_2.

From the manner in which the frequency surface has been derived...it is evident that:

For a given value of σ_2, the distribution of σ_1 is independent of σ. [Fisher's italics.] On the other hand, it is clear...that for a given value of σ_1 the distribution of σ_2 does involve σ. In other words, if, in seeking information as to the value of σ, we first determine σ_1, then we can still further improve our estimate by determining σ_2; but if we had first determined σ_2, the frequency curve for σ_1 being entirely independent of σ, the actual value of σ_1 can give us no further information as to the value of σ. The whole of the information to be obtained from σ_1 is included in that supplied by a knowledge of σ_2.

This remarkable property of σ_2, as the methods which we have used to determine the frequency surface demonstrate, follows from the distribution of frequency density in concentric spheres over each of which σ_2 is constant. It therefore holds equally if σ_3 or any other derivate be substituted for σ_1. If this is so, then it must be admitted that:

The whole of the information respecting σ, which a sample provides, is summed up in the value of σ_2. [Fisher's italics.]

Fisher went on to observe that this property of σ_2 is quite dependent on the assumption that the population is normal, and showed that indeed σ_1 is preferable to σ_2, at least in large samples, for estimating the scale parameter of the double exponential distribution, providing both estimators are appropriately rescaled. He closed the paper by proposing that in actual situations the sample measure of kurtosis, β_2, be calculated, and

If this is near 3 the Mean Square Error will be required; if, on the other hand, it approaches 6, its value for the double exponential curve, it may be that σ_1 is a more suitable measure of dispersion.

By 1922, Fisher had named sufficiency, related it to maximum likelihood, stated the factorization theorem, and applied the concept in a variety of situations. The theory of estimation had taken a giant leap forward; the concept of sufficiency, one of Fisher's most original contributions, had been born.

3. LAPLACE AND THE METHOD OF SITUATION

In the last part of the Second Supplement (1818) to his monumental *Théorie Analytique des Probabilités*, Laplace presented an investigation which is strikingly similar to Fisher's work of 1920. Laplace considered the problem we would now refer to as linear regression through the origin. In his notation, we have a system of n equations,

$$p_i y - a_i + x_i = 0 \quad (i = 1, ..., n),$$

where the p_i's and a_i's are known, and y and the x_i's are unknown, the x_i's being the errors of observation. The problem is to estimate y. Laplace had earlier in the volume discussed the method of least squares, for which he here used the name 'most advantageous method'; the purpose of this section of the Second Supplement was to discuss an alternative method of estimation introduced by Boscovich in 1757, which Laplace called the 'method of situation'.

Laplace began by repeating the definition of the 'method of situation' and a computational algorithm which had appeared in his *Mécanique Céleste* in 1799. Briefly, the method is: estimate y by that value which minimizes

$$\sum_{i=1}^{n} |p_i y - a_i|. \tag{1}$$

If all the p_i are positive and the observations are indexed in such a way that

$$\frac{a_1}{p_1} \geqslant \frac{a_2}{p_2} \geqslant ... \geqslant \frac{a_n}{p_n},$$

and if the integer r is defined to be such that

$$p_1 + ... + p_{r-1} < p_r + ... + p_n, \quad p_1 + ... + p_r > p_{r+1} + ... + p_n,$$

then the value of y which minimizes (1) is given by

$$y = a_r / p_r. \tag{2}$$

Of course, in the special case where all $p_i \equiv 1$, this simply gives the median of the a_i's. Edgeworth (1923) has called the solution (2) a weighted median; we use the symbol y_{MS}, MS for 'method of situation', for the expression (2) in order to distinguish it from the least squares or 'most advantageous' estimator,

$$y_{LS} = \frac{\Sigma p_i a_i}{\Sigma p_i^2}.$$

After proving that y_{MS} given by (2) does in fact minimize (1), Laplace proceeded to investigate the distribution of y_{MS}. He assumed that the observational errors x_i all obeyed the same probability density $\phi(x)$, but he assumed nothing more about ϕ than that it was an even function, $\phi(x) = \phi(-x)$. His later analysis assumes implicitly that ϕ is twice differentiable at zero and $\phi(0) > 0$. Laplace then correctly derived the density of $y_{\mathrm{MS}} - y$, and showed that as n increases, this density approaches the normal density with mean zero and variance $\{4\phi^2(0)\sum p_i^2\}^{-1}$. In the special case where all p_i's are 1 this agrees with the now standard results for the sample median; indeed, it seems likely that Laplace was the first to derive the asymptotic distribution of a single order statistic.

Earlier in the volume Laplace had derived the asymptotic distribution of the least squares estimator y_{LS}; he now compared y_{MS} and y_{LS} as estimators of y on the basis of the variances of their asymptotic distributions, and concluded that the 'method of situation' was preferable to the 'most advantageous method' if and only if

$$\{2\phi(0)\}^2 > \left\{\int_{-\infty}^{\infty} x^2 \phi(x)\, dx\right\}^{-1}.$$

He noted that y_{LS} was to be preferred if ϕ is a normal density.

If Laplace had halted his investigation at this point and gone on to other matters, it would still have been a notable work. For the first time, two reasonable methods of estimation had been compared for general populations and the precise conditions under which one would be preferable to another spelled out. In addition, he had presented the first large sample theory for a single order statistic, using an argument that would still be considered modern today, an argument which can be easily extended to nonsymmetrical error distributions and sample percentiles other than the median. But Laplace did not stop here.

As Fisher did in a similar situation one hundred years later, Laplace, having examined the distributions of y_{MS} and y_{LS} separately, went on to consider their joint distribution. His object in doing this was to show how the two estimators could be combined to provide a new estimator which would be better than either:

In combining the results of these two methods, one can obtain a result whose probability law of error will be more rapidly decreasing.

He accomplished this as follows. First, he presented an expression for the joint density of $y_{\mathrm{MS}} - y$ and $y_{\mathrm{LS}} - y$ in terms of the inversion of a characteristic function. Next, he developed the integrand as a series. Finally, he was able to perform the required integration by taking n large, showing that the joint asymptotic distribution of $\zeta = y_{\mathrm{MS}} - y$ and $\zeta' = y_{\mathrm{LS}} - y$ had a density proportional to

$$\exp\left[-\frac{k}{2k''}\zeta'^2 \Sigma p_i^2 - \frac{k''}{k}\frac{\left\{\zeta\frac{\phi(0)}{k} - \zeta'\frac{k'}{k''}\right\}^2}{2\left(\frac{k''}{k} - \frac{k'^2}{k^2}\right)}\Sigma p_i^2\right], \tag{3}$$

where

$$k = \tfrac{1}{2}, \quad k' = \int_0^{\infty} x\phi(x)\, dx, \quad k'' = \int_0^{\infty} x^2 \phi(x)\, dx.$$

We would now recognize this as the density of the bivariate normal distribution with mean $(0, 0)'$ and covariance matrix $(\Sigma p_i^2)^{-1}\{\sigma_{jk}\}$, where

$$\sigma_{jk} = \{2\phi(0)\}^{-(4-j-k)}\int_{-\infty}^{\infty} |x|^{j+k-2}\phi(x)\, dx.$$

Laplace then used this result to find, in terms of ϕ, the value of C for which the asymptotic variance of $y_{LS} - (y_{LS} - y_{MS})\,C$ is smallest. The appropriate value of C is, again in Laplace's notation,

$$C = \frac{\dfrac{\phi(0)}{k}\left\{\dfrac{\phi(0)}{k} - \dfrac{k'}{k''}\right\}}{\dfrac{k}{k''} - \dfrac{k'^2}{k''^2} + \left\{\dfrac{\phi(0)}{k} - \dfrac{k'}{k''}\right\}^2}.$$

Laplace wrote, in closing the Second Supplement,

the result of the most advantageous method [i.e. y_{LS}] must therefore be diminished by the quantity $[(y_{LS} - y_{MS})\,C]$; and the probability of an error u using this corrected result will be proportional to the preceding exponential. [The correct expression had been given.] The importance of this new result will be increased if $(\phi(0)/k) - (k'/k'')$ is not zero; there is accordingly an advantage to correcting the most advantageous method in this way. When one does not know the distribution of the errors of observation this correction is not feasible; but it is remarkable that in the case where this probability is proportional to e^{-hx^2}; that is, when $\phi(x) \propto e^{-hx^2}$, the quantity

$$\frac{\phi(0)}{k} - \frac{k'}{k''}$$

will be zero. Then the result of the most advantageous method will receive no correction from the result of the method of situation, and the probability law of an error is unchanged.

Thus we see that the consideration of the joint distribution of y_{LS} and y_{MS} led Laplace to the 'remarkable' conclusion that not only is y_{LS} a better estimator of y than y_{MS} when the errors are normally distributed, but y_{LS} is better than any other linear combination of the two. Earlier in the Second Supplement, Laplace had performed a similar but less complicated calculation to also show that y_{LS} could not be improved by linear combination with any linear unbiased estimator of y. He thus clearly realized that, in this limited sense, neither y_{MS} nor any linear estimator could add information about y to y_{LS}.

4. Conclusion

While it is true that Laplace had clearly come across and described one aspect of sufficiency, at least as it related to his problem, it must be admitted that he did not go so far as Fisher, and he did not isolate the concept. Both Laplace and Fisher took the important step of considering the joint distribution of competing estimators, rather than merely looking at their distributions separately. Both were led by this step to realize that for normally distributed data, one of their estimators could add nothing to the other. But Laplace stopped where Fisher continued, and did not abstract that element of the problem which was responsible for this state of affairs.

Ironically, whereas Fisher considered only normally distributed data for the greater part of his paper, Laplace considered a more general class of distributions, and this greater generality apparently kept him from realizing the special nature of the density (3) when $\phi(x) \propto \exp(-hx^2)$. It is tempting to speculate that if Laplace had lived in an age where the normal distribution was considered as 'normal' as it was by Fisher, he might have looked at the density (3) when $\phi(0)/k = k'/k''$, realize that the conditional distribution of y_{MS} given y_{LS} did not change with y, and hit upon the concept of sufficiency, as Fisher did a century later. However, such speculation is worse than idle. For it not only serves to mask the fact that it is the generality of Laplace's work which both motivated his achievement and made it important, but also serves to diminish unfairly the magnitude of Fisher's great accomplishment, leaping from a particular fact to the general concept of sufficiency with such speed and force that the momentum carried him on to develop a whole new theory of estimation.

Actually, the great difference between the historical contexts of these works makes a real comparison impossible. Laplace's work took place in the infancy of mathematical statistics; the method of least squares had only been published 13 years before and the problem of point estimation was fuzzily understood at best. By Fisher's time, the issues had become much clearer, through the work of Karl Pearson and others. Yet, the similarities between the two men's work, despite their different circumstances, sheds some light on the creative processes of great minds.

BIBLIOGRAPHICAL NOTES

Laplace's *Deuxième Supplément a la Théorie Analytique des Probabilités* was published separately in 1818 dated 'Février 1818' and appeared as an appendix to the *Théorie Analytique* with the publication of the third edition in 1820. The first part of this Supplement, which is not discussed here, originally appeared in *Connaissance des Temps pour l'an 1820* (published 1818), pp. 422–43, bearing the legend 'Lu à l'Academie des Sciences, le 4 août 1817'. This would apparently date the work discussed here some time in the latter part of 1817.

Boscovich's description of the method of situation and the relation between his work and Laplace's have been discussed by Eisenhart (1961). While the method itself and a geometrical version of the algorithm presented here are due to Boscovich, the analysis of the probabilistic properties of the estimator seems to have been original with Laplace.

Boscovich, and Laplace in *Mécanique Céleste* (Première Partie, Livre III, No. 40), had considered the more general model $p_i y + z - a_i + x_i = 0$, and determined the estimate of the parameter z from that of y by the requirement that the sum of the residuals be zero.

Fisher does not appear to have been aware of Laplace's Second Supplement. He nowhere refers to it, and the only explicit reference to *Théorie Analytique* in his collected papers is to the second edition (1814), which did not contain the Second Supplement. He does refer to the third edition in his books, in connexion with Laplace's use of Bayes's theorem.

While the first paper of Fisher's to use the name sufficiency was his 1922 paper 'On the mathematical foundations of theoretical statistics', the name did appear in *Nature* for 24 November 1921, in the abstract of Fisher's 17 November presentation to the Royal Society.

This research was initiated in the Department of Statistics, University of Wisconsin, under the partial support of the Office of Naval Research and completed while the author was on leave in the Department of Statistics, University of Chicago, under the partial support of the National Science Foundation.

REFERENCES

EDDINGTON, A. S. (1914). *Stellar Movements and the Structure of the Universe*. London: Macmillan.

EDGEWORTH, F. Y. (1923). On the use of medians for reducing observations relating to several quantities. *Phil. Magazine* **46** (6th series), 1074–88.

EISENHART, C. (1961). Boscovich and the combination of observations. Chapter 7 in *R. J. Boscovich, Studies of His Life and Work*, ed. L. L. Whyte. London: Allen and Unwin. Reissued (1963) by the Fordham University Press, New York.

FISHER, R. A. (1920). A mathematical examination of the methods of determining the accuracy of an observation by the mean error, and by the mean square error. *Monthly Notices R. Astronomical Soc.* **80**, 758–70. Reprinted (1950) in Fisher's *Contributions to Mathematical Statistics*. New York: Wiley.

FISHER, R. A. (1922). On the mathematical foundations of theoretical statistics. *Phil. Trans. R. Soc. Lond. A* **222**, 309–68. Reprinted (1950) in Fisher's *Contributions to Mathematical Statistics*. New York: Wiley. Abstract in *Nature, Lond.* **108** (1921), 421.

GAUSS, C. F. (1816). Bestimmung der Genauigkeit der Beobachtungen. In *Carl Friedrich Gauss Werke* Band 4, pp. 109–17, Göttingen: Königlichen Gesellschaft der Wissenschaften (1880).

LAPLACE, P. S. DE (1818). *Deuxième Supplément a la Théorie Analytique des Probabilités*. Paris: Courcier. Reprinted (1847) in *Oeuvres de Laplace* **7**, pp. 569–623. Paris: Imprimerie Royale; (1886) in *Oeuvres Complètes de Laplace* **7**, pp. 531–80. Paris: Gauthier-Villars. The first Part of this Supplement, which is not discussed here, originally appeared (1818) in *Connaissance des Temps pour l'an 1820*, pp. 422–43.

[*Received December 1972. Revised February 1973*]

Biometrika (1972), **59**, 2, *p.* 239
Printed in Great Britain

Studies in the History of Probability and Statistics. XXIX

The discovery of the method of least squares

By R. L. PLACKETT

University of Newcastle upon Tyne

Summary

The circumstances in which the discovery of the method of least squares took place and the course of the ensuing controversy are examined in detail with the aid of correspondence. Some conclusions are drawn about the attitudes of the main participants and the nature of historical research in statistics.

Some key words: Calculus of probability; Gauss; History of statistics; Legendre; Method of least squares; Priority in scientific discovery.

1. Introduction

The technique of combining independent observations on a single quantity by forming their arithmetic mean had appeared by the end of the seventeenth century (Plackett, 1958). Early in the eighteenth century, a generalized version in which weights are assigned to the observations was introduced by Cotes. A clear account of his work is included in an unpublished National Bureau of Standards report by C. Eisenhart. Subsequent developments in the analysis of data were concerned with methods of estimation from linear models when the number of unknown parameters was two or more. A method devised independently by Euler and Mayer consisted of subdividing the observations into as many groups as there are known parameters, and then equating the observed total for each group to its theoretical value (Eisenhart, 1964). The subdivision is arbitrary and the subjective element was removed by Boscovich. Suppose that we have observations y_i on $\alpha + \beta x_i$ ($i = 1, ..., n$). Call $y_i - \alpha - \beta x_i$ the ith deviation and $y_i - \hat{\alpha} - \hat{\beta} x_i$ the ith residual, where $\hat{\alpha}$ and $\hat{\beta}$ are estimates of α and β. Boscovich proposed to estimate α and β by quantities which satisfy the following conditions: (*a*) the sum of the deviations is zero; and (*b*) the sum of the absolute values of the deviations is a minimum. He gave a geometrical method of solution, and applied it to the lengths of five meridian arcs. A study of the work of Boscovich on the combination of observations, and its influence on Laplace, is made by Eisenhart (1961). Another method was proposed by Euler and Lambert, according to which the estimates of α and β should be the quantities which minimize the absolute value of the largest deviation (Sheynin, 1966). An algorithm for finding the minimax residual was given in 1783 by Laplace, and in 1789 he simplified his earlier procedure. The second memoir also contains a simplified version of what Laplace describes as the ingenious method which Boscovich has given for the purpose of determining the most probable values. Both methods are applied to geodetic data, and in each case a careful analysis is made by comparing the largest residuals with the errors to which the observations are susceptible.

The method of least squares was discovered independently by Gauss and Legendre during the period which followed. Although Gauss had been using the method since about

1795, the first explicit account was published in 1805 by Legendre. Four years later Gauss gave his version, in the course of which he referred to his earlier work. This remark led to much controversy concerning the priority for the discovery. The technical innovations of the period have already been fully reviewed (Seal, 1967). We shall examine the germination of Gauss's ideas on the subject in the intervals before his publications, his reactions to the work of Laplace and Legendre and the impact on all concerned of the questions of priority then raised. The material on which the analysis is based is presented in the next section. Readers who are mainly interested in the discussion can turn there at once and refer back for further information as required.

2. MATERIAL

Galle (1924) states that the basic ideas of the method of least squares occurred to Gauss in the autumn of 1794 when he read about the treatment of a surplus of observations in the first volume of Lambert (1765). Gauss was 17 years old at the time, attending the Collegium Carolinum in Brunswick and preparing for his university studies. No evidence is cited in support of Galle's statement, but in the correspondence and other material which is reproduced below Gauss mentions 1794 on several occasions as the year of the discovery. He arrived at Göttingen in October 1795. A list of the books which he borrowed from the university library during his student years has been compiled by Dunnington (1955, pp. 398–404). The first item is volumes I and II of Richardson's novel *Clarissa*, then popular throughout Europe, and the second (24 October 1795) is the three volumes of Lambert's work.

From 1796 to 1814, Gauss recorded brief and often enigmatic summaries of his work in a mathematical diary. An entry for 1798 reads as follows:

Calculus probabilitatis contra La Place defensus. Gott. Jun. 17.

Here # signifies that Gauss attached some importance to this entry, although much less than for others where there is multiple vertical and horizontal scoring. The entry is mentioned in letters from Gauss to Olbers on 24 January 1812, and from Gauss to Laplace on 30 January 1812. On the basis of these references, Klein and Schlesinger argue convincingly (*Werke*, X/1, 533) that the entry is concerned with the method of Boscovich and Laplace. They also suggest that Gauss had just become acquainted with Laplace's memoir of 1789. Such may indeed be the case, but although the list of books which Gauss borrowed from the library of the University of Göttingen in 1798 before 17 June contains several volumes of *Mémoires de l'Académie de Paris*, those for 1789 receive no specific mention.

An exchange of letters between Gauss and Schumacher in 1832 identifies the following two items as applications of the method of least squares. The German word 'Modulen' which occurs in the original of the first item has been translated as 'units of length'. Gauss is using 2 toises ($= \frac{1}{1000}$ league $\simeq 3\cdot898$ metres) as his unit of length (Delambre, 1814, vol. 3, p. 566). The first item (Gauss, 1799) takes the form of a letter to the editor, von Zach, who has added two footnotes, and the second (Gauss, 1800) is a set of corrections.

Allow me to point out a printer's error in the July issue of the A.G.E. Page xxxv of the introduction, in the account of the arc between the Panthéon and Évaux, must read 76145.74 instead of 76545.74. The sum is correct and the error cannot be in any other place.* I discovered this error when I applied my method, a specimen of which I have given you,† to determine the ellipse simply from these four measured arcs, and found the ellipticity to be 1/150; after correction of that error I found 1/187, and 2565006 units of length in the whole quadrant (namely without consideration of the degree in Peru).

The difference between 1/150 and 1/187 is certainly not important in this case, because the end-points lie too close together.

Brunswick, 24 Aug. 1799. C.F. Gauss

* This printers error is confirmed, and may also be recognized by the decimal degree figure set beside it, $2^D.66868$ – v.Z.

† Here at another time – v.Z.

Corrections to Volume 4 of the Allg. Geogr. Ephemer.

At the very same place p. 378 nr. 3 line 9, instead of ellipticity 1/150 must be meant 1/50. In line 12, instead of the words *'The difference between 1/150 and 1/187 is certainly not important in this case....'* greater intelligibility can be achieved as follows. *'The difference between 1/150, the ellipiticity which was found by the French surveyors* (A.G.E. volume 4, p. xxxvii of the Introduction and p. 42) *and 1/187, which I have found, is certainly not important in this case'*.

The method of least squares was first named and published by Legendre in *Nouvelles méthodes pour la détermination des orbites des comètes,* which appeared in 1805 (Merriman, 1877). Gauss became aware of Legendre's work not long after publication and took an early opportunity to study the details. We give first an extract from Gauss (1806, p. 184), which again takes the form of a letter to von Zach, and then the relevant portion of a letter to Gauss's friend, the physician and astronomer Olbers (Schilling & Kramer, 1900).

Brunswick, 8 July 1806

...I have not yet seen Legendre's work, which you take this occasion to mention. I intentionally did not go out of my way to do so, in order not to disrupt the sequence of my own ideas while working on my method. It was through a couple of words which De la Lande let fall in the recent *Histoire de l'Astronomie* 1805, *méthode des moindres quarrés,* that I tumbled to the idea that a principle which I had made use of for twelve years in many a calculation, and which, too, I intended to employ in my book, although it admittedly does not form an essential part of my method – that this principle had been employed by Legendre as well....

Gauss to Olbers. Brunswick, 30 July 1806.

...Hr. v. Zach writes to me further that you have offered to review Legendre's work on the orbits of comets. I will, therefore, with pleasure send to you at Bremen the copy which Hr. v. Zach sent me; but will you please allow me first to keep it for a few weeks yet. From a preliminary inspection it appears to me to contain much that is very beautiful. Much of what was original in my method, particularly in its first form, I find again also in this book. It seems to be my fate to compete with Legendre in almost all my theoretical works. So it is in the higher arithmetic, in the researches on transcendental functions connected with the rectification of the ellipse, in the fundamentals of geometry, and now again here. Thus, for example, the principle which I have used since 1794, that the sum of squares must be minimized for the best representation of several quantities which cannot all be represented exactly, is also used in Legendre's work and is most thoroughly developed. He does not seem to know your work....

Eight months later Gauss made the same point concerning prior use and he also referred somewhat more explicitly to the matter concealed behind the entry in his diary for 17 June 1798.

Gauss to Olbers. Brunswick, 24 March 1807.

...I am occupied at present with treating the problem: 'To determine the most probable values of a number of unknown quantities from a *larger* number of observations depending on them', on the basis of the calculus of probability. The principle that the sum of squares of the differences between the calculated and observed quantities must be minimized, I have employed for many years; I mentioned it to you long ago, and it is now also advanced by Legendre. In this connection also, the principle is preferable to that of Laplace, according to which the sum of all differences must be zero, and the sum of the same differences, each taken with positive sign, should be a minimum. It can be shown that this is not admissible on the basis of the calculus of probability, but leads to contradictions....

Gauss completed in 1806 a German version of the book he had been preparing, but conditions were such that he had difficulty in finding a publisher. Eventually, Perthes agreed to publish provided that the work was translated into Latin. Thus it was that *Theoria Motus Corporum Coelestium* appeared in 1809. Section 186 contains a simple but penetrating comparison of the method of least squares with its competitors in which the idea of admissibility briefly appears. The following is the translation published by Davis (1857), with a few corrections.

On the other hand, the principle that the sum of the squares of the differences between the observed and computed quantities must be as small as possible may, in the following manner, be considered independently of the calculus of probabilities.

When the number of unknown quantities is equal to the number of the observed quantities depending on them, the former may be so determined as exactly to satisfy the latter. But when the number of the former is less than that of the latter, an absolutely exact agreement cannot be determined, in so far as the observations do not enjoy absolute accuracy. In this case care must be taken to establish the best possible agreement, or to diminish as far as practicable the differences. This idea, however, from its nature, involves something vague. For, although a system of values for the unknown quantities which makes *all* the differences respectively less than another system, is without doubt to be preferred to the latter, still the choice between two systems, one of which presents a better agreement in some observations, the other in others, is left in a measure to our judgement, and clearly innumerable different principles can be proposed by which the former condition is satisfied. Denoting the differences between observation and calculation by Δ, Δ', Δ'' etc., the first condition will be satisfied not only if $\Delta\Delta + \Delta'\Delta' + \Delta''\Delta'' + $ etc., is as small as possible (which is our principle), but also if $\Delta^4 + \Delta'^4 + \Delta''^4 + $ etc., or $\Delta^6 + \Delta'^6 + \Delta''^6 + $ etc., or in general, if the sum of any of the powers with an even exponent becomes as small as possible. But of all these principles ours is the most simple; by the others we shall be led into the most complicated calculations.

On the other hand our principle, which we have made use of since the year 1795, has lately been published by LEGENDRE in the work *Nouvelles méthodes pour la détermination des orbites des comètes*, Paris 1806, where several other properties of this principle have been explained, which, for the sake of brevity, we here omit.

If we were to adopt a power with an infinitely great, even exponent, we should be led to that system in which the greatest differences become as small as possible.

LAPLACE made use of another principle for the solution of linear equations, the number of which is greater than the number of unknown quantities, which had been previously proposed by BOSCOVICH, namely that the differences themselves, but all of them taken positively, should make up as small a sum as possible. It can be easily shown, that a system of values of unknown quantities, derived from this principle alone, must necessarily* exactly satisfy as many equations out of the number proposed, as there are unknown quantities, so that the remaining equations come under consideration only so far as they help to *determine the choice*: if, therefore, the equation $V = M$, for example, is of the number of those which are not satisfied, the system of values found according to this principle would in no respect be changed, even if any other value N had been observed instead of M, provided that, denoting the computed value by n, the differences $M - n$, $N - n$, were affected by the same signs. On the other hand, LAPLACE qualifies in some measure this principle by adding a new condition: he requires, namely, that the sum of the differences, the signs remaining unchanged, be equal to zero. Hence it follows, that the number of equations exactly represented may be less by unity than the number of unknown quantities; but what we have said before will still hold good if there are at least two unknown quantities.

* Except the special cases in which the problem remains, to some extent, indeterminate.

In fact, both of the conditions on the differences are due to Boscovich. There appears to be a slip of the pen in the last sentence, with 'less by unity' in place of 'more by unity'.

The *Theoria Motus* became known to Legendre through Sophie Germain, whose career is summarized by Dunnington (1955, p. 68). A letter from Legendre to Gauss soon followed. Only a short extract is given in *Werke*, X/1, 380, but the original is in the Gauss archives at Göttingen. The interest of this letter, and the part it plays in the relationship between Legendre and Gauss, are such as to justify the complete translation which follows.

Legendre to Gauss. Paris, 31 *May* 1809.

I imagine Sir, that Mademoiselle Germain will have delivered the message which she kindly agreed to take, which was to thank you very much for the paper which you were kind enough to send me on the summation of a number of series. Your writings on a subject of which I have always been fond could not fail to interest me deeply, and I noted the fecundity of your genius which led you to discover a fourth or fifth demonstration of the proposition to which I have given the name of the law of reciprocity between two prime numbers.

A few days ago Mlle. Germain received from Germany your Theoria Motus Corporum Coelestium; she has passed on extracts, and from the little I have managed to read, I can see that this work is worthy of your reputation, and that you have vindicated analysis from the reproach which might have been levelled at it, of not providing astronomers with practical methods of resolving with the necessary exactitude the problem of determining the orbit of a planet by three observations.

In my researches into determining the orbits of comets, I too had the aim of demonstrating that analysis, properly handled, was capable of providing solutions which were as quick as, and more reliable than, those arrived at by synthetic methods, such as those of Olbers. Your object, however, is more extensive, and you have endowed science with a method which will become very useful in the art of astronomy, especially if new planets continue to be discovered.

It was with pleasure that I saw that in the course of your meditations you had hit on the same method which I had called *Méthode des moindres quarrés* in my memoir on comets. The idea for this method did not call for an effort of genius; however, when I observe how imperfect and full of difficulties were the methods which had been employed previously with the same end in view, especially that of M. La Place, which you are justified in attacking, I confess to you that I do attach some value to this little find. I will therefore not conceal from you, Sir, that I felt some regret to see that in citing my memoir p. 221 you say *principium nostrum quo jam inde ab anno* 1795 *usi sumus* etc. There is no discovery that one cannot claim for oneself by saying that one had found the same thing some years previously; but if one does not supply the evidence by citing the place where one has published it, this assertion becomes pointless and serves only to do a disservice to the true author of the discovery. In Mathematics it often happens that one discovers the same things that have been discovered by others and which are well known; this has happened to me a number of times, but I have never mentioned it and I have never called *principium nostrum* a principle which someone else had published before me. You have treasures enough of your own, Sir, to have no need to envy anyone; and I am perfectly satisfied, besides, that I have reason to complain of the expression only and by no means of the intention.

I will make the point as well, Sir, taking quite the opposite standpoint, that you have inadvertently attributed to M. de La Place something which belongs to Euler. You say on page 212, 'per theorema elegans *primo ab ill. La Place inventium* integrale $\int e^{-t^2} dt$ á $t = -\infty$ ad $t = +\infty$, $= \sqrt{\pi}$'. This theorem was discovered a long while beforehand by Euler who gave in general the integral $\int dx \left(l\frac{1}{x} \right)^{\frac{1}{2}(2n+1)}$ from $x = 0$ to $x = 1$. Therefore one has only to set $e^{-t^2} = x$ to change $\int e^{-t^2} dt$ into $\frac{1}{2} \int dx \left(l\frac{1}{x} \right)^{-\frac{1}{2}}$. See Com. petr. vol. v, Nov. Comm. vol. xvi, Nova acta petr. vol. v, etc.

I have the honour to be, Sir, your obedient servant.   Le Gendre

With regard to the final paragraph, De Moivre derived the normal function in November 1733, using a constant which Stirling almost at once showed to be $\sqrt{\pi}$ (K. Pearson, 1924). However, the force of this letter for Gauss resided in the paragraph immediately preceding. When writing to Olbers later in the year, Gauss took the opportunity to seek his support.

Gauss to Olbers.   Göttingen, 4 October 1809.

...Do you still remember, dearest friend, that on my first visit to Bremen in 1803 I talked with you about the principle which I used to represent observations most exactly, namely that the sum of squares of the differences must be minimized when the observations have equal weights? That we discussed the matter in Rehburg in 1804 of that I still clearly recollect all the circumstances. It is important to me to know this. The reason for the question can wait...

A formal letter from Laplace to Gauss two years later indicates that the question of priority was still under active discussion. The letter may have accompanied the memoirs mentioned by Gauss in his reply (30 January 1812).

Laplace to Gauss.   15 November 1811.

M. Gauss says in his work on elliptical movement that he was conversant with it before M. Le Gendre had published it; I would greatly like to know whether before this publication anything was printed in Germany concerning this method and I request M. Gauss to have the kindness to inform me about it.

Soon afterwards, Gauss put his case at length to Olbers and Laplace in letters separated by less than a week. In the first letter Gauss again asks Olbers for support but still withholds the reason. The second letter opens with a problem in metric number theory, first solved in 1928 by Kuzmin (Uspensky, 1937, Appendix III; Gnedenko, 1957). No reply from Laplace is recorded, and all his papers were destroyed by a fire at the Château de Mailloc in 1925 (Smith, 1927; K. Pearson, 1929, p. 215).

Gauss to Olbers.   Göttingen, 24 January 1812.

...I hear that Hr. Delambre has given *ad modum suum,* a very prolix theory about *moindres carrés* in the French *Moniteur*; I do not read the *Moniteur*, which also only comes here in yearly deliveries. Perhaps you will find an opportunity sometime, to attest *publicly* that I already stated the essential ideas to you at our first personal meeting in 1803. I find among my papers that in June 1798, when the method was one which I had *long* applied, I first saw Laplace's method and indicated its incompatibility with the principles of the calculus of probability in a short diary-notebook about my mathematical occupations. In Autumn 1802 I entered in my astronomical notebook the eighth set of elements of Ceres, found by the method of least squares. The papers have now been lost, in which I applied that method in earlier years, e.g. in Spring 1799 on Meyberg's table of the equation of time. The only thing which is surprising is that this principle, which suggests itself so readily that no particular value at all can be placed on the idea alone, was not already applied 50 or 100 years earlier by others e.g. Euler or Lambert or Halley or Tobias Mayer, although it may very easily be that the last, for example, has applied that sort of thing without announcing it, just as every calculator necessarily invents a collection of devices and methods which he propagates by word of mouth only as occasion offers....

Gauss to Laplace.   Göttingen, 30 January 1812.

I express my deep thanks for the two memoirs which you did me the honour of sending me and which I received a few days ago. The functions which you deal with therein, as well as the questions of probability, on which you are preparing a large work, have a great attraction for me, although I have worked little myself on the latter. I am reminded however of a curious problem which I worked at 12 years ago, but which I did not succeed in solving to my satisfaction. Perhaps you would care to study it for a few moments: in this case I am sure that you will find a complete solution. Here it is. Let M be an unknown quantity between the limits 0 and 1, for which all values are either equally probable or varying according to a given law: and that we suppose it expressed as a continued fraction

$$M = \cfrac{1}{a' + \cfrac{1}{a'' + \text{etc.}}}.$$

What is the probability, that in stopping the expansion at a finite term, $a^{(n)}$, the following fraction

$$\cfrac{1}{a^{(n+1)} + \cfrac{1}{a^{(n+2)} + \text{etc.}}}$$

lies between the limits 0 and x? I denote it by $P(n, x)$ and I obtain for it supposing that all values of M are equally probable

$$P(0, x) = x;$$

$P(1, x)$ is a transcendental function depending on the function

$$1 + \tfrac{1}{2} + \tfrac{1}{3} + \ldots + 1/x$$

which Euler calls inexplicable and on which I have given several results in a memoir presented to our scientific society which will soon be printed. But in the cases when n is larger, the exact value of $P(n, x)$ seems intractable. However I have found by very simple arguments that for infinite n we have

$$P(n, x) = \frac{\log (1+x)}{\log 2}.$$

But the efforts which I made at the time of my researches to determine

$$P(n, x) - \frac{\log (1+x)}{\log 2}$$

for a very large, but not infinite, value of n were unfruitful.

· · ·

I have used the method of least squares since the year 1795 and I find in my papers, that the month of June 1798 is the time when I reconciled it with the principles of the calculus of probabilities: a note about this is contained in a diary which I kept about my mathematical work since the year 1796, and which I showed at that time to Mr. De Lindenau.

However my frequent applications of this method only date from the year 1802, since then I use it as you might say every day in my astronomical calculations on the new planets. As I had intended since then to assemble all the methods which I have used in one extensive work (which I began in 1805 and of which the manuscript originally in German, was completed in 1806, but which at the request of Mr Perthes I afterwards translated into Latin: printing began in 1807 and was finished only in 1809), I am in no hurry to publish an isolated fragment, therefore Mr. Legendre has preceded me. Nevertheless I had already communicated this same method, well before the publication of Mr Legendre's work, to several people among other to Mr. Olbers in 1803 who must certainly remember it. Therefore, in my *theory of the motions of planets*, I was able to discuss the method of least squares, which I have applied thousands of times during the last 7 years, and for which I had developed the theory, in section 3 of book II of this work, in German at least, well before having seen Mr Legendre's work – I say, could I have discussed this principle, which I had made known to several of my friends already in 1803 as being likely to form part of a work which I was preparing, – as a method derived from Mr. Legendre? I had no idea that Mr. Legendre would have been capable of attaching so much value to an idea so simple that, rather than being astonished that it had not been thought of a hundred years ago, he should feel annoyed at my saying that I had used it before he did. In fact, it would be very easy to prove it to everyone by evidence which could not be refuted, if it were worth the trouble. But I thought that all those who know me would believe it on my word alone, just as I would have believed it with all my heart if Mr Legendre had stated that he had already been conversant with the method before 1795. I have many things in my papers, which I may perhaps lose the chance of being first to publish; but so be it, I prefer to let things ripen.

. . .

Continue, Sir, to honour me with your good will, which I rank among the things most essential to my happiness.

Ch. Fr. Gauss.

Olbers readily agreed to support Gauss and some four years later his promise was fulfilled by the addition of a footnote to a paper on the period of a star (Olbers, 1816, p. 192).

Olbers to Gauss.   Bremen, 10 *March* 1812.

. . . I have now received the November issue of *M.C.*, and with it your beautiful elimination method for *moindres carrés*. I can attest publicly at the first opportunity, and will do so with pleasure, that you had already told me the basic principle in 1803. I remember this quite well as if it had happened today. There must also be something concerning it written down among my papers. For I noted it then, together with your interpolation formula, which you communicated to me at that time. . . .

A letter to Olbers in 1819 describes the progress of the work which freed the method of least squares from assumptions of normality and culminated in the monographs presented to the Royal Society of Göttingen in 1821, 1823 and 1826.

Gauss to Olbers.   Göttingen, 22 *February* 1819.

. . . I am also occupied at present with a new basis for the so-called method of least squares. In my first basis I supposed that the probability of an observational error x was represented by e^{-hhxx}, in which event that method gives the most probable result with complete rigour in all cases. When the law of error is unknown, it is *impossible* to state the most probable results from observations *already made*. Laplace has considered the matter from a different angle and chosen a principle, which leads to the method of least squares, and which is quite independent of the law of error, when the number of observations is indefinitely large.

With a moderate number of observations, however, one remains quite in the dark if the law of error is unknown, and Laplace himself has also nothing better to say in this case, than that the method of least squares may also be applied here because it affords convenient calculation. I have now found that, by the choice of a principle somewhat different from that of Laplace (and indeed, as cannot be denied, one such that its assumption can be justified at least as well as that of Laplace, and which, in my opinion, must strike *anyone without a previous predilection* as more natural than Laplace's) – all those advantages are retained, namely that in all cases for every error-law the method of least squares will be the most advantageous, and the comparison of the precision of the results with that of the observations, which I had based in my *Theoria* on the error-law e^{-hhxx}, remains generally valid. At the same time there is the advantage, that everything can be demonstrated and worked out by very clear, simple, analytical developments, which is by no means the case with Laplace's principle and treatment, thus, for instance, the generalization of his conclusions from two unknown parameters to any number does not yet appear to have the justification necessary....

In 1827 Legendre received a letter from Jacobi on the subject of elliptic functions. Legendre was then 75 whereas Jacobi was 23. Notwithstanding the disparity in age, a correspondence was established and lasted about five years (Legendre, 1875). Here is a brief extract from the inaugural letter, followed by the closing section of Legendre's reply.

Jacobi to Legendre. Königsberg in Prussia, 5 August 1827.

...It is only very recently that these researches have taken shape. However they are not the only investigations in Germany with the same object. Mr. Gauss, being informed of them, told me that he had already developed in 1808 the cases of 3 sections, 5 sections, and 7 sections, and discovered at the same time the new scales of modules which are related to them. It seems to me that this news is quite interesting...

Legendre to Jacobi. Paris, 30 November 1827.

...How can Mr. Gauss have dared to tell you that the greater part of your theorems were known to him and that he discovered them as early as 1808?...This extreme impertinence is incredible on the part of a man who has sufficient personal merit to have no need of appropriating the discoveries of others.... But this is the same man who, in 1801, wished to attribute to himself the discovery of the law of reciprocity published in 1785 and who wanted to appropriate in 1809 the method of least squares published in 1805. – Other examples will be found in other places, but a man of honour should refrain from imitating them.

Meanwhile the controversy originated by the publication of the *Theoria Motus* continued to reverberate in Germany also (Peters, 1860–5).

Schumacher to Gauss. Altona, 30 November 1831.

...I believe I have already said to you that Zach printed a letter from you in the G.E. (1799 October, p. 378), in which you evidently mentioned the method of least squares, which you had thus already communicated to Zach at the time. You write of the French survey 'I discovered this error when I applied my method, a specimen of which I have given you' etc. Zach noted at that place 'Here at another time' but the other time has never come. Because you gave the results of your calculations, it seems to me that it is easy to show that these were derived by the method of least squares. Besides, Zach is still alive, and has surely preserved your letter. Do you not find it worth the trouble to establish the matter beyond doubt once and for all, even in the face of the polite doubts of the French, which I find particularly offensive?

Gauss to Schumacher. Göttingen, 3 December 1831.

...The place you mention in Zach's A.G.E. is well-known to me; the application of the method of least squares mentioned there concerns an extract from Ulugh Beigh's table of the equation of time, printed earlier in the same journal, which had led to a number of quite curious results. I had communicated these results to Zach with the remark that in connection with them I had employed a method of my own, which I had used for years, of combining quantities involving random errors in a consistent way free from arbitrariness. However, I did not inform him of the nature of the method. I believe I have written to you once before, that in no event will I discuss this passage, where the method was publicly indicated for the first time, also that I do not wish one of my friends to do it with my assent. This would amount to recognizing that my announcement (Th.M.C.C.) that I had used this method many times since 1794 is in need of justification, and with that I shall never agree. When Olbers attested, that I communicated the whole method to him in 1802 [*sic*], this was certainly well meant; but if he had asked me beforehand, I would have disapproved it strongly....

After 1827 Gauss published nothing further about the theoretical basis, but he lectured on the method of least squares from 1835 (Dunnington, 1955, p. 409), and anticipated ideas of decision theory in a letter to Bessel. A section of the following extract has been previously translated by Edgeworth (1908, pp. 386–7).

Gauss to Bessel.   Göttingen, 28 February 1839.

...I have read with great interest your paper in Astronomische Nachrichten, and on how the law for the probability of errors of observation arising simultaneously from several sources, approaches the formula $e^{-xx/hh}$; yet, speaking sincerely, this interest was less concerned with the thing itself than with your exposition. For the former has been familiar to me for many years, though I myself have never arrived at carrying out the development completely. Moreover, the fact that I later abandoned the metaphysics used in the T.M.C.C. for the method of least squares also happened rather for a reason which I myself have not mentioned publicly. Namely, I must consider it less important in every way to determine that value of an unknown parameter for which the probability is largest, although still infinitely small, rather than that value, by relying on which one is playing the least disadvantageous game; or if fa denotes the probability that the unknown x has the value a, then it is less important that fa should be a maximum than that $\int fxF(x-a)\,dx$, taken over all possible values of x, should be a minimum, where for F is chosen a function which is always positive and which always increases with increasing arguments in a suitable way. Choosing the square for this function is purely arbitrary and this arbitrariness lies in the nature of the subject. Without the known extraordinarily large advantages afforded by choosing the square, any other function satisfying those conditions could be chosen, and moreover is chosen in quite exceptional cases. But I do not know whether I have expressed myself sufficiently clearly....

The last letter which refers to the discovery of the method of least squares has essentially the same theme in the relevant portion as the letter from Gauss to Laplace in 1812, but the expression of ideas is now calm and even light-hearted. Gauss had been looking through the papers of Tobias Mayer and was led by easy stages to a reconsideration of the whole affair.

Gauss to Schumacher.   Göttingen, 6 July 1840.

...Among other papers by Mayer (where incidentally, I have found nothing suitable for dissemination), which are still extant and now in my possession, I can find a few quarto sheets of rough notes or draft calculations which contain a small number of passages which are plainly numerical calculations comparing his theory with observations. As far as I can recall at the moment there are about 4 to 6 such passages.

. . .

These sketch notes, incidentally, have been somewhat of a disappointment to me in another respect. You know that I have never laid great store myself upon the procedures which I have been using since 1794, and which have subsequently been given the name Méthode des moindres quarrés. Please understand me correctly, I do not mean in relation to the great benefits which they yield, that is clear enough; but that is not how I measure the value of things. Rather it was for this reason, or to this extent, that I did not rate it very highly, that from the very beginning the idea seemed to me to be so natural, so accessible, that I did not doubt in the slightest that many people, who had to deal with numerical calculations, must of their own accord have arrived at such a device, and used it, without thinking it worthwhile to make much fuss about a thing so natural. To be precise, I had Tobias Mayer particularly in mind, and I remember very clearly that, when I used to discuss my method with other people (as, for example, happened really frequently during my student days, 1795–1798), I often expressed the view, that I would lay a hundred to one that Tobias Mayer had already used the same method in his calculations. I now know from these papers that I should have lost my bet. In the event they contain eliminations, e.g., of three unknown quantities from four or five equations, but in such a way that the most commonplace calculator would have done it, without any trace of a more refined style.
    At all events, I pass this on in confidence. It has really pained me to have my opinion of Mayer somewhat lowered; but what good would it do to publicize it. I loathe minxit in patrios cineres....

Gauss died in 1855 and in 1856 his friend Sartorius published the biography which has remained an important source of information for other biographers. The section which refers to the method of least squares (pp. 42–3) confirms many of the points already made, and ends with a couple of significant ~~questions~~ quotations.

...One time, mentioning the dispute, he expressed himself to us in the words 'The method of least squares is not the greatest of my discoveries'. Another time, to several listeners, he stressed only the words 'They might have believed me.'...

3. DISCUSSION

The principle that publication establishes priority is supported by reason as much as by precedent. A mathematician or scientist attaches great importance to his reputation as a scholar, both in respect of personal esteem and because it determines his livelihood. His reputation is based on the work which he has achieved, and consequently he needs to establish that such work is indeed his own and not derived from others. The publication of his work by any method which is generally accepted, such as a doctoral dissertation, a book, or in a recognized journal, clearly defines the point in time when his discoveries become known to others. Thus the originator of a piece of research can be assessed in such a way that justice is seen to be done. These arguments are given because some commentators on Gauss take another view. For example, Dunnington (1955, p. 19) writes 'according to the custom then in vogue Legendre gained the right of priority'. In fact, the custom has changed little since the time of the dispute between Newton and Leibniz about the discovery of the differential calculus.

Sartorius had no hesitation in assuming that practical need, the observation of nature itself, led Gauss to the method of least squares. The year in which he first applied the method is given variously as 1794 and 1795. Gauss then turned to the problem of establishing a relationship between the principle of minimizing a sum of squares and the calculus of probability. According to a contemporary account of his lecture to the Royal Society of Göttingen on 15 February 1826 (*Werke*, 4, 98), he formulated in 1797 the problem of selecting from all possible combinations of the observations that one which minimized the uncertainty of the results. However, he found that no progress was possible in terms of the most probable values of the unknown quantities when the distribution of errors was unknown. He therefore decided to follow the opposite path first, and look for the distribution which gave the arithmetic mean as the most probable value. Both at this stage and subsequently, the work of Laplace was a stimulus which led Gauss to improve upon the results of his great contemporary. Indeed, Laplace is cast in the role of an adversary from whom the calculus of probability must be protected, and with whom Gauss wrestles in his extraordinary letter to Olbers in 1819. The fact that Gauss completed his theoretical researches on the method of least squares in the same year that Laplace died is doubtless a coincidence, but his letter to Bessel in 1839 tempts us to speculate on what further advances he might otherwise have made.

From first to last, Gauss regarded the method of least squares as his own discovery: he refers to 'meine Methode' both in 1799 and in 1840. When he first saw Legendre's work in July 1806, Gauss commented very favourably and expressed regret about the competition between them. But after Legendre made known his disapproval of the references in the *Theoria Motus* to 'principium nostrum', there was a change of emphasis. On a number of subsequent occasions, Gauss remarked that the idea of minimizing the sum of squared differences was essentially simple and likely to have been used previously. According to his 1840 letter to Schumacher, this had always been his opinion, but he does not mention in it his 1806 letter to Olbers. Possibly he considered that the controversy was disproportionate in size compared with what was involved, and that another aspect should be stressed. The name which Legendre gave, 'méthode des moindres quarrés', must have been generally accepted almost at once and, is, of course, still in use today, suitably translated. According to Dunnington (1955, p. 113), 'by adopting this nomenclature Gauss showed that he did not feel hurt because he was anticipated', but the fact that we find 'der sogennanten Methode der Kleinsten Quadrate', and 'dem später der Name Méthode des moindres quarrés beigelegt ist' suggests a certain amount of antipathy towards this description of 'meine Methode'.

We now turn to Legendre. A full account of his career would be out of place but some aspects of his character are important in a study of his relationship with Gauss. His attitude towards the question of priority is well expressed by the following translated extract from *Mémoires de l'Académie des Sciences* for 1786 (Élie de Beaumont, trans. Alexander, 1867):

> I shall not conclude this article without giving notice that the greater part of the propositions contained therein have been discovered by M. Euler and published in the 7th volume of the *Nouveau Mémoires de Petersbourg* and in some other works, a fact of which I was ignorant when I was engaged in these researches.

According to his biographer in *Nouvelle Biographie Générale* (1859, vol. 30, cols. 385–388), Legendre had studied the works of Euler so assiduously that it could be said he knew them by heart. He always spoke his mind, even when his interests were adversely affected. At the time of the French Revolution he was forced to hide himself, and he was unable to resume public instruction before December 1795 (Élie de Beaumont, 1867). In 1824 he lost his pension because he refused to vote in favour of a candidate for the National Institute, proposed by the government. Immediately after the publication of the first two volumes of his *Traité des fonctions elliptiques et des intégrales eulériennes* in 1825 and 1826, the theory was transformed by Abel and Jacobi. He generously recognized the merits of this work and incorporated their discoveries in a third volume, which appeared shortly before his death (Nielsen, 1929).

Legendre was thus a man of integrity. The full text of his letter to Gauss, which has not been previously published, contains a judicious mixture of praise and criticism, whereas the portion reproduced in Gauss's collected works is critical only. An account of the least squares controversy is given by Bell (1939, pp. 294–5) in his well-known set of biographical sketches of mathematicians. Unfortunately, he makes mistakes about names and dates, and there are reasons for rejecting his conclusions about Legendre. According to Bell, Legendre (i) practically accused Gauss of dishonesty, (ii) regarded the method of least squares as his own ewe lamb, and (iii) passed on his unjustified suspicions to Jacobi and so prevented Jacobi from coming to cordial terms with Gauss. Now Bell was almost certainly unaware of all that Legendre's letter contained. But even if we confine attention to the section which refers to principium nostrum, the emphasis is on two points only: the reason why a claim of priority has to be established by publication; and the convention that a rediscovery is passed over in silence. The suggestion about a ewe lamb is quite out of character, as is shown by Legendre's treatment of the discoveries of Abel and Jacobi. As regards the alleged subversion of Jacobi, who was quite capable of forming his own opinions, the comments made by Gauss on his discoveries must already have been very disheartening. The caustic tone of Legendre's letter of 1827 can reasonably be interpreted as encouragement for a young man of great promise whom he considered to be treated unfairly. But the entire course of Bell's account is so tendentious that it scarcely merits a detailed refutation.

For his part, Gauss was less than wholehearted in accepting the principle that publication establishes priority. This is illustrated by his use of the term 'principium nostrum', the passage with Jacobi, the letter to Bessel, and numerous other instances (Dunnington, 1955). His procedure was to wait until all the aspects of a problem had been rigorously explored before publishing, and treat casually what others did meanwhile. As he wrote to Laplace, 'j'aime mieux faire mûrir les choses'. He published nothing on non-Euclidean geometry, and his work on elliptic functions became fully known only after his death.

Another consequence of Legendre's critical remarks was that Gauss felt concerned to ensure that his announcement of earlier work was accepted as truthful, and his concern was increased after the note from Laplace. He asked Olbers to attest publicly that he had previously communicated the matter to him, but after Olbers had done so, Gauss expressed disapproval because he considered that his announcement did not need justification. Presumably the opportunity for reflection which came with the passage of time, and a dislike of controversy, had led Gauss to have second thoughts about his request. But his disappointment with the response to his announcement is evident in the remark 'Man hätte mir wohl glauben können.'

The last word may appropriately be left with Laplace (1820, p. 353), although he was writing long before the controversy died away.

M. Legendre eut l'idée simple de considérer la somme des carrés des erreurs des observations, et de la rendre un minimum, ce qui fournit directement autant d'équations finales, qu'il y a d'élements à corriger. Ce savant géomètre est le premier qui ait publié cette méthode; mais on doit à M. Gauss la justice d'observer qu'il avait eu, plusieurs années avant cette publication, la même idée dont il faisait un usage habituel, et qu'il avait communiquée à plusieurs astronomes.

We must distinguish the history of statistics and probability from the search for priorities in published work. The two are not identical, as Seal (1967) appears to suggest, and some important differences are as follows. First, a historical account should include if possible the process of challenge and discovery which precedes each publication, and which is illustrated by the episode described above. Secondly, we must attempt to decide whether a publication had any real influence at the time, or whether it sank at once into oblivion and was only rediscovered many years later after the same results had been found independently by other investigators. For example, the work of Adrain on the method of least squares was contemporary with that of Gauss, Laplace and Legendre, but it had no effect whatsoever on the development of the subject. Finally, published work must be interpreted so as to include correspondence, and unpublished work may also be important. Many scientists whose influence has been decisive left behind unpublished manuscripts, and wrote hundreds of letters to their contemporaries. Fortunately, all such material relating to Gauss has been collected and published as the result of immense labours of scholarship. Such are the grounds for believing that the detailed description of published scientific papers in chronological order of appearance gives only a partial and possibly a biased interpretation of the history of statistics and probability.

My debt to the editors of Gauss's collected works and correspondence has already been mentioned. I am grateful to the State and University Library of Lower Saxony for permission to publish a translation of the letter from Legendre to Gauss. They retain the copyright of the manuscript. I express my deep thanks to Mr R. M. White, who made the translation, and to Dr A. Fletcher who gave advice on units of measurement. Other translations have greatly benefited from the help kindly given by Mr White, Dr Fletcher and Professor D. A. West. The comments of Professor G. A. Barnard, Professor G. Waldo Dunnington, Dr C.

Eisenhart, Dr M. G. Kendall, Professor W. Kruskal, Professor E. S. Pearson, Dr H. L. Seal and Dr O. B. Sheynin on an earlier version of this paper have been of much assistance. I am responsible for any errors which remain.

REFERENCES

BELL, E. T. (1939). *Men of Mathematics*. London: Gollancz.

DELAMBRE, J. B. J. (1814). *Astronomie Théorique et Pratique*, 3 vols. Paris.

DUNNINGTON, G. W. (1955). *Carl Friedrich Gauss: Titan of Science*. New York: Exposition.

EDGEWORTH, F. Y. (1908). On the probable errors of frequency constants. *J. R. Statist. Soc.* **71**, 381–97.

EISENHART, C. (1961). Boscovich and the combination of observations. *Roger Joseph Boscovich*, ed. L. L. Whyte, pp. 200–12. London: Allen and Unwin.

EISENHART, C. (1964). The meaning of 'least' in least squares. *J. Wash. Acad. Sci.* **54**, 24–33.

ÉLIE DE BEAUMONT, L. (1867). Memoir of Legendre (trans. Alexander). *Ann. Rep. Smithsonian Institution for* 1867, pp. 137–57.

GALLE, A. (1924). *Über die Geodätischen Arbeiten von Gauss*. C. F. Gauss *Werke*, 11 (2). Berlin: Springer.

GAUSS, C. F. (1799). Vermischte Nachrichten no. 3. *Allgemeine Geographische Ephemeridenz* **4**, 378.

GAUSS, C. F. (1800). *Monatliche Correspondenz.... 1*, 193.

GAUSS, C. F. (1806). II Comet vom Jahr 1805. *Monatliche Correspondenz.... 14*, 181–6.

GAUSS, C. F. (1866–1933). *Werke*, 12 vols. Leipzig.

GAUSS, C. F. (1857). *Theory of the Motion of the Heavenly Bodies....* Trans. C. H. Davis. Boston: Little and Brown.

GNEDENKO, B. W. (1957). Über die Arbeiten von C. F. Gauss zur Wahrscheinichkeitsrechnung. *C. F. Gauss Gedenkband anlasslich des* 100. *Todestages am* 23 *Februar* 1955, ed. H. Reichardt, pp. 193–204. Leipzig: Teubner.

LAMBERT, J. H. (1765). *Beyträge zum Gebrauche der Mathematik und deren Anwendung*. 3 vols. Berlin.

LAPLACE, P. S. (1783). Mémoire sur la figure de la terre. *Histoire de l'Académie Royale des Sciences. Année* 1783. *Avec les Mémoires de Mathématique & de Physique, pour la même Année. Mémoires*, pp. 17–46. Paris, 1786.

LAPLACE, P. S. (1789). *Sur quelques points du systéme du monde. Historie de l'Académie des Sciences. Année* 1789. *Avec. les Mémoires...Mémoires*, pp. 1–87. Paris, 1793.

LAPLACE, P. S. (1820). *Théorie Analytique des Probabilités*, 3rd edition. Paris: Courcier.

LEGENDRE, A. M. (1875). Correspondance mathématique entre Legendre et Jacobi. *J. für die reine und angewandte Mathematik* **80**, 205–79.

MERRIMAN, M. (1877). A list of writings relating to the method of least squares, with historical and critical notes. *Trans. Connecticut Acad. Art. Sci.* **4**, 151–232.

NIELSEN, N. (1929). *Géomètres Français sous la Revolution*. Copenhagen: Levin and Munksgaard.

OLBERS, W. (1816). Ueber den veränderlichen Stern im Halse des Schwans. *Z. für Astronomie und verwandte Wissenschaften* **2**, 181–98.

PEARSON, K. (1924). Historical note on the origin of the normal curve of errors. *Biometrika* **16**, 402–4.

PEARSON, K. (1929). Laplace. *Biometrika* **21**, 202–16.

PETERS, C. A. (Ed.) (1860–65). *Briefwechsel zwischen C. F. Gauss und H. C. Schumacher*, 6 vols. Altona.

PLACKETT, R. L. (1958). The principle of the arithmetic mean. *Biometrika* **45**, 130–5.

SARTORIUS, W. (1856). *Gauss zum Gedächtniss*. Leipzig.

SCHILLING, C. & KRAMER, I. (Eds.) (1900). *Briefwechsel zwischen Olbers und Gauss*. Berlin.

SEAL, H. L. (1967) Studies in the History of Probability and Statistics. XV. The historical development of the Gauss linear model. *Biometrika* **54**, 1–24.

SHEYNIN, O. B. (1966). Origin of the theory of errors. *Nature, Lond.* **211**, 1003–4.

SMITH, E. S. (1927). The tomb of Laplace. *Nature, Lond.* **119**, 493–4.

USPENSKY, J. V. (1937). *Introduction to Mathematical Probability*. New York: McGraw-Hill.

[*Received December* 1971. *Revised January* 1972]

DEVELOPMENT OF THE NOTION OF STATISTICAL DEPENDENCE*

H.O. Lancaster

(received 21 June, 1971)

1. Early work

Naturally, the study of the distribution of single random variables had preceded that of pairs or sets of random variables. However,
it is a surprise to find that the first steps towards a general theory
of dependence were taken as a result of Francis Galton's interest in
genetics. The explanation seems to be that the astronomers were accu-
stomed to assume mutual independence of the errors in their measurements
of the observables and they were not expecting the observables, for example, the positions of the stars, to be related to one another by any
mathematical laws or for any physical reason. In biology, the observables such as height of father and height of son were obviously mutually
dependent and yet not mutually determined - a type of problem not met
by the astronomers.

However, the astronomers had obtained the multivariate normal
distribution without carrying through all its implications. Laplace
(1811), Plana (1813), Gauss (1823) and Bravais (1846) all derived normal
correlations as the joint distribution of linear forms in independently
distributed normal variables but did not define a coefficient of correlation. They expressed the density function as a constant times the
exponent of a general quadratic form. Bravais (1846) in his study on
the errors in the position of a point gave the normal density function
as

$$(1.1) \qquad \frac{K}{\pi} e^{-(ax^2 + 2exy + by^2)} dxdy.$$

---

* Invited address delivered at the sixth New Zealand Mathematics
Colloquium, held at Wellington, 17-19 May, 1971.

The ellipses of equal probability density are then given by

(1.2)
$$ax^2 + 2exy + by^2 = D$$

as stated on his page 272. For each ellipse, the horizontal tangents could be drawn and each point of contact lay on a straight line through the origin, a diameter of the ellipse conjugate to the x axis. This is the line of the modes of the conditional distributions of X for given $Y = y$, since at every other point on this horizontal line the quadratic form, $ax^2 + 2exy + by^2$, corresponds to an ellipse with a greater value of the D of equation (1.2). This is also the line of the conditional means of X but Bravais did not make this point. Bravais (1846) also determined the principal axes and gave the equation in the form,

(1.3)
$$\frac{x^2}{\alpha_0} + \frac{y^2}{\alpha_1} = D.$$

He also determined ϖ the probability inside the axes of equal probability in the form,

(1.4)
$$\frac{d\varpi}{ds} = \frac{K}{\pi} e^{-\frac{Ks}{\pi}},$$

and

(1.5)
$$\varpi = 1 - e^{-\frac{Ks_1}{\pi}}.$$

These equations (1.4) and (1.5) can be regarded as a statement of the distribution of χ^2 with two degrees of freedom or even as an elementary form of the Pearson lemma. Bravais (1846) determined also the principal axes and the value of the constant in the equation for the density in three dimensions of random variables X, Y and Z by successively integrating out the variables. He did not display the bivariate density as the product of a marginal density and a conditional density.

K. Pearson (1920) suggested that the interpretations of K. Pearson (1895) and many later authors were incorrect; for Bravais (1846) was not proposing a theory of the joint distribution of the observables but only of the errors. As already mentioned, Francis Galton was the first

worker to study the mutual dependence of one variable on another as they
appear in the physical world, namely, the relations between heights and
other attributes in human parents and offspring. He had found it difficult to obtain sufficient data in human subjects and so was examining
the produce of seeds of the sweet pea. Galton (1877) observed the phenomenon of 'reversion' whereby the mean of the produce of large seeds was
always closer to the 'average ancestral type' than the parents and similarly for the produce of small seeds; he gives biological reasons, such
as differential mortality, for this phenomenon. In later papers, he used
'regression' rather than 'reversion' to the mean or 'mediocrity'. The
reversion coefficient may be written as λ and $\hat{y} = \lambda x$ is the line of reversion or regression of Y on X. If the variance of Y, the offspring, is·
to be equal to the variance of the parent, X, and if *homoscedasticity*,
that is, a uniform conditional variance, be assumed then the variance of
Y can be partitioned into the conditional variance and the variance of
conditional means

$$(1.6) \qquad \text{var } Y = (1 - r^2)\text{var } Y + \text{var } \hat{y}$$
$$= (1 - r^2)\text{var } Y + \lambda^2 \text{ var } X,$$

and $\lambda = r$ if var X = var Y, so that the variance is to be preserved from
generation to generation. In this earlier paper, Galton (1877) did not
seem to realise that there are two regression lines. However, Galton
(1886a) gave a diagram displaying them both; it is of interest that he
obtained, by smoothing an observed distribution of heights of offspring,
the ellipses of equal density such as Bravais (1846) had determined in
the theoretical distribution. Galton (1886b) enlisted the help of Dickson
(1886), who wrote the joint distribution as a product of the marginal
distribution of Y and the conditional distribution of X.

Now it appears that Dickson (1886) could well have taken notice of
several other themes in the multivariate normal distribution. Thus
Bienaymé (1852) had generalized the central limit theorems of Laplace
and of Fourier to several variables. However, he had then made a linear
transformation of his variables to obtain the distribution of the theoretical χ^2 for he was not interested in the joint distributions of the

4

variables or errors. Todhunter (1869) also had examined errors jointly
normally distributed and in an appendix to this note, Cayley (1869) had
given the general expression for the constant multiplier of the density,
constant × exp - $\frac{1}{2}x^{T}Cx$. Further, Todhunter (1869) had obtained the
characteristic function for the normal correlation.

Herschel (1850) in an extended review of Quetelet's (1846)
Lettres à S.A.R. le duc ... had shown how the normal distribution could
be derived as the only possible distribution of errors in two dimensions
such that it was possible to obtain two distinct pairs of independent
random variables, related by an orthogonal transformation. Indeed,
Herschel (1850) sketched a proof of the proposition, that if X is independent of Y and if also $X \cos\theta + Y \sin\theta$ is independent of $-X \sin\theta +$
$Y \cos\theta$, then X is normal and Y is normal. This characterization was bitterly attacked by Ellis (1850). However, the hypotheses were stated more
exactly by Boole (1854), who made the explicit assumptions that the variables X and Y were mutually independent and that the density was dependent solely on the distance from the origin; a functional equation resulted,

(1.7) $$f(x^2 + y^2)f(0) = f(x^2)f(y^2)$$

which had a solution of the form,

(1.8) $$f(x^2) = (2\pi)^{-\frac{1}{2}}\exp(-\frac{1}{2}x^2),$$

since $\int f(x^2)dx = 1$, and so the distribution was normal. Boole (1854)
also clearly recognized that there could not be a universal law of
errors.

Assumptions closely related to those of Herschel (1850) were used
explicitly by Maxwell (1860, 1867) in his derivation of the kinetic theory of gases. He deduced the normal distribution of the velocities of
the gas molecules by assuming that the components of the velocities of
the molecules along any orthogonal set of axes are mutually independent.
As is noted at page 122 of Plummer (1940) the assumption of independence
of the components along three orthogonal axes was disputed and subsequ-

296

ently replaced by an analysis of the dynamical conditions; Plummer (1940) praises Herschel (1850) for realising that some assumptions have to be made to reach a solution and for making simple and explicit assumptions, effective in leading to a solution. Many modern writers have overlooked the point in Maxwell's discussion and have assumed both the mutual independence about arbitrary orthogonal axes and the normality and so have multiplied their hypotheses unnecessarily.

The multinomial distribution is perhaps the simplest of the joint distributions and was early studied. Bienaymé (1838) gave the asymptotic distribution of linear forms in the cell frequencies of the multinomial suitably standardised and also the asymptotic expression for the probability of the set of cell frequencies. The exponent in this expression is very close to Pearson's χ^2 ; in fact it is the sum $\sum$(observed-expected)^2/(observed). This contribution has been almost entirely overlooked by subsequent authors although Meyer (1874) mentions it.

2.   The development of the notion of dependence by Karl Pearson

The best introduction to the next era of development is the statement of K. Pearson (1920), who wrote
'It will be seen from what has gone before that in 1892 the next steps to be taken were clearly indicated. They were, I think,
(a)   The abolition of the median and quartile processes as too inexact for accurate statistics.
(b)   The replacement of the laborious processes of dividing by the quartiles and averaging the deduced values of r, by a direct and if possible "best" method of finding r.
(c)   The determination of the probable errors of r as found by the "best" and other methods.
(d)   The expression of the multiple correlation surface in an adequate and simple form.

These problems were solved by Dr Sheppard or myself before the end of 1897.

Closely associated with these problems arose the question of

6

generalising correlation. Why should the distribution be Gaussian,
why should the regression curve be linear?

As early as 1893 I dealt with quite a number of correlation tables
for long series and was able to demonstrate

(i)   by applying Galton's process of drawing contours of equal frequency that most smooth and definite systems of contours can arise from
long series, obviously mathematical families of curves, which are (a)
ovaloid, not ellipsoid, and (b) which do not possess - like the normal
surface contours - more than one axis of symmetry.

(ii) that regression curves can be quite smooth mathematical curves
differing widely from straight lines

(iii) that in cases wherein (i) and (ii) hold, homoscedasticity is not
the rule.

I obtained differential equations to such systems, but for more
than 25 years while often returning to them, have failed to obtain their
integration.

This seems to me the desideratum of the theory of correlation at
the present time: the discovery of an appropriate system of surfaces,
which will give bi-variate skew frequency. We want to free ourselves
from the limitations of the normal surface, as we have from the normal
curve of errors.'

To Karl Pearson and W.F.R. Weldon, the importance of Galton's
work for future studies of heredity and evolution was clear. Weldon
(1890 and 1892) began to accumulate empirical distributions in one and
two dimensions in order to compare variations in biological variables in
space and time. To explain the non-normality of one of Weldon's empirical distributions Pearson (1894) solved the problem of how to dissect a
mixture of normal distributions into its component normal curves and so
began his great series on the *Contributions to the mathematical theory
of evolution*. Pearson (1895b, 1901 and 1916a) introduced his system of
curves, which have since found many applications in statistics, as the
solutions to a hypergeometric differential equation. This series of

298

curves appears naturally in the solution of differential equations giving the limiting or equilibrium forms of distribution for certain diffusion processes and determining bivariate densities which have
'diagonal' or homogenous polynomial expansions.

K. Pearson (1895b) constructed other joint distributions by sampling experiments with card packs and roulette wheels; he noted that if
the material 'obeys a law of skew distribution, the theory of correlation
as developed by Galton and Dickson requires considerable modification.'
Pearson (1896) gave some historical notes on the history of multivariate
normality, in which he assigned rather more importance, as Pearson (1920)
was later to remark, to the memoir of Bravais (1846) than it deserved in
the particular context, for Bravais (1846) had considered neither regression nor conditional distributions and had not paid any special attention
to the coefficient of correlation. In the same article, Pearson (1896)
introduced the multivariate normal distribution by making a linear transformation from mutually independent centred normal variables, the elements
of a vector, $\underset{\sim}{\xi}$, to the elements of a vector, $\underset{\sim}{\eta}$.

$$(2.1) \qquad\qquad \underset{\sim}{\eta} = \underset{\sim}{A}\underset{\sim}{\xi},$$

where A is $m \times n$, $m \leq n$ and of rank m, a transformation which had already
been used by Laplace (1810) and Plana (1813) and also by Todhunter (1869).
By integrating out superfluous variables, Pearson (1896) obtained the
joint density function,

$$(2.2) \qquad\qquad f(\eta_1, \eta_2, \ldots, \eta_m) = \text{constant } \exp(-\tfrac{1}{2}\chi^2),$$

where the expression χ^2 appears for the first time in the literature as
a quadratic form in the variables. If $m = 2$, there was linear regression
between the two variables and the conditional variance of one variable
for fixed values of the other was reduced in the ratio, $(1 - \rho^2)$, where
ρ was the coefficient of correlation. Pearson (1896) gave the general
formula for the bivariate density,

$$(2.3) \qquad f(x,y) = \text{constant} \times \exp\left[-\tfrac{1}{2}(g_1 x^2 + 2hxy + g_2 y^2)\right],$$

and derived from it the standard form as we now know it. He then

8

showed how to estimate r, the coefficient of correlation in a sample
by maximum likelihood. He extended the analysis to normal distributions
in three and higher dimensions. The reader cannot but wish that he had
separated out the purely mathematical or distributional theory from the
genetical applications, for the mixture of the two aspects is very confusing. It appears to the modern reader of Pearson (1896), Edgeworth
(1892) and Sheppard (1898) that Pearson had a much better understanding
of the mathematics and the applications than either of the others,
although this view has not been generally held. Pearson (1896) obtained
the general formula for the multivariate normal, namely

$$(2.4) \qquad f(x_1, x_2, \ldots, x_p) = (2\pi)^{-\frac{1}{2}p} |\underset{\sim}{R}|^{-\frac{1}{2}} \exp(-\tfrac{1}{2}\underset{\sim}{x}^T \underset{\sim}{R}^{-1} \underset{\sim}{x}) ,$$

a result already known to Todhunter (1869) and Cayley (1869).

It should be noted that at this time, 1896, few joint distributions, not simply products of the marginal distributions, were known and
the lack of such examples and a corresponding theory led such capable
mathematicians as H.W. Watson (1891) into serious difficulties. There
was also lacking a clear notion of multivariate measure or distribution,
free of the notion of sample. Thus we find the multiplier N appearing
in formulae, as in the first formula of Section 5 of Pearson (1896) in,
what seems to us, a rather irrelevant manner. Perhaps, only after the
axiomatizations of probability of Kolmogorov (1933), for example, were
clear statements possible. Appropriate practical examples were also
lacking. Meteorological or vital statistical examples lacked a mathematical model to explain their form. A simple model was possible in
genetics. Galton, Weldon and Pearson all believed that correlation or
lack of independence in the distribution of attributes in relatives was
usually brought about by the possession of random elements in common,
that is, the presence or absence of genetic factors. To imitate such
a genetic model, Weldon (1906) carried out dice throwing experiments
which yielded a bivariate binomial distribution, with marginal distrib-
utions of the form, $(\tfrac{1}{2} + \tfrac{1}{2})^{12}$, and six random variables held in common.
Pearson (1895b) too was carrying out experiments with teetotums (roul-
ette wheels) and card packs to obtain empirical joint distributions.

Some novel applications of the theory were now possible. Pearson and
Filon (1898) obtained the joint distribution of the errors in estimates
with the aid of some plausible simplifying assumptions such as joint
normality, thus extending the theory in which mutual independence had
been assumed. They were also able to consider the effect of sampling
on one subset of the variables on the joint distribution of the complementary subset. Now that Pearson (1895b) had described a number of different frequency curves, it became for the first time practical to ask
whether empirical distributions were better fitted by one theoretical
distribution or another. Sheppard (1898b) proposed to make such comparisons by treating the difference, observed-expected, as a binomial
variable, for example, for each of the cells of a contingency table.
As Sheppard (1929) was later to remark, he had paid insufficient attention in 1898 to the fact that these binomial variables were correlated.
Pearson (1900a) with his knowledge of the mutivariate normal theory was
able to attain the solution, the Pearson χ^2, which could be used as a
criterion to test goodness of fit. Pearson (1900a) made the assumption
that variables having marginal normal distributions were jointly normal.
In the asymtotically normal distribution arising from the multinomial,
this assumption can be justified by the use of the generalised central
limit theorem of Bernstein (1926). A closely related theme had been
developed by Bienaymé (1838) who had obtained the asymptotically normal
distribution of linear forms in the cell contents of a multinomial distribution. Later, Sheppard (1898) independently obtained this result.
The idea is useful for Frechet (1951) has since defined multivariate
normality by the property that every linear form in the variables is
normal.

Pearson (1904) considered multivariate distributions in qualitative
variables, for which he introduced the term, 'contingency' for any measure of the total deviation of the classification from independent probability. These measures might in some cases be independent of the ordering;
he was, therefore, making a new departure freeing the theory of the necessity to give primacy to the product moment correlation. For bivariate
distributions, he defined ϕ^2 , the integral of the square of the like-

lihood ratio of the bivariate density to the product of the marginal
densities, taken with respect to the product measure. He concluded
that he had generalized the work of Yule (1900, 1901 and 1903).

Pearson (1905) in effect defined *correlation* in a general sense
to mean that the distribution of Y was not the same for every value of
X, that is $P(Y \in B | X \in A)$ was not constant for every choice of the set A.
He considered the conditional means and the variance of the conditional
mean and the *correlation ratio*.

Pearson and Heron (1913) were concerned mainly with the estimation
of indexes of correlation rather than with the introduction of new theory. Pearson introduced a notion here and in other articles, that in a
partitioned marginal space of 'discrete' variables, the variable could
be regarded as approximating in some way to an underlying continuous
variable. In the discrete space formed from the space of a continuous
random variable, there is indeed a linear form in the indicator variables of the cells of the condensed space which has maximum correlation
with the continuous marginal variable and this maximum correlation tends
to unity as the partition is appropriately refined.

The introduction of a new system of distributions in one dimension
was a great step forward, but K. Pearson was much less fortunate in his
generalizations of the idea to two-dimensional distributions; indeed,
Pearson (1923a) noted that, after the introduction of his system of curves to graduate empirical frequency distributions in one dimension, he
had attempted to describe a system of bivariate distributions as the solutions of a differential equation but that much work had failed to provide such a system. Certain particular results had been obtained; in
general, a bivariate distribution could not be represented as a product
distribution by a rotation of axes, for example, if the density was referred to the 'principal inertial system of the contour system' of the
bivariate density, a set of independent random variables was not thereby
obtained, as it would be in the multivariate normal distributions.
E.C. Rhodes (1923) and L.N.G. Filon had obtained some special forms of
bivariate densities, which they had not published at the time of the

completion of their work. Pearson (1923a) was rather concerned that the correlation between such variables appeared to be determined by the marginal distributions. In Pearson (1923a and b) some surfaces given by analytic expressions were examined for properties such as linear regression. If homoscedasticity were imposed, the normal distribution was characterized. Pearson (1925) attempted to extend to two dimensions the Charlier method of approximating to the normal in one dimension. He wished to obtain bivariate surfaces for arbitrary values of the sample size and moments and mixed moments up to the 4th order, the fifteen-constant bivariate density. However, we cannot regard this effort as productive as there are too many parameters to be fitted and it is possible that negative values of the density may result.

The most successful attempt to obtain bivariate densities was, according to Pearson (1923a), made by S. Narumi (1923, a, b and c). Narumi (1923a) had begun by considering a 'regression function', by which he meant any functions of the form $\hat{y}_x \equiv \hat{y}(x)$ and $\hat{x}_y \equiv \hat{x}(y)$, such as the mode of the conditional distribution for given x, the conditional mean, a linear function in the independent variable. 'Homoscedasticity' was reinterpreted as the same form of conditional distribution for each value of the independent variable. He then considered various differential equations. It is of interest that if the centres of location of the conditional distributions are situated on straight lines and if the conditional distributions have the same form, then there are but few possibilities for either the marginal variables are mutually independent, the distribution is jointly normal or the joint distribution is degenerate. It is clear from Narumi (1921b) that this approach leads to very heavy algebra and functional equations, difficult to solve except in the normal distribution. The principal extension of the theory due to Narumi, Rhodes and Pearson may be said to be the densities of the form, $f(\underset{\sim}{x}) = (1 \pm \underset{\sim}{x}^T \underset{\sim}{A} \underset{\sim}{x})^\alpha$, the bivariate hypergeometric distribution and the multivariate hypergeometric distribution and the multivariate binomial distribution. Bernstein (1927) obtained the independence distributions, the joint normal distribution and some other special distributions as a result of his investigations in this problem.

REFERENCES

Bernstein, S.N. (1926). *Sur l'extension du théorème limite du calcul des probabilités aux sommes de quantités dépendantes*, Math. Ann. 97, 1-59.

Bernstein, S. (1927). *Fondements géométriques de la théorie des corrélations*, Metron 7(2), 1-27.

Bienaymé, J. (1838). *Sur la probabilité des résultats moyens des observations; démonstration directe de la règle de Laplace*, Mémor. Sav. Étrangers Acad. Sci. Paris 5, 513-558.

Bienaymé, J. (1852). *Sur la probabilité des erreurs d'après la méthode des moindres carrés*, J. Math. Pures Appl. 17, 33-78.

Boole, G. (1854). *On a general method in the theory of probabilities*, Phil. Mag. (4) 8, (reprinted in *Studies in probability and statistics* (1952), Watts and Co., London).

Bravais, A. (1846). *Analyse mathématique sur les probabilités des erreurs de situation d'un point*, Mém. de l'Instit. de France 9, 255-332.

Cayley, A. (1869). See Todhunter (1869).

Dickson, J.D.H. (1886). *Appendix to Galton (1886)*, Proc. Roy. Soc. 40, 63-73.

Edgeworth, F.Y. (1892). *The law of error and correlated averages*, Phil. Mag. (5) 34, 429-438 and 518-526.

Ellis, R.L. (1850). *Remarks on an alleged proof of the 'Method of Least Squares' contained in a late number of the Edinburgh Review*, Phil. Mag. (3) 37, 321-8 and 462.

Fréchet, M. (1951). *Généralisations de la loi de probabilité de Laplace*, Ann. Instit. Henri Poincaré 12, 1-29.

Galton, F. (1886a). *Regression towards mediocrity in hereditary stature*, J. Anthrop. Inst. 15, 246-263.

Galton, F. (1886b). *Family likeness in stature*, Proc. Roy. Soc. 40, 42-63.

Galton, F. (1877). *Typical laws of heredity*, Proc. Royal Instit. Gt. Britain 8, 282-301.

Galton, F. (1908) *Memories of my life*, E.P. Dutton and Co., New York.

Gauss, C.F. (1823). *Anwendung der Wahrscheinlichkeitsrechnung auf eine Aufgabe der praktischen Geometrie*, Astron. Nachr. 1, 81-88.

Herschel, J. (Anonymously) (1850). A review of *Lettres à S.A.R. le duc règnant de Saxe-Cobourg et Gotha sur la théorie des probabilités appliquée aux sciences morales et politiques*, by M.A. Quetelet (1846), Edinburgh Review 92, 1-57.

Kolmogorov, A.N. (1933), *Grundbegriffe der Wahrscheinlichkeitsrechnung*, Ergebnisse der Math. 2(3), 196-262.

Laplace, P.S. (1811). *Mémoires sur les intégrales définies et leur application aux probabilités*, Mémoires de l'Institut Impérial de France, 1811, 279-347.

Meyer, A. (1874). *Calcul des probabilités de A. Meyer publié sur les manuscrits de l'auteur par F. Folie*, Mém. Soc. Liège, 6(2), x+446 pp.

Maxwell, J.C. (1860). *Illustrations of the dynamical theory of gases*, Phil. Mag. (4) 19, 19-32 and (4) 20, 21-37.

Maxwell, J.C. (1867). *On the dynamical theory of gases*, Philos. Trans. Roy. Soc. Lond. 157, 49-88.

Narumi, S. (1923a, b and c). *On the general forms of bivariate frequency distributions which are mathematically possible when regression and variation are subjected to limiting conditions. Parts I and II*, Biometrika 15, 77-88 and 209-221.

Pearson, K. (1894). *Contribution to the theory of evolution*, Philos. Trans. Roy. Soc. Ser. A. 185, 71-110.

Pearson, K. (1895a). *Note on regression and inheritance in the case of two parents*, Proc. Roy. Soc. 58, 240-241.

Pearson, K. (1895b). *Contributions to the mathematical theory of evolution. II. Skew variation in homogeneous material*, Philos. Trans. Roy. Soc. Ser. A. 186, 343-414.

Pearson, K. (1896). *Mathematical contributions to the theory of evolution. III. Regression, neredity and panmixia*, Philos. Trans. Roy. Soc. Ser. A. 187, 253-318.

Pearson, K. (1900a). *On the criterion that a given system of deviations from the probable in the case of a correlated system of variables is such that it can be reasonably supposed to have arisen from random sampling*, Philos. Mag. 50, 157-175.

Pearson, K. (1901). *Mathematical contributions to the theory of evolution. X. Supplement to a memoir on skew variation*, Philos. Trans. Roy. Soc. Ser. A. 197, 443-459.

Pearson, K. (1904). *Mathematical contributions to the theory of evolution. XIII. On the theory of contingency and its relations to association and normal correlation*, Drapers' Company Research Memoirs, Biometric Series 1. 35pp.

Pearson, K. (1905). *Mathematical contributions to the theory of evolution. XIV. On the general theory of skew correlation and non-linear regression.* Drapers' Company Research Memoirs, Biometric Series. ii+54.

Pearson, K. (1916a). *Mathematical contributions to the theory of evolution. XIX. Second supplement to a memoir on skew variation*, Philos. Trans. Roy. Soc. Ser. A. 216, 429-457.

Pearson, K. (1920). *Notes on the history of correlation*, Biometrika 13, 25-45.

Pearson, K. (1923a). *Notes on skew frequency surfaces*, Biometrika 15, 222-30.

Pearson, K. (1925). *The fifteen constant bivariate frequency surface,* Biometrika 17, 268-313.

Pearson, K. and Filon, L.N.G. (1898). *Mathematical contributions to the theory of evolution. IV. On the probable errors of frequency constants and on the influence of random selection on variation and correlation,* Philos. Trans. Roy. Soc. Ser.A. 186, 343-414.

Pearson, K. and Heron, D. (1913). *On theories of association,* Biometrika 9, 159-315.

Plana, G.A.A. (1813). *Mémoire sur divers problèmes de probabilité,* Mém. Accad. Imper. Turin 20 (années 1811-1812), 355-408.

Plummer, H.C. (1940). *Probability and frequency,* Macmillan and Co., London, xi+277.

Quetelet, A. (1846). *Lettres à S.A.R. le duc regnant de Saxe-Cobourg et Gotha sur la théorie des probabilités appliqué aux sciences morales et politiques,* Hayez, Bruxelles, iv+450pp. (English transl. by O.G. Downes (1849). Charles and Edward Cayton, London.)

Rhodes, E.C. (1923). *On a certain skew correlation surface,* Biometrika 14, 355-377.

Sheppard, W.F. (1898a). *On the application of the theory of error to cases of normal distribution and normal correlation,* Philos. Trans. Roy. Soc. Ser. A. 192, 101-167.

Sheppard, W.F. (1898b). *On the geometrical treatment of the 'Normal Curve' of statistics with special reference to correlation and to the theory of errors,* Proc. Roy. Soc. 62, 170-173.

Sheppard, W.F. (1929). *The fit of formulae for discrepant observations,* Philos. Trans. Roy. Soc. Ser. A. 228, 115-150.

Todhunter, I. (1869). *On the method of least squares,* Trans. Camb. Philos. Soc. 11, 219-238.

Watson, H.W. (1891). *Observations on the law of facility of errors.* Proc. Birmingham Phil. Soc. 7, 289-318.

Weldon, W.F.R. (1890). *The variation occurring in the Decapod Crustacea. - 1. Crangon vulgaris,* Proc. Roy. Soc. 47, 445-453.

Weldon, W.F.R. (1892). *Certain correlated variations in Crangon vulgaris,* Proc. Roy. Soc. 51, 2-21.

Weldon, W.F.R. (1906). *Inheritance in animals and plants,* pp. 81-109 in T.B. Strong, *Lectures on the method of science,* Clarendon Press, Oxford, viii+249.

Yule, G.U. (1900). *On the association of attributes in statistics,* Philos. Trans. Roy. Soc. Ser.A. 194, 257-319.

Yule, G.U. (1901). *On the theory of consistence of logical class frequencies and its geometrical representation,* Philos. Trans. Roy. Soc. Ser A. 197, 91-134.

Yule, G.U. (1903) *Notes on the theory of association of attributes in statistics,* Biometrika 2, 121-134.

University of Sydney

FLORENCE NIGHTINGALE AS STATISTICIAN.

BY EDWIN W. KOPF   (1916)

The somewhat legendary accounts of this remarkable woman contain but few references to that part of her life and work which should appeal to the students of the history of modern social statistics.   More or less is understood of her radical innovations in the nursing care of the sick in institutions and especially in military hospitals; a definite idea exists of her capacities as reformer, administrator, and nurse.   Comparatively little is known of her, however, as a constructive compiler and interpreter of descriptive social statistics.   One biographer alone, Sir Edward T. Cook, speaks with sympathy and understanding of her as a statistician.   He calls her a "passionate statistician."*

The activities of Miss Nightingale in statistics may be classed under several broad categories.   We may think of her in terms of her forty years of thought and achievement in the Indian question; in safeguarding the health of the British soldier; in reorganizing civil and military hospital administration at home and abroad; and, in this latter regard, of her pioneer services to the profession of nursing.   Her keen intellect, applied to these major projects of her career, comprehended the utility of the statistical method as a means of developing a basis of established fact for social reform.

In early life, Miss Nightingale showed peculiar aptitude for collecting and methodically recording current historical facts. Her observations during the travels of the Nightingales in Europe over the period 1837–1839 are a curious mixture of comment and criticism on the then existing laws, land systems, social conditions, and benevolent institutions.   Throughout her life she collected an immense number of pamphlets, reports, and returns which she skillfully analyzed with telling effect in her campaigns for hospital and sanitary reform.   Following her nursing apprenticeship with the Fliedners at

---

*The writer is indebted to Sir Edward T. Cook's "Life of Florence Nightingale" for abstracts from Miss Nightingale's private papers, and for much of the material in this paper.

Kaiserswerth, she undertook further training at the *Maison de la Providence* in Paris; here she proceeded to collect hospital reports, returns, statistical forms, and general information on hospital construction and sanitation. Among her papers then were elaborately tabulated analyses of hospital organizations and nursing systems and their end results. These inquiries extended to both France and Germany. She seems also to have addressed circulars of inquiry on the same subjects to representative hospitals in the United Kingdom.

Miss Nightingale was profoundly influenced by the works of Adolphe Quetelet, the Belgian astronomer, meteorologist, and statistician. Perhaps her practice of methodically recording the facts of her botanical researches led her to one of Quetelet's laws of flowering plants. The common lilac flowers, he averred, when the sum of the squares of the mean daily temperatures, counted from the end of the frosts, equals 4264° centigrade. While this "law" delighted her, she regarded it as a lesser example of Quetelet's researches and statistical conclusions. She was fascinated most by Quetelet's "*Sur l'Homme et le Développement de ses Facultés*," published in 1835, in which he outlined his conception of statistical method as applied to the life of man. From Quetelet, Miss Nightingale learned much of the science and art which describes human society in terms of numbers. From him she learned the methods, general aims, and results of qualified inquiry into social facts and forces.

MILITARY AND SANITARY STATISTICS OF THE CRIMEAN WAR.

The discipline of Quetelet's new science of social inquiry was to have its first influence upon the military and sanitary statistics of the Crimean War. Miss Nightingale found the medical records of the Scutari hospitals in lamentable condition. Even the number of deaths was not accurately recorded. The three separate registers then maintained gave each a totally different account of the deaths among the military forces. None of the statistical records was kept in uniform manner. She was able to introduce an orderly plan of recording the principal sickness and mortality data of the military hospital establishments which came within the sphere of her influence.

Miss Nightingale's experience in the Crimea filled her with an ardent desire to remedy the scandalous neglect of sanitary precautions in the Army; her study of the available data convinced her that the greater number of deaths in hospitals need not have occurred at all. During the first seven months of the Crimean campaign, a mortality of 60 per cent. per annum from disease alone occurred, a rate of mortality which exceeded even that of the Great Plague in London, and a higher ratio than the case mortality of cholera. Miss Nightingale's vigorous use of these facts resulted in a series of reforms, which in turn reduced this terrible rate of mortality. She observed, also, that if sanitary neglect prevailed in the Army afield, it probably affected the Army at home in considerable degree likewise. She compared the mortality in civil life with the mortality in army barracks. Between the ages of 25 and 35 she found that the mortality among soldiers was nearly double that in civil life. In writing to Sir John McNeill she said: "it is as criminal to have a mortality of 17, 19, and 20 per thousand in the Line, Artillery and Guards, when that in civil life is only 11 per 1,000, as it would be to take 1,100 men out upon Salisbury Plain and shoot them."

Her further observations on the Chatham military hospitals were: "This disgraceful state of our Chatham Hospitals is only one more symptom of a system, which, in the Crimea, put to death 16,000 men—the finest experiment modern history has seen upon a large scale, viz., as to what given number may be put to death at will by the sole agency of bad food and bad air." Among her private notes of 1856 her biographer found this: "I stand at the altar of the murdered men, and while I live, I fight their cause." The one weapon upon which she placed most dependence was her collection of sanitary statistics.

HEALTH, EFFICIENCY, AND HOSPITAL ADMINISTRATION OF THE BRITISH ARMY.

The results of her personal studies of army medical statistics were embodied in a report, from the first intended as a confidential communication to the War Office and the Army Medical Department. There had been considerable delay in the formation of the Royal Commission on the health of the Army which she had requested in her November, 1856, interview with Lord Panmure. The Royal Warrant establishing the Commission was not issued until May 5, 1857. During this exasperating period of delay Miss Nightingale held in reserve her array of statistics, until, having begun her agitation with the sovereign and continuing through the politicians, she was almost ready to plead her cause with the people. In three months from the day the Royal Warrant was issued, the Commission presented its report. In the meantime, Lord Panmure had asked Miss Nightingale for her "Notes Affecting the Health, Efficiency and Hospital Administration of the British Army." These notes are the least known of her works, because they were never officially published. It has never become known how much of the final Report of the Royal Commission was actually the work of Miss Nightingale. Printed at her private expense and circulated among influential people, her "Notes" made a profound impression. They have been termed "a treasury of authentic fact . . . affording a complete elucidation of the causes which had brought about failure, and showing the means by which the country could best hope to safeguard the truly sacred task of providing for the health of its troops in future wars." Another of her friends who read the proof said: "It has so much the character of good, sincere, enlightened conversation on a subject which is thoroughly understood and appreciated, and so little the appearance of having been 'got up' or of pretension of any kind, literary or artistic." Another reader said: "I regard it as a gift to the Army, and to the country altogether priceless."

The preface to the Notes gave the keynote. Hospitals were shown to be but part of wider programs involving the general health and efficiency of the Army. This was emphasized by the fact that those who fell before Sebastopol by disease were above seven times the number who fell by the enemy. The introductory chapter gave the history of the health of the British Armies in previous campaigns. Six of the twenty sections of the Notes dealt with the medical history of the Crimean War. Two other sections discussed the mortality of armies in peace and war and the necessity for a statistical department of the army. There were also numerous appendices, supplementary notes, and graphic illustrations and diagrams.

PIONEER IN GRAPHIC ILLUSTRATION OF STATISTICS.

It must be remembered that these Notes were written by Miss Nightingale in the short space of six months, and while in delicate health. In the preparation of her report she had but little assistance; the gathering of the data was facilitated, however, by the friendly coöperation of many broadminded men in the public service. Dr. Farr, for instance, aided materially in the preparation of the comparisons between the mortality of civil and army life and in editing the graphical illustrations which he especially commended. These graphical diagrams were at that time somewhat of an innovation in statistics, and had no significant precedent save in the statistical works of A. M. Guerry, a contemporary of Quetelet. The Report of the Commission, containing some thirty-three written answers by Miss Nightingale to leading questions by the Commission, together with her original tabulation of the appalling morbidity and mortality statistics of the British Army, was issued to the public in January, 1858. The graphical illustrations in her own Notes portrayed, by means of shaded or colored squares, circles and wedges, (1) the deaths due to preventable causes in the hospitals during the Crimean War and (2) the rate of mortality in the British Army at home. "Our soldiers enlist," as she put it, "to death in the barracks." She reprinted this graphic section and distributed it, with a brief memorandum, to leading members of Parliament and to medical and commanding officers throughout the country, in India, and in the Colonies. "It is our flank march upon the enemy," she said.

STATISTICAL DEPARTMENT OF BRITISH ARMY FOUNDED.

The chief product of the Commission's work of interest from the statistical standpoint was the report of the subcommittee on Army Medical Statistics. This committee, consisting of Mr. Sidney Herbert, Sir A. Tulloch, and Dr. Farr, reported in June, 1858, and published its "First Annual Statistical Report on the Health of the Army" in March, 1861. The compilations were directed by Dr. Thomas Graham Balfour, under whose leadership British army statistics became the best and most useful obtainable in Europe.

The facts published in the "Notes" did not go unchallenged. A pamphlet appeared anonymously calling them in question. Miss Nightingale immediately prepared a reply. This second note was entitled "A Contribution to the Sanitary History of the British Army during the late War with Russia," and constituted a scathing and eloquent account of the preventable mortality which she had witnessed in the East. The graphic charts of the "Notes" were reproduced.

CONSTRUCTION, ORGANIZATION, AND MANAGEMENT OF CIVIL HOSPITALS.

The opposition to the recommendations of the subcommission on Army Barracks stimulated Miss Nightingale to prepare a more extended discussion of hospital construction, organization, and management. From her extensive experience in and study of hospital systems in Germany, France, and Ireland and in the Crimea, she prepared two addresses on hospital construction and sanitation for the Liverpool meeting of the National Association for the Promotion of Social Science. These papers were reprinted as "Notes on Hospitals." These "Notes" in three editions, the last in 1863, revolutionized ideas of hospital construction.

It was pointed out that the hospital statistics then available gave little information of real value on the proportion of recoveries, of deaths, and the average duration of hospital treatment for different diseases, duly qualified by sex and age. A common agreement on the number and nature of statistical data to be tabulated was recommended. A unique feature of this Liverpool address was a mortality table for hospital nurses and attendants showing the greatly increased prevalence of communicable diseases among this class of hospital employees, as compared with the mortality from the same causes in civil life. The deplorable existence of "hospital gangrene" and "hospital septicemia" in that day of defective hospital sanitation and construction was effectively portrayed by these mortality statistics.

A brief inquiry into the precedent circumstances will be of interest. When Miss Nightingale returned from the Crimea she directed much thought and attention to hospital statistics as an adjunct to administration of institutions for the care of the sick. She found a complete lack of scientific coördination. The statistics were not kept along uniform lines. Each hospital followed its own nomenclature and classification of diseases. The available data had never been tabulated upon forms which would render the statistics of one hospital comparable with those of another. The data had little value for advancing medical knowledge or as an adjunct to hospital management. With the assistance of Dr. Farr, and of other friendly physicians, she drew up a standard list of diseases (largely a selection from the d'Espine-Farr System) and a set of model hospital statistical forms. She had her model forms printed in 1859 and persuaded some of the London hospitals to adopt them experimentally. She and Dr. Farr studied the tabulated results, which had sufficient value to show how large a field of qualified statistical inquiry had been opened by the introduction of her forms.

Miss Nightingale's skill in so effectively employing the statistical method in army sanitary reform had led to her election, in 1858, to fellowship in the Royal Statistical Society. On October 16, 1874, the American Statistical Association elected her an honorary member. A photographic reproduction of her letter of acceptance appears on the following page.

INTERNATIONAL STATISTICAL CONGRESS OF 1860.

This growing association with the leaders of thought in the statistical world of her time enabled her to take an active part in drawing up the program of the second section of the International Statistical Congress, held at London in 1860. This Section dealt with sanitary statistics. Miss Nightingale and Dr. Farr incorporated the forms for uniform hospital statistics, which had been experimentally introduced into a group of London hospitals in 1859, in a paper read for her before the Section by Dr. McMillian. Additions and recommendations,

London Jany 30 1875

Sir,
    I request that you will be
good enough to thank the

American Statistical Association
        in my name
for the honour which they
have done me   in electing
me an Honorary Member
& to assure them that I am
      ever their faithful Servant

            Florence Nightingale

I trust that I may be allowed to
send a Blue Book, being the last
"Annual Sanitary Report" of the
India Office: & a Pamphlet "Life
or Death in India" by myself: as
a small contribution to their Library:
                        F.N.
To the
Corresponding Sec^y

chiefly by Dr. Berg of Sweden and Dr. Neumann of Berlin, were concurred in by the author in a letter to the Earl of Shaftesbury, President of the Section.

This paper on uniform classifications and forms for hospital statistics was afterward widely circulated among physicians and hospital officials. Large quantities of the forms were supplied to hospitals in various parts of the country. The Paris hospitals took up the plan. Guy's Hospital, London, prepared a statistical analysis of its experience for the years 1854 to 1861; St. Thomas' for the years 1857 to 1860; and St. Bartholomew's, for 1860. At a meeting held at Guy's Hospital on June 21, 1861, it was unanimously agreed to adopt a uniform plan of registration, that each hospital should publish its own statistics annually, and that the forms devised by Miss Nightingale should be used so far as practicable.

ELEMENTS OF HOSPITAL MEDICAL STATISTICS.

Miss Nightingale then prepared the detailed paper on "Hospital Statistics and Hospital Plans" for the Dublin meeting of the National Association for the Promotion of Social Science in 1861.' In this paper she emphasized the seven primary tabulation elements of hospital sickness statistics, which were:

(1) Number of patients remaining in hospital on first day of year.

(2) Patients admitted during year.

(3) Patients recovered or relieved during the year.

(4) Patients discharged as incurable, unrelieved, for irregularities or at own request.

(5) Patients died during year.

(6) Patients remaining at end of year.

(7) Mean duration of cases in days and fractions of a day.

These tabulation "elements" were to be compiled as seven separate tables, each showing diseases classified thereunder by sex and age (by single years under five and by five year periods thereafter). The additional consideration of diseases contracted in hospital while under treatment was also provided for. These extensions of the original paper were suggested by Drs. Berg and Neumann. Miss Nightingale held that these supplementary tables would bring out the fact of the then scandalous prevalence of "hospital diseases" such as gangrene and septicemia.

The statistics of the various hospitals adopting her forms were published in the *Journal* of the Royal Statistical Society for September, 1862. Miss Nightingale's system of uniform hospital statistics was never generally successful over any considerable period of time. The plan, but partly realized in the requirements of the King's Hospital Fund, demands for its complete and effective operation a more intelligent appreciation of and a finer enthusiasm for statistical facts than is afforded even by present day voluntary and competitive hospital systems in metropolitan districts.

A further example of Miss Nightingale's use of the statistical method in hospital economy was her study of the questions relating to the possible removal of St. Thomas' Hospital at the instance of the Southeastern Railway. The railway company proposed the removal of the hospital to provide for an extension of the right-of-way from London Bridge to Charing Cross. She analysed the origins of cases served by the hospital, tabulated the proportions of cases within certain radial distances, and showed the probable effect upon patients of the removal of the hospital to the several possible sites suggested. This method of fitting hospital accommodation to the needs of populations has only recently been revived. It represents a legitimate application of demographic principles to the study of the relief of dire human needs.

The statistics of surgical operations from the standpoint of hospital cost and practical end results were next considered by Miss Nightingale. In a commentary on St. Bartholomew's statistical report, Miss Nightingale outlined the minimum requirements of a report form for the nature and result of surgical operations. The subject was further developed in a paper read for her before the Berlin meeting of the International Statistical Congress in 1863.

Before the close of the London meeting of the International Statistical Congress, 1860, Miss Nightingale addressed a letter to Lord Shaftesbury. The letter was read to the whole Congress and adopted by it as a resolution. The resolution impressed upon governments the prime necessity for publishing more extensive and numerous abstracts of the statistical information in their possession.

MISS NIGHTINGALE AND THE CENSUS OF ENGLAND, 1861.

Miss Nightingale made a determined effort to extend the scope and application of the Census of 1861, largely in the direction of collecting statistics which would serve as a foundation for sanitary reform. Her aim was twofold: one was to enumerate the sick and infirm on Census Day. To those who denied that it could be accomplished, she pointed out that it had been done elsewhere, notably in Ireland. Her second ambition was to obtain complete data on the housing of the population; this, too, had been practicable in Ireland, she urged. In pressing her point with the Census officials she said: "The connection between the *health* and the *dwellings* of the population is one of the most important that exists. The 'diseases' can be approximated also. In all the more important—such as smallpox, fevers, measles, heart disease, etc. —all those which affect the *national* health, there will be very little error. Where there *is* error in these things, the error is uniform . . . and corrects itself. . . . " These few remarks still serve as prolegomena to any future census plans for England or any other country.

The Census Bill came up late in the session and not much comment can be found on the foregoing suggestion for a development of Census inquiry. The only critical comment made in the debate proceeded from Lord Ellenborough, who, far from considering the innovations of sickness and housing statistics, proposed to exclude most of the inquiries already suggested. Miss Nightingale, in her conception of census methods and results, was far ahead of her day and generation. Subsequent censuses, chiefly in the United States, in Tasmania, and in one or two other countries, have included sickness and housing inquiries.

SCOPE AND USE OF STATISTICS.

Miss Nightingale's activities in furthering statistical progress were the outgrowth of her deep conviction, variously expressed in her several papers, that the social and moral sciences are in method and substance statistical sciences. In her several papers on metaphysical topics, she asserts that statistics were to her almost a religious exercise. Her conception of theology was that its true function was to ascertain the "character of God." Statistics, she mused, discovered and codified law in the social sphere and thereby revealed certain aspects of "the character of God." Doubtless, in these speculations she was profoundly influenced by the studies of Quetelet in moral statistics, as typified in his "*Recherches statistiques sur le Royaume de Pays-Bas.*"

Statistics as an element in political education also appealed with peculiar force to Miss Nightingale. In a letter to Benjamin Jowett she said: "The Cabinet Ministers, the army of their subordinates . . . have for the most part received a university education, but no education in statistical method. We legislate without knowing what we are doing. The War Office has some of the finest statistics in the world. What comes of them? Little or nothing. Why? Because the Heads do not know how to make anything of them. Our Indian statistics are really better than those of England. Of these no use is made in administration. What we want is not so much (or at least not at present) an accumulation of facts, as to teach men who are to govern the country the use of statistical facts."

She proposed a number of leading questions which she desired to see investigated by the statistical method: What had been the result of twenty years of compulsory education? What is the effect of town life on offspring in number and in health? What are the contributions of the several social classes to the population of the next generation? In proposing these inquiries she anticipated by more than a generation the work of the eugenists and biometricians of the Galton-Pearson school. Her friend, Adolphe Quetelet, had inaugurated studies of this character. Both he and Dr. Farr had hoped that she would pursue her inquiries more extensively. In conversation with Benjamin Jowett she proposed to found at Oxford a Professorship or Lectureship in Applied Statistics. Mr. Jowett seems to have discussed the matter with Mr. Arthur Balfour and Professor Alfred Marshall. Miss Nightingale, on her part, consulted Mr. Francis Galton, who responded earnestly and worked out a detailed plan. There was more or less discussion over the matter, but in the press of other affairs, the proposition was discarded. It is certainly of very great interest in this connection to observe that many years afterward such an enterprise was undertaken by the University College in London; probably, however, not directly issuing from Miss Nightingale's suggestion.

VITALITY OF ABORIGINAL RACES.

Following her disquisitions on army, medical, and civil hospital statistics, came her statistical investigations into colonial questions: the first into the vitality of native or aboriginal races in the Colonies; the second into the sanitary condition and material welfare of the population and military establishment of India. In a paper before the National Association for the Promotion of Social Science, meeting at Edinburgh in 1863, she discussed the gradual disappearance of the native races when brought into contact with the influences of civilization. The paper was suggested by Sir George Grey, who, at that time, was deeply engaged with questions of Colonial policy. The preliminary inquiry related to the probable effect of European school usages and school education generally, upon the health of children, of parents, and of races which had not heretofore been brought under any system of education. With the assistance of the Duke of Newcastle, she prepared a form which was sent by the Colonial office to the governors of the various colonies. From the returns of 143 schools she deduced the mortality of school children by age period and by sex, and further classified the statistics by causes of death. A second inquiry into the statistics of colonial hospitals gave important information on the causes of high institutional mortality among the native races. The numbers involved were small, and the results were necessarily considerably in error; but, in the main, the conclusions as to the neglect of sanitary precautions and the change in the living habits of native races were sound.

A further paper on the aboriginal races in Australia was read before the same Association at York, England, in 1864. This essay contained copious quotations from correspondence with colonial governors over points raised by the first paper. The Colonial Office circulated the reprint of the Edinburgh paper widely. Miss Nightingale has been considered a pioneer in work for arresting the decline of native races, so far as such work has been possible.

BRITISH ARMY IN INDIA.

After the death of Sidney Herbert, Miss Nightingale devoted the larger portion of her time and attention to the Indian question. Her earlier years of service to the British Army at home were not more significant than her later years of endeavor for the army and people in India. The greater proportion of the lives of soldiers lost in India was not chargeable to battle, but to disease caused by insanitation and general ignorance of military tropical hygiene. In 1859, it was found that the average annual death rate of the British armies in India had been 69 per 1,000 since the year 1817.* In recent times the figure has been 5 per 1,000. The changes in living and working accommodations which brought about this reduction in army mortality are directly traceable to the recommendations of the Royal Commission, appointed in 1859, which reported in 1863. An unrecorded fact, however, is that Miss Nightingale's suggestion led directly to the appointment of the Commission and that the greater part of the Report was her handiwork. The suggestions upon which permanent reforms in army sanitation in India were based were also her work.†

For eight months, during the latter part of 1858 and the earlier part of 1859, Miss Nightingale importuned Lord Stanley for the appointment of a Royal Sanitary Commission which would do for the armies in India what had been done for the armies at home. She had contemplated for two years before,

---

* While this figure included battle casualties of the campaigns of the forty years ending with 1856, the mortality figure was still excessive. It was stated that the registered mortality among British troops in India was six times that of Englishmen of the same ages at home. Again, in an earlier investigation into mortality from disease among troops in the East Indies (including for the most part British India) death rates varying from 40 to 98 per 1,000 of mean strength over the period 1840–1848 were found.

† British soldiers in India today live in barracks which surpass in comfort and sanitation any that can be found in other countries. Every regiment, battery, and depot has its regimental institute, a sort of soldiers' club, library, reading and recreation room, a temperance association, and a theatre. The use of alcoholic beverages is discouraged and every encouragement is given to useful employment for the men.

the appointment of a Commission to investigate the entire question of the armies in British India.   The mutiny of almost the whole of the Bengal native army and the contingents in northern India in 1857, which had filled the minds of the British population with thoughts of vengeance and repression against the native army, had only served to fix her attention upon sanitary and other administrative reform on behalf of the soldiers.   Her analysis of the statistics of army mortality in India convinced her that there was murder committed not by the Sepoys alone.   To her mind, it was murder to doom British soldiers to death by neglect of the most elementary sanitary precautions.

Anticipating the appointment of the Commission, she began collecting, tabulating, and interpreting data she derived from circulars of inquiry which she had drafted and sent to all the stations in India.   The inquiry form lacked little in requisite completeness and precision of detail.   In the meantime, Miss Nightingale and Dr. Farr searched the sickness and mortality records of the India Office.

The report of the Indian Sanitary Commission when issued in 1863 comprised in all 2,028 pages, mostly in small print. The greater part of the statistical work in the Report bore clear evidence of Miss Nightingale's influence.   Her inquiry blank, in the first instance, provided the vehicle for the transmission of much of the evidence.   The replies to her questions occupy the whole of the second volume of the Report.   In October, 1861, the Commission requested her to submit her interpretation of these Stational Reports.   Her observations upon these reports occupied 23 pages and are the most remarkable of her published works.   Her unusual treatment of the subject by the addition of illustrative wood-cuts describing Indian hospitals and barracks, made the Treasury demur, however, at the cost, but Miss Nightingale was permitted to pay for the printing out of her private purse.   Copies of these observations were sent to the queen and to influential members of the government.   Sir John McNeill wrote: "The picture is terrible, but it is all true.   There is no one statement from beginning to end that I feel disposed to question and there are many which my own observation and experience enable me to confirm." In detail, her notes related to the camp diseases which follow the selection of poor sites, defective disposal of human wastes, overcrowding in barracks, lack of suitable occupation and exercise, dietaries and defective hospital arrangements. The sources of statistical data for the armies in India were also criticised. In addition to her observations, Miss Nightingale prepared with Dr. Sutherland an abstract of the returns upon which her "Observations" were founded.

Moreover, when the Report was published she moved with her characteristic decision to secure for it the newspaper and periodical publicity which it deserved. She contributed a popular résumé of the Royal Commission's Report to the National Association for the Promotion of Social Science, meeting at Edinburgh in 1863, entitled "How People may Live and not Die in India." The paper was republished in 1864, with a preface and an account of what the Commission had actually achieved in sanitary works and measures. In 1868, the Secretary of State printed her résumé of the Indian Sanitary Question from 1859 to 1867 and her memorandum of advice and suggestions on the entire situation as it then stood. The dispatch from Sir Stafford Northcote under date of April 23, 1863 (drafted by Miss Nightingale), was printed at the same time. This dispatch resulted in the first of the annual series of Indian Sanitary Reports. In the reports for 1868 and 1869 appear two of Miss Nightingale's contributions; in the first an Introduction of eight pages and in the second her paper on "Sanitary Progress in India."

Her statistical enterprises of this period are well summarized in a few phrases which are here quoted from one of her letters: "I am all in the arithmetical line now . . . I find that every year . . . there are in the Home Army, 729 men alive every year who would have been dead but for Sidney Herbert's measures, and 5,184 men always on active duty who would have been 'constantly sick in bed.' In India the difference is still more striking. Taken on the last two years, the death rate of Bombay is lower than that of London,* the healthiest city in Europe. And the death rate of Calcutta is lower than that of Liverpool and Manchester. But this is not

---

*These figures related only to conditions prevailing at the time this letter was written.

the greatest victory. The Municipal Commissioner of Bombay writes that the 'huddled native masses clamorously invoke the aid of the Health Department if but one death from cholera occurs; whereas formerly half of them might be swept away and the other half think it all right.'"

In 1873, the National Association for the Promotion of Social Science invited her to contribute a paper on the ten years of progress in India since her "How People may Live and not Die in India" paper appeared. Her title for this later paper was "How some People have Lived and Not Died in India." The India Office reprinted the paper in its 1874 bluebook. The salient fact developed in this report was that the death rate in the Indian Army had been reduced from 69 per 1,000 to 18 per 1,000. This summary of ten years of sanitary progress in India she was qualified to prepare in consequence of her editorial work on the annuals issued by the sanitary department of the India Office.

In 1877, Miss Nightingale published two letters on famine in India and followed these by an article in the *Nineteenth Century* magazine. This article, "The People in India," gave the principal facts about the Indian famines and proceeded further to describe in considerable detail the evils of usury in the Bombay Deccan. Beginning with 1874, Miss Nightingale collected statistics of irrigation in India, and of the effect of irrigation on the life and health of the people. These data, the appendix of the second part of an unpublished work on the Indian Land Question and Irrigation Systems, were afterward partially used in several isolated papers. She thought much on education in India. There had been a neglect of elementary education. The exception was found in the system of village schools established by Lieutenant-Governor James Thomason, of what is now the Agra Province. The report of the Indian Education Commission of 1883 directed attention to the essential difficulties residing in the language, credal, race, and traditional differences of the populations of the several provinces. The two chief difficulties in the way of a diffusion of education among the masses in India are the large agricultural population, among whom it is in all countries difficult to advance any system of education, and the existence of a hereditary class, whose object has been to maintain their monoply of learning as the chief buttress of their social supremacy.   Questions such as these occupied Miss Nightingale's attention in the years 1881 and 1882.

The succeeding years were taken up in turn by Army Hospital Service reform, district nursing organization, nursing education and Indian finance problems.

In 1891 Miss Nightingale finally laid aside her ambition to found at Oxford a Professorship in Applied Statistics. This relinquishment of active interest in the progress of statistics was indicative only of her failing physical powers. The gradual failure of sight, memory, and mental apprehension proceeded during the last fifteen years of her life.   Her death occurred on August 13, 1910.

FLORENCE NIGHTINGALE AS STATISTICIAN.

Florence Nightingale may well be assigned a position in the history of social statistics next to those occupied by Quetelet and Farr.   Her ardent, genuine sympathy for the sick and distressed was greatly augmented by a positive genius for marshalling definite knowledge of the forces which make for disease and suffering.   The same intellect which sharply separated the formulae, procedures, and practical methods of nursing from its abiding principles as one of the humanities, also discerned the statistical facts of sickness and other forms of disharmony between the individual and his environment.   In hospital care of the sick, as an instance, Miss Nightingale replaced the astigmatic case viewpoint with one embodying a grasp of total situations.   This is one function of statistics.

Her earnest perception of this truth—an essential in the equipment of the statistician—was firmly supported by her control over laborious detail and by her scrupulous care in testing the statistical foundations of her premises.   The interpretations she placed upon the facts developed in her researches show a careful regard for the competent counsel which she so often consulted.   In all these respects Miss Nightingale exhibited the prime qualities of one thoroughly versed in the art of preparing and reflectively analysing social data.

Studies in the History of Probability and Statistics. XXV.
On the history of some statistical laws of distribution
BY O. B. SHEYNIN
(1971)
Institute for the History of Natural Science and Technology, Academy of Sciences, Moscow

SUMMARY

The history of the χ^2 distribution and the contributions by Herschel, Maxwell and Boltzmann are discussed. The use of the beta distribution by Bayes is also described.

1. THE USE OF THE χ^2 DISTRIBUTION IN PHYSICAL THEORY

The deduction of the χ^2 distribution is due to Abbe (1863); see also Sheynin (1966). While considering the accuracy of archery, Herschel (1869) almost arrived at this distribution for two degrees of freedom, a fact mentioned to me in 1967 by Professor W. Kruskal in a private communication. For three degrees of freedom similar results were arrived at by Maxwell (1860) somewhat prior to Herschel. Boltzmann arrived at this distribution at first (1878) for two and three degrees of freedom and then (1881) in the general case.

Thus, these works should find their place in the history of the χ^2 distribution beside those of Laplace, Bienaymé and Ellis (Lancaster, 1966), and, of course, beside those of Abbe (Helmert, 1876); see also David (1957) and Pearson (1900).

Considering the probability of an arrow deviating by z from the centre of a circular target Herschel (1869, pp. 506–7) starts from an obvious formula

$$\text{pr}(z|\phi) \propto e^{-kz^2}\,dz,$$

where ϕ is the polar angle of the point of hit, the polar coordinate system being (z, ϕ), and k is to be estimated from observations. Then, since the circumference of a circle is proportional to z,

$$\text{pr}(z \leqslant z_0) \propto \int_0^{z_0} z\,e^{-kz^2}\,dz. \tag{1}$$

Herschel could have arrived at the χ^2 distribution for two degrees of freedom had he considered instead of (1) the probability that $z^2 \leqslant z_0$.

But the introduction of z^2 does not seem sufficiently natural. More natural, however, would have been a similar procedure of Maxwell (1860, p. 381), this being his classical paper where he arrived at the normal distribution of velocities of molecules

$$f(x) = \frac{1}{\alpha\sqrt{\pi}}\,e^{-x^2/\alpha^2} \quad (-\infty < x < +\infty). \tag{2}$$

It is not our intention to discuss his deduction of (2); what we notice is that Maxwell, starting from (2) and giving no intermediate explanation, observes that

$$\frac{4N}{\alpha^3\sqrt{\pi}}\,v^2\,e^{-v^2/\alpha^2}\,dv \tag{3}$$

molecules out of N, the whole number of them, have space velocities belonging to the segment $[v, v+dv]$. Reconstruction of (3) is easy enough: starting from (2), we have a similar distribution for the simultaneous probability of $\{\xi < x,\ \eta < y,\ \zeta < z\}$ and

$$\text{pr}(\text{velocity} \in [v, v+dv]) = \int_0^{2\pi} d\phi \int_0^{\pi} \sin\Theta\,d\Theta \int_v^{v+dv} z^2\,e^{-z^2/\alpha^2}\,dz,$$

this being equivalent to (3). Here ξ, η and ζ are the components of velocity in each of the three dimensions.

Introducing probability proper, i.e. eliminating N, and studying the square of the velocity rather than the velocity itself, Maxwell could have found $\text{pr}(\xi^2 + \eta^2 + \zeta^2 \in [v, v+dv])$ which would have been the χ^2 distribution for three degrees of freedom. This transformation would have been quite natural, for instance, in studying the kinetic energy (K.E.) of molecules. Maxwell, however, restricted himself to remarking that the mean value of v^2 is equal to $\frac{3}{2}\alpha^2$.

On the other hand, Boltzmann (1878, p. 252) studied just this distribution of the K.E. and arrived at the χ^2 distribution for two, and then (p. 257), for three degrees of freedom. Only in the first instance did he mention his proceeding from the normal law

$$\frac{k}{\pi} e^{-k(u^2+v^2)} \, du \, dv \tag{4}$$

and explain that the integration had been made with respect to the polar angle.

More interesting is that Boltzmann (1881, p. 576), also with no explanations given, writes down the distribution of the K.E. of molecules in the general case of a dimensions, i.e.

$$\mathrm{pr}\,(\tfrac{1}{2}mv^2 \in [\chi, \chi+d\chi]) = \frac{h^{\frac{1}{2}a}}{\Gamma(\tfrac{1}{2}a)} e^{-h\chi} \chi^{\frac{1}{2}a-1} d\chi. \tag{5}$$

The meaning of h, first introduced in 1878, is not given in either article but it should be equal to $2k/m$, where k is the same as in (4) and m, as in (5), is the mass of the molecule.

In deducing (5), Boltzmann may have used the discontinuity, Dirichlet, factor as both Abbe and Helmert did. But then, he would possibly have said so. And at least two other methods for the calculation of the a dimensional integral of the exponential function were open to him.

The lack of references to Abbe and Helmert testify to the fact that Boltzmann did not know their works.

None of the nineteenth-century authors mentioned thought of compiling a table of the χ^2 distribution. The mode of using statistical considerations in physics proper could be seen in Maxwell (1860) who was evidently satisfied with deducing several distributions and giving a corresponding overall numerical estimate of the behaviour of molecules.

In error theory, the criteria for estimating systematic influences, bias, in observations developed by Abbe and Helmert were nonparametric and the χ^2 distribution itself was not used.

2. The beta distribution

This distribution is due to Bayes (1764). The beta function had been repeatedly considered by several scholars prior to Bayes, but it was his probabilistic problem that led to the calculation of

$$\frac{\int_b^c x^p(1-x)^q \, dx}{\int_0^1 x^p(1-x)^q \, dx} = I_c(p+1, q+1) - I_b(p+1, q+1), \tag{6}$$

where $0 \leqslant b < c \leqslant 1$, p and q are large positive integers and I is the symbol for the incomplete beta function. Of course, Bayes used neither these terms nor the symbols.

For the denominator of (6) he easily deduced

$$\mathrm{B}(p+1, q+1) = \frac{1}{(q+1)\dbinom{p+q+1}{p}}$$

and, for the estimation of the whole fraction (6), used (Bayes, 1765) a supplementary curve

$$y = c\left(1 - \frac{n^2 x^2}{pq}\right)^{\frac{1}{2}n} \quad (n = p+q).$$

A more detailed description, including an improvement of Bayes's estimate due to Price (see Sheynin, 1969) and, of course, including the related work of Laplace (*Théor. anal. prob.*, chap. 6 of book 2) would be of interest for the history of mathematical calculus. We, however, shall restrict ourselves to probabilistic problems proper.

The integral limit theorem in De Moivre could be written as

$$\mathrm{pr}\left[-z \leqslant \frac{\mu/n - p}{\{(pq)/n\}^{\frac{1}{2}}} \leqslant z\right] \to \frac{2}{(2\pi)^{\frac{1}{2}}} \int_0^z e^{-\frac{1}{2}x^2} dx, \tag{7}$$

where pq/n is the dispersion of μ/n, the relative frequency, the notation being that in standard use (which is not the case with Bayes).

Neither Bayes nor Price used this limit theorem while the latter observed, moreover (in the covering letter to Bayes, 1764) that it is not 'rigorously exact' for the case of a finite n. Price also noticed that the problem being solved by Bayes is the converse to that of De Moivre's. We shall now study this point.

If, notwithstanding the opinion of Price (and, obviously, Bayes), the Bayesian formula is to be considered in the limiting case, $n \to \infty$, it would give

$$\mathrm{pr}\left\{-z \leqslant \frac{\bar{p}-a}{(pq/n^{\frac{3}{2}})^{\frac{1}{2}}} \leqslant z\right\} \to \frac{2}{(2\pi)^{\frac{1}{2}}} \int_0^z e^{-\frac{1}{2}x^2}\,dx, \tag{8}$$

where $a = p/n$ and $\bar{p}$ is the estimate of p, the classical probability in the binomial scheme. In a somewhat different form (8) is to be found in a commentary by Timerding to the German translation of Bayes (1908). Lacking in this commentary, however, is the understanding that, as a first approximation,

$$a = E(\bar{p}), \quad pq/n^{\frac{3}{2}} = \mathrm{var}\,(\bar{p}).$$

Neither Bayes nor Price possessed any notion about dispersion but, what is remarkable, they both obviously understood the practical worthlessness of using (7) for the converse problem of deducing p from the relative frequency of the event studied.

It is generally known that the so-called Bayes's formula

$$\mathrm{pr}\,(A_i|B) = \frac{\mathrm{pr}\,(B|A_i)\,\mathrm{pr}\,(A_i)}{\sum\limits_{j=1}^{n}\mathrm{pr}\,(B|A_j)\,\mathrm{pr}\,(A_j)} \tag{9}$$

is not to be found in Bayes. Expressed in words only, (9) is contained in Laplace (*Essai philosophique*) and it is possibly Cournot (1843, § 88, p. 158) to whom this incorrect expression (Bayes's formula) is due.

Acknowledgement is due to Professor H. O. Lancaster for reprints of his own paper (1966) and that of David (1957) and to Professor W. Kruskal for permitting us to mention his own discovery of Herschel's connexion with the χ^2.

REFERENCES

ABBE, E. (1863). Über die Gesetzmässigkeit der Vertheilung der Fehler bei Beobachtungsreihen. *Ges. Abh.* **2**, 55–81.

BAYES, T. (1764). An essay towards solving a problem in the doctrine of chances. Reprinted in *Biometrika*, 1958, **45**, 296–315.

BAYES, T. (1765). A demonstration of the second rule in the essay, etc. *Phil. Trans. R. Soc.* **54**, 296–325. [This includes a covering note and subsequent remarks by Price.]

BAYES, T. (1908). *Versuch zur Lösung eines Problems der Wahrscheinlichkeitsrechnung*. Herausgeber, H. E. Timerding. Ostwald Klassiker No. 169. Leipzig: Engelmann.

BOLTZMANN, L. (1878). Weitere Bemerkungen über einige Probleme der mechanischen Wärmetheorie. *Wiss. Abh.* **2**, 250–88.

BOLTZMANN, L. (1881). Über einige das Wärmegleichgewicht betreffende Sätze. *Wiss. Abh.* **2**, 572–81.

COURNOT, A. A. (1843). *Exposition de la théorie des chances et des probabilités*. Paris: Hachette.

DAVID, H. A. (1957). Some notes on the statistical papers of F. R. Helmert. *Bull. Stat. Soc. N.S.W.* **19**, 25–8.

HELMERT, F. R. (1876). Über die Wahrscheinlichkeit der Potenzsummen der Beobachtungsfehler etc. *Z. Math. und Phys.* **21**, 192–218.

HERSCHEL, J. F. W. (1869). *Familiar Lectures on Scientific Subjects*, pp. 496–507. New York: Routledge.

LANCASTER, H. O. (1966). Forerunners of the Pearson χ^2. *Aust. J. Statist.* **8**, 117–26.

MAXWELL, J. C. (1860). Illustrations of the dynamical theory of gases. *Scient. Papers*, vol. **1**, 377–410. Paris, 1927: Libr. scient. Hermann.

PEARSON, K. (1900). On a criterion that a given system of deviations from the probable in the case of a correlated system of variables is such that it can be reasonably supposed to have arisen from random sampling. *Phil. Mag.* **50**, 157–75.

SHEYNIN, O. B. (1966). Origin of the theory of errors. *Nature, Lond.* **211**, 1003–4.

SHEYNIN, O. B. (1969). On the work of T. Bayes in the theory of probability (Russian). *Proc. 12th Conf. Inst. Hist. Nat. Sci. and Technology, Hist. of Math. and Mech. Sect.* 40–57. Moscow: Viniti publ.

[*Received August* 1970. *Revised October* 1970]

Some key words: History of normal, chi-squared and beta distributions; History of kinetic theory and Bayes's formula.

Biometrika (1971), **58**, 2, *p.* 369
Printed in Great Britain

Studies in the History of Probability and Statistics. XXVI

The work of Ernst Abbe

By M. G. KENDALL

Scientific Control Systems Ltd

SUMMARY

A brief biographical note on E. Abbe is given, followed by an abstract of his paper deriving the χ^2 distributions and the distribution of the serial correlation coefficient.

1. INTRODUCTION

Mr O. B. Sheynin has recently called attention to a most remarkable paper by Ernst Abbe, presented in 1863, in which Abbe derives not only the χ^2 distribution, but R. L. Anderson's (1942) distribution of the serial correlation coefficient. The paper itself is rather long, but an abstract is given as § 2 of this note. The paper is a superbly competent piece of work and perhaps the most remarkable anticipation of later studies of distribution theory that has yet come to light.

Abbe was born in 1840, the son of a master spinner at Eisenach. His childhood was one of privation; his father worked on his feet 16 hours every day with no break for meals. Ernst, however, won scholarships and was helped through his studies by his father's employer. He received his doctorate at Jena at the age of 21 with a dissertation on thermodynamics. Two years later he was appointed professor, the paper under notice here being a dissertation for the acquisition of Venia Docendi in the Faculty of Philosophy.

What further contributions Abbe might have made to distribution theory must remain a matter of speculation. In 1866, at the age of 26, he was approached by Carl Zeiss with some optical problems and the remainder of his life was spent in theoretical and practical research in optics and astronomy. He is, in fact, an outstanding example of scientific genius combined with practical inventiveness. Moreover, when he became a wealthy man in later life, having been taken into partnership by Zeiss, he introduced into industrial relations measures which were not to become general for another half-century, the eight-hour day, paid vacations, sick-benefit, pensions, severance pay, annual bonuses among others.

The dissertation, an abstract of which follows, is published in Abbe's collected works (1906). For the biographical details I am indebted to the extensive account of his life and works by Volkmann (1966). I cannot find that Abbe ever returned to the subject of the dissertation, but he was never eager to publish his work and his private papers, if they still survive, might possibly contain something of interest.

2. ABSTRACT OF ERNST ABBE'S PAPER 'UBER DI GESETZMÄSSIGKEIT IN DER VERTHEILUNG BEI BEOBACHTUNGSREIHEN'

2·1. *Formulation of problem*

The introduction refers to Gauss's theory of errors of observation:

One relies partly on the sum of squares of errors or on the mean error derived from it, by comparing it with the precision of such observations obtained independently; partly one also takes into account the way in which the errors are distributed.

Abbe defines the normal distribution in terms of a precision constant h in the form

$$w = \frac{h}{\sqrt{\pi}} \exp(-h^2 x^2). \tag{1}$$

He defines the sum of squares

$$\Delta = \sum_{j=1}^{n} x_j^2 \tag{2}$$

and the sum of squares of first differences

$$\sum_{j=1}^{n-1} (x_j - x_{j+1})^2,$$

but goes on to make the definition circular by defining

$$\theta = \sum_{j=1}^{n} (x_j - x_{j+1})^2 \quad (x_{n+1} = x_1). \tag{3}$$

The problem then is:

What is, for any number of observations, the probability for the occurrence of a system of errors in which the function Δ has a certain value (or a value lying between certain limits) and in which simultaneously the function θ has a certain other value (or a value lying between certain limits)?

2·2. *Distribution of sum of squares*

The probability of a system of errors with a sum of squares between 0 and Δ is

$$\Phi(\Delta) = \frac{h^n}{\pi^{\frac{1}{2}n}} \int \exp\left(-h^2 \sum_{j=1}^{n} x_j^2\right) dx_1 \dots dx_n \tag{4}$$

integrated over values for which

$$0 < \sum_{j=1}^{n} x_j^2 < \Delta. \tag{5}$$

Abbe considers the contour integral

$$\frac{1}{2\pi} \int_{-\infty}^{\infty} \frac{\exp\{\sigma(a+i\phi)\}}{a+i\phi} d\phi = \begin{cases} 1 & (\sigma > 0), \\ 0 & (\sigma < 0). \end{cases} \tag{6}$$

We put

$$\sigma = \Delta - \sum_{j=1}^{n} x_j^2, \tag{7}$$

and substitute this 'discontinuity factor' in (4) to get

$$\Phi(\Delta) = \frac{1}{2\pi} \frac{h^n}{\pi^{\frac{1}{2}n}} \int_{-\infty}^{\infty} d\phi \frac{\exp\{\Delta(a+i\phi)\}}{a+i\phi} \int \exp\{-\Sigma(h^2+a+i\phi) x_j^2\} dx_1 \dots dx_n. \tag{8}$$

The second integral is easily evaluated to give

$$\Phi(\Delta) = \frac{h^n}{2\pi} \int_{-\infty}^{\infty} d\phi \frac{\exp\{\Delta(a+i\phi)\}}{(a+i\phi)(h^2+a+i\phi)^{\frac{1}{2}n}}. \tag{9}$$

The integral can be further reduced if we put

$$\frac{1}{(h^2+a+i\phi)^{\frac{1}{2}n}} = \frac{1}{\Gamma(\frac{1}{2}n)} \int_{-\infty}^{\infty} dy \exp\{-(h^2+a+i\phi)y\} y^{\frac{1}{2}n-1}.$$

The integral with respect to ϕ can be evaluated from equation (7) to yield

$$\Phi(\Delta) = \frac{h^n}{\Gamma(\frac{1}{2}n)} \int_0^{\Delta} e^{-h^2 y} y^{\frac{1}{2}n-1} dy. \tag{10}$$

Abbe remarks of this integral 'wie auch nach andern Methoden gefunden wird' but he does not indicate these other methods.

If we now define the mean error (mittler Fehler) δ by $\delta^2 = \Delta/n$, we find for what nowadays would be called the frequency function of δ, equivalent to that of χ^2,

$$f(\delta) = \frac{2n^{\frac{1}{2}n} h^n}{\Gamma(\frac{1}{2}n)} \exp\left(-nh^2\delta^2\right)\delta^{n-1}. \tag{11}$$

It is to be noted that Abbe derives the distribution by considering errors about a parent zero mean. He does not consider errors about the sample mean.

2·3. *Mean square successive differences*

Abbe obtains an expression for the sum θ of (3) by a similar technique. Putting in (7)

$$\sigma = \theta - \sum_{j=1}^{n} (x_j + x_{j+1})^2, \tag{12}$$

he finds the integral

$$\Psi(\theta) = \frac{h^n}{2\pi\,\pi^{\frac{1}{2}n}} \int_{-\infty}^{\infty} d\phi\, \frac{\exp\{\theta(a+i\phi)\}}{a+i\phi} \oint_{-\infty}^{\infty} dx_1...dx_n \exp\left\{-h^2\sum_{j=1}^{n} x_j^2 - (a+i\phi)\,\Sigma(x_j - x_{j+1})^2\right\}. \tag{13}$$

The last part of this expression no longer permits of separation of the variable. However, by an orthogonal transformation we can reduce $\Sigma(x_j - x_{j+1})^2$ to the form $\Sigma\lambda_j\,\xi_j^2$ while transforming Σx_j^2 to $\Sigma\xi_j^2$. The Jacobian of the transformation is unity. Hence (13) reduces to

$$\Psi(\theta) = \frac{1}{2\pi}\frac{h^2}{\pi^{\frac{1}{2}n}} \int_{-\infty}^{\infty} \frac{d\phi}{a+i\phi} \exp\{\theta(a+i\phi)\} \prod_{j=1}^{n} \int_{-\infty}^{\infty} d\xi_j \exp\left[-\{h^2 + \lambda_j(a+i\phi)\}\xi_j^2\right]. \tag{14}$$

The terms on the right can now be integrated to give

$$\Psi(\theta) = \frac{h^2}{2\pi} \int_{-\infty}^{\infty} d\phi\, \frac{\exp\{\theta(a+i\phi)\}}{(a+i\phi)\{h^2 + \lambda_1(a+i\phi)\}^{\frac{1}{2}}...\{h^2 + \lambda_n(a+i\phi)\}^{\frac{1}{2}}}. \tag{15}$$

Abbe then determines the coefficients λ by showing that they are the roots of the circulant

$$\begin{vmatrix} 2-\lambda & -1 & 0 & 0 & ... & -1 \\ -1 & 2-\lambda & -1 & 0 & ... & 0 \\ 0 & -1 & 2-\lambda & -1 & ... & 0 \\ . & . & . & . & . & . \\ . & . & . & . & . & . \\ . & . & . & . & . & . \\ -1 & 0 & 0 & 0 & ... & 2-\lambda \end{vmatrix} = 0, \tag{16}$$

which gives

$$\lambda_k = 2\left\{1 - \cos\left(\frac{2\pi k}{n}\right)\right\} = 4\sin^2\left(\frac{\pi k}{n}\right). \tag{17}$$

He points out that one root is zero and that the others occur in pairs except perhaps one. Writing $\nu = \frac{1}{2}(n-1)$ for n odd and $\frac{1}{2}(n-2)$ for n even, we find

$$\Psi(\theta) = \frac{h^{n-1}}{2\pi} \int_{-\infty}^{\infty} d\phi\, \frac{\exp\{\theta(a+i\phi)\}}{(a+i\phi)\{h^2 + \lambda_1(a+i\phi)\}...\{h^2 + \lambda_\nu(a+i\phi)\}}.$$

This expression can be evaluated by a contour integration. The poles are at

$$\phi = ai,\ \left(a + \frac{h^2}{\lambda_1}\right)i,\ ...,\ \left(a + \frac{h^2}{\lambda_\nu}\right)i.$$

For complex ϕ with positive imaginary part the integrand vanishes at infinity. Hence by the theory of residues the contribution from the pole at $(a+h^2/\lambda_j)i$ is

$$-\frac{2\pi\lambda_1\ldots\lambda_j^{\nu-1}}{h^{n-1}}\frac{\exp(-\theta h^2/\lambda_j)}{(\lambda_j-\lambda_1)\ldots(\lambda_j-\lambda_{j-1})(\lambda_j-\lambda_{j+1})\ldots(\lambda_j-\lambda_r)}, \tag{18}$$

and from (17)

$$\Psi'(\theta) = 1 - \sum_{j=1}^{\nu}\frac{\lambda_j^{\nu-1}\exp(-\theta h^2/\lambda_j)}{(\lambda_j-\lambda_1)\ldots(\lambda_j-\lambda_\nu)}. \tag{19}$$

When n is even, part of the integrand can be simplified by writing

$$\frac{1}{2(\tfrac{1}{4}h^2+a+i\phi)^{\frac{1}{2}}} = \frac{1}{2\sqrt{\pi}}\int_0^\infty \frac{du}{\sqrt{u}}\exp(-\tfrac{1}{4}h^2+a+i\phi). \tag{20}$$

We then find

$$\Psi'(\theta) = \frac{h}{2\sqrt{\pi}}\int_0^\infty \frac{du}{\sqrt{u}}e^{-\tfrac{1}{4}h^2 u}\left[1 - \sum_{j=1}^{\nu}\frac{\lambda_j^{\nu-1}\exp\{-(\theta-\mu)\lambda_j\}}{(\lambda_j-\lambda_1)\ldots(\lambda_j-\lambda_\nu)}\right]. \tag{21}$$

The formulae can be written in a somewhat different form if we replace the λ's by the trigonometrical expressions of (17). Abbe displays them explicitly. He concludes the section by saying that the frequency function of θ can be obtained by differentiating $\Psi'(\theta)$.

2·4. *Distribution of ratio*

Abbe then proceeds to consider the

way in which the relative frequency of a system of errors depends on the relationship which exists between the values of the two functions Δ and θ.

In modern terminology this is equivalent to the distribution of the first serial correlation coefficient circularly defined. He considers the probability of a system of errors for which $\theta \leqslant \mu\Delta$. By the same technique he derives expressions similar to (20) and (21) for the distribution function

$$X(\mu) = 1 - \frac{\sum_j (\lambda_j-\mu)^{\frac{1}{2}(n-3)}(1-\mu/\lambda_j)}{(\lambda_j-\lambda_1)\ldots(\lambda_j-\lambda_\nu)},$$

where the summation is restricted to those terms for which $\lambda_j > \mu$. For the case n even he says that we can proceed in the same way but does not give the formula explicitly. This formula is equivalent to one of the main results of R. L. Anderson who, of course, carried the subject a great deal further. Abbe then goes on to point out that the distribution of μ has the property of being independent of the precision constant h, an early example of Studentization.

It would seem that Abbe pursued this work by some kind of numerical studies of the function $X(\mu)$. He points out that it is symmetrical and says

a further discussion of the numerical value of the integral for different values of the argument shows—this is stated here without proof—that the function $X(\mu)$ reaches an absolute maximum for the value of 2 (strictly so for even n and with close approximation for odd n) and decreases from this point on both sides very quickly to zero, if the number n is at all large.

In conclusion he says that a test of μ depends on the calculation of the function $X(\mu)$ and therefore demands a presentation of this function in a form which facilitates its numerical evaluation.

Such a form can be achieved easily in several ways, at least if we presuppose a high value of n—the only case which is of practical interest . . . but we cannot carry through this development here because the necessary groundwork would exceed the scope of this paper.

REFERENCES

ABBE, E. (1906). *Gesammelte Abhandlungen*, vol. II. Jena: Gustav Fischer Verlag.
ANDERSON, R. L. (1942). Distribution of the serial correlation coefficient. *Ann. Math. Statist.* **13**, 1–13.
VOLKMANN, J. (1966). Ernst Abbe and his work. *Appl. Optics* **5**, 1720–31.

[*Received January* 1971. *Revised February* 1971]

Some key words: History of chi-squared distribution; Distribution of first serial correlation coefficient; Distribution of quadratic forms and of ratios.

Int. Stat. Rev., Vol. 41, No. 1, 1973, pp. 59–68/*Longman Group Ltd*/*Printed in Great Britain*

Entropy, Probability and Information

M. G. Kendall[1]

Summary

A historical account is given of the development of the concepts of entropy, communication theory and their relation to statistical information. There are many misunderstandings about the nature of information as often defined in statistical theory and an attempt is made to remove them.

1. By the middle of the nineteenth century heat was accepted as a mode of molecular motion. The law which asserts the fact, the so-called First Law of Thermodynamics, has the important implication that in a conservative system wherein the energy remains constant, there is a loss of *usable* energy as temperatures within the system approach equality. The idea was given formal expression by Clausius (1822–1888) in a paper of 1854 (see, for example, his book of 1887). Clausius introduced the concept of entropy and stated what is now known as the Second Law of Thermodynamics: no process is possible whose sole result is the transfer of heat from a colder to a hotter body.

In terms of the cycle of a heat engine, if Q is the quantity of heat and T is the temperature at which heat is supplied to the system, then

$$\frac{dQ}{T} = 0 \text{ for a reversible cycle}$$

$$> 0 \text{ for an irreversible cycle.} \tag{1}$$

The quantity $dS = dQ/T$ is an element of the entropy S. For an isolated system the entropy cannot decrease.

2. In 1868 Maxwell (1831–1879) published his famous paper on the dynamical theory of gases, showing that under certain conditions the distribution of velocities of N molecules whose component velocities are ξ, η, ζ is given by

$$dN = Nk \exp \left\{ -\xi^2 - \eta^2 - \zeta^2 \right\} d\xi d\eta d\zeta. \tag{2}$$

So far as I know, this was the first contribution to thermodynamics of a probabilistic kind. Maxwell emphasized the statistical nature of the approach by pointing out at a later date that the Second Law was a macro phenomenon and might be falsified in the small.[2]

3. Stimulated by Maxwell's work, Boltzman (1844–1906) took the next step, still in the theory of heat, about 1877. It was he who linked the thermodynamic concept of entropy

---

[1] Formerly Chairman, Scientific Control Systems (Holdings) Ltd.

[2] For example, with what is usually known as Maxwell's Demon. If we imagine a cylinder of gas divided into two halves by a septum with a shutter in it, and a dexterous Demon who can operate the shutter so as to allow fast-moving particles to enter one chamber but not the other, heat will accumulate in that chamber and the Second Law will be falsified. Maxwell, so far as I can ascertain, did not use the word "demon" but speaks of his *Theory of Heat* of "a being whose faculties are so sharpened that he can follow every molecule in its course". Szilard (1929) pointed out the obvious that if the observer is part of the system the Second Law may be preserved.

with the statistical concept of disorder. He specifically makes an analogy between the distribution of molecular velocities in a gas and that obtained by the random drawing of balls from a bag. This gives him a multinomial distribution, to which he approximates by the Stirling formula

$$n! \doteq (2n\pi)^{\frac{1}{2}} (n/e)^n.$$

The logarithm of the resulting probabilities gives him terms of type $n \log_e n$, a form which continually recurs in modern theories of communication. "The theorem that H (the negative entropy) decreases through molecular collisions amounts to this, that the distribution of molecular velocities in a gas occasioned by collision approaches the most probable" (Boltzmann, 1895, p. 42).

4. Entropy as defined in a thermodynamical context arises naturally as an additive quantity. Probabilities are multiplicative. It can be shown as a matter of mathematics that *if* the entropy S is a function of the probability P of a state, then S must be proportional to $\log P$. (The probabilities concern the velocities of molecules, not their position in space.) When we come to consider information as a function of probability the same kind of relationship will apply. However, "information" is a subtle concept and we need to consider its nature.

5. Boltzmann had some trouble with a paradox arising from the deterministic nature of his mechanics which assumed, *inter alia*, that molecules were perfectly elastic. If we stop the system at some point and reverse all the velocities, it will retrace its history and the entropy will accordingly decrease. What makes us think that one sequence rather than the other is occurring? The same point was taken by Eddington in the 1920s in his discussion of "Time's Arrow", and it is implicit in one form of Maxwell's demon. The subtle problems associated with the nature of time and its direction remain, in my opinion, unsolved. We must assume that, at least in our part of the cosmos, entropy is increasing. What the cosmological implications may be, whether this involves the Heat Death of the Universe, are fortunately not questions which affect the theories of probability and information.

6. Before leaving thermodynamics, we may notice a much more sophisticated definition of entropy given by Carathéodory in 1909 in the form: "In the neighbourhood of any arbitrary state J of a thermally isolated system there are states J' which are inaccessible from J." This formulation does not depend on the theory of heat engines. There does not seem to have been an attempt to axiomatize information theory in a similar way.

7. The application of probability theory to physical systems involving particles and states requires a specification of the basic reference set. Statistical mechanics recognizes at least three such sets. In the Maxwell-Boltzmann system we have r *distinguishable* particles distributed over $n\,(>r)$ cells (states, energy levels, etc.) so that each of the n^r arrangements is equally probable. In Bose-Einstein statistics the particles are indistinguishable and the $\binom{n+r-1}{r}$ distinguishable arrangements are equally probable. In Fermi-Dirac statistics not more than one particle can appear in any one cell and there are $\binom{n}{r}$ equally probable arrangements. Communication theory has not, up to now, found it necessary to incorporate these more specialized reference sets.

8. We now come to the foundation of the mathematical theory of "information", which may be regarded as beginning with the work of R. V. L. Hartley (1928). The background of the subject is no longer thermodynamics but electrical communications. Hartley was writing for electrical engineers and his paper is distinguished by some unusual features, notably its extreme clarity and the fact that he gives no references to previous work. In particular, he does not use the word "entropy", nor does he speak in terms of probability. One of his main results, a foreshadowing of a famous theorem of Shannon's, is that there is a limit in any communication channel to the speed at which "information" can be transmitted; the narrower the bandwidth the longer must be the duration.

9. However, the word "information" as used in communication theory has a highly specialized meaning somewhat remote from what is ordinarily understood by the word. Hartley distinguishes between an objective and a psychological content of "information". In the terminology of later writers, the "information" content has nothing to do with the semantics of the message. If I send by telegram a copy of Jabberwocky to an illiterate Chinese peasant, the "information" conveyed is complete provided that there are no transmission errors. This is so far from the meaning which is ordinarily ascribed to phrases like "the communication of information" that one is not surprised that the work of Hartley and his successors has been widely misunderstood outside the realm of communications engineering. I refer to the point again later, but in the first instance let us examine what it was that Hartley did mean.

10. We suppose that we have a set of symbols called an alphabet. The simplest possible alphabet consists of two symbols, A and B, or 0 and 1 or dot and dash; and, as is well known, every alphabet can be reduced to binary form, as in the Morse code. There is no compelling reason in logic alone for using a binary system. However, all digital electronic computers work in binary because the determination of one way out of a two-way state (holes in a card, polarity of a magnet, direction of current, etc.) is much less liable to instrumental error than in more elaborate mechanisms. It is sometimes convenient to conduct a theoretical discussion in terms of binary although most of the argument applies generally to an alphabet of any number of letters.

11. A message of length n in an alphabet of S letters can occur in S^n ways. "In estimating the capacity of the physical system," says Hartley, "we should ignore the question of interpretation, make each selection perfectly arbitrary, and base our result on the possibility of the receiver's distinguishing the result of selecting any one symbol from that of selecting any other." From this viewpoint the amount of "information" in any one correctly transmitted message is measured by the number of all the possible messages it might have been, namely S^n. Equivalently, if the receiver was in doubt as to which of the S^n messages he would get, the more symbols accurately received the greater the reduction in his uncertainty and hence the greater the "information". What we are concerned with here is clearly related to the correctness of the transmission, not the meaning ascribed to the symbols. If some extraterrestrial intelligence sent us a series of signals which were correctly received (i.e. were uncontaminated by error) the "information" would be complete, notwithstanding that we had not the slightest inkling of what they meant.

12. Hartley proceeds to point out that the (exponential) number of sequences is not suitable for use directly as a measure of "information". It is the number of selected symbols, namely n, which must be measured against the capacity of the transmission system. Those familiar with the game known as "Twenty Questions" will at once grasp the point. By affirmative or negative answers to the questions concerning a binary sequence: "Is the first symbol A?", "Is the second symbol A?" and so on we can identify the message completely with n questions. Hartley then writes the "amount of information" H as

$$H = Kn \tag{3}$$

where K is some constant depending on the number S of symbols. He shows that K is the logarithm of the number of symbols and so, in the binary code,

$$H = n \log_e 2. \tag{4}$$

It is customary to take $\log_e 2$ as the unit of "information" so that the amount of "information" is n units. The phrase "binary digit" has been abbreviated to "bit" and one speaks of the "information" content in bits. Except when speaking to specialists where there is no possibility of misunderstanding one should avoid expressions such as "a bit of information" which has a totally different meaning in colloquial speech.

If the logarithm of 2^n is taken to base 10 the resultant information is sometimes said to be expressed in hartleys.

13. Hartley's work was comparatively unnoticed until Shannon took it up and extended it in a paper (1949) which may be regarded as the effective starting-point of the current interest in the subject. Shannon divides a communication system into five parts: information source, transmitter, channel, receiver and destination. For present purposes it is the information source which is of main interest. It now becomes explicit that "information" is a statistical property of the set of possible messages, not of an individual message. Whereas, in Hartley's simplified exposition, each symbol can occur at any point with equal probability, Shannon generalizes to an alphabet and a language in which (a) certain sequences are inadmissible (such as three Z's in a row in English), (b) the symbols occur with different relative frequencies (the reason for representing E with a single dot in Morse) and (c) the successive occurrences of symbols are not independent, so that the message *en clair* can be regarded as a stochastic process (and particularly as a Markoff process with probabilities $p(i, j)$ that symbol i will be followed by symbol j). If the probability of occurrence of symbol i in an alphabet of S letters is $p_i (i = 1, 2, ..., S)$ Shannon lays down the following requirements for a measure H of "information" produced: it should be continuous in the p_i; if all the p_i are equal so that $p_i = 1/S$, H should be a monotonic increasing function of S; and if a selection of symbols is broken down into successive selections, the original H should be a (weighted) sum of the individual values of H – an additivity postulate. He then shows that the only measure conforming to these requirements is

$$H = -K \sum_{i=1}^{S} p_i \log p_i, \tag{5}$$

where K is a positive quantity.

Shannon notes the analogy with entropy in Boltzmann's formulation and describes the sum on the right in equation (5) as the entropy of the set of possible symbols (events). The corresponding entropy of an information source is this quantity if the symbols are independent. But if there are serial correlations in the permissible messages, the probability of a transmission having reached state i is P_i and the probability from that state of producing symbol j is $p_i(j)$, the entropy is defined as

$$- \sum_{i, j} P_i p_i(j). \tag{6}$$

14. The ratio of the entropy of a source to the maximum which it could have with the same set of symbols is its relative entropy, and is a measure of the efficiency of the source. One of the problems of communication is to encode the source so as to improve its efficiency. The complementary quantity equal to one minus the relative entropy is called the redundancy. Basing himself on counts of relative frequency of letters, diagrams, etc. Shannon remarks that English is about 50 per cent redundant.

15. Reviewing information theory in 1966, Gilbert remarked that mathematicians found Shannon's paper a gold mine, physicists were interested in a new interpretation of entropy as information and there were many other applications, to linguists, music, cryptography and gambling. But, he says, "Information theory was a glamour science for many years. It was popularly held that information theory held the key to progress in remote fields to which in fact it did not apply. Interesting evidence of this remains in some of the editorials which appeared in the *Transactions of the Professional Group on Information Theory* between 1955 and 1959: the editors seem appalled by a flood of worthless interdisciplinary papers submitted in the name of information theory." The theory of "information" has indeed added little beyond Shannon's work to the communication of information as ordinarily understood.

16. More recently Hart (1971) has come to the same conclusion in considering means of concentration employed in statistical economics. He refers to previous work on the entropy function as a measure of concentration of output or employment in industry and compares some of the proposed methods with the more traditional measures of classical statistics such as Lorenz curves and Gini concentration. His general conclusion, based on both theory and practical application to a number of UK industries is that "the redundancy and entropy measures of business concentration are inappropriate".

17. We now have to consider what all this has to do with information as understood in the theories of statistics and probability. The development of statistical data, especially on the electronic computer, has put at the statistician's disposal an enormous amount of material which is usually spoken of as "information". In the computer world it is customary to speak of organized EDP (Electronic Data Processing) as an information system, and in particular management information systems are an important part of any enterprise, whether on the computer or not. Such "information" is sought for and evaluated on the basis of its semantic content. Most people would be surprised to learn that any other attitude is possible.

18. However, "information" is not an abstract quantity to be considered apart from the situation to which it relates. One might say that the telephone directory is a mine of information; but it is of no use to anyone who wants to know how to play, say, the Sicilian Defence at chess. Information only assumes any kind of reality when we specify what the information is about. In current speech we usually take this field of application for granted, or as understood from the context. We may then speak of a source containing a lot or only a little information, meaning that it contains a lot of (or a few) statements of fact about its subject. We acquire a rather vague notion of *amount* of information.

19. But information in this sense is not necessarily additive. Suppose I have a bag containing balls of unknown colour and I want to know whether one or more are red. If I take one from the bag and it proves to be red, no further drawing adds any information to the answer. Any crossword addict knows that there is a diminishing return of information as more letters of an unknown word are identified. In the case of a thermo-system it is reasonable to suppose (and indeed follows almost by definition) that entropy is additive. In the case of a communications source it is not implausible to require that the "information" in a set of symbols is the sum of contributions from each symbol (although, so far as I can see, there is no logical necessity for any additivity postulate). But in information concerning a subject, as the expression is ordinarily understood, there may be very good reasons why additivity is repugnant to common sense requirements.

20. The use of the word "information" in a specialized sense in the theory of statistics dates back to a measure proposed by Fisher (1934, 1935). If t is an unbiased estimator of a parameter θ in a frequency distribution $f(x, \theta)$ and L is the likelihood of a sample of n independent observations, then under certain regularity conditions

$$\operatorname{var} t \geqq 1/E\left(\frac{\partial \log L}{\partial \theta}\right)^2 = -1/E\left(\frac{\partial^2 \log L}{\partial \theta^2}\right). \tag{7}$$

The quantity I defined as

$$I = E\left(\frac{\partial \log L}{\partial \theta}\right)^2 \tag{8}$$

was defined as the "amount of information" in the sample (presumably information about θ). For random samples I is proportional to the sample size n. Asymptotically the inequality of equation (7) becomes an equality for maximum likelihood statistics so the "information" about θ is inversely proportional to the variance of the ML estimator. The regularity conditions justifying statements such as these are critical; Pitman (1936) points out, for example, that I is unsatisfactory if the range of the distribution depends on the parameter under estimate.

21. Kempthorne and some of his co-workers, notably Papaionnou (1971) have recently asked some rather searching questions about "information": how it is to be defined, for example, and what purpose is served by setting up a measure of it. They regard "information" essentially as an indefinable like the straight line of Euclidean geometry and, in common with other writers, set down the axioms which a measure of information should obey. Possibly this is the best approach to a study of information theory, but I have some reservations about the axioms, and particularly the additivity postulate. However, there are certain features which seem to be common to all modern treatments:

(1) In some sense or other, additional observations should, in general, reduce the uncertainty about a statistical distribution.
(2) If this reduction can be quantified, it can be identified in some form with "amount of information".
(3) The amount of information cannot be negative.
(4) If the data are condensed the amount of information cannot be increased (and in general will be reduced; there is "loss of information").
(5) If a random variable x is regarded as providing information about a random variable y, then y provides the same amount of information about x.
(6) In particular, if x and y are independent neither conveys any information about the other.

22. In recent work we may discern three main lines of attack.

The first, mainly pursued by Russian writers, is completely divorced from any physical theory and can be regarded as an axiomatization, a rigorization and a generalization of the mathematics implicit in Shannon's papers. A good deal of this work is incorporated in a series of *Mathematische Forschungsberichte* (volumes 4, 6, 10, 17, 18) under the title of *Arbeiten zur Informationstheorie* (see references under *Arbeiten*). Much of this is pure mathematics concerned, for example, with extensions from discontinuous to continuous variables. So far as I can see, it does not enlarge our understanding of the nature of information or provide any working tools for the statistician.

23. The second line of development, which closely accompanies the first, is the concept of the amount of "information" contained in one random variable about another random variable. A basic paper on the subject by Gel'fand and Yaglom (1959) is available in English.

If a variable ξ takes values x_1, x_2, ..., x_s and another variable η takes values y_1, y_2, ..., y_t with probabilities p_1, ..., p_s and q_1, ..., q_t; and if the probability of x_i, y_j is r_{ij} the amount of "information" about the variable ξ contained in the variable η (or more precisely, the average amount of "information" about the variable ξ conveyed by specifying the value of η) is measured by

$$J(\xi, \eta) = \sum_{i=1}^{s} \sum_{j=1}^{t} r_{ij} \log \frac{r_{ij}}{p_i q_j}. \tag{9}$$

It is not difficult to show that J has certain properties which we should require of any measure of "information". It is essentially positive unless ξ and η are independent. It is symmetrical in ξ, η. If $\xi = \eta$, J reduces to the entropy of ξ. In other cases $J(\xi, \eta) \leq J(\eta, \eta)$. If some values of ξ are averages or amalgamations of subsets "information" is lost; and so on. For example, if ξ, η are distributed in the bivariate normal form with correlation ρ

$$J(\xi, \eta) = -\tfrac{1}{2} \log(1-\rho^2). \tag{10}$$

Applications to regression analysis are obviously possible. If ξ is a multinormal distribution with dispersion determinant A, η is multinormal with determinant B and ξ, η together have determinant C, then

$$J(\xi, \eta) = \tfrac{1}{2} \log \frac{AB}{C}. \tag{11}$$

24. The third line of development is best accounted for by Kullback's book (1959), which brings the theory of "information" and the orthodox theory of statistics into line.

Consider a discrimination between two hypotheses H_1 and H_2, both concerned with the frequency distributions f_1 and f_2 of a variable x. Let a value of x be given. Then by Bayes' rule connecting prior with posterior probabilities

$$P(H_i \mid x) = \frac{P(H_i) f_i(x)}{P(H_1) f_1(x) + P(H_2) f_2(x)}, \quad i = 1, 2. \tag{12}$$

Hence

$$\log \frac{f_1(x)}{f_2(x)} = \log \frac{P(H_1 \mid x)}{P(H_2 \mid x)} - \log \frac{P(H_1)}{P(H_2)}. \tag{13}$$

Verbally expressed, this equation implies that the logarithm of the likelihood ratio f_1/f_2 is the logarithm of the odds in favour of H_1 after the event less the logarithm before the event. It is defined by Kullback as the "information" in x for discrimination in favour of H_1 against H_2. The mean "information" $I(1; 2)$ is defined as

$$I(1; 2) = \int_{-\infty}^{\infty} \{\log f_1 | f_2\} f_1 dx. \tag{14}$$

Likewise the mean "information" in x for discrimination in favour of H_2 against H_1 is

$$I(2; 1) = \int_{-\infty}^{\infty} \log(f_2/f_1) f_2 dx. \tag{15}$$

The quantity J defined as

$$J(1; 2) = I(1; 2) - J(2; 1)$$

$$= \int_{-\infty}^{\infty} (f_1 - f_2) \log(f_1/f_2) dx \tag{16}$$

is called the *divergence* between the frequency functions. Those who dislike Bayesian type arguments may begin from equations (14)–(16), which do not incorporate prior probabilities.

The divergence may be regarded as a measure of the "distance" between two frequency distributions. For example, if f_1 is $N(\mu_1, 1)$ and f_2 is $N(\mu_2, 1)$, J becomes $\frac{1}{2}(\mu_2 - \mu_1)^2$. However, J does not obey the triangular inequality of ordinary distance functions.

25. This approach can be linked with the entropy of communication theory in the following way. Suppose we have S mutually exclusive and exhaustive hypotheses and that we can determine which is true from one observation (e.g. with an alphabet of S letters, one observation tells us which occurs). In generalization of equation (12) the prior probability is p_i and the amount of "information" is $-\log p_i$. The mean "information" is then

$$- \sum_{i=1}^{S} p_i \log p_i$$

which is the entropy of the set of H's.

26. A link can also be established with Fisher's measure of "information". Consider a small variation in the parameter θ, H_1 referring to θ and H_2 so $\theta + \Delta\theta$. Then

$$I(\theta, \theta + \Delta\theta) = \int f(x, \theta) \log \{f(x, \theta)/f(n, \theta + \Delta\theta)\} \, dx$$

$$= - \int f(x, \theta)\{\log f(x, \theta + \Delta\theta) - \log f(x, \theta)\} \, dx$$

$$= - \int f(x, \theta) \left\{ \Delta\theta \frac{\partial \log f}{\partial \theta} + \frac{1}{2}(\Delta\theta)^2 \frac{\partial^2 \log f}{\partial \theta^2} \right\} dx$$

$$+ 0(\Delta\theta)^3 \qquad (17)$$

Under the usual regularity conditions the integral of $\partial \log f/\partial \theta$ is zero and the integral of $-\partial^2 \log f/\partial \theta^2$ is I, the Fisher information. Hence for small deviations

$$I(\theta, \theta + \Delta\theta) = I(\Delta\theta)^2 \qquad (18)$$

and similarly

$$J(\theta, \theta + \Delta\theta) = I(\Delta\theta)^2 \qquad (19)$$

The results are easily extended to the simultaneous estimation of several parameters.

27. Kullback goes on to prove that I has the usual properties of "information". For a function of two independent variables x, y

$$I(1; 2; x, y) = I(1; 2; x) + I(1; 2; y) \qquad (20)$$

and in the case of dependence

$$= I(1; 2; x) + I(1; 2; x \mid y). \qquad (21)$$

"Information" in this sense is lost under grouping of variate values or transformation to new values unless the statistic of transformation is sufficient.

28. The quantity of J equation (16) is justly denoted by that letter because it was introduced by Jeffreys (1946) in a different context. He was interested in prior probability distributions of parameters which were invariant, and in particular in invariant expressions for the difference between two distributions of chance. J is one of the expressions he derives; and he notes that it has the form of the square of a distance in curvilinear coordinates.

29. The failure of the Fisher measure of information in cases where the regularity conditions are violated has led to a search for alternative measures. Papaionnou and Kempthorne (1971) point out one further disadvantage, namely that I does not readily generalize to the case where multivariate parameters are concerned. The obvious generalization of (8) to such a case is to consider the $k \times k$ matrix (k being the number of parameters)

$$I_{ij} = \left[E \left(\frac{\partial \log L}{\partial \theta_i} \right) \left(\frac{\partial \log L}{\partial \theta_j} \right) \right], \quad i, j = 1, 2, ..., k \tag{22}$$

The diagonal elements, of course, can be regarded as the univariate measures of information for each parameter separately. The off-diagonal elements have no obvious interpretation. Papaionnou and Kempthorne suggest forming a measure of information from the trace of the matrix and show that this obeys the usual requirements. So does the determinant of the matrix except that it does not obey the additivity axiom.

30. Other measures have been proposed. Many of them are variants of the entropy suitably defined for continuous distributions. That of Kullback and Leibler (1951) is effectively the one set out in equation (15). A different kind of measure of divergence between two populations was suggested by Bhattacharya (1943)

$$p(f_1, f_2) = \int \{ f_1(x) f_2(x) \}^{\frac{1}{2}} dx \tag{23}$$

where x is p-dimensional. The negative of the logarithm of this function can also be considered as a measure of information.

31. There is now a fairly extensive literature on "information" in the narrower statistical sense – a bibliography is given in Papaionnou and Kempthorne. The subject links up in various ways with the classical theory of statistics; for example, with the concept of minimal sufficiency statistics (for which there should be no loss of information), with measures of divergence between two distributions (which can be considered apart from questions of "information") and with Bayesian-type inference (the difference between prior and posterior distributions being, in some sense, additional information). In my view, however, it remains an open question whether these various measures tell us any more (I had almost written "give us more information") about a statistical situation than the measures from which they are derived. But perhaps it is not worth while arguing over questions of terminology. The objectives of estimation, inference and experimental design are fairly plain and provided that we understand what is implied by the use of the word "information" in statistical contexts it may well provide a useful quantification of our ideas.

References

Arbeiten zur Informationstheorie

    I (Chintschen, A. J., Faddejev, D. K., Kolmogoroff, A. N., Renyi, A. and Balatoni, J.) Dritte Auflage 1967.
    II Fel'fand, I. M. and Jaglom, A. M. 1958
    III Kolmogoroff, A. N. and Tichimirow, W. M. 1960
    IV Dobruschin, R. L. 1963
    V Pinsker, M. S. 1963

    V.E.B. Deutscher Verlag der Wissenschaften, Berlin.

Bhattacharya, A. (1943). On a measure of divergence between two statistical populations defined by their probability distributions. *Calcutta Math. Soc.* **35**, 99.
Boltzmann, L. (1895). *Vorlesungen über Gastheorie*. Ier Theil, Leipzig, Barth.
Carathéodory, C. (1909). Untersuchungen über die Grundlagen der Thermodynamik. *Math. Annalen*, **67**, 355.
Clausius, R. J. E. (1887). *Die Mechanische Wärmetheorie*. Braunchweig.
Fisher, R. A. (1934). Probability, likelihood and quantity of information in the logic of uncertain inference. *Proc. Roy. Soc.* A, **146**, 1.

Fisher, R. A. (1935). The logic of inductive inference. *J. Roy. Statist. Soc.* **98**, 39.

Gel'fand, I. M., Yaglom, A. M. (1959). Calculation of the amount of information about a random function contained in another random function. *Am. Math. Soc. Trans. Series 2*, A, **12**, 199.

Gilbert, E. N. (1966). Information theory after 18 years. *Science*, **152**, 320.

Hart, P. E. (1971). Entropy and other measures of concentration. *J. Roy. Stat. Soc.* A, **134**, 73.

Hartley, R. V. L. (1928). Transmission of information. *Bell System Technical Journal*, 7, 535.

Jeffreys, H. (1946). An invariant form for the prior probability in estimation problems. *Proc. Roy. Soc.* A, **186**, 453.

Kullback, S. (1959). *Information Theory and Statistics*. New York, Wiley.

Kullback, S., Leibler, R. A. (1951). On information and sufficiency. *Am. Math. Stats.* **22**, 79.

Maxwell, J. C. (1868). On the dynamical theory of gases. *Phil. Mag. IV series*, **35**, 129 and 185.

Papaionnou, P. C., Kempthorne, O. (1971). On statistical information theory and related measures of information. Aerospace Res. Lab. ARL. 71–0059. Ohio.

Pitman, E. J. G. (1936). Sufficient statistics and intrinsix accuracy. *Proc. Camb. Phil. Soc.* **32**, 567.

Shannon, C. E., Weaver, W. (1949). *The Mathematical Theory of Communication*. University of Illinois Press.

Szilard, L. (1929). Uber die Entropieverminderung in linearen thermodynamischen System bei Eingriffen intelligenter Wesen. *Zeitschrift für Physik*, **53**, 840.

Résumé

L'auteur rappèle brièvement l'histoire de l'idée d'entropie dans la théorie de la chaleur et trace sa connection avec les idées analogues dans la théorie de communication et celle d'"information". Il offre un commentaire critique sur l'idée d'information dans le sens restreint que l'on le trouve dans la théorie statistique.

A History of Random Processes

I. Brownian Movement from Brown to Perrin

STEPHEN G. BRUSH

(1968)

Contents

1. ROBERT BROWN's Observations and Interpretations Thereof 2
2. Miscellaneous Observations and Qualitative Experiments, 1840—1878 7
3. Criticisms of the Molecular-impact Theory 10
4. EINSTEIN's Theory of Brownian Movement 14
5. SMOLUCHOWSKI's Theory of Brownian Movement 24
6. PERRIN's Experiments and the Reality of Atoms 30

Modern science differs from Newtonian science in its emphasis on processes. Psychologists study the development of personality from infancy through adolescence rather than merely analyzing the faculties of the adult human; biologists deal with evolution and life cycles of organisms rather than concentrating on the classification of species and descriptive anatomy; chemists and physicists are interested more in the reactions of molecules, atoms and particles than in the static properties of substances; and astronomers have turned away from mapping the heavens and analyzing the cyclic motions of planets to speculating about the evolution of stars and of the universe. The new concerns of science have also stimulated new developments in mathematics, notably the theory of random ("stochastic") processes.

In this paper, which is the first of a series on random processes in modern science and mathematics, we examine the history of observations and theories of Brownian movement. Starting with ROBERT BROWN — who was apparently the first to recognize, in 1828, that the irregular movements of all kinds of small particles in fluids have a physical rather than a biological cause — we review various unsuccessful attempts to explain these movements. This survey is a prelude to the more detailed discussion of the theories of EINSTEIN and SMOLUCHOWSKI.

It is surprising that Brownian movement played almost no role in physics until 1905, and was generally ignored even by the physicists who developed the kinetic theory of gases, though it is now frequently remarked that Brownian movement is the best illustration of the existence of random molecular motions. Moreover, the existence of Brownian movement itself has been interpreted to imply that the Second Law of Thermodynamics does not have absolute validity on the microscopic level; yet this argument was almost completely overlooked at the time when the validity and consequences of the Second Law were being hotly disputed in the 1890's.

Early attempts to apply kinetic theory to Brownian movement employed the theorem that all particles have equal average kinetic energy in a state of thermal equilibrium (Waterston-Maxwell equipartition theorem). Observers attempted to

measure the speed of particles in Brownian movement, but none of the results
came anywhere near the theoretical prediction. Only after EINSTEIN had developed
his theory was the reason for the discrepancy realized; three-quarters of a century
of experimentation produced almost no useful results, simply because no theorist
had told the experimentalists what quantity should be measured!

EINSTEIN's theory of Brownian movement combined two postulates that
seemed to have nothing to do with each other (perhaps this was characteristic of
EINSTEIN). He took a formula from hydrodynamics, for the force on a sphere
moving through a viscous fluid; and another formula, from the theory of solutions,
for the osmotic pressure of dissolved molecules. Inserting these (apparently incompatible) physical characterizations into his description of the random motion
of a particle, he arrived at his famous result for the mean-square displacement
of the particle. SMOLUCHOWSKI, shortly afterwards, published a more compre-
hensible derivation of a similar result, using a theoretical model taken over from
gas theory.

JEAN PERRIN quickly attempted to establish EINSTEIN's theory by experimental test of the displacement formula and of another formula for the vertical
distribution of particles in a fluid. PERRIN also claimed that the confirmation of
EINSTEIN's theory proved the real existence of the atom, previously considered
a merely hypothetical entity. He was remarkably successful in putting over this
argument, and in fact the anti-atomistic "Energetics" movement never recovered
from this blow.

1. Robert Brown's Observations and Interpretations Thereof

The modern reader of BROWN's papers may be puzzled by his use of the word
"Molecule." The significance of this term for biologists in the early nineteenth
century was strongly influenced by the theories of the French naturalist GEORGES
LOUIS LECLERC, Comte DE BUFFON (1707—1786). BUFFON taught that all plants
and animals developed out of a basic supply of "organic molecules"; these
molecules were like interchangeable parts, and their formation into an organism
of a particular kind was simply guided by an "interior mold."[1] BUFFON's ideas
seemed to be supported by microscopic observations of JOHN NEEDHAM, LAZZARO
SPALLANZANI, and others in the eighteenth century. While these naturalists did
not necessarily agree with BUFFON or with each other about the generation and
constitution of organisms, they did report having seen tiny particles from organic
substances that appeared to be self-animated when placed in a fluid. Later in
the nineteenth century, the doctrine of organic molecules was to be swallowed
up by the cell theory, but in the 1820's it was still being defended by some biologists.[2] BROWN's discovery, therefore, was not his observation of the motion of

[1] G. L. L. BUFFON, *Histoire Naturelle, Générale et Particulière, avec la Description
du Cabinet du Roy*, Tome Second, Chapitre IV & VI; Tome Quatrième, "Le Boeuf"
(p. 437ff.) (Paris: L'Imprimerie Royale, 1749, 1753). See also P. FLOURENS *Histoire
des Travaux et des Idées de Buffon*, 2. ed., p. 67ff. (Paris: Hachette, 1850).
[2] On the history of Brownian movement before BROWN, see F. I. F. MEYER,
"Historische-physiologische Untersuchungen über selbstbewegliche Molecüle der
Materie," p. 327—498 in volume 4 of *Robert Brown's Vermischte Botanische Schriften*,
ed. C. G. NEES VON ESENBECK (Leipzig: F. Fleischer & Nürnberg: L. Schrag, 1825—
1834).

microscopic particles in fluids; that observation had been made many times before; instead, it was his emancipation from the previously current notion that such movements had a specifically organic character. What BROWN showed was that almost any kind of matter, organic or inorganic, can be broken into fine particles that exhibit the same kind of dancing motion; thus he removed the subject from the realm of biology into the realm of physics.

ROBERT BROWN (1773—1858) was one of England's greatest botanists. He is best known for his discovery of the nuclei of plant cells, as well as for classifying a large number of unfamiliar plants which he brought back from an Australian expedition in 1801—1805. The first volume of his *Prodromus Florae Novae Hollandiae et Insulae Van Diemen* was published in 1810, and in the same year he became librarian to JOSEPH BANKS, President of the Royal Society. Though offered a university chair he preferred to stay with BANKS where he had the use of valuable collections. The library and collections were bequeathed to him by BANKS on the latter's death in 1820, with the condition that they were to go the British Museum after BROWN's death. BROWN arranged for the British Museum to take over the collections in 1827, with himself as keeper of the botanical department on an assured stipend.[3]

In 1828, BROWN wrote a pamphlet entitled "A brief account of microscopical observations made in the months of June, July and August, 1827, on the particles contained in the pollen of plants; and on the general existence of active molecules in organic and inorganic bodies." This work is a bibliographic curiosity, for it was never officially "published" even though it was printed in the *Edinburgh Journal of Science* in 1828 and reprinted elsewhere.[4] It was originally intended only for private circulation, and BROWN avoided the formal manner of presentation customarily found in articles written for scientific journals. Although BROWN's informal style pleases those of us who like to see how one step leads to another in a scientific investigation, his inclusion of preliminary conjectures led to some later misunderstandings about his views on the cause of the movement he observed.

The researches originated, BROWN tells us, in an attempt to find the mode of action of pollen in the process of impregnation. The first plant examined was *Clarckia pulchella*, whose pollen contains particles varying from $1/4000^{th}$ to $1/5000^{th}$ of an inch in length.

> While examining the form of these particles immersed in water, I observed many of them very evidently in motion; their motion consisting not only of a change of place in the fluid, manifested by alterations in their relative positions, but also not infrequently by a change of form of the particle itself .... In a few instances the particle was seen to turn on its longer axis. These motions were such as to satisfy me, after frequently repeated observation, that they arose neither from currents in the fluid, nor from its gradual evaporation, but belonged to the particle itself.

[3] See the article "Robert Brown" by J. B. FARMER in *Makers of British Botany* (ed. F. W. OLIVER), Cambridge University Press, 1913.

[4] ROBERT BROWN, Edinburgh New Philosophical Journal 5, 358—371 (1828); Annales des Sciences Naturelles (Paris) 14, 341—362 (1828); FRORIEP's Notizen aus dem Gebiete der Natur- und Heilkunde 22, 161—170 (1828); OKEN's Isis 21, 1006—1012 (1828); Philosophical Magazine 4, 161—173 (1828); POGGENDORFF's Annalen der Physik 14, 294—313 (1828); *The Miscellaneous Botanical Works of Robert Brown* published for the Ray Society by R. HARDWICKE, vol. 1, p. 465—479 (1866).

1*

Brown then examined particles (or "Molecules" as he now started to call them) from several other plants, not only living ones but also some that had been preserved in an herbarium for not less than a century. He observed similar movements in all cases. At this point he recorded rather recklessly his first guess about the origin of the motion:

> Reflecting on all the facts with which I had now become acquainted, I was disposed to believe that the minute spherical particles or Molecules of apparently uniform size ... were in reality the supposed constituent or elementary Molecules of organic bodies, first so considered by Buffon and Needham, then by Wrisberg with greater precision, and very recently by Dr. Milne Edwards, who has revived the doctrine ... I now therefore expected to find these molecules in all organic bodies: and accordingly on examining the various animal and vegetable tissues, whether living or dead, they were always found to exist ....

But after studying several mineralized vegetable remains, Brown began to suspect that moving molecules could also be obtained from inorganic sources. It turned out that practically every conceivable substance, from a piece of window glass to a fragment of the Sphinx, could be made to yield particles that moved in water.

Brown's memoir attracted considerable attention, and several other scientists reported observations of a similar kind. But there was almost universal condemnation of Brown, at least on the Continent, for what was thought to be his opinion that the molecules are self-animated. All kinds of physical explanations for the motion were suggested: unequal temperatures in the strongly illuminated water, evaporation, air currents, heat flow, capillarity, motions caused by the hands of the observer, and so forth.[5]

Michael Faraday gave a Friday evening lecture on Brownian movement at the Royal Institution on February 21, 1829, in which he defended Brown. According to Faraday, Brown's experiments were carefully done and sufficed to show that the movements could not be explained by any of the causes so far suggested. In fact, Brown had simply admitted that he could not account for the motions; but by using the term "molecule" (which Faraday was careful to distinguish from "ultimate atoms") Brown had laid himself open to misunderstanding, "because the subject connects itself so readily with general molecular philosophy that all *think* he must have meant this or that ...."[6]

---

[5] G. W. Muncke, "Ueber Robert Brown's microscopische Beobachtungen ..." Poggendorff's Annalen der Physik, 17, 159—176 (1829). Francois Raspail, "Observations et expériences propres à démontrer que les granules qui sortent dans l'explosion du grain de pollen, bien loin d'être les analogues des animalcules spermatiques, comme Gleichen l'avait pensé le premier, ne sont pas même des corps organisés", Mem. Soc. Hist. Nat., Paris, 4, 347—362 (1828); Edinburgh Journal of Science 10, 96—108 (1828). C. A. S. Schultze, *Mikroskopische Untersuchungen ueber des Herrn Robert Brown Entdeckung lebender, selbst in Feuer unzerstoerbarer Teilchen in all Koerpern, und ueber Erzeugung der Monaden*, Carlsruhe und Freiburg; in der Herder'-schen Kunst- und Buchhandlung, 1828. C. M. Marx, "Ueber Molecul-Bewegungen," Schweigger's Journal für Chemie und Physik 61, 121—122 (1831). Franz Unger, "Ueber die Bewegung der Molecüle," Flora 15, 713—717 (1832). R. E. Grant, "On the influence of light on the motions of infusoria," Edinburgh Journal of Science 10, 346—350 (1829). Robert Bakewell, "An account of Mr. Needham's original discovery of the action of the pollen of plants, with observations on the supposed existence of Active Molecules in Mineral Substances," Mag. Nat. Hist. 2, 1—9 (1829).

[6] Bence Jones, *The Life and Letters of Faraday*, London: Longmans, Green & Co., 1870, Vol. 1, p. 403.

The physicist DAVID BREWSTER (who was the editor of the *Edinburgh Journal of Science*, in which BROWN's original memoir had been printed) thought that physical causes would probably be found sufficient to explain the motion. But even if a complete physical explanation were not yet possible, he thought it quite improper to attribute the motion to animal life. There was nothing surprising in the fact that molecules should have their own characteristic motion:

Why should not the molecules of the hardest solids have their orbits, their centres of attraction, and the same varied movements which are observed in planetary and nebulous matter? The existence of such movements has already been recognized in mineral and other bodies. A piece of sugar melted by heat, and without any regular arrangement of its particles, will in process of time gradually change its character, and convert itself into regular crystals .... In these changes the molecules must have turned round their axes, and taken up new positions within the solid .... Before another century passes away, the laws of such movements will probably be determined; and when the molecular world shall thus have surrendered her strongholds, we may look for a new extension of the power of man over the products of inorganic nature.[7]

BREWSTER is the intellectual ancestor of the modern physicist who is supremely confident that all problems in chemistry, and perhaps even biology, can "in principle," be reduced to the solution of the appropriate Schrödinger equation; we need only build a big enough computer to solve that equation. At the same time, BREWSTER was here representing rather accurately the attitude of nineteenth-century physicists toward Brownian movement: it was a phenomenon that would certainly find its ultimate explanation in a future theory of molecular motion, but was not worth the trouble of a detailed investigation at the present time. The leading natural philosophers of the day had more important problems to be solved first.

In a second memoir, "Additional remarks on active molecules," BROWN replied to some of his critics and reported further experiments.[8] He disclaimed the view that the molecules are animated, admitting that some readers may have misunderstood him because he had "communicated the facts in the same order in which they occurred, accompanied by the views which presented themselves in the different stages of the investigation." He now wished to prove that the motion of particles in a fluid cannot be due, as others had suggested, to "that intestine motion which may be supposed to accompany its evaporation." To accomplish this proof, he mixed water containing particles with almond-oil. After being shaken, the mixture contained drops of water ranging from $1/50^{th}$ to $1/200^{th}$ of an inch in diameter. Being surrounded by almond-oil, these drops of water do not evaporate for a considerable time. Some of the drops contained only a single particle. "But in all the drops thus formed and protected, the motion of the particles takes place with undiminished activity, while the principal causes assigned for that motion, namely, evaporation, and their mutual attraction and repulsion, are either materially reduced or absolutely null."

[7] D. BREWSTER, "Observations relative to the motions of the molecules of bodies," Edinburgh Journal of Science 10, 215—220 (1829).

[8] ROBERT BROWN, "Additional remarks on active molecules," Edinburgh Journal of Science 1, 314—320 (1829); Annales des Sciences Naturelles (Paris) 19, 104—110 (1830); Edinburgh New Philosophical Journal 8, 41—46 (1830); Froriep's Notizen aus dem Gebiete der Natur- und Heilkunde 25, 305—310 (1829); Philosophical Magazine 6, 161—166 (1829); *The Miscellaneous Botanical Works of Robert Brown*, vol. 1, p. 479—486.

Brown was aware that the evaporation of liquids was considered by physicists to be somehow connected with "intestine motions," but he misunderstood the connection; he thought that if evaporation were suppressed by some external cause, then the intestine motion would also have to stop.

The other cause of motion that Brown thought he had excluded by this experiment — mutual attraction or repulsion of the particles — was occasionally proposed later in the nineteenth century. No one seems to have noticed Brown's refutation of this explanation: the fact that a single particle in a drop of water will exhibit the same motion as it does when other particles are present.

In this second memoir, Brown also referred to the previous observations of Leeuwenhock, Stephen Grant, Needham, Buffon, and Spallanzani, but he said that all these writers confused "Molecular" motion with animalcular motion. He also cited Gleichen, Wriberg, Müller, and James Drummond. He noted that in 1819, Bywater of Liverpool had published an account of microscopical observations "in which it is stated that not only organic tissues, but also inorganic substances consist of what he terms animated or irritated particles." But, according to Brown, Bywater was susceptible to "optical illusions." Thus Brown disposed of a possible claimant to the discovery of the generality of Molecular motion.[9]

It would take us too far from the main subject of this paper to pursue the effect of Brown's writings on biological speculations about "organic molecules" later in the nineteenth century. We merely note in passing that by the 1870's, at least, it was becoming common for authors of books on the microscope to include warnings about Brownian movement, in case observers should mistake it for the motion of living beings and attempt to build fantastic theories on it.[10,11]

Before leaving Brown, I think it may not be superfluous to quote the recollections of Charles Darwin from the 1830's:

> I saw a good deal of Robert Brown, "facile Princeps Botanicorum," as he was called by Humboldt. He seemed to me to be chiefly remarkable for the minuteness

---

[9] The only publication of Bywater which I have been able to find is an article "On Animalcules, particularly on the Polyopes," Phil. Mag. **49**, 283—288 (1817). In this article Bywater does not mention movements in any substances other than those derived from animals and plants. The British Museum catalogue lists a book, *Physiological Fragments; or, Sketches of various subjects intimately connected with the study of Physiology* (London: Baldwin, Cradock & Joy, 1819) which is probably the one Brown refers to; this was reissued in 1824 with an addendum, *Supplementary Observations, to shew that vital and chemical energies are of the same nature and both derived from solar light* (London: R. Hunter). Brown says he has not seen the original memoir of 1819 but only a later edition published in 1828.

[10] See for example C. P. Robin, *Traité du Microscope, son mode d'emploi*, Paris: J. B. Baillière & fils, 1871, p. 526; W. B. Carpenter, *The Microscope and its revelations*, London: J. Churchill, fifth edition 1875, p. 199.

[11] On biological speculations, see E. Rádl, section "Spekulationen über kleinere Lebensteilchen als die Zelle," p. 386—389 in *Geschichte der Biologische Theorien*, II. Teil, Leipzig: Engelmann, 1909. (The English translation of Rádl's book, *The History of Biological Theories*, published by Oxford University Press in 1930, lacks many of the references for this section.)

The history of "abiogenesis," "biogenesis,' 'organic molecules," and Pasteur's germ theory is discussed by T. H. Huxley in his Presidential Address to the British Association in 1870: Report of the British Association, **40**, lxxiii—lxxxix (1870).

of his observations and their perfect accuracy. His knowledge was extraordinarily great, and much died with him, owing to his excessive fear of ever making a mistake. He poured out his knowledge to me in the most unreserved manner, yet was strangely jealous on some points. I called on him two or three times before the voyage of the *Beagle* [1831], and on one occasion he asked me to look through a microscope and describe what I saw. This I did, and believe now that it was the marvelous currents of protoplasm in some vegetable cell. I then asked him what I had seen; but he answered me, "That is my little secret."[12]

2. Miscellaneous Observations and Qualitative Explanations, 1840—1878

After the initial flurry of excitement caused by BROWN's publications in 1828—29, interest in Brownian movement dropped off to almost nothing for about thirty years.[13] But with the development of thermodynamics and the revival of the kinetic theory of gases in the 1850's, there was a new stimulus for researches into the relation between heat and microscopic motion. This was certainly a factor in the explanations proposed for Brownian movement later in the nineteenth century. However, until NÄGELI's paper in 1879 (see next section), there was no serious attempt to develop a quantitative theory of Brownian movement, based on the mechanical-atomic theory of heat. What is perhaps most significant about the history of this period is the absence of any publications on Brownian movement by the kinetic theorists — CLAUSIUS, MAXWELL and BOLTZMANN.

Notions of the old caloric theory of heat were still evident in the discussion of the British biologists GRIFFITH & HENFREY, in their *Micrographic Dictionary* published in 1856:

Heat is the only agent which affects [the motion]; this causes the motion to become more rapid. Hence it might be attributed to the various impulses which each particle receives from the radiant heat emitted by those adjacent. Or, as it takes place when the temperature is uniform, may it not arise from the physical repulsion of the molecules, uninterfered with by gravitation, hence free to move? The effect of heat would then be explicable, because this increases the natural repulsion of the particles of matter, as in the conversion of water into vapour ....[14]

In 1858, JULES REGNAULD (Professor of Physics at the École de Pharmacie in Paris, and later Professor of Pharmacology) did some experiments which convinced him that Brownian movement was due to the absorption of heat by the particles from rays of light falling on the suspension; the transmission of this heat to the surrounding fluid sets up currents which move the particles.[15]

[12] *Charles Darwin: His Life told in an autobiographical Chapter, and in a selected series of his published letters* (edited by his son, FRANCIS DARWIN) (London: Murray, 1892, reprinted by Schuman, New York, 1950) p. 46.

[13] J. D. BOTTO, "Observations microscopiques sur les mouvements des globules végétaux suspendus dans un menstrue," Mem. della R. Accad. delle scienze di Torino [II] **2**, 457—471 (1840).

E. W. WEBER, "Mikroskopische Beobachtungen sehr gesetzmässigen Bewegungen, welche die Bildung von Niederschlägen harziger Körper aus Weingeist begleiten," Berichte über die Verhandlungen, K. Sächsischen Gesellschaft der Wissenschaften, Mathematisch-Physikalische Klasse, Leipzig, 57—67 (1854).

[14] J. W. GRIFFITH & A. HENFREY, article "Molecular motion" in *Micrographic Dictionary* (London: J. Voorst, 1856), p. 428; third edition, 1875, vol. 1, p. 498.

[15] J. REGNAULD, "Etudes relatives au phénomène désigné sous le nom du mouvement Brownien," Journ. Pharm. Chim. [3] **34**, 141 (1858).

CHRISTIAN WIENER (Professor of Descriptive Geometry and Geodesy at Karls-ruhe), reporting on his experiments in 1863, gave various arguments to show that Brownian movement cannot be due to external causes but must be attributed to internal motions in the fluid. WIENER's concept of atomic motion, however, was derived from the period before CLAUSIUS and MAXWELL. He believed that matter consists not only of material atoms but also of aether atoms; heat is the kinetic energy of both kinds of atoms. The essential difference between solids and liquids, he said, is that in solids the direction of vibration of the molecules is opposite to that of the aether atoms, whereas in liquids it is the same; heat of melting is the energy needed to reverse the relative directions of these two vibrations. WIENER's explanation of Brownian movement is closely related to these ideas about aether vibrations, and he says that the wavelength of red light is about the same as the diameter of the smallest groups of molecules that move together in the liquid.[16]

WIENER's theory was criticized, in 1894, by MEADE BACHE, who said:

The theory of Herr WIENER, that the movements are due to the action of the red-wave of light and heat is refuted by the single fact that, as I have proved by experiment, one may interpose at pleasure between the source of light or heat and the particles, either a violet glass or a red glass, without being able to observe the slightest alteration in the movements.[17]

Nevertheless, WIENER is credited by some later writers[18] with being the first to discover that Brownian movement is due to molecular motions of molecules in the liquid.

GIOVANNI CANTONI, an Italian physicist, also attributed Brownian movement to thermal motions in the liquid, and considered that this phenomenon provides a "beautiful and direct experimental demonstration of the fundamental principles of the mechanical theory of heat."[19] The honor of having discovered the true cause of Brownian movement has also been claimed for him.[20]

In 1868, J. B. DANCER, in Manchester, said:

The cause of the phenomenon is not yet satisfactorily accounted for. Some have imagined that it is the physical repulsion of the particles when uninfluenced by gravitation. The author ... thinks that the movement may possibly be connected with the absorption and radiation of heat.[21]

W. STANLEY JEVONS, the British writer on political economy and scientific method, claimed that Brownian movement is an electrical phenomenon and is related to osmosis, "for if a liquid is capable of impelling a particle in a given

---

[16] CHR. WIENER, "Erklärung des atomistischen Wesens des flüssigen Körperzustandes und Bestätigung desselben durch die sogenannten Molecularbewegungen," POGGENDORFF's Annalen der Physik **118**, 79—94 (1863).

[17] R. MEADE BACHE, "The secret of the Brownian movements," Proceedings of the American Philosophical Society **33**, 163—177 (1894).

[18] THE SVEDBERG, "The Brownien Movements" Ion **1**, 373—402 (1909); J. PERRIN, "Mouvement brownien et réalité moléculaire," Ann. Chimie et Physique [8] **18**, 1—114 (1909).

[19] GIOVANNI CANTONI, "Su alcune condizioni fisiche dell'affinità, e sul moto browniano," Reale Istituto Lombardo di scienze e lettere (Milano) Rendiconti [2] **1**, 56—67 (1868). See also "Sul moto brauniano," *ibid.* **22**, 152—155 (Sem. 1, 1889).

[20] IC. GUARESCHI, "Nota sulla storia del movimento browniano," Isis **1**, 47—52 (1913).

[21] J. B. DANCER, "Remarks on molecular activity as shown under the microscope," Proceedings of the Manchester Literary and Philosophical Society **7**, 162—164 (1868).

direction, the particle if fixed is capable of impelling the liquid in an opposite direction by an equal force."[22] Lacking VAN'T HOFF's theory of osmotic pressure, however, JEVONS was unable to quantify this suggestion or relate it to diffusion theory in the way that EINSTEIN did in 1905 (see section 4, below). Ignoring DANCER's protest that "the results of many experiments point to heat as a probable case,"[23] JEVONS elaborated his electrical explanation and even proposed a new name, *pedesis*, for the effect. (BROWN, after all, was not the first to see dancing particles under the microscope; he was merely a good publicist.) JEVONS suggested that pedesis plays an important role in sewage treatment, geological processes, and so forth, by maintaining particles in suspension; it is also involved in the detergent action of soap. Yet, despite a brief reference to pedesis as a random process that might be treated by probability theory, JEVONS made no real progress toward a quantitative theory.[24] Only one other scientist, WILLIAM RAMSAY, adopted his term "pedesis."[25]

During the 1870's, the opinion that Brownian movement is somehow related to heat was frequently expressed. But there was still considerable confusion and vagueness in such explanations, stemming mainly from the lingering belief that motion is connected only with a *change* of temperature. It was understandable that the absorption of radiation coming from outside the fluid might cause unequal heating within it, and thereby set up currents which would produce motion of the particles. An atomic explanation, on the other hand, would have to account for the fact that the particles move even in fluids of uniform temperature, simply because of bombardments by individual molecules. The suggestion was occasionally made that several molecules might move together to cause the visible motion of the particle, and that such cooperative motions could be viewed as a possible fluctuation from the average state of motion.[26] One might think that such a suggestion would be countered by the argument that the Second Law of Thermodynamics denies the possibility of such conversions of thermal energy into mechanical energy if there is no temperature-difference in the system. While this argument may have operated unconsciously or privately to discourage explanations based on fluctuations, there is no explicit reference to it in print until later on.[27] It certainly did not prevent scientists like JOSEPH DELSAULX, in Brussels, from talking about the "Thermodynamic origin of Brownian motions."[28]

To summarize the situation in 1878: first, the phenomenon of Brownian movement was becoming widely known, as can be seen from the many references

[22] W. S. JEVONS, "On the so-called molecular movements of microscopic particles," Proceedings of the Manchester Literary and Philosophical Society, 9, 78—82 (1870).

[23] J. B. DANCER, *ibid.* p. 82.

[24] W. S. JEVONS, "On the movement of microscopic particles suspended in liquids," Journal of Science and Annals of Astronomy, 8, 167 (1878); "On the pedetic action of soap," Jornal of Science 8, 514 (1878).

[25] See note 34 below.

[26] E. BUDDE, "Ueber Untersuchungen in Betreff der Brown'schen Molecularbewegung." Sitzungsberichte Niederrheinischen Gesellschaft für Natur- und Heilkunde 27, 108—110 (1870); J. THIRION; "Les mouvements moléculaires," Rev. Quest. Sci. 7, 5—55 (1880).

[27] GOUY, "Note sur le mouvement brownien," Journal de Physique [2] 7, 561—564 (1888).

[28] J. DELSAULX, Monthly Microscopic Journal 18, 1—6 (1877).

to it in scientific journals and even in fiction.[29] Second, while a minority of scientists still attributed its cause to electrical effects, osmosis, or surface tension, most seemed to think that it must be connected with thermal molecular motions. Third, there was still no quantitative theory that could be tested against experiment.

3. Criticisms of the Molecular-impact Theory

In 1879, the German botanist KARL NÄGELI published a long memoir in which he attempted to disprove the suggestion that Brownian movement is caused by the collisions of the particles with molecules in the surrounding fluid.[30] NÄGELI was an authority on microscopic methods of observation, and also had some knowledge of physics which he frequently tried to apply to biology.[31] The immediate stimulus for this particular work was a question that had come up in a discussion at the Munich Academy of Sciences on the spreading of fungus infections by the wind. NÄGELI, motivated partly by the need to defend his own theory of fungus action, developed a general theory of the motions of small particles in air, in which the role of collisions could be compared with the role of intermolecular forces. The main part of his argument applied to sun-motes being bombarded by air molecules, but he claimed that it extended also to Brownian movement of particles in liquids.

NÄGELI had one great advantage over all the scientists who had previously written on Brownian movement: he was familiar with the estimates of molecular masses and speeds that had been obtained from the kinetic theory of gases.[32] He could therefore argue as follows:

Since the idea that the molecules of a gas travel past each other with large velocities has entered physics and, because of its irrefutable proof, has found general agreement, one might also suppose that the "dancing motion" of sun-motes is caused by the frequent and variously directed impulses which they receive from gas molecules. ... [But] because of their greater weight, they [sun-motes] are like bodies completely at rest among the air molecules flying hither and yon, and there can be no question of a dancing or quivering of the sun-motes resulting from molecular collisions.

This can easily be proved by a calculation of the number and energy of molecular collisions which a particle of definite size under definite conditions experiences in

---

[29] See for example GEORGE ELIOT'S *Middlemarch* (1872), Book I, Chapter XVII, p. 181, in which ROBERT BROWN'S memoir is offered to the Rev. Mr. FAREBROTHER by the surgeon LYDGATE.

[30] KARL NÄGELI, "Über die Bewegungen kleinster Körperchen," Sitzungsberichte der mathematisch-physikalischen Classe der K. Bayerischen Akademie der Wissenschaften zu München **9**, 389—453 (1879).

[31] KARL NÄGELI & S. SCHWENDENER, *Das Mikroskop, Theorie und Anwendung Desselben*, Leipzig: Engelmann, 1867, 2. aufl. 1877; English translation, *The microscope in Theory and Practice*, New York: Macmillan, 1892. KARL NÄGELI, *Mechanisch-physiologische Theorie der Abstammungslehre*, München und Leipzig: Oldenbourg 1884. (There is a section on p. 729—735 of the last-named book dealing with heat and molecular motion.) For a modern evaluation of NÄGELI'S research see J. S. WILKIE, "Nägeli's work on the fine structure of living matter," Annals of Science **17**, 27—62 (1961) and earlier papers.

[32] J. LOSCHMIDT, "Zur Grösse der Luftmolecüle," Sitzungsberichte der kaiserlichen Akademie der Wissenschaften in Wien, Klasse II, **52**, 395—413 (1865); similar estimates were published by LOTHAR MEYER, L. LORENZ, G. JOHNSTONE STONEY, and WILLIAM THOMSON during 1867—70.

the air. Such a calculation has a firm base, since, thanks to the mechanical theory of gases, one has a rather accurate idea of the weight and velocity of a gas molecule ... .

We assume that in 1 cc of gas at $0°$ at a pressure of 760 mm of mercury, there are 21 trillion molecules, so that the oxygen molecule has a weight of 7- and the nitrogen molecule a weight of 6 hundred-thousand-trillionth gram. The former moves with an average velocity of 461 meters/sec, the latter 492 meters/sec, so that the kinetic energy ($\frac{1}{2} m v^2$) is equally large for each.

The gas molecules behave like completely elastic spheres in their mutual interactions. Their collisions with the mote-particles may or may not be elastic likewise ... .[33]

The particle acquires after such a collision, if it is perfectly elastic, a velocity $\frac{2av}{a+b}$, where v is the velocity of the gas molecule, $a =$ mass of gas molecule, $b =$ mass of particle. Since the smallest fission fungus still weighs about 300 million times as much as a gas molecule, it acquires a velocity of only 0.002 mm/sec. If the collisions are not elastic, the effect is still smaller.

The motion which a sun-mote, and on the whole any particle found in the air, can acquire by the collisions of an individual gas molecule or a multitude of such molecules is therefore so extraordinarily small, and the number of simultaneous collisions against the particle from all sides is so extraordinarily large, that the particle behaves just as if it were completely at rest.

As for Brownian movement, WIENER and EXNER showed that its cause is to be sought in the liquid and is to be ascribed to the internal motions peculiar to the liquid state. But should this be understood in the sense that it is the collisions themselves of the liquid molecules moving in different directions, and not the molecular forces which cause the particles visible in the microscope to dance, then such an assumption would be no better founded than the analogous supposition for the dance of sun-motes.

It is not possible, said NÄGELI, to give an exact value of the molecular velocity in liquids, but it must be less than in gases, so that the same arguments apply even more strongly. NÄGELI concluded that the cause of particle motion must lie not in thermal molecular motions but in attractive and repulsive forces.

The British chemist WILLIAM RAMSAY, writing three years after NÄGELI but with no apparent knowledge of his paper, used similar arguments against the idea that Brownian movement could result from the impacts of individual molecules. The active particles have masses at least 125 million times as great as that of a water molecule, according to RAMSAY's estimate.

If molecules do not coalesce and move as a whole, then they would appear to have no possible power of giving motion to a mass so much larger than themselves. But that molecules have arrangement is probable, owing to the power which some liquids possess of rotating the plane of polarized light.

CLERK-MAXWELL supposed for some time that the attraction of two molecules varies inversely as the fifth power of the distance. If attraction at distance 2 is 1, attraction at distance 1 would be 64. Why do not all molecules therefore coalesce? Probably, because their own proper motion, of which heat represents the higher harmonics, causes them to fly apart again. The wavelength of that motion is not so minute, and although we possess no means of ascertaining the amplitude of such vibrations, still their rate is so prodigious as to give rise to an almost incredible impact.[34]

The last paragraph quoted reveals how little most scientists really understood of the kinetic theory of gases in the 1880's. For example, MAXWELL had assumed

---

[33] NÄGELI, *op. cit.* (note 30), my translation.

[34] WILLIAM RAMSAY, "On Brownian or pedetic motion," Proceedings of the Bristol Naturalists' Society **3**, 299—302 (1882).

that molecules *repel* each other with an inverse fifth power force, not attract.[35] On the other hand, RAMSAY was going far beyond any established results of kinetic theory when he suggested here, and again in 1892, that water molecules might move together "in complex groups of considerable mass, and of some stability" which could produce Brownian movement.[36]

The same idea was brought forward a few years later by the French physicist LÉON GOUY. He admitted that while one could not explain Brownian movement simply by invoking "the uncoordinated movements of the molecules, which one often regards as constituting thermal movement," "one may conceive that molecular movements in liquids are partly coordinated for spaces comparable to 1 micron. ... The existence of Brownian movement seems to show that in reality something similar to this takes place."[37] This might seem like circular reasoning, but GOUY was fortunate enough to be addressing physicists, in the language of physics, at a time when they were ready to pay attention, and thus GOUY got a certain amount of credit for "discovering" the cause of Brownian movement.[38] (Some of this he deserved for his experimental work, in which he showed that external influences such as strong magnetic fields have no effect on the movements, so that the cause must be sought in factors internal to the liquid.) Of more historical significance than his qualitative discussion of the cause of Brownian movement is his suggestion that it might offer an exception to the Second Law of Thermodynamics:

Whatever idea one may have as to the cause that produces [the movement], it is no less certain that work is expended on these particles, and one can conceive a mechanism by which a portion of this work might become available. Imagine, for example, that one of these solid particles is suspended by a thread of diameter very small compared to its own, from a rachet wheel; impulses in a certain direction make the wheel turn, and we can recover the work. This mechanism is clearly unrealisable, but there is no theoretical reason to prevent it from functioning. Work could be produced at the expense of the heat of the surrounding medium, in opposition to CARNOT's principle. It appears that one can then make precise the meaning of HELMHOLTZ's reservations about this principle, in the case of living tissues; this principle would then be exact only for the gross mechanisms that we know how to make, and it would cease to be applicable when the *receptor* organ has dimensions comparable to 1 micron.[39]

[35] JAMES CLERK MAXWELL, "On the dynamical theory of gases," Philosophical Transactions of the Royal Society of London," **157**, 49—88 (1867), reprinted in vol. 2 of my *Kinetic Theory* (see note 46, below).

[36] WILLIAM RAMSAY, "Pedetic motion in relation to colloidal solutions," Chemical News **65**, 90—91 (1892); Proceedings of the Chemical Society **8**, 17—19 (1894).

[37] See ref. 26; see also the reprint with notes by J. THIRION in Rev. Quest. Sci. [3] **15**, 251—258 (1909) which includes a shorter paper on the same subject by GOUY.

[38] J. PERRIN, "L'agitation moléculaire et le mouvement brownien," Comptes Rendus Acad. Sci. Paris **146**, 967—970 (1908); G. L. DE HAAS-LORENTZ, *Die Brownsche Bewegung und einige Verwandte Erscheinungen*, Braunschweig: F. Vieweg u. Sohn, 1913, p. 9.

[39] GOUY cites HELMHOLTZ, *Sur la thermodynamique des théorèmes chimiques*, (Academie de Berlin, 1882), Traduit par M. G. Chaperon dans le Journal de Physique, 1884. WILLIAM THOMSON, who first proposed the generalized "Dissipation of energy" version of the Second Law, suggested that organic life might be excepted from its domain of validity:

"Any restoration of mechanical energy, without more than an equivalent of dissipation, is impossible in inanimate material processes, and is probably never

GOUY'S remarks found a responsive reader in HENRI POINCARÉ, who pointed out their significance to his audience at the Congress of Arts and Science in St. Louis (1904):

The biologist, armed with his microscope, long ago noticed in his preparations disorderly movements of little particles in suspension: this is the Brownian movement; he first thought this was a vital phenomenon, but he soon saw that the inanimate bodies danced with no less ardor than the others; then he turned the matter over to the physicists. Unhappily, the physicists remained long uninterested in this question; the light is focused to illuminate the microscopic preparation, thought they; with light goes heat; hence inequalities of temperature and interior currents produce the movements in the liquid of which we speak.

M. GOUY, however, looked more closely, and he saw, or thought he saw, that this explanation is untenable, that the movements become more brisk as the particles are smaller, but that they are not influenced by the mode of illumination.

If, then, these movements never cease, or rather are reborn without ceasing, without borrowing anything from an external source of energy, what ought we to believe? To be sure, we should not renounce our belief in the conservation of energy, but we see under our eyes now motion transformed into heat by friction, now heat changed inversely into motion, and that without loss since the movement lasts forever. This is the contrary of the principle of CARNOT.

If this be so, to see the world return backward, we no longer have need of the infinitely subtle eye of MAXWELL's demon; our microscope suffices us. Bodies too large, those, for example, which are a tenth of a millimeter, are hit from all sides by moving atoms, but they do not budge, because these shocks are very numerous and the law of chance makes them compensate each other: but the smaller particles receive too few shocks for this compensation to take place with certainty and are incessantly knocked about.[40]

POINCARÉ's interest in Brownian movement presents us with a new historical problem: why didn't he go ahead and work out the quantitative theory of it himself? The only evidence I have found that throws any light on this question is contained in a thesis by L. BACHELIER, presented to the Faculté des Sciences de Paris in 1900. The thesis is dedicated to POINCARÉ, who was also one of the examiners, and is mainly concerned with the theory of speculation on the Paris stock market. With this particular application apparently foremost in his mind, BACHELIER developed a theory of stochastic processes similar to the modern theory of Brownian movement (the "Wiener process"). At the end of the thesis there is printed in one paragraph the "Seconde Thèse: Propositions données par la Faculté" which reads as follows:

Résistance d'une masse liquide indefinie pourvue de frottements intérieurs, regis par les formules de NAVIER, aux petits mouvements varies de translation d'une sphere solide, immergee dans cette masse et adherente à la couche fluide qui la touche.[41]

---

effected by means of organized matter, either endowed with vegetable life or subjected to the will of an animated creature" (Phil. Mag. **4**, 306 (1852)).

In 1862, he agreed with his brother JAMES THOMSON that plants and animals might possess a "vital principle" which could reverse the dissipation of energy; see JAMES THOMSON's *Collected Papers in Physics and Engineering* (ed. J. LARMOR) Cambridge University Press, 1912, p. lv.

[40] JULES HENRI POINCARÉ, "The principles of mathematical Physics," pp. 604—622 in *Congress of Arts and Science, Universal Exposition, St. Louis, 1904*, Volume I (Boston & New York: Houghton, Mifflin & Co., 1905). The quotation is from p. 610.

[41] L. BACHELIER, Thèses présentées à la Faculté des Sciences de Paris pour obtenir le Grade de Docteur ès Sciences Mathématiques (Paris: Gouthier-Villars, 1900). The Première Thèse, "Théorie de la Speculation," was also published in Ann. Sci. Ecole

But there is no hint in BACHELIER's thesis that the problem of stock-market speculation has anything to do with the problem of the resistance of a viscous fluid to the movement of a solid sphere.

4. Einstein's Theory of Brownian Movement

In his "Autobiographical Notes," written for the Schilpp collection *Albert Einstein: Philosopher-Scientist* (1949), EINSTEIN indicated the motivation for his work on the theory of Brownian movement and its relation to the state of physics at the beginning of this century:

Not acquainted with the earlier investigations of BOLTZMANN and GIBBS, which had appeared earlier and actually exhausted the subject, I developed the statistical mechanics and the molecular-kinetic theory of thermodynamics which was based on the former. My major aim in this was to find facts which would guarantee as much as possible the existence of atoms of definite finite size. In the midst of this I discovered that, according to atomistic theory, there would have to be a movement of suspended microscopic particles open to observation, without knowing that observations concerning the Brownian motion were already long familiar.[42]

Since there is some doubt about the accuracy of these statements, it should be noted that EINSTEIN himself recognized at the beginning of this essay that "Every reminiscence is colored by today's being what it is, and therefore by a deceptive point of view."

After summarizing his theory of Brownian movement, EINSTEIN continued his reminiscences with the following remarks:

The agreement of these considerations with experience together with PLANCK's determination of the true molecular size from the law of radiation (for high temperatures) convinced the skeptics, who were quite numerous at that time (OSTWALD, MACH) of the reality of atoms. The antipathy of these scholars towards atomic theory can indubitably be traced back to their positivistic philosophical attitude. This is an interesting example of the fact that even scholars of audacious spirit and fine instinct can be obstructed in the interpretation of facts by philosophical prejudices. The prejudice — which has by no means died out in the meantime — consists in the faith that facts by themselves can and should yield scientific knowledge without free conceptual construction. Such a misconception is possible only because one does not easily become aware of the free choice of such concepts, which, through verification and long usage, appear to be immediately connected with the empirical material.

EINSTEIN then went on to discuss the application of Brownian motion theory to radiation, a topic we will come to in a later paper in this series.

Not only did EINSTEIN's theory provide a decisive breakthrough in the understanding of the phenomenon of Brownian motion; it also, in the opinion of MAX BORN, did "more than any other work to convince physicists of the reality of atoms and molecules, of the kinetic theory of heat, and of the fundamental part

Normale Supérieure [3] **17**, 21—86 (1900). For a recent discussion of the relation between BACHELIER's work and Brownian movement, see M. F. M. OSBORNE, "Reply to 'comments on Brownian motion in the stock market,'" Operations Research **7**, 807—811 (1959).

[42] ALBERT EINSTEIN, "Autobiographical notes," pp. 1—95 in *Albert Einstein: Philosopher-Scientist* (ed. P. A. SCHILPP) Library of Living Philosophers, Tudor Pub. Co., 1949, reprinted by Harper & Brothers, New York, 1959. The quoted passage is from p. 46—49 (German with English translation on facing pages).

of probability in the natural laws."[43] It will therefore be worthwhile to digress somewhat from the history of Brownian movement itself to recall some of the background of late nineteenth-century theoretical physics underlying EINSTEIN's theory.

Kinetic theories of matter, identifying heat qualitatively or quantitatively with molecular motion, had been frequently proposed since the seventeenth century.[44] At the time of ROBERT BROWN's work, vague qualitative notions about thermal molecular motion were common enough, but HERAPATH's enthusiastic though inaccurate attempts to advance to something like the modern kinetic theory of gases had been rebuffed by the Royal Society in 1821.[45] The kinetic theory was revived after the general acceptance of the law of conservation of energy in the middle of the nineteenth century; kinetic theory was then thought of as the molecular extension of the "mechanical theory of heat" or thermodynamics. The theory was developed extensively by CLAUSIUS, MAXWELL, and BOLTZMANN.[46]

EINSTEIN, contrary to his statement at the beginning of the above quotation, was familiar with BOLTZMANN's treatise, Vorlesungen über Gastheorie;[47] he cites it on the fourth page of his own first paper on kinetic theory in 1902.[48] Two aspects of BOLTZMANN's work had a special attraction for EINSTEIN: first, the formulation of "statistical mechanics" (as we now call it, following GIBBS) by means of generalized Lagrange-Hamilton dynamics; and second, the emphasis on statistical fluctuation phenomena. EINSTEIN's purpose, as he says, was not only to connect thermodynamics with general mechanics, but also to deal with phenomena explicitly involving the atomic structure of matter. Moreover, he was very much interested in putting PLANCK's quantum theory into the language of statistical mechanics, and the papers on Brownian movement contain several references to the radiation distribution law. Perhaps the best way to characterize EINSTEIN's work on kinetic theory is to say that he wanted to remove the restriction to *gases*, which seemed to limit the applicability of most of the work of CLAUSIUS, MAXWELL, and BOLTZMANN, and develop a theory sufficiently general to deal with liquids, solids, and radiation.

---

[43] MAX BORN, "Einstein's statistical theories," pp. 161—177 in SCHLIPP, *op. cit.*; quotation from p. 166.

[44] See S. G. BRUSH, *Kinetic Theory, Volume 1, The Nature of Gases and Heat* (New York: Pergamon Press, 1965) and works cited therein; G. R. TALBOT & A. J. PACEY, "Some early kinetic theories of gases: Herapath and his predecessors," British Journal of History of Science **3**, 133—149 (1966).

[45] See S. G. BRUSH, "The Royal Society's first rejection of the kinetic theory of gases (1821), John Herapath versus Humphry Davy" Notes and Records of the Royal Society of London, **18**, 161—180 (1963).

[46] S. G. BRUSH, *Kinetic Theory, Volume 1* (see note 44) and *Volume 2, Irreversible Processes* (New York: Pergamon Press, 1966). "The Development of the Kinetic Theory of Gases. III. Clausius" and "... IV. Maxwell" Annals of Science, **14**, 185—196, 243—255 (1958); "Development of the kinetic theory of gases. V. Equation of State" and "... VI. Viscosity" American Journal of Physics **29**, 593—605 (1961), **30**, 269—281 (1962).

[47] L. BOLTZMANN, *Vorlesungen über Gastheorie* (Leipzig: J. A. Barth, I. Teil, 1896, II. Teil, 1898); *Lectures on Gas Theory* (English translation with introduction, notes, and bibliography by S. G. BRUSH) (Berkeley: University of California Press, 1964).

[48] A. EINSTEIN, "Kinetische Theorie des Wärmegleichgewichts und des zweiten Hauptsatzes der Thermodynamik, "Annalen der Physik (4) **9**, 417 (1902).

EINSTEIN was certainly aware of the attacks on kinetic theory which had been made by MACH, OSTWALD, and their followers in the 1890's. In its simplest form, the issue was: why should we base our theories of matter on atomic hypotheses, when a phenomenological description such as thermodynamics contains all the necessary information about a physical system and also avoids the various paradoxes and inconsistencies that plague atomic theory? EINSTEIN spoke directly to this point in his first paper on Brownian movement[49] when he contrasted the predictions of classical thermodynamics and of the molecular-kinetic theory of heat. According to the former, small suspended particles in a liquid will not exert any force on a semi-permeable membrane placed in the liquid, although it is known that molecules of a dissolved non-electrolyte will exert an osmotic pressure in such a situation. On the other hand, the molecular-kinetic theory maintains that "a dissolved molecule is differentiated from a suspended body *solely* by its dimensions" and a certain number of suspended particles must produce the same osmotic pressure as the same number of molecules in solution. Thus, in EINSTEIN's view, classical thermodynamics introduces an artificial distinction between suspended particles and dissolved molecules, based on the fact that the latter are too small for us to see and so we don't recognize them as being the same kind of entity.

It is obvious that any theory of Brownian movement must deal with the motion of particles in liquids. Unfortunately, at the time EINSTEIN first began to work on this problem, there was no adequate quantitative theory of liquids developed from the molecular viewpoint, comparable to the kinetic theory of gases. In fact, the kinetic theory of liquids is such a difficult subject that even today it has not advanced as far as gas theory had gone by 1905. On the other hand, there was in existence, late in the nineteenth century, a sizable body of theoretical research on solids and liquids treated as continua. Much of this research had originally been done for the purpose of developing theories of propagation of light through the ether, but the substantive mathematical results turned out to be more relevant to the properties of "real" liquids and solids. In particular, there was a well-established formula — the "Stokes formula" — for the force resisting the motion of a sphere through a liquid.

The origin of the Stokes formula itself is worth noting. In 1845, STOKES had derived his general equations for the flow of fluids, taking account of "internal friction" (viscosity).[50] The original application of these equations was to the

---

[49] A. EINSTEIN, "Ueber die von der molekular-kinetischen Theorie der Wärme geforderte Bewegung von in ruhenden Flüssigkeiten suspendierten Teilchen," Annalen der Physik [4] **17**, 549—560 (1905) (see also publications cited in notes 62 and 63 below).

[50] G. G. STOKES, "On the theories of the internal friction of fluids in motion, and of the equilibrium and motion of elastic solids," [1845] Trans. Cambridge Phil. Soc. **8**, 287—319 (1849), reprinted in Stokes' *Mathematical and Physical Papers* vol. I, pp. 75—129 (Cambridge University Press, 1880). See also his review of earlier work of CHALLIS, GREEN, AIRY, BARRÉ DE SAINT-VENANT, NAVIER, POISSON, and others, in: "Report on recent researches in Hydrodynamics," Report of the British Association **16**, 1—20 (1846).

It should be noted that in the work discussed here STOKES used an equation less general than what is now called the "Navier-Stokes equation," in that he assumed that the terms involving the dependence of pressure on the rate of change of density could be dropped. This is tantamount to ignoring what is now called the "bulk viscosity."

computation of air-resistance corrections to results obtained with swinging pendulums. This subject was discussed in another paper in 1850; he remarked that

> On account of the inconvenience and expense attending experiments in a vacuum
> apparatus, the observations are usually made in air, and then it becomes necessary
> to apply a small correction, in order to reduce the observed result to what would
> have been observed had the pendulum been swung in a vacuum.[51]

STOKES quoted some experiments of Colonel SABINE (1829), which according
to SABINE might indicate "an inherent property in the elastic fluids, analogous
to that of viscidity in liquids, of resistance to the motion of bodies passing through
them ...."[52] However, STOKES argued that previous theories of the motion of
solids through viscous fluids had erroneously assumed that the fluid simply glides
past the surface of the solid, ignoring the tangential action between the surface
and the fluid. As evidence for such tangential action — i.e., for the assumption
that the layer of fluid in contact with the surface does not move past it — STOKES
quoted an experiment of Sir JAMES SOUTH:

> ... on attaching a piece of gold leaf to the bottom of a pendulum, so as to stick
> out in a direction perpendicular to the surface, and then setting the pendulum in
> motion, Sir JAMES SOUTH found that the gold leaf retained its perpendicular position
> just as if the pendulum had been at rest; and it was not until the gold leaf carried
> by the pendulum had been removed to some distance from the surface, that it began
> to lag behind. This experiment shews clearly the existence of a tangential action
> between the pendulum and the air, and between one layer of air and another.[53]

STOKES was able to solve his hydrodynamic equations in the special case of
a sphere moving uniformly in a fluid; he found that, provided terms involving
the square of the velocity may be neglected, a force K imparts to a sphere of
radius P a velocity $K/6\pi k P$, where k is the viscosity of the fluid. (We are now
following EINSTEIN's notation.) The surprising feature of this result to STOKES
was that the resistance to an object moving with a given speed is proportional
to its radius rather than to its surface area. One consequence of this fact is that
"the resistance to a minute globule of water falling through the air with its
terminal velocity depends almost wholly on the internal friction of air .... The
terminal velocity thus obtained is so small in the case of small globules such as
those of which we may conceive a cloud to be composed, that the apparent
suspension of the clouds does not seem to present any difficulty."[54]

Even before his long memoir containing these results had appeared, STOKES
had already published a shorter paper "On the Constitution of the Luminiferous
ether" in which he assumed that the ether close to the surface of the earth is
at rest relative to it, and then applied the hydrodynamical theory which he had
developed for this case.[55] Throughout the nineteenth century, this hypothesis

---

[51] G. G. STOKES, "On the effect of the internal friction of fluids on the motion
of pendulums," [1850] Trans. Cambridge Phil. Soc. **9**, [8]—[106] (1856); *Papers*,
vol. III, pp. 1—141 (Cambridge University Press, 1901); quotation from *Papers*, III, 1.

[52] *Papers*, III, 3, quoted from p. 232 of SABINE's paper "On the reduction to a
vacuum of the vibrations of an invariable pendulum," Phil. Trans. **119**, 207—239
(1829).

[53] *Papers*, III, 7. The *Dictionary of National Biography* has an interesting article
on Sir JAMES SOUTH.

[54] *Papers*, III, 10.

[55] G. G. STOKES, Phil. Mag. [3] **32**, 343—349 (1848); *Papers* II, 8—13 (1883).

was elaborated and criticized by theoretical physicists in discussions of the ether.[56] It is therefore not surprising that EINSTEIN was familiar with the Stokes formula; although the only reference he gives is to KIRCHHOFF's *Vorlesungen über Mechanik*,[57] he may have been led to that source by his reading of LORENTZ's papers.[58] The formula was also used indirectly by J. J. THOMSON in his determination of the electrical charge on ions produced by Röntgen rays, in 1898.[59]

The other cornerstone of EINSTEIN's theory was J. H. VAN'T HOFF's theory of osmotic pressure in solutions.[60] But this theory applied (or was thought to apply) only to dissolved molecules which are about the same size as the molecules of the liquid. Nevertheless, VAN'T HOFF's formula for osmotic pressure was one of the very few quantitative results known to be valid for liquids.

Here were two theories dealing with particles in liquids: the hydrodynamic theory of STOKES, based on the assumption that the liquid is a continuous medium which sticks to a solid surface moving through it with not too high a speed; and the osmotic theory of VAN'T HOFF, based on the assumption that the particle is itself a molecule mixed in with a molecular liquid. To put it another way: the equations of hydrodynamics are valid in situations where solid boundaries or suspended particles are acted on by a steady force originating in the liquid, and turbulence or random molecular motion in the liquid has no significant effect on the motion of the particle. (OSBORNE REYNOLDS and others in the 1880's had already shown the need to avoid this restriction, and to take account of surface effects.[61]) The osmotic theory, on the other hand, is valid in situations where *all* the motion of the dissolved particle must be attributed to random molecular motion. Thus the two theories seemed to have mutually exclusive regions of validity; it required the reckless genius of an EINSTEIN to ignore that difficulty and boldly combine them. How much that move upset EINSTEIN's contemporaries is well illustrated by the voluminous appendix, filled with justifications and

---

[56] E. T. WHITTAKER, *A History of the Theories of Aether and Electricity* (London: Nelson & Sons, 1951).

[57] G. R. KIRCHHOFF, *Vorlesungen über Mechanik* [*Vorlesungen über Mathematische Physik*, erster Band], Leipzig: B. G. Teubner, vierte Auflage, 1897; Sechsundzwanzigste Vorlesung, p. 380.

[58] See for example H. A. LORENTZ, "Ein allgemeiner Satz, die Bewegung einer reibenden Flüssigkeit betreffend, nebst einigen Anwendungen desselben," in *Abhandlungen über theoretische Physik* (Leipzig und Berlin: B. G. Teubner, 1907) p. 23—42, based on a paper originally in Zittingsverlag Akad. v. Wet., Amsterdam, **5**, 168 (1896). LORENTZ cited KIRCHHOFF's *Mechanik* for the derivation of the Stokes formula.

[59] J. J. THOMSON, "On the charge of electricity carried by the ions produced by Röntgen rays," Phil. Mag. [5] **46**, 528—545 (1898); this work is mentioned in LORENTZ's paper cited in note 58.

[60] J. H. VAN'T HOFF, "Lois de l'équilibre chimique dans l'état dilué gazeux ou dissous," Kongliga Svenska Vetenskaps — Academiens Handlingar, Stockholm, xxi, no. 17, p. 1—58 (1884—5); for further details see J. R. PARTINGTON, *A History of Chemistry*, Vol. 4, p. 654 ff. (London: Macmillan, 1964).

[61] O. REYNOLDS, "An experimental investigation of the circumstances which determine whether the motion of water shall be direct or sinuous, and of the law of resistance in parallel channels," Phil. Trans. **174**, 935—987 (1884). See H. ROUSE & S. INCE, *History of Hydraulics* (Iowa Institute of Hydraulic Research, 1957, reprinted by Dover Publications, New York, 1963).

apologies, which was published with the reprint of EINSTEIN's papers on Brownian movement in OSTWALD's *Klassiker*.[62]

Since the details of EINSTEIN's theory are easily accessible in a paperback reprint of the English translation of the *Klassiker* edition,[63] I will give here only a summary. EINSTEIN argued that the suspended particles visible in the microscope should exert an osmotic pressure against a semi-permeable membrane, just as the molecules of a dissolved substance do. In fact, the pressure is just that which would be exerted by an ideal gas of the same number of point-atoms in the same space, provided that one can ignore interactions between the particles. To justify this assertion, EINSTEIN carried out a short calculation with the help of his own statistical mechanical formulation. This calculation probably obscured the argument for most readers at the time, since the use of what we now call the "partition function" or "phase integral" had not yet become familiar. What the formalism does, or should do, in such an investigation, is to free one from the need for calculating pressures and thermodynamic functions by considering collisions of particles with the wall of the container, and computing the time for a particle with a certain speed to go back and forth between two opposite walls. Clearly the microscopic particles do not bounce back and forth between parallel walls, moving in straight lines in the way that one thinks of gas molecules doing in elementary kinetic-theory derivations. The point is that these particles are nevertheless in statistical thermal equilibrium with the molecules of the fluid, and this is what determines their average pressure and energy.

EINSTEIN could therefore use the ideal-gas equation for the osmotic pressure:

$$p = \frac{RT}{V*} \frac{n}{N} = \frac{RT}{N} v \qquad (4.1)$$

where $T =$ absolute (KELVIN) temperature, $n =$ number of suspended particles in a volume $V*$ (which is partitioned out of a larger volume V), $N =$ AVOGADRO's number, and the concentration is $v = n/V$.

The basis of EINSTEIN's description of Brownian movement is the notion that the suspended particles are "diffusing" through the liquid, in such a way that dynamical equilibrium is always maintained between the osmotic force originating in a concentration gradient, and the viscous force which must retard the motion of a particle according to hydrodynamics. The osmotic force tends to push the particles from regions of high concentration to regions of low concentration; the magnitude of the force is the same as the pressure-gradient in the direction of motion. One way to develop the argument (not quite the same as EINSTEIN's, for reasons noted below) would be to write down the equation relating force to pressure-gradient,

$$Kv = \frac{\partial p}{\partial x} \qquad (4.2)$$

[62] ALBERT EINSTEIN, *Untersuchungen über die Theorie der Browenschen Bewegungen*, hrsg. R. FÜRTH (Leipzig: Akademische Verlagsgesellschaft 1922).

[63] ALBERT EINSTEIN, *Investigations on the theory of the Brownian Movement* (translated by A. D. COWPER), Methuen & Co., 1926, reprinted by Dover Publications, Inc., 1956.

and then substitute the value for pressure given by van't Hoff's ideal gas law (4.1) on the right hand side, thereby arriving at the equation

$$K\nu = \frac{RT}{N}\,\frac{\partial \nu}{\partial x}\,.\tag{4.3}$$

We could then use the Stokes formula for the velocity of a particle moving through a viscous medium,

$$v = \frac{K}{6\pi k P}\tag{4.4}$$

to eliminate the force K from the left-hand-side of equation (4.3), and obtain

$$6\pi k P v\nu = \frac{RT}{N}\,\frac{\partial \nu}{\partial x}\,.\tag{4.5}$$

We would then have a relation between several quantities that presumably can be measured: the radius of the particle (P), its velocity (v), the concentration of particles (ν), the concentration gradient $\partial \nu/\partial x$, Avogadro's number (N), the gas constant (R), and the temperature (T).

However, Einstein did not propose (4.5) as the basic prediction of the theory to be tested by experiment. In fact, he did not even follow the sequence indicated above, deriving (4.3) from (4.2), although as Fürth suggested in his notes[62] this would have been the most direct route. Instead, Einstein gave a derivation of (4.3) which superficially seems to be independent of any formula for the pressure such as (4.1) — actually, both (4.3) and (4.1) are consequences of the same statistical mechanical formulation — and proceeded instead from expressions for energy and entropy. Again, the effect was probably to obscure the argument for readers who did not have Einstein's own familiarity with statistical mechanics.

Another reason why Einstein did not propose the equation (4.5), even though it is perfectly consistent with his development up to this point, is that he wanted to use a mathematical description of "diffusion" which avoided ascribing a well-defined instantaneous velocity to a particle. Perhaps this is why he did not even introduce a symbol for velocity as we have done in equation (4.4), but immediately translated that equation into an expression for the number of particles passing unit area per unit of time, $\nu K/6\pi k P$. He could then equate this expression to the conventional formula for the rate of diffusion, namely, $-D(\partial \nu/\partial x)$, where D is the "coefficient of diffusion." The velocity of an individual particle is never mentioned in the paper after this paragraph.

Einstein's first two basic equations are then the relation between osmotic force and concentration gradient,

$$K\nu = \frac{RT}{N}\,\frac{\partial \nu}{\partial x}\tag{4.3}$$

and the relation between diffusion rate and the flow of particles through a viscous medium resulting from this same force,

$$\frac{K\nu}{6\pi k P} = D\,\frac{\partial \nu}{\partial x}\,.\tag{4.6}$$

It is perhaps a further sign of Einstein's preference for abstract formulations that he writes both of these equations with two terms on the left hand side equated to zero on the right hand side.

By combining (4.3) and (4.6) EINSTEIN obtained the following expression for the diffusion constant:

$$D = \frac{RT}{N} \frac{1}{6\pi k P}.$$ (4.7)

EINSTEIN now turned to "a closer consideration of the irregular movements which arise from thermal molecular movement,"[64] or what we would now call the mathematical description of a certain type of stochastic process. Hardly any of the mathematics is original with EINSTEIN; what is new is the attempt to describe particle motions that are in principle still deterministic (on the molecular level) by a certain mode of probabilistic analysis.

Let us assume, EINSTEIN said, that the motion of each particle is independent of the others, and moreover that "the movements of one and the same particle after different intervals of time [are] mutually independent processes, so long as we think of these intervals of time as being chosen not too small." To be more specific, assume that the movements of a particle in two consecutive intervals of time τ are independent. Let the number of particles which experience, in the time interval τ, a displacement which lies between Δ and $\Delta + d\Delta$, be expressed in the form

$$dn = n\,\phi(\Delta)\,d\Delta.$$ (4.8)

(That is, $\phi(\Delta)$ is *defined* by equation (4.8), in the traditional indirect manner of theoretical physics.) Further, let the concentration ν be regarded as a function of space and time, $f(x, t)$. The values of this function after the time-interval τ has elapsed can be computed in terms of the distribution function for displacements, $\phi(\Delta)$; that is, $f(x, t+\tau)$ will depend on the values of $f(x+\Delta, t)$ for all possible values of Δ, weighted by $\phi(\Delta)$. If τ is very small (note that this contradicts the original assumption about τ), and if only small values of Δ need be taken into account, then one can derive a differential equation which relates the time-variations of f to its space-variations:

$$\frac{\partial f}{\partial t} = D\,\frac{\partial^2 f}{\partial x^2}$$ (4.9)

where

$$D = \frac{1}{\tau} \int_{-\infty}^{+\infty} \frac{\Delta^2}{2}\,\phi(\Delta)\,d\Delta.$$ (4.10)

(This formula comes from an expansion of $f(x+\Delta, t)$ in powers of Δ.)

In solving equation (4.9), we may as well (as EINSTEIN remarks) take advantage of our assumption that the particles move independently of each other, and therefore interpret x not as the actual space coordinate of a particle but rather as the distance it has moved from its starting place at $t=0$. The solution of (4.9) is then:

$$f(x, t) = \frac{n}{\sqrt{4\pi D}}\,\frac{e^{-x^2/4Dt}}{\sqrt{t}}.$$ (4.11)

---

[64] *Ibid.*, p. 12.

EINSTEIN could now obtain immediately the root-mean-square displacement of
a particle in the x-direction,

$$\lambda_x = \sqrt{\overline{x^2}} = \sqrt{2Dt}. \qquad (4.12)$$

Thus *the mean displacement is proportional to the square root of the time.*

All that is left is to substitute into (4.12) the expression previously found for
the diffusion constant [equation (4.7)]:

$$\lambda_x = \sqrt{t} \sqrt{\frac{RT}{N} \frac{1}{3\pi k P}}. \qquad (4.13)$$

This is EINSTEIN's final result in this paper, and he noted that it can be used
either to predict the mean displacement if one assumes values for N, R, T, K,
and P, or else to calculate N if everything else is known.

In this first paper, EINSTEIN did not emphasize very strongly the significance
of his result that λ_x is proportional to the square root of the time, and in fact it
is quite probable that most early readers of the paper gave up in bewilderment
before they got to this result. EINSTEIN wrote two more papers, published in the
Zeitschrift für Elektrochemie in 1907 and 1908, calling his results to the attention
of experimentalists and attempting to explain them more simply.[65] In the first
of these papers he pointed out that according to the molecular theory of heat
(*i.e.*, the Waterston-Maxwell equipartition theorem) the mean square velocity of
a suspended particle should be determined by the equation

$$\frac{m}{2} \overline{v^2} = \frac{3}{2} \frac{RT}{N}. \qquad (4.14)$$

For the colloidal platinum solutions investigated by SVEDBERG,[66] in which the
mass of the particles is about 2.5×10^{-15} g, this equation indicates a root-mean-
square velocity of 8.6 cm/sec. However, EINSTEIN showed that there is no possibility of observing this velocity, because of the very rapid viscous damping,
which one can calculate from the Stokes formula. The velocity of such a particle
would drop to $\frac{1}{10}$ of its initial value in about 3.3×10^{-7} sec. "But, at the same
time," said EINSTEIN, "we must assume that the particle gets new impulses to
movement during this time by some process that is the inverse of viscosity, so
that it retains a velocity which on an average is equal to $\sqrt{\overline{v^2}}$. But since we must
imagine that the direction and magnitude of these impulses are (approximately)
independent of the original direction of motion and velocity of the particle, we
must conclude that the velocity and direction of motion of the particle will be
already very greatly altered in the extraordinarily short time θ [$=3.3 \times 10^{-7}$ sec],
and, indeed, in a totally irregular manner. It is therefore impossible — at least
for ultramicroscopic particles — to ascertain $\sqrt{\overline{v^2}}$ by observation."[67]

According to EINSTEIN's theory, the mean velocity in an interval τ will be
inversely proportional to $\sqrt{\tau}$; that is, it increases without limit as the time interval
becomes smaller. Hence any attempt to measure the "instantaneous" velocity

[65] *Ibid.*, p. 63—67, 68—85. (The translation is marred by several misprints which
have not been corrected in the reprint.)

[66] T. SVEDBERG, "Über die Eigenbewegung der Teilchen in kolloidalen Lösungen,"
Zeitschrift für Elektrochemie **12**, 853—860, 909—910 (1906).

[67] A. EINSTEIN, *op. cit.* (note 63) p. 66.

of particles in Brownian movement will give erratic and meaningless results. It is for just this reason that all the efforts of experimentalists, who knew nothing more of kinetic theory than the equipartition theorem, had failed to lead to any definite conclusion about the average speeds of suspended particles. They were simply measuring the wrong thing until EINSTEIN pointed out that only the ratio of mean square displacement to time could be expected to have any theoretical significance. One can hardly find a better example in the history of science of the complete failure of experiment and observation, unguided (until 1905) by theory, to unearth the simple laws governing a phenomenon.

The peculiar nature of random motion governed by a diffusion equation had reared its head in physics once before. In 1854, WILLIAM THOMSON (later Lord KELVIN) had applied the diffusion equation (*i.e.*, FOURIER'S equation for heat conduction) in his theoretical studies of the motion of electricity in telegraph lines.[68] After going through almost exactly the same mathematical analysis that EINSTEIN was to make 50 years later, THOMSON wrote:

> We may infer that the retardations of signals are proportional to the squares of the distances, and not to the distances simply; and hence different observers, believing they have found a "velocity of electric propagation," may well have obtained widely discrepant results; and the apparent velocity would, *caeteris paribus*, be the less, the greater the length of wire used in the observation.[69]

In an article on the "Velocity of electricity" published in *Nichol's Cyclopedia* in 1856, THOMSON quoted nine different results for the "velocity of electricity" ranging from 1,430 to 288,000 miles per second, and pointed out that the diversity of values measured under different conditions can probably be attributed to the fact that the actual time required for an electric impulse to get from one place to another is proportional to the square of the distance.[70] The validity of THOMSON's "law of squares" was of considerable economic importance at the time because the technical problems involved in laying the Atlantic Cable were just then being thrashed out. THOMSON had to defend his theoretical prediction against experiments that appeared to contradict it; those engineers who were enthusiastic about pushing ahead with the cable did not like the law of squares. THOMSON argued that the transmission of messages would be very slow at large distances unless the cable is made very thick, and said:

> Capitalists ought to require a very "matter-of-fact" proof of the attainability of a sufficient rapidity in the communication of actual messages, by whatever cable may be proposed, before sinking so large an amount of property in the Atlantic, as would be involved in any cable of ordinary or of extraordinary great lateral dimensions to form an electric communication between Britain and America.[71]

Apparently the scientists who attempted to measure the velocity of particles in Brownian movement later in the nineteenth century had not followed the dispute about THOMSON's law of squares in the electric telegraph problem, and

---

[68] WILLIAM THOMSON, "On the theory of the electric telegraph" (extracts from letters to Prof. STOKES, October and November 1854), Proceedings of the Royal Society of London, 7, 382—399 (1855), reprinted in his *Mathematical and Physical Papers*, Vol. II, p. 61—76 (Cambridge University Press, 1884).

[69] THOMSON, *Mathematical and Physical Papers*, p. 65.

[70] *Ibid.*, p. 131—137.

[71] *Ibid.*, p. 92—102 (reprinted from Athenaeum, 1856).

they obtained a similar collection of wildly varying results, none of them in agreement with the equipartition theorem.[72]

Einstein was also well aware that his formula for the mean square displacement could not be applicable to very short time intervals. The formula (4.12) seems to imply an infinite instantaneous velocity. But, as Einstein noted in a second paper written in 1905, "we have implicitly assumed in our development that the events during the time t are to be looked upon as phenomena independent of the events in the time immediately preceding. But this assumption becomes harder to justify the smaller the time t is chosen."[73] One can ignore this objection and retain the assumption that events at any time are completely independent of those at any other time; one then has what might be called a mathematical idealization of Brownian movement, sometimes called a "Wiener process".[74] The most common method for modifying the assumptions at short times to give a physically reasonable result leads to the "Ornstein-Uhlenbeck process."[75] Both processes are still of considerable interest in mathematical physics and other applications, and there is really no need to choose between them except for a particular purpose.

Another important result obtained by Einstein in his second paper is a formula for the probability distribution of the vertical distance of a particle of density ϱ and volume v from the bottom of a container:

$$dW = \text{const}\, e^{-(N/RT)\,v\,(\varrho-\varrho_0)\,g x}\,dx. \tag{4.15}$$

Since this formula follows directly from the Maxwell-Boltzmann distribution law,[76] we will not discuss its derivation. The formula was used by Perrin in some of his experimental work,[77] to determine N, and thus played an important role in establishing the "real existence" of the atom (see section 6).

5. Smoluchowski's Theory of Brownian Movement

Shortly after the publication of Einstein's first paper on Brownian movement in 1905, there appeared a paper by the Polish physicist Marian von Smolu-

---

[72] See for example the table of results of Regnauld, Wiener, Ramsay, R. Exner, and Zsigmondy, in T. Svedberg, "The Brownian Movements," Ion 1, 373—402 (1909).

[73] Einstein, *Investigations on the theory of the Brownian Movement*, p. 34.

[74] Norbert Wiener, "The average of an analytic functional," Proc. Nat. Acad. Sci. USA 7, 253—260 (1921); "The average of an analytic functional and the Brownian motion," *ibid.* 294—298, and many other papers. The modern theory is summarized by E. B. Dynkin, *Markov Processes* (translated from Russian) (Berlin: Springer-Verlag, 1965).

[75] E. G. Uhlenbeck & L. S. Ornstein, "On the theory of the Brownian motion," Phys. Rev. [2] 36, 823—841 (1930).

[76] L. Boltzmann, "Studien über das Gleichgewicht der lebendigen Kraft zwischen bewegten materiellen Punkten," Sitzungsberichte der kaiserlichen Akademie der Wissenschaften in Wien, Klasse II, 58, 517—560 (1868); J. C. Maxwell, "On the final state of a system of molecules in motion subject to forces of any kind," Nature 8, 537—538 (1873).

[77] J. Perrin, "L'agitation moléculaire et le mouvement brownien," Comptes Rendus Acad. Sci. (Paris) 146, 967—970 (1908), reprinted in *Oeuvres Scientifiques de Jean Perrin*, Paris: Centre National de la Recherche Scientifique, 1950.

CHOWSKI on the same subject.[78] SMOLUCHOWSKI began by citing EINSTEIN'S work, and said that EINSTEIN's results "agree completely with those that I obtained several years ago by following a completely different line of thought." However, he thought that his own method is "more direct, simpler, and more convincing than EINSTEIN's." This judgment is of course a matter of taste, but it is probably true that most physicists and chemists at that time found it easier to follow SMOLUCHOWSKI's arguments, based on combinatorics and the mean-free-path approximation of kinetic theory, than EINSTEIN's, based on abstract statistical mechanics and the diffusion equation. SMOLUCHOWSKI also made a much greater effort to review the experimental results that had some bearing on the theory, as well as to justify his theoretical assumptions.

According to SMOLUCHOWSKI, observations show that the motion is more rapid, if the diameter of the particles is smaller, though few reliable measurements have been made. Different observers have reported contradictory results concerning the influence of the medium, but it is clear that the motion is liveliest in fluids of the smallest viscosity. The motion is almost entirely independent of external influences. SMOLUCHOWSKI then cited and criticized several theoretical explanations of Brownian movement that had been proposed:

... there follows immediately from [the independence of external influences] the untenability of any theories based on an external energy source; and especially any supposition that one has here convection streams originating from temperature inequalities. The inadmissibility of these latter explanations follows moreover from simple considerations of another kind. The motions would have to stop completely in water at a temperature of 4°, whereas actually they continue down to the freezing point with scarcely diminished strength (MEADE BACHE). The reduction of the thickness of the liquid layer to a small fraction of a millimeter by placing it on a cover glass would be expected to diminish the mobility greatly, whereas there is no evidence of this effect. Calculation shows that in this case a temperature drop of the order of magnitude of 100,000° in 1 cm would be necessary to create a convection stream of the observed velocity. It is known that such streams occur in containers of larger dimensions, but these collective motions of a larger number of particles are completely different from the irregular vibrating Brownian motions.

(Here SMOLUCHOWSKI attacked one of the most common misconceptions about the molecular-kinetic explanation of Brownian movement: the idea that a large number of molecules near the suspended particle must move in unison in order to produce the observed motion. SMOLUCHOWSKI showed, as we will see, that such cooperative motions need not be postulated arbitrarily, but rather that equivalent *fluctuations* must be expected as a natural consequence of the randomness of molecular motions.)

It may be noted [he continued] that the maximum temperature difference produced in the neighborhood of a spherical completely black particle exposed to direct sunlight is $ca/k = (1/300)°$ (assuming: radiation intensity $c = 1/30$, radius $a = 10^{-4}$ cm, thermal conductivity $k = 10^{-3}$ (water)). This is in agreement with the previous remarks

---

[78] M. R. VON SMOLAN SMOLUCHOWSKI, "Zarys kinetycznej teorji ruchów Browna i roztworów metnych," Rozprawy i Sprawozdania z Posiedzeń Wydziału Matematyczno-Przyrodniczego Akademii Umiejetności (Krakow), A **46**, 257—282 (1906); German translation in Annalen der Physik [4], **21**, 756—780 (1906), reprinted in *Abhandlungen über die Brownsche Bewegung und verwandte Erscheinungen*, Leipzig: Akademische Verlagsgesellschaft m.b.H., 1923. The quotations in the text are my translation from the German version.

about the impossibility of REGNAULD's explanation on the basis of the origin of
convection streams in the neighborhood of a particle as a result of absorption of
radiation at its surface.

The independence of the Brownian phenomenon of the intensity of illumination
also contradicts KOLÁČEK's and QUINCKE's theories, which find in it an analogy to
radiometer motions, or to the various phenomena of periodic capillary motions investigated by QUINCKE, respectively. It seems very difficult to understand how a
continuous radiation can give rise to the periodic expansion of warmer over colder
fluid layers at the surface of each particle, assumed by QUINCKE, and how there can
be any connection between the extraordinary phenomenon of periodic capillary motion
that occurs in certain cases (oil in soap solution, alcohol in salt solution, etc.) and the
very general phenomenon of Brownian motion which is independent of the substance.
It is indeed very probably that a sufficiently strong radiation can give rise to motions,
but these would be completely different from Brownian motions.

Similarly, SMOLUCHOWSKI rejected explanations based on intermolecular repulsive forces, impurities, and so forth.

Having gone through the ritual of demolishing all other explanations of the
phenomenon (a procedure which EINSTEIN did not consider necessary), SMOLU-
CHOWSKI turned to the kinetic theory (which he had clearly already decided to
use). At this point it should be noted that SMOLUCHOWSKI had already done
some substantial research in the kinetic theory of rarified gases — in particular,
the problem of the temperature-discontinuity at a solid surface — and had just
published a paper extending the work of JEANS on the "persistence of velocities"
in collisions of gas molecules.[79] JEANS had been trying to improve on the elementary
mean-free-path calculations of kinetic theory, in which it was usually assumed
that after a collision, a molecule "forgets" its previous motion and simply assumes
the average velocity and direction of motion characteristic of the place in the gas
where the collision occurred. That assumption had been the basis of MAXWELL's
original calculation of the viscosity of a gas in 1859, though MAXWELL later
abandoned it when he developed his more general theory based on transfer
equations.[80] Nevertheless, the mean-free-path theory was easier to apply in many
calculations, since MAXWELL's equations could only be solved in the special case
of inverse fifth-power repulsive forces. SMOLUCHOWSKI was thus carrying on a
type of research previously pursued by CLAUSIUS, MAXWELL, O. E. MEYER, TAIT,
and JEANS, in which one tries to describe the effect of collisions on the path of
a molecule and thus on the properties of a gas.[81] EINSTEIN, on the other hand,
was working along lines first opened up by BOLTZMANN, MAXWELL (in his later
papers), and GIBBS; there, the objective was to deduce more general results from
a postulated probability distribution for configurations of the entire system of
molecules (described by generalized coordinates), without making specific as-

[79] M. R. VON SMOLAN SMOLUCHOWSKI, "O średniej swobodnej drodze czasteczek
gazu i o jej zeiazku teorja dyfuzji," Rozprawy i Sprawozdania z Posiedzen Wydzialu
Matematyczno-Przyrodniczego Akademii Umiejetnosci (Krakow), A 46, 129—139
(1906); French translation reprinted in *Pisma Marjana Smoluchowskiego*, Cracovie:
Académie Polonaise des Sciences et des Lettres.

[80] J. C. MAXWELL, papers reprinted in S. G. BRUSH, *Kinetic Theory*, Volumes 1
and 2.

[81] See S. G. BRUSH, "Development of the kinetic theory of gases. VI. Viscosity,"
American Journal of Physics 30, 269—281 (1962). J. H. JEANS, *The Dynamical Theory
of Gases*, Cambridge University Press, 1904; 4th edition (1925) reprinted by Dover
Publications.

sumptions about the intermolecular forces and collisions that determine transitions from one configuration to another. The two methodologies — kinetic theory and statistical mechanics — lead to similar results in one region of application, the equilibrium properties of gases, but kinetic theory could be extended to the calculation of transport properties of gases, whereas statistical mechanics could be extended to the calculation of equilibrium properties of liquids and solids. The explanation of Brownian motion involved a subject which neither kinetic theory nor statistical mechanics could claim to have conquered: transport properties and fluctuations in liquids. It was therefore valuable to have both viewpoints brought to bear on the problem.

SMOLUCHOWSKI recognized the reason why previous attempts to apply kinetic theory to Brownian movement had failed: they had been based either on a naive invocation of the equipartition theorem, or on a consideration of individual collisions of the molecules with the suspended particles. He mentioned NÄGELI,[31] who thought that he could refute the kinetic explanation "by indicating the smallness of the velocity produced by a collision. Thus a molecule of water, colliding with a particle of diameter 10^{-4} cm (and of density 1) would impart to it a velocity of only 3×10^{-6} cm/sec, which is much less than the order of magnitude of Brownian motion. In actuality the successive impulses would combine with each other, but NÄGELI thought that on the average they must cancel out, since they act in all directions of space, and that the end result could not be noticeably greater."[82] SMOLUCHOWSKI showed the fallacy of this argument, and at the same time hinted at a possible alternative derivation of EINSTEIN's displacement formula, by a simple combinatorial calculation. Suppose a gambling game consists of a sequence of random events — for example, throws of dice — in which there is an equal probability of winning or losing each time. Then the probability that in n throws there will be m favorable and $n - m$ unfavorable ones (hence a net gain of $2m - n$) is

$$p_{n,m} = \frac{n!}{2^n\, m!\,(n-m)!} = \frac{1}{2^n} \binom{n}{m}. \tag{5.1}$$

The average positive or negative deviation from the value zero (i.e., the average of the absolute value of the net total gain or loss in n throws) is

$$v = 2 \sum_{m-n/2}^{n} (2m - n)\, p_{n,m} = \frac{n}{2^n} \binom{n}{n/2}. \tag{5.2}$$

For large n, this reduces to

$$v = \sqrt{\frac{2n}{\pi}}. \tag{5.3}$$

According to equation (5.3), the *velocity* acquired by a suspended particle as a result of random impacts of molecules will be proportional to the square root of the number of impacts. Thus, if there are 10^{16} impacts per second, each of which transfers a velocity component in the X direction of $\pm 10^{-6}$ cm/sec, then each particle will acquire a velocity of 100 cm/sec after one second.

While this calculation shows the error in NÄGELI's argument, it does not by itself lead to the correct result, according to SMOLUCHOWSKI. It is not true, for

[82] SMOLUCHOWSKI, *Abhandlungen über die Brownsche Bewegung*, p. 7.

example, that each collision changes the velocity of the particle by the same amount on the average; instead, the probability of increases in the velocity becomes less, the greater the velocity itself. One expects that in the equilibrium state the particles will have an average velocity given by the equipartition theorem,

$$C = c \sqrt{\frac{m}{M}}. \tag{5.4}$$

(c = average velocity of the molecules, m = mass of molecules, M = mass of particles). However, the velocity calculated from this formula is generally much larger than what is actually observed (1,000 times larger, in one case). This paradox can easily be explained if we remember that according to kinetic theory the particle will be changing its direction of motion as a result of molecular impacts more than 10^{16} times per second, so that we cannot observe the instantaneous velocity but only the displacement over a large number of segments of a zigzag path. This is of course just Einstein's explanation, but it gains concreteness with the help of the kinetic-theory picture.

In constructing his quantitative theory, Smoluchowski argued that the magnitude of the velocity of the particle will always fluctuate around its equilibrium value given by (5.4), but its direction will change by a small amount at each impact of a molecule. The average change in direction was assumed to be $3\,mc/4M$, according to "the laws of collisions of elastic spheres."[83] Therefore Smoluchowski assumed that the magnitude of the velocity of the particle is always constant, but its direction changes at each impact by an amount $3\,mc/4MC$ in a random direction. He also assumed that the molecular impacts occur at equal time intervals, so that the path of the particle is a chain made up of segments of equal lengths.

In all this it is perfectly clear that Smoluchowski is simply adapting the mean-free-path description of the path of a gas molecule, with the single difference that here the persistence of motion after a collision is almost complete, whereas in the case of a gas molecule moving among other gas molecules there is relatively little persistence of motion.

Smoluchowski thus reduced the theory of Brownian movement to the mathematical problem: find the mean square end-to-end distance of a chain composed of n segments, each of length l, each rotated in a randomly chosen direction from the direction of the preceding one by a small angle ε. The general solution, which Smoluchowski obtained by solving a recursion equation involving an integral over n trigonometric functions, is

$$\Delta_n^2 = l^2 \left\{ \frac{2n}{\delta} + 1 - n - 2\,\frac{(1-\delta)^2 - (1-\delta)^{n+2}}{\delta^2} \right\} \tag{5.5}$$

where $\delta = 1 - \cos\varepsilon \approx \varepsilon^2/2$. In the limit when $n\,\delta$ is small, this reduces to

$$\Delta = n\,l \left(1 - \frac{n\,\delta}{6}\right). \tag{5.6}$$

---

[83] *Ibid.*, p. 9. In his attempt to reproduce this result, Fürth obtains the result .806 mc/M (*Ibid.*, p. 116—117).

(This is almost the same as straight-line motion, with a small correction for curvature.) When n is large, the root-mean-square displacement reduces to

$$\Delta = l \sqrt{\frac{2n}{\delta}}. \qquad (5.7)$$

Substituting $\varepsilon = 3\,mc/4\,MC\ \big(= 3C/4c$ according to equation (5.4)$\big)$ and $l = C/n$ for the length of each segment (if n segments are traversed in one second), SMOLU-CHOWSKI arrived at the result

$$\Delta = \frac{8}{3}\,\frac{c}{\sqrt{n}}. \qquad (5.8)$$

In order to put his result into a form directly comparable with EINSTEIN's — that is, to express the average displacement in terms of the size of the particle and the viscosity and temperature of the medium — SMOLUCHOWSKI used another result from the kinetic theory of gases. The number of collisions per second experienced by a molecule of radius R, at rest in a gas containing N point molecules in unit volume moving with average velocity c, is[70]

$$n = N R^2 \pi c. \qquad (5.9)$$

The average change in the velocity C caused by each collision is $2mC/3M$. Hence the resisting force of the medium is

$$S = \frac{2\pi}{3}\,R^2 \varrho\, c = \frac{2}{3}\,m\,n. \qquad (5.10)$$

So the number of collisions, n, can be related to the resisting force, S, by the formula[84]

$$n = \frac{3}{2}\,\frac{S}{m}. \qquad (5.11)$$

If we now assume that the Stokes formula for the resisting force is applicable,

$$S = 6\pi\mu R,$$

we can substitute $n = 9\pi\mu R/m$ into equation (5.8) and obtain the result

$$\Delta = \frac{8}{9\sqrt{\pi}}\,\frac{c\sqrt{m}}{\sqrt{\mu R}}. \qquad (5.12)$$

Recalling that EINSTEIN's λ_x in equation (4.13) is the average component of displacement in the x-direction, and thus corresponds to SMOLUCHOWSKI's Δ divided by $\sqrt{3}$, and remembering that $RT/N = mc^2/3$, we see that SMOLUCHOWSKI's result is smaller than EINSTEIN's by a factor of $\sqrt{\frac{27}{64}}$ but is otherwise identical.

The slight discrepancy in the numerical factor is perhaps not surprising in view of the various approximations used by EINSTEIN and SMOLUCHOWSKI; as FÜRTH pointed out in his notes to SMOLUCHOWSKI's paper in the *Klassiker* reprint, SMOLUCHOWSKI himself adopted EINSTEIN's formula in his later papers.[85]

---

[84] It should be noted that SMOLUCHOWSKI did not write down explicitly the step corresponding to our equation (5.11), and that his equation (23) has a misprint, ϱ in place of S, so his derivation is somewhat puzzling at first sight.

[85] *Ibid.*, p. 118—119.

6. Perrin's Experiments and the Reality of Atoms

The French physicist JEAN PERRIN (1870—1942) occupies a pivotal position
in the history of Brownian movement. He is generally credited with having
established the Einstein-Smoluchowski theory by his experiments, but just as
important was his role as propagandist for atomism. In addition, his emphasis
on the analogy of Brownian-movement paths with non-differentiable functions
(previously studied by RIEMANN & WEIERSTRASS) seems to have stimulated some
of the later research on functional integrals, especially that of WIENER.[86]

PERRIN's earlier work had been in the field of cathode rays and X-rays. In
his first paper (1895), he found evidence that cathode rays are negatively charged
particles, and in 1896 he was awarded the JOULE prize of the Royal Society of
London for his experiments on cathode rays. After an interval of several years,
during which he was occupied with organizing a course in physical chemistry at
the Sorbonne, he returned to the laboratory in 1903 and started to work on
contact electricity and colloidal solutions.[87,88]

Two works published during this pedagogical interlude give some indication
of his familiarity with contemporary theoretical physics. In a short contribution
to a symposium on molecular hypotheses at Paris in 1901, he proposed a "nucleo-
planetary" model of atomic structure, with a positively-charged "sun" surrounded
by many smaller negatively charged "planets." The periods of rotation of these
planets might correspond to different wavelengths in the emission spectrum.[89]
In 1903, he published the first volume of a textbook on physical chemistry. In
the preface to this book, he reviewed the status of molecular hypotheses. While
conceding that science should not base itself on atomism if that meant simply
reducing the visible to the invisible or unknowable, he maintained that atomic
hypotheses could be legitimate if they dealt with sensations that were at least
possible even if they had not yet been realized. He suggested, as an analogy,
that the germ theory of disease might have been developed and successfully tested
before the invention of the microscope; the microbes would have been hypothetical entities, yet, as we know now, they could eventually be observed. PERRIN
was therefore receptive to atomistic theories, all the more so because he seemed
to think that the alternative, "energetics," had degenerated into a pseudo-
religious cult.[90]

---

[86] NORBERT WIENER, *I am a Mathematician* (Garden City, New York: Doubleday
& Co., 1956), p. 33f.

[87] For biographical information see FERNAND LOT, *Jean Perrin*, Editions Seghers,
1963; ALBERT RANC, *Jean Perrin, un grand savant au service du socialisme*, Paris:
Editions de la Liberté, 1945; LOUIS DE BROGLIE, *Hommage National à Jean Perrin*,
Institut de France, Académie des Sciences, Paris, 1962. "Prix Gaston Planté,"
Comptes Rendus Acad. Sci. Paris **149**, 1207—1210 (1909).

[88] A selection of PERRIN's papers was reprinted in *Oeuvres Scientifiques de Jean
Perrin*, Centre National de la Recherche Scientifique, Paris, 1950; hereafter cited as
Oeuvres.

[89] J. PERRIN, "La structure nucléo-planétaire des atomes" (Conclusion d'une conference sur les *Hypothèses moléculaires* faite le 16 fevrier 1901 aux *Amis de l'Université
de Paris*) Revue Scientifique, **15**, p. 449 (1901); *Oeuvres*, 165—167.

[90] J. PERRIN, *Traité de Chimie Physique. Les Principes*. Paris: Gauthier-Villars,
1903.

PERRIN's interest in Brownian movement was first clearly shown in a lecture he gave to the Société de Philosophie in 1906. In this lecture he discussed the meaning and limits of validity of the Second Law of Thermodynamics, and raised the question of its compatibility with molecular hypotheses.[91] BOLTZMANN had maintained that the Second Law is perfectly consistent with kinetic theory as long as one demands only that the law be *statistically* valid; fluctuations that correspond to entropy decreases can occur, but so rarely that it is extremely improbable that they would be observed in any actual experiment. Critics of the kinetic theory, such as the mathematician, ERNST ZERMELO, argued that this "merely" statistical validity was not good enough; irreversibility is a fundamental property of natural processes, and any molecular hypothesis — or perhaps all conceivable molecular hypotheses based on Newtonian mechanics — that permits any exceptions must be wrong.[92] Referring to this criticism (which he attributes to LIPPMANN), PERRIN said:

If one recalls all the beautiful discoveries that we owe to molecular hypotheses, he would hesitate to support this radical opinion. But since one has no direct proof of the existence of molecules, he can only go by esthetic reasons if he has not succeeded in proving by experimental arguments that the second law does not have the character of absolute rigor, in the name of which one would sacrifice the molecular theories. We are going to try to show that such arguments exist.

Briefly, we are going to show that sufficiently careful observation reveals that at every instant, in a mass of fluid, there is an irregular spontaneous agitation which cannot be reconciled with CARNOT's principle except just on the condition of admitting that his principle has the probabilitistic character suggested to us by molecular hypotheses.[93]

He then discussed the phenomena of Brownian movement, together with a qualitative kinetic explanation, and then proposed a method for violating the Second Law. Suppose one starts with a liquid containing particles only in its lower layers. As a result of Brownian movement, these particles will tend to move upwards. By inserting in the liquid a piston made of a membrane permeable to the liquid but not to the particles, one could obtain mechanical work from this upward motion. This is just the concept of "osmotic pressure" of the particles in Brownian movement that EINSTEIN had used in his theoretical paper the previous year, though at this point PERRIN did not mention EINSTEIN.

In another popularisation, written at about the same time for the *Revue du Mois*, PERRIN drew an analogy between the physical discontinuity of matter and the mathematical properties of curves without tangents.[94] He pointed out that in teaching the concept of limit, we usually draw a curve and calculate the average velocity of a point moving on the curve from the quotient of its finite displacement and corresponding finite time interval. Then we say to the student: "You understand, don't you, that when this distance tends to zero, the average velocity tends to a limit." If the student is not too bright or too critical he will be intimidated by this demonstration and agree that of course the curve has an instantaneous velocity or tangent at any point.

[91] J. PERRIN, "Le contenu essential des principes de la thermodynamique," Bull. Soc. Philosophie 6, 81—111 (1906); *Oeuvres*, 57—80.
[92] See S. G. BRUSH, *Kinetic Theory, Volume 2, Irreversible Processes*. New York: Pergamon Press, 1966.
[93] PERRIN, *Oeuvres*, p. 68.
[94] J. PERRIN, "La discontinuité de la Matière," Revue du Mois, 1, 323—343 (1906).

But, said Perrin, the mathematicians of the preceding century have finally recognized that it is futile to attempt to prove rigorously, by such geometric arguments, that every continuous function has a derivative.

But they still thought the only interesting functions were the ones that can be differentiated. Now, however, an important school, developing with rigor the notion of continuity, has created a new mathematics, within which the old theory of functions is only the study (profound, to be sure) of a group of singular cases. It is curves with derivatives that are now the exceptions; or, if one prefers the geometrical language, curves with no tangent at any point become the rule, while the familiar regular curves become some kind of curiosities, doubtless interesting, but still very special.

There are still those who consider themselves men of good sense, who would say that these developments, while interesting to mathematicians, have nothing to do with the real world. But they are quite wrong, Perrin asserted, for one has only to look in the microscope to see that ordinary matter, which appeared perfectly continuous, homogeneous, and static to the naked eye, is really discontinuous, heterogeneous, and dynamic. In short, we have no reason to think that the properties of matter vary in a regular way as we go from one point of space to another. The ultramicroscope, recently developed by Siedentopf, Zsygmondy, Cotton, and Mouton, reveals the fine structure of matter more clearly than was ever possible before. Again, Perrin gave a qualitative discussion of Brownian movement, and suggested that such phenomena support the kinetic theory, but now he suggests that his interest in the subject is more than that of an expositor and commentator on the state of other people's work:

I will say only, announcing the results of reasonings that will be detailed in a subsequent article of M. Langevin, that the properties of fluids imply that molecules have diameters of about ten-millionth of a millimeter, and masses so small that a hundred billion molecules of water weigh scarcely a thousandth of a milligram.[95]

Langevin's simplified version of the Einstein theory was presented to the Académie des Sciences in Paris on March 9, 1908.[96] Perrin wrote later, "ever since I became, through M. Langevin, acquainted with the theory, it has been my aim to apply to it the test of experiment."[97] In the previous year, Seddig[98] and Svedberg[99] had attempted to verify Einstein's formula [equation (4.13)], but Perrin did not think their results were conclusive.[100]

[95] Ibid., p. 339.

[96] P. Langevin, "Sur la theorie du mouvement brownien," Comptes Rendus Acad. Sci. Paris, 146, 530—533 (1908); reprinted in Oeuvres Scientifiques de Paul Langevin, Centre National de la Recherche Scientifique, Paris, 1950.

[97] Jean Perrin, Atoms (translated by D. L. Hammick) London: Constable & Company Ltd., second English edition, 1923, p. 114.

[98] Max Seddig, "Abhängigkeit der Brownschen Molekularbewegung von der Temperatur," Sitzungsber. der Ges. zur Beförderung der gesammten Naturwissenschaften (Marburg), 182—188 (1907); "Ueber die sogenannte Brownsche Molekularbewegung und deren Abhängigkeit von der Temperatur," Naturwissenschaftliche Rundschau 23, 377—379 (1908).

[99] Theodor Svedberg, Studien zur Lehre von der Kolloiden Lösungen, Upsala, 1907. See also Die Existenz der Moleküle: experimentelle Studien, Leipzig: Akademische Verlagsgesellschaft m.b.H. "Neuere Untersuchungen über die Brownsche Bewegung," Jahrbuch Radioakt. 10, 467—515 (1913).

[100] Jean Perrin, Atoms, p. 120, note 1.

In his first series of experiments, PERRIN attempted to verify the formula (4.15) for the equilibrium distribution of the particles, which he wrote in the form:

$$2.3 \log \frac{n_0}{n} = \frac{1}{k} \cdot mgh \left(1 - \frac{1}{\varrho}\right). \tag{6.1}$$

It will be recalled that this equation can be derived directly from the Maxwell-Boltzmann distribution law; the derivation does not depend on the more dubious assumptions which EINSTEIN used in obtaining his displacement formula, such as the validity of Stokes law for very small particles. However, PERRIN did use the Stokes law to determine the mass of the particles in an associated experiment. He found that by taking the value of AVOGADRO's number that could then be estimated from kinetic theory, *i.e.* 7×10^{23}, he could fit his data very well; the agreement was even better if he chose $N = 6.7 \times 10^{23}$. He concluded:

> The average kinetic energy of a granule of the colloid is therefore equal to that of a molecule. This is, established by experiment, the hypothesis that EINSTEIN and LANGEVIN have indicated as equivalent to that of M. GOUY (theorem of equipartition of kinetic energies). At the same time, the kinetic theory of fluids seems to gain some support, and molecules become a little more tangible.[101]

At the meeting of the Académie des Sciences on May 18, 1908, one week after the presentation of PERRIN's first report, VICTOR HENRI reported a cinematographic study of Brownian movement, conducted in order to check EINSTEIN's displacement formula.[102] SVEDBERG[99] had found that the displacements are 6 or 7 times larger than those calculated from EINSTEIN's formula. HENRI's experimental results were four times larger than the theoretical prediction, although the proportionality of Δ^2 to t was confirmed. HENRI suggested that perhaps STOKES' law does not apply to such small particles.

Since the use of STOKES' law appeared to be the weakest link in EINSTEIN's chain of deductions, PERRIN next carried out a direct experimental test of the law for small particles of gamboge. He concluded that STOKES' law is valid, at least for the average displacement of a particle in a short time, for particles as small as $\frac{1}{10}$ of a micron.[103]

Returning to the experiment on distribution of particles, PERRIN presented more data and suggested that the Brownian movement might offer a new and more precise way of determining AVOGADRO's number.[104] He followed up this suggestion in a note presented on October 5, 1908, in which he reviewed four methods of determining N: (1) from kinetic theory; (2) from electrolysis, combined with measurements of the electronic charge; (3) by PLANCK and LORENTZ, from PLANCK's "beautiful electromagnetic theory of black-body radiation;" (4) from Brownian movement. PERRIN's best value of N has now become 71×10^{22}; the corresponding value of the electronic charge would be 4.1×10^{-10} esu.[105]

[101] See note 77.

[102] VICTOR HENRI, "Étude cinématographique des mouvements browniens," Comptes Rendus Acad. Sci. Paris, **146**, 1024—1026 (1908).

[103] JEAN PERRIN, "La loi de Stokes et le mouvement brownien," Comptes Rendus Acad. Sci. Paris **147**, 475—476 (7 Sept. 1908).

[104] JEAN PERRIN, "L'origine du mouvement brownien," Comptes Rendus Acad. Sci. Paris **147**, 530—532 (21 Sept. 1908).

[105] JEAN PERRIN, "Grandeur des molécules et charge de l'électron," Comptes Rendus Acad. Sci. Paris **147**, 594—596 (1908).

On November 30, 1908, CHAUDESAIGES, a student working in PERRIN's laboratory, reported on a new test of the Einstein displacement formula, using spherical grains of gamboge whose radius had been precisely measured.[106] CHAUDESAIGES
found that EINSTEIN's formula is completely exact, not only with respect to the
proportionality constant, provided that one takes $N = 64 \times 10^{22}$. The distribution
of displacements was stated to follow the law of errors, as expected.

PERRIN conducted some further experiments, to verify EINSTEIN's formula[107]
for rotational motion,[108] and to obtain more accurate values of N and atomic
parameters such as the electronic charge.[109] In 1909, he was awarded the Prix
Gaston Planté for his work on cathode rays, X-rays, and especially Brownian
movement.[87] For the next few years, he seems to have devoted much of his time
to popularizing the significance of his work on Brownian movement, in particular
the idea that atoms have now been proved to exist.[110] In this he was surprisingly
successful. In fact, the willingness of scientists to believe in the "reality" of
atoms after 1908, in contrast to previous insistence on their "hypothetical"
character, is quite amazing.

The evidence provided by the Brownian movement experiments of PERRIN
and others seems rather flimsy, compared to what was already available from
other sources. The fact that one could determine AVOGADRO's number and the
charge on the electron by one more method seems hardly sufficient to justify
such profound metaphysical conclusions. Several independent methods of determining these parameters had been known since 1870 or before,[32] to say nothing
of the many successes of kinetic theory in predicting the properties of gases.
Perhaps it was the novelty of EINSTEIN's deduction of the displacement formula,
tying together seemingly unrelated properties of liquids, that startled scientists
into conceding that the evidence for atomism had now become irrefutable. But
the evidence for the quantitative validity of the displacement formula was not
yet very good. PERRIN had to explain away the discordant results of VICTOR
HENRI in order to be able to claim that CHAUDESAIGE's measurements constituted
a verification of the theory:

> It is necessary to admit that some unknown complication or some systematic
> source of error has falsified the results of VICTOR HENRI, for the measurements that
> I am going to summarize leave no doubt of the rigorous exactitude of the formula
> proposed by EINSTEIN.[111]

The following statements are typical of the reactions of scientists to PERRIN's
work on Brownian movement. NERNST, in the sixth edition of his *Theoretische
Chemie* in 1909, added a new section on kinetic theory and heat, in which he dis-

[106] CHAUDESAIGES, "Le mouvement brownien et la formule d'Einstein," Comptes
Rendus Acad. Sci. **147**, 1044—1046 (30 Nov. 1908).

[107] ALBERT EINSTEIN, *Investigations on the theory of the Brownian Movement*, p. 33
[from Annalen der Physik, 1906].

[108] JEAN PERRIN, "Le mouvement brownien de rotation," Comptes Rendus Acad.
Sci. Paris **149**, 549 (1909).

[109] PERRIN & DABROWSKI, "Mouvement brownien et constantes moléculaires,"
Comptes Rendus Acad. Sci. Paris, **149**, 477—479 (1909).

[110] See the bibliography in *Oeuvres*, p. vii—xii.

[111] JEAN PERRIN, "Mouvement brownien et réalité moléculaire," Annales de Chimie
et de Physique [8] **18**, 1—114 (1909); *Oeuvres*, 171—239 (quotation from p. 214).
See also *Atoms*, p. 121.

cussed Brownian movement in connection with the limits of validity of the Second Law. He concluded the section with the remark:

In view of the *ocular* confirmation of the picture which the kinetic theory provides us of the world of molecules, one must admit that this theory begins to lose its hypothetical character.[112]

ARRHENIUS, in a lecture in Paris on March 13, 1911, reviewed the work of PERRIN and SVEDBERG and said:

After this, it does not seem possible to doubt that the molecular theory entertained by the philosophers of antiquity, LEUCIPPUS and DEMOCRITOS, has attained the truth, at least in essentials.[113]

The most dramatic reaction was that of OSTWALD, because he had been an outspoken critic of atomism as late as 1906. In his Ingersoll lecture at Harvard, he had said

... as I have been maintaining for the last ten years, the matter-and-motion theory (or scientific materialism) has outgrown itself and must be replaced by another theory, to which the name *Energetics* has been given .... The question as to the identity or non-identity of the different portions of water is without meaning, since there is no means of singling out the individual parts of the water and identifying them ... atoms are only hypothetical things ... .[114]

In the preface of the fourth edition of his *Grundriss der allgemeinen Chemie*, written in 1909, OSTWALD completely reversed himself:

I have convinced myself that we have recently come into possession of experimental proof of the discrete or grainy nature of matter, for which the atomic hypothesis had vainly sought for centuries, even millenia. The isolation and counting of gas ions on the one hand — which the exhaustive and excellent work of J. J. THOMSON has crowned with complete success — and the agreement of Brownian movements with the predictions of the kinetic hypothesis on the other hand, which has been shown by a series of researchers, most completely by J. PERRIN — this evidence now justifies even the most cautious scientist in speaking of the *experimental* proof of the atomistic nature of space-filling matter. What has up to now been called the atomistic hypothesis is thereby raised to the level of a well-founded theory, which therefore deserves its place in any textbook intended as an introduction to the scientific subject of general chemistry.[115]

There was only one major dissent from the scientific consensus on the reality of atoms after 1908. ERNST MACH, in 1909, reprinted his essay on the principle of conservation of work, in which he wrote, reviewing the history of the subject:

... it was concluded that, if heat can be transformed into mechanical work, heat consists in mechanical processes — in motion. This conclusion, which has spread over the whole cultivated world like wildfire, had, as an effect, a huge mass of literature on this subject, and now people are everywhere eagerly bent on explaining heat by means of motions; they determine the velocities, the average distances, and the paths of the molecules, and there is hardly a single problem which could not, people say, be completely solved in this way by means of sufficiently long calculations and of different hypotheses. No wonder that in all this clamour the voice of one of the most

---

[112] W. NERNST, *Theoretische Chemie*, Stuttgart: Enke, Sechste Auflage 1909, p. 212.

[113] S. ARRHÉNIUS, *Conférences sur quelques thèmes choisis de la chimie physique pure et appliquée, faites à l'Université de Paris du 6 au 13 Mars 1911*. Paris: Librairie Scientifique A. Hermann et fils, 1912. The quotation is from page 12.

[114] WILHELM OSTWALD, *Individuality and Immortality*, Boston: Houghton, Mifflin & Co., 1906. Quotations from pages 7, 40 and 41.

[115] WILHELM OSTWALD, *Grundriss der allgemeinen Chemie*. Leipzig: Verlag von Wilhelm Engelmann, 4. Aufl. 1909. Quotation from the "Vorbericht."

eminent, that of the great founder of the mechanical theory of heat, J. R. MAYER, is unheard: "Just as little as, from the connexion between the tendency to fall *(Fall-kraft)* and motion, we can conclude that the essence of this tendency is motion, just so little does this conclusion hold for heat. Rather might we conclude the opposite, that, in order to become heat, motion — whether simple or vibrating, like light or radiant heat — must cease to be motion." (*Mechanik der Wärme*, Stuttgart, 1867, p. 9) ....[116]

If, then, we are astonished at the discovery that heat is motion, we are astonished at something which has never been discovered. It is quite irrelevant for scientific purposes whether we think of heat as a substance or not.[117]

ALBERT EINSTEIN received a copy of this essay and wrote to MACH as follows (9. 8. 1909), enclosing copies of some of his own works:

Especially I ask you to glance at the paper on Brownian movement, since here there is a motion that one must believe to be "heat motion."[118]

But MACH was not convinced by this argument, and continued to assert that atomism was merely a hypothesis, though perhaps a useful one.[119]

It took a long time for Brownian movement to work itself into the mainstream of physics. But when it was finally recognized as a phenomenon worthy of serious study, the consequences were striking. In the next paper of this series, we will see how theories of Brownian movement interacted with the quantum theories of the 20th century.

Acknowledgments. This research was supported in part by the U. S. Atomic Energy Commission, while the author was employed at the Lawrence Radiation Laboratory, Livermore, California. Final revision of the paper was helped by useful suggestions from Drs. A. M. BORK, M. BUNGE, and L. SWENSON.

[116] E. MACH, *Die Geschichte und die Wurzel des Satzes von der Erhaltung der Arbeit*. Vortrag ... 2. unveränderter Abdruck nach der in Prag 1872 ersch. 1. Aufl. Leipzig: J. A. Barth, 1909. English translation, *History and Root of the Principle of the Conservation of Energy* [sic]. Chicago: The Open Court Pub. Co., 1911. Quotation from page 37 of the English translation.

[117] *Ibid.*, p. 47.

[118] FRIEDRICH HERNECK, "Zum Briefwechsel Albert Einsteins mit Ernst Mach," Forsch. Fortschr. 37, 239—243 (1963).

[119] See S. G. BRUSH, "Mach and Atomism," Synthese 18, 192—215 (1968).

Department of History and
Institute for Fluid Dynamics
and Applied Mathematics
University of Maryland
College Park

(Received December 16, 1967)

BRANCHING PROCESSES SINCE 1873

Address delivered on the occasion of the Centenary

DAVID G. KENDALL
(1966)

1. *Introduction*

The term "branching process" appears to have been coined by A. N. Kolmogorov and N. A. Dmitriev [**30**] in 1947 to describe the stochastic processes which arise when the theory of probability is introduced into population mathematics, but the subject is much older than one might suppose from this fact, and goes back nearly one hundred years.

The early history of the theory of branching processes centres round the figure of the Reverend Henry William Watson, clergyman, mathematician and alpinist. An account of his activities brings to life the kind of world in which our Society was founded. Watson was born in 1827 and entered Trinity College, Cambridge (by way of King's College, London) in 1846. In 1850 he was second Wrangler (second to Besant; the third and fourth were Wostenholme and Hayward). Watson was very active as a founder of societies; in particular he helped to found the Alpine Club in 1857. One would like to be able to claim that he was a founder-member of our own Society, but the evidence for this is exceedingly frail, and he may in fact never have been a member at any time.† The three other mathematicians just mentioned certainly did join the Society, and if Watson did not do so then this was one of the few steps in life which Hayward and he failed to take in common, for both became fellows of their Cambridge colleges (Hayward was at St. John's), both were early members of the Alpine Club, both were assistant masters at Harrow School, and both became Fellows of the Royal Society. They married the sisters Emily and Marianne Rowe. They died, in 1903, still closely associated, within three weeks of each other.

I have been unable to establish a definite link between Watson and the founding of the Society. However, Mr. T. S. Blakeney of the Alpine Club tells me that there might be a link between Watson and another notable event of 1865, the first ascent of the Matterhorn. In 1855 Watson was a member of one of two parties who together made the second ascent of the Dufourspitze, the highest point of Monte Rosa. It is just possible that a guide in the other party may have been the elder one of the two Taugwalders who were with Whymper on that tragic occasion ten years later, and who alone with him survived it.

Hayward spent most of his life at Harrow, but Watson in 1865 became Rector of Berkswell with Barston, near Coventry, where he spent the rest

---

† The evidence, such as it is, is footnoted in §5 of this paper.

[JOURNAL LONDON MATH. SOC., 41 (1966), 385–406]

of his working life.† This was by no means the end of his mathematics, however; from his country living Watson published some half-dozen books, of which *A Treatise on the Kinetic Theory of Gases* (1876) was the most notable.

What will chiefly concern us here is a correspondence in 1873 with Francis Galton, from which there eventually emerged the fundamental Galton-Watson-Haldane-Steffensen " Criticality Theorem", which was to be the source of the modern theory of branching processes.

I had hoped to be able to present you with a portrait of Watson in an Alpine setting, but the best I can do is to show this slide‡ of the party which made the first ascent of the Dufourspitze (a fortnight earlier than Watson's). The kind of pioneering vigour which Watson possessed is more evident here than in the portrait of Watson in Trinity College Library (Plate I).

2. *The extinction of surnames*

For some years Galton had been interested in the decay of families of "men of note"; peers, judges, and the like. De Candolle (1806–1893) appears to have been the first [5] to point out that the stock of ordinary family names is continually being eroded, that the rate of this erosion should be calculable, and that until we know what the rate of erosion is we cannot decide whether the extinction of peerages, for example, is an indication of diminished fertility. Galton accordingly gave to the problem a precise mathematical formulation, and communicated it in 1873 [**16**] to what he called "a well-known mathematical periodical of a high class, the *Educational Times*". This periodical, it is pleasing to note, was published by C. F. Hodgson and Sons. It was subtitled "Journal of the College of Preceptors", and from 1867 it printed regular and detailed reports of the transactions at meetings of the London Mathematical Society.

PROBLEM 4001: A large nation, of whom we will only concern ourselves with the adult males, N in number, and who each bear separate surnames, colonise a district. Their law of population is such that, in each generation, a_0 per cent of the adult males have no male children who reach adult life; a_1 have one such male child; a_2 have two; and so on up to a_5 who have five.

Find (1) what proportion of the surnames will have become extinct after r generations; and (2) how many instances there will be of the same surname being held by m persons.

---

† See [**19**] for an interesting account of how Watson came to be at Berkswell.
‡ Not reproduced here. See the *Alpine Journal* for 1916, facing page 291.

The only solution received did not please Galton, who declared that the man had made " a frightful mull of it ".   He therefore persuaded Watson to take up the matter.   It is not clear how their acquaintanceship began, but Galton had already consulted Watson a year or so earlier about a device for obtaining useful work from the energy of waves, and in particular from the relative tossing and pitching of two hulks.

Watson attacked the problem with the aid of generating functions,† and recognised that its complexity was essentially that associated with functional iteration, although he did not refer to and almost certainly did not know about Schröder's discussion [40] of this topic in the *Mathematische Annalen* of 1871.   In fact, if we write p_k for Galton's a_k (removing the restriction $k \leqslant 5$) and if we put

$$f(s) = \sum_0^\infty p_k s^k \quad (0 \leqslant s \leqslant 1),   \tag{1}$$

and

$$f_1(s) = f(s), \quad f_{n+1}(s) = f\big(f_n(s)\big) = f_n\big(f(s)\big) \quad (n = 1, 2, \ldots),   \tag{2}$$

then the power-series for f_n will have as coefficients the terms of the probability-distribution for the total number of males in the n-th generation (the original male constituting the zero-th generation).   Watson observed that the probability q_n of extinction by the n-th generation satisfies the equations

$$q_1 = p_0,$$

$$q_{n+1} = f(q_n),$$

and that if $q_n \to q$ when $n \to \infty$, then

$$q = f(q).   \tag{3}$$

This last equation always has the root $q = 1$, and from this fact Watson deduced, incorrectly, the inevitability of the extinction of the male line.

Here the matter rested.   Galton vigorously pressed Watson to continue with the problem, but he was busy with other things, and in a series of letters to de Candolle we find Galton trying to persuade him to enlist the aid of the Swiss mathematicians.   What Galton would particularly have liked was a " reasonable " form for the generating function f which behaved well under iteration.   Had Watson read Schröder, he would have experimented with the linear fractional family, have discovered the error in his treatment of the equation (3), and so have anticipated Haldane and Steffensen by sixty years.   As it was, the problem remained dormant.   In 1924 Karl Pearson published Volume II [35] of his *Life, Letters, and Labours of Francis Galton*, reprinting some of the correspondence I have mentioned

---

† For Watson's solution see references [45], [17], [18] and [35].

and summarising Watson's analysis (with which he appears to have found nothing amiss). No new progress was made until the period 1927–1930, when the problem re-appeared independently in this country and in Denmark; this time its investigators were to meet with more success.

R. A. Fisher in 1922 [13] had touched on what is essentially Galton's problem in a genetical context, but he did not follow up the matter until 1930 [14, 15]; by then J. B. S. Haldane [20] had pursued Fisher's ideas and in 1927 had roughly sketched a correct statement of the criticality theorem which lies at the heart of the matter (we will give a formal statement of this in a moment). These papers contain no references to the Galton–Watson work, although I find it hard to believe that they were not at least partially and indirectly stimulated by what had gone before. It should be mentioned that these papers of Fisher and Haldane contained a great deal else of mathematical interest which T. E. Harris [24] has now linked up with the general theory. In particular Fisher gives a lot of interesting detail in the special case when $f(s) = \exp\{c(s-1)\}$.

While we cannot be sure that there is a not a line of descent from Galton–Watson to Fisher–Haldane, it is virtually certain that the next "outbreak" of the problem (in 1929) was genuinely independent of the others. This time the enquiry was prompted by the personal predicament of a Danish telephone engineer.

Agner Krarup Erlang was a member, through his mother, of the famous Krarup family, the name of which was about to become extinct. This led him to submit what was essentially Galton's problem in its original surname context as a challenge to the readers of the *Matematisk Tiddskrift* in 1929, together with a partial solution of his own which has remained unpublished. Erlang had arrived at Watson's equation (3) by another route, for he wrote it as

$$q = p_0 + p_1 q + p_2 q^2 + \cdots,$$

and interpreted this as the assertion that the chance q of extinction is equal to (i) the chance of no male in the first generation, plus (ii) the chance of one male in the first generation, followed by ultimate extinction, plus (iii) the chance of two males in the first generation, followed by the ultimate extinction of both sub-families, and so on.

Erlang realised, as Watson had not, that the equation (3) can have two roots in the relevant interval [0, 1], and that in fact there will be (exactly) one root in [0, 1] in addition to the root $q = 1$ if and only if the expected number of sons per parent,

$$m = \sum_0^\infty k p_k,$$

is greater than unity.† Erlang died before his challenge [9] was published (his obituary appears beside it), but there is reason to believe from some remarks of Steffensen [43] that he had conjectured what is in fact the basic theorem of the subject : *it is always the smallest root of* (3) *which is the appropriate one*; *thus extinction is almost certain for sub-critical populations with* $m < 1$ *and for critical populations with* $m = 1$, *but there is always a positive chance of survival for super-critical populations with* $1 < m \leqslant \infty$.

It is quite clear from [20] that this theorem was known to Haldane in 1927 ; he virtually stated it then, and by remarks and references to the work of Koenigs showed that he knew how to prove it. But for a clear statement and detailed proof of the theorem we have to wait until 1930 when J. F. Steffensen set out his solution [42] to Erlang's problem in the *Matematisk Tidsskrift*. On seeing this note W. P. Elderton told Steffensen about the earlier attempts of Galton and Watson, of which Steffensen (and presumably therefore also Erlang) had been ignorant. Incidentally this suggests that not only Erlang and Steffensen but also Elderton were unaware of the parallel developments in the field of genetics. From this point onwards it is not profitable to follow the history of the subject in complete detail, for the basic theorem was to be re-discovered over and over again, especially during the war period, and no doubt we have not yet seen its last re-discovery.

Elderton also remarked that the probabilities p_k might in practice prove to be in geometric progression, and Steffensen was quick to seize on this idea, realising that if we put

$$p_0 = \alpha, \ \ p_k = (1 - \alpha)(1 - \beta)\beta^{k-1} \ \ \ (k = 1, 2, \ldots) \tag{5}$$

then f will be a linear fractional function and the iterations can be made explicit.

Meanwhile A. J. Lotka had seen Steffensen's note in the *Matematisk Tidsskrift* and he immediately [32] applied the branching-process theorem to the data contained in the 1920 United States census of white males, obtaining $q = 0 \cdot 88$ as the probability of the termination of the male line of descent from a new-born male. (In a subsequent calculation, according to Harris [24], he reduced this value to $q = 0 \cdot 82$.) He also [33] made the extremely interesting observation that for the population under study Elderton's suggested law (5) fits the facts fairly well, with $\alpha = 0 \cdot 4813$ and $\beta = 0 \cdot 5586$. If this is used instead of the empirical values of the p_k, we shall have

$$q = \alpha/\beta = 0 \cdot 862, \ \text{and} \ m = \frac{1 - \alpha}{1 - \beta} = 1 \cdot 175 > 1.$$

---

† We here (as always) exclude the uninteresting case $f(s) \equiv s$.

The generating function f corresponding to the use of (5) is

$$f(s) = 0 \cdot 4813 + \frac{0 \cdot 2290s}{1 - 0 \cdot 5586s},$$

and it is a simple matter to iterate this as far as may be desired by first reducing it to the canonical form [**41**] for a hyperbolic projectivity. If now Z_n denotes the number of males in the n-th generation, then

$$E(s^{Z_n}) = f_n(s)$$

and

$$E(e^{-uZ_n/m^n}) = f_n(e^{-u/m^n}) \rightarrow q + \frac{(1-q)^2}{(1-q)+u} \quad (0 \leqslant u \leqslant \infty).$$

From this, using the fact that $P(Z_n > 0) = q_n \rightarrow q$, and invoking a standard limit theorem, we deduce that

$$\lim_{n \to \infty} P(Z_n/m^n \geqslant x \mid Z_n > 0) = e^{-(1-q)x} \quad (x > 0). \tag{6}$$

That is, *the ultimate sizes of groups of males inheriting a common surname from a common ancestor may be expected to have an exponential distribution, when $1 < m < \infty$.* So far as I am aware, no attempt has ever been made to check this prediction; the statistics of the distribution of surnames appear never to have been collected in a suitable form.

If the probabilities p_1, p_2, ... do not follow a geometric law then there is a similar limit theorem for the distribution of Z_n/m^n (given that $Z_n > 0$) whenever $1 < m < \infty$, but the form of the limit distribution will be different.

In fact a much stronger result holds. It is clear that

$$E(Z_{n+1} \mid Z_n, Z_{n-1}, ..., Z_0) = mZ_n,$$

and so, if we write $W_n = Z_n/m^n$, we shall have

$$E(W_{n+1} \mid W_n, W_{n-1}, ..., W_0) = W_n;$$

that is, the stochastic process $\{W_n : n = 0, 1, 2, ...\}$ forms a *martingale*. Martingales have many interesting properties of which the most famous, due to J. L. Doob [**6**], goes as follows: *if the random variables W_n constituting the martingale satisfy $W_n \geqslant 0$ for all n, then with probability one the sequence $W_0, W_1, W_2, ...$ will converge to a finite limit W (a random variable).* It is then a relatively simple matter to show in the present case that this limiting random variable W will have a distribution uniquely characterised by the formulae

$$\left.\begin{array}{l} E(e^{-uW}) = \phi(u) \quad (0 \leqslant u \leqslant \infty), \\[2mm] \phi(mu) = f\big(\phi(u)\big), \\[2mm] \phi'(0) = -1, \end{array}\right\} \tag{7}$$

provided† that $\Sigma k^2 p_k < \infty$.

---

† In proving the last of the relations (7) "almost sure" convergence has to be reinforced by mean-square convergence, and this is the only reason for the condition on the second moment.

These relations show that W has a non-degenerate exponential distribution (perhaps supplemented by a mass at $W = 0$) if and only if f has the linear fractional form.

What this tells us is that if a large number N of males all having different surnames colonise a district, and if (females being available as and when required) they each propagate with a finite average replacement rate $m > 1$ (with $\Sigma k^2 p_k < \infty$), then after a long time has elapsed about qN of the surnames will have disappeared, while the remainder $(1-q)N$ will persist forever. If we restrict attention to those surnames which by chance survive, the number $Z_n^{(\sigma)}$ of representatives carrying any particular surname σ in the n-th generation will tend to infinity in such a way that

$$Z_n^{(\sigma)} = C^{(\sigma)} m^n + o(m^n) \quad (n \to \infty), \tag{8}$$

where $C^{(\sigma)}$ is constant for each surname σ. The different values of $C^{(\sigma)}$ will be distributed as independent random variables over $(0, \infty)$ with a distribution determined by

$$E(e^{-uC}) = \frac{\phi(u) - q}{1 - q} \quad (0 \leqslant u \leqslant \infty), \tag{9}$$

where ϕ is determined from f by the relations (7).

This is the final solution to the Galton–Watson problem. It was evolved between 1944 and 1950, mainly by the efforts of D. Hawkins and S. Ulam [26], T. E. Harris [21, 22, 23], and A. M. Yaglom [46]. The relevance of martingale theory was pointed out by Doob. The pathologies excluded by the assumption $\Sigma k^2 p_k < \infty$ were noticed and investigated by N. Levinson [31].

3. Later work

In the last section we have reviewed the development of branching process theory up to the point at which the original question formulated by Galton could be said to have been definitively answered. Here we shall very briefly mention a few of the further applications and theoretical developments which have resulted from the intensive cultivation of the subject during the last twenty years; a comprehensive account could not be given in this short space, nor would one be necessary, for there is now an excellent book on the whole subject by T. E. Harris [24] which gives a virtually complete exposition up to 1963.

One straightforward generalisation leads to the *multitype* Galton–Watson process; there are now objects (not necessarily living organisms) of K different types, and the process is characterised by a collection $f^i(s_1, \ldots, s_K)$ $(i = 1, 2, \ldots, K)$ of generating functions for which the coefficient of

$$s_1^{r_1} s_2^{r_2} \ldots s_K^{r_K}$$

in f^i is the probability that an individual of type i will in the first generation produce r_j individuals of the j-th type $(j = 1, 2, ..., K)$. The expected growth rate m is now replaced by a $(K \times K)$ matrix M of which the (i, j)-th element is the expected number of first-generation progeny of the j-th type to a parent of type i, and the theory takes its simplest form when we have *positive regularity* in the sense that some iterate of the matrix M has all its elements positive. The matrix M will then by a theorem of Perron have a real positive latent root ρ which is algebraically simple and which exceeds all other latent roots in modulus; the corresponding left latent vector $\mathbf{v}$ can be taken to have positive elements. If $\mathbf{Z}_n$ is the row-vector whose elements enumerate the individuals of types $1, 2, ..., K$ in the n-th generation, then a theorem of C. J. Everett and Ulam [10], and Harris [23], shows that with probability one

$$\mathbf{Z}_n = C\rho^n \mathbf{v} + \mathbf{o}(\rho^n) \quad (n \to \infty), \tag{10}$$

where C is a scalar random variable whose distribution depends on the type of the initial ancestor. (The proof of this theorem involves the assumption that the numbers of the various types in the first generation have finite quadratic moments. In these circumstances it can further be shown, with probability one, that C will not vanish unless $\mathbf{Z}_n$ is zero for all sufficiently large n.)

This implies that if the population does not become extinct then the ratios of the components of $\mathbf{Z}_n$ converge with probability one to the *fixed* ratios between the components of the vector $\mathbf{v}$. Thus, while the ultimate size of the population depends on the value assumed by the random variable C, the ultimate proportions between the numbers of the various types converge to values which are fully determined by the structure of the system *and which involve no chance element at all*. This remarkable theorem has recently been applied by J. H. Pollard [36] to a description of the growth of the female human population of Australia, the types $j = 1, 2, ..., K (= 60)$ corresponding to the sixty one-year age-groups in the age-range from 0 to 60 years. Pollard's work was stimulated by some highly original investigations by P. H. Leslie†.

Many other limit theorems are known both for this case and for numerous variants in which the discrete time variable n or the discrete " type " variable j or both are replaced by continuous variables. These results are of great technical interest, but on the present occasion it will be more suitable to mention instead one or two of the many applications. Those in which the participating individuals are nucleons instead of biological organisms spring at once to mind; the basic criticality theorem here has an immediate application familiar to literally every schoolboy.

---

† See, for example, *Biometrika*, 35 (1948), 213–245, and subsequent articles in that journal.

A biological application which may be new to some of the present audience presents itself when we consider the growth of an epidemic. Here in the simplest case Z_t is the number of infectious persons at large in the (closed, homogeneously mixing) population at time t. If X_t is the current number of susceptibles, then the simplest possible assumption is that the group enumerated by Z_t will be subject to a gross " birth rate " (associated with infections) which is a multiple of $X_t Z_t$ and to a gross " death rate " (associated with true death, recovery or any other sort of removal) which is a multiple of Z_t alone.

At the outset of the epidemic X_t will be large and will vary only slightly; as an approximation we may equate it to X_0. We will then have net birth and death rates per head per unit of time which can be taken to be

$$\lambda = X_0, \text{ and } \mu = \xi \text{ (constant)}.$$

The criticality theorem for this continuous-time branching process goes as follows (c.f. §4):

if $\lambda \leqslant \mu$,   then the number Z_t of infectious persons eventually becomes and remains zero, with probability one;

if $\lambda > \mu$,   then the number Z_t of infectious persons becomes zero with probability $(\mu/\lambda)^{Z_0}$, and otherwise grows indefinitely (ultimately exponentially).

This means that if the initial number of susceptibles X_0 does not exceed the " threshold " ξ, then the epidemic will be of small dimensions and will rapidly die out, whereas if X_0 does exceed ξ then we have two quite different possibilities, realised respectively with probabilities

$$(\xi/X_0)^{Z_0} \text{ and } 1 - (\xi/X_0)^{Z_0}.$$

These are (i) the epidemic may die out before more than a small fraction of the population has been infected; and (ii) the number of infectious persons may build up to a point at which the above approximation fails and the classical " deterministic " theory of epidemics can then be appealed to, to describe the subsequent course of the outbreak.

There is also an application to the theory of queues, created by Erlang in 1908 and intensively developed during the last fifteen years. If we look at a queueing system from the point of view of the person responsible for providing service, then we shall chiefly be interested in the duration of the " busy periods " during which service is continuous. Now suppose that customers arrive at random (i.e., in a Poisson process). Let a customer who arrives to find the system free (and who therefore initiates a busy period) occupy it for a time T_0. During his service-period a number Z_1 of further customers will arrive (conditionally, given T_0, Z_1 is a Poisson variable with parameter λT_0, if λ is the arrival rate); let $f(s)$ generate the absolute probability distribution of Z_1. Then it will be seen that if we call these Z_1 customers the male children of the initiating customer, the evolution of the busy period is equivalent to the history of a branching process with generating function f, and so by the criticality theorem the system will again come to rest with probability one if and only if $m = f'(1) \leqslant 1$; i.e., if and only if $\lambda E(T_0) \leqslant 1$; i.e., if and only if the rate λ of the input does not exceed the capacity $1/E(T_0)$ of the system.

The literature on busy periods dates from 1942 (when it was begun by a remarkable paper of Borel). For a time I wondered whether Erlang's preoccupation with the criticality theorem at the end of his life may have been associated with some unrecorded work of his on busy periods in queueing theory, but my friends in Denmark say that there is no evidence for this, and that Erlang's motive was almost certainly the one mentioned in §2.

4. *Some new properties of the simple birth process*

One of the simplest branching processes in continuous time is the simple birth-and-death process; this† is a Markovian process in which the current random variable Z_t (which enumerates the population at time t) changes only by jumps of amount $+1$ or -1, and in which these two sorts of jump occur with respective probabilities $\lambda Z_t \, dt + o(dt)$ and $\mu Z_t \, dt + o(dt)$ for the time interval $(t, t + dt)$. The system can be made more complicated, and perhaps more realistic, in a great variety of ways.

Thus one can let the parameters λ and μ depend on the current value of Z_t, instead of being constants. Pathologies arise when $\lambda(Z)$ increases too rapidly with Z, because it may then become possible for the system to make an infinity of transitions in a finite time; these phenomena are now fairly thoroughly understood from the work of Feller [12], S. Karlin and J. L. McGregor [27], and G. E. H. Reuter [38, 39]. There is an instructive analogy with diffusion theory which has been explored by Feller, while Karlin and McGregor have perfected a technique first introduced by W. Lederman and Reuter‡ which yields the spectral expansion of the transition probabilities by an appeal to classical work on moment problems and orthogonal polynomials.

Other modifications which have been studied include versions which allow λ and μ to depend on the time t (Kendall, [28]), which take into account the ages of the individuals and discuss stochastic fluctuations in the age-distribution (Kendall, [29]), or which replace the negative-exponential life-times by arbitrarily distributed ones (R. Bellman and T. E. Harris; see Harris, [24]). It is also possible to introduce different types of organism

---

† For further details and references see Bartlett's book [1]. The first serious study of the birth-and-death process was made by W. Feller [11], but on the present occasion it is especially proper to refer to the pioneer paper [34] by A. G. McKendrick.

‡ See *Phil. Trans. Roy. Soc. London* (A), 246 (1954), 321–369.

(see again Harris [24]); when the assemblage of types attains the power of the continuum this extension links up with the work on the age-distribution. These multi-type systems have been extensively studied in the Soviet Union.

In the simplest case, when λ and μ are constants, elementary methods show that

$$F_t(s) = E(s^{Z_t} \mid Z_0 = 1) = \frac{\mu(1-\theta) - (\mu - \lambda\theta) s}{(\lambda - \mu\theta) - \lambda(1-\theta) s} \quad (0 \leqslant s \leqslant 1), \qquad (11)$$

where

$$\theta = e^{-(\lambda - \mu) t}.$$

It is worthy of notice that if we observe such a system at regular times $t = n\tau$ $(n = 0, 1, 2, \ldots)$, then we obtain a discrete-time branching process of the kind discussed in §2, where now $f_n(s) = F_{n\tau}(s)$, and the function $f \equiv F_\tau$ has the linear fractional form which Lotka found useful in his study of human populations.

We can use (11) to investigate the limiting distribution of the reduced random variable $W_t = Z_t e^{-(\lambda - \mu) t}$ when $\lambda > \mu$ and $t \to \infty$. We easily find that

$$\lim_{t \to \infty} E(e^{-uW_t}) = \frac{\mu}{\lambda} + \left(1 - \frac{\mu}{\lambda}\right) \frac{(1 - \mu/\lambda)}{(1 - \mu/\lambda) + u} \quad (0 \leqslant u \leqslant \infty), \qquad (12)$$

so that

$$\lim_{t \to \infty} P(Z_t \geqslant x e^{(\lambda - \mu) t} \mid Z_t > 0) = e^{-(1 - \mu/\lambda) x} \quad (x > 0). \qquad (13)$$

The parallel with (6) is complete because μ/λ is now the chance of extinction.

It is also easily verified that $\{W_t : t \geqslant 0\}$ is a continuous parameter nonnegative martingale, and so (if, as we may, we take a separable version $\{\hat{Z}_t : t \geqslant 0\}$) we can assert in virtue of Doob's theorem that $\hat{W}_t = \hat{Z}_t e^{-(\lambda - \mu) t}$ converges with probability one to a finite limit W, as $t \to \infty$. Here W is a random variable which vanishes with probability μ/λ and which otherwise is distributed over the interval $(0, \infty)$ in accordance with the exponential distribution indicated at the right-hand side of (13). Thus for almost all sample trajectories ω of the (separable) process we can write

$$\hat{Z}_t(\omega) = W(\omega) e^{(\lambda - \mu) t} + o(e^{(\lambda - \mu) t}) \quad \text{as } t \to \infty, \qquad (14)$$

where the " constant " $W(\omega)$ varies randomly from one realisation ω of the process to the next in the manner just described.

All this is quite well known and with much more information of the same kind will be found in Harris's book [24]. What will interest us here is the possibility of saying more about the remainder term $o(e^{(\lambda - \mu) t})$ in (14), when in addition to prescribing Z_0 we also prescribe the ultimate behaviour of the system, i.e., the value of W. We shall find that when the sample paths are tied down in this manner at both extremities of the time scale, it then becomes possible to make interesting assertions about the conditional fluctuations at intermediate times. The problem is more complicated and the results are less attractive when births and deaths both occur than when deaths are excluded, and so in the interests of elegance we shall in the remainder of this account confine our attention to the birth process alone; *i.e.*, we shall put $\mu = 0$. In this case (11) simplifies to

$$E(s^{Z_t} \mid Z_0 = 1) = \frac{s}{e^{\lambda t} - s(e^{\lambda t} - 1)}, \qquad (15)$$

where of course the expression on the right-hand side must be raised to the power Z_0 if we want the system to start in a state other than $Z_0 = 1$. Iteration of (15) now shows that the joint generating function

$$E(s_1^{Z_{t_1}} s_2^{Z_{t_2}} \dots s_k^{Z_{t_k}} s^{Z_t} \mid Z_0 = 1) \qquad (0 < t_1 < t_2 < \dots < t_k < t)$$

is equal to

$$\frac{s s_k s_{k-1} \dots s_1}{\theta - (\theta - \theta_k)s - (\theta_k - \theta_{k-1})s s_k - \dots - (\theta_1 - 1)s s_k s_{k-1} \dots s_1}, \qquad (16)$$

where θ denotes $e^{\lambda t}$ and θ_j denotes $e^{\lambda t_j}$ $(j = 1, 2, \dots, k)$; we shall have occasion to quote this formula in a moment.

We now prove the following result, which is believed to be new.

THEOREM I.   (i) *If* $\{\hat{Z}_t : t \geqslant 0\}$ *is a separable version of the simple birth process* $\{Z_t : t \geqslant 0\}$, *then* $\hat{W}_t = \hat{Z}_t e^{-\lambda t}$ *converges with probability one as* $t \to \infty$ *to a positive finite random variable* W *having the distribution*

$$\frac{1}{(r-1)!} e^{-w} w^{r-1} dw \qquad (0 < w < \infty) \qquad (17)$$

when $Z_0 = r \geqslant 1$.

(ii) *The process* $\{Z_t : t \geqslant 0\}$ *possesses a conditional distribution with* $W = \lim Z_n e^{-\lambda n}$ *as conditioning random variable; this may be described informally by saying that*

$$\{Z_{\lambda^{-1} \log[1 + (t/W)]} : t \geqslant 0\} \qquad (18)$$

is a Poisson process with unit parameter when W *is given.*

Note. A formal statement of part (ii) of the theorem will be given in the course of the proof. [See the two lines following formula (21).]

Proof. The argument takes its simplest form when $Z_0 = 1$, and only this portion of the proof will be set out in detail here. The extension to values of $Z_0 = r > 1$ is straightforward but tedious; it amounts to expressing at rather greater length the following simple idea. We can think of the process Z_t as the sum of r independent components $Z_t^{(i)}$, each being a simple birth process with $Z_0^{(i)} = 1$ $(i = 1, 2, \dots, r)$. Formula (17) follows from the fact that $W = \Sigma W^{(i)}$ and from the convolution properties of the negative exponential distribution. Similarly the proof of the extension of the second part of the theorem consists essentially of the observations that each $Z^{(i)}_{\lambda^{-1}\log(1+t)}$ is (conditionally) a Poisson process with parameter $W^{(i)}$, and that the sum of r independent Poisson processes is again a Poisson process, the parameters being additive.

In view of these remarks we have nothing to prove in part (i) of the theorem, for this result with $Z_0 = 1$ is a special case of (12)–(14) above. We are therefore left with the problem of constructing a proof of part (ii) in the simplest case when $Z_0 = 1$. In order to do this we must first express the result to be proved in more precise language.

Let $\mathscr{B}$ denote the sigma-algebra of Borel subsets of $(0, \infty)$, and let $\mathscr{B}_1$ consist of all the half-lines (c, ∞) for $0 \leqslant c \leqslant \infty$. Then $\mathscr{B}_1$ is a π-system in the sense of Dynkin [8] (*i.e.*, it is closed under the formation of binary intersections), and it is a generating π-system in the sense that $\mathscr{B}$ is the smallest sigma-algebra extending $\mathscr{B}_1$. Similarly let $\mathscr{A}_1$ denote the system of the sets A_1 of mappings $\omega \equiv Z_.$ from $[0, \infty)$ into† $\{1, 2, ..., \infty\}$, with $Z_0 = 1$, such that either either A_1 is void, or A_1 contains all such mappings, or A_1 consists of exactly those mappings satisfying a set of conditions like

$$Z_{t_r} = j_r \quad (r = 1, 2, ..., k),$$

where k is a positive integer, the j's are positive integers or ∞, and $0 < t_1 < t_2 < ... < t_k$. Once again $\mathscr{A}_1$ is a π-system, and now we define $\mathscr{A}$ to be the sigma-algebra generated by $\mathscr{A}_1$; when we speak of a " measurable " set of mappings $Z_.$, we shall mean an element of $\mathscr{A}$.

In seeking the conditional distribution of the sample paths given W and given that $Z_0 = 1$, we have to try to construct a function $Q_.(.)$ from $(0, \infty) \times \mathscr{A}$ to the reals such that

(a) *for each fixed w in $0 < w < \infty$, $Q_w(.)$ is a probability measure on $\mathscr{A}$;*

(b) *for each fixed A in $\mathscr{A}$, $Q_.(A)$ is a Borel measurable function; and*

(c) *if A is in $\mathscr{A}$ and B is in $\mathscr{B}$, then*

$$\int_{W \in B} Q_W(A)\, dP = P(A \cap \{W \in B\}), \qquad (19)$$

where P on $\mathscr{A}$ is the probability measure for the birth process.

Here it is important to note that $W = \lim \hat{Z}_t e^{-\lambda t}$ was defined in (i) in terms of the separable version $\{\hat{Z}_t : t \geqslant 0\}$ of the original process $\{Z_t : t \geqslant 0\}$, so that $\hat{Z}_t(\omega) = \hat{Z}_t(Z_.)$ is not necessarily equal to $Z_t(\omega)$. This produces no difficulty, however, because $\lim Z_n(\omega) e^{-\lambda n}$ (where now $n \to \infty$ through integer values) exists and is equal to $W(\omega)$ with probability one, and we can use this alternative definition of W in (19) whenever it suits us.

---

† The compactification point ∞ is added for separability purposes.

Now let Q denote the probability measure on $\mathscr{A}$ which is associated with the Poisson process with unit parameter (started at $t = 0$ with $Z_0 = 1$). We define an operator V_w for $0 < w < \infty$ by

$$(V_w Z)_t = Z_{\lambda^{-1}\log[1+(t/w)]} \quad (0 \leqslant t < \infty), \tag{20}$$

so that, for each w, V_w maps the set of all mappings $Z_.$ onto itself in a one–one manner. Obviously both V_w and its inverse carry elements of $\mathscr{A}$ to elements of $\mathscr{A}$. A detailed proof of this last fact consists in noting that V_w (for example) maps $\mathscr{A}_1$ into $\mathscr{A}_1 \subseteq \mathscr{A}$, so that $\mathscr{A}_1 \subseteq V_w^{-1}\mathscr{A}$, and that the system $V_w^{-1}\mathscr{A}$ is a λ-system in the sense of Dynkin. [A λ-system is a system of sets which contains the whole space and is closed under the formation of (i) binary disjoint unions, (ii) differences $C\setminus D$ in the case when $C \supseteq D$, and (iii) limits of increasing sequences; it is a true theorem that if a λ-system covers a generating π-system then it also covers the generated sigma-algebra.] These considerations make it permissible to define

$$Q_w(A) \equiv Q(V_w A) \quad (0 < w < \infty, \ A \in \mathscr{A}), \tag{21}$$

and *the precise formulation of part (ii) of the theorem is that $Q_w(A)$ enjoys the three required properties* (a), (b), *and* (c).

That (a) holds is immediate, and the proof of (b) is not much harder. We have only to note (as an elementary calculation readily shows) that $Q_.(A_1)$ is a Borel measurable function whenever A_1 is in $\mathscr{A}_1$, and that the system of sets $C \in \mathscr{A}$ for which $Q_.(C)$ is Borel measurable is a λ-system. The substantial part of the proof of the theorem thus consists in establishing property (c).

Here it is enough to prove that

$$\phi(A_1, B_1) \equiv \int_{B_1} Q_w(A_1) e^{-w} dw = P(A_1 \cap \{W \in B_1\}) \equiv \psi(A_1, B_1) \tag{22}$$

for all A_1 in $\mathscr{A}_1$ and B_1 in $\mathscr{B}_1$, for we can then use the π/λ-theorem twice, first to show that $\phi(A_1, B) = \psi(A_1, B)$ for all Borel sets B, and then to show that $\phi(A, B) = \psi(A, B)$ for all A in $\mathscr{A}$ and B in $\mathscr{B}$, and this is equivalent to (19) in virtue of part (i) of the theorem. We therefore turn to the proof of (22).

We have to show that

$$\int_c^\infty Q(Z_{\tau_1} = j_1, \ Z_{\tau_2} = j_2, \ ..., \ Z_{\tau_k} = j_k) e^{-w} dw$$

$$= P(Z_{t_1} = j_1, \ Z_{t_2} = j_2, \ ..., \ Z_{t_k} = j_k, \ W > c),$$

where

$$\tau_j = w(e^{\lambda t_j} - 1) \quad (j = 1, 2, ..., k),$$

for all finite $c \geqslant 0$ and for all finite positive integers j (we can exclude as trivial the cases $c = \infty$, some $j = \infty$), and in order to do this it will be sufficient to prove the corresponding identity between generating functions, namely,

$$\int_0^\infty E_Q(s_1{}^{Z\tau_1}s_2{}^{Z\tau_2}\ldots s_k{}^{Z\tau_k})\,e^{-(1+u)\,w}\,dw = E_P(s_1{}^{Z_{t_1}}s_2{}^{Z_{t_2}}\ldots s_k{}^{Z_{t_k}}e^{-uW}),\qquad(24)$$

where $0 \leqslant s_j < 1$ $(j = 1, 2, \ldots, k)$, $0 < u < \infty$, and E_Q and E_P denote integration with respect to the Q- and P- measures respectively.

Now the right-hand side of (24) is equal to

$$\frac{s_k s_{k-1}\ldots s_1}{u + \theta_k - (\theta_k - \theta_{k-1})\,s_k - (\theta_{k-1} - \theta_{k-2})\,s_k s_{k-1} - \ldots - (\theta_1 - 1)\,s_k s_{k-1}\ldots s_1}\,,\quad(25)$$

where $\theta_j = e^{\lambda t_j}$ $(j = 1, 2, \ldots, k)$. This is proved by putting $s = \exp(-ue^{-\lambda t})$ in (16) and letting $t \to \infty$ through integer values; we here make use of the fact that $\lim_{n\to\infty} Z_n e^{-\lambda n} = W$ with probability one.

On the other hand a straightforward calculation with the Poisson process shows that the expectation occurring in the integrand on the left-hand side of (24) is

$$s_k s_{k-1}\ldots s_1 \exp\bigl\{-\tau_1(1 - s_1 s_2\ldots s_k)$$
$$- (\tau_2 - \tau_1)(1 - s_2 s_3\ldots s_k) - \ldots - (\tau_k - \tau_{k-1})(1 - s_k)\bigr\},$$

and if we now replace τ_j by the formula for it involving t_j and w, and perform the w-integration, we again obtain (25). The proof of the theorem is thus complete.

Theorem I opens up a great many possibilities which we may exploit in another place, and we shall only present one of these in detail here. If we invoke the law of the iterated logarithm in the form appropriate for the Poisson process, (c.f. V. Strassen [44], and P. Hartman and A. Wintner [25]), we are led to the following conclusion which throws a great deal of light on the behaviour of the remainder term in (14) in the case $\mu = 0$ (the birth process).

THEOREM II. *If W is the random variable defined in part (i) of Theorem I, then with probability one we can write*

$$\log \hat{Z}_t = \lambda t + \log W + \zeta(t)\,e^{-\lambda t/2}\sqrt{\left(\frac{2 \log \lambda t}{W}\right)},\qquad(26)$$

where the cluster-set of limit points of $\zeta(t)$ as $t \to \infty$ consists exactly of the segment $[-1, 1]$.

Proof. (i) Let A be the set of mappings Z such that

$$\limsup_{n\to\infty}\ \limsup_{N\to\infty}\ \frac{Z_{\lambda^{-1}\log(1+n)} - n Z_N e^{-\lambda N}}{\sqrt{\{2n Z_N e^{-\lambda N}\log\log n\}}} = +1.$$

This is an $\mathscr{A}$-measurable set; if we put $B=(0,\infty)$, then (19) and (21) show that

$$P(A)=\int_0^\infty Q(V_w A)\,e^{-w}\,dw.$$

Now a mapping $Z.$ belongs to $V_w A$ if and only if $V_w^{-1} Z.$ belongs to A, and

$$(V_w^{-1} Z)_\tau = Z_{w(e^{\lambda\tau}-1)},$$

so that $V_w A$ consists of the mappings $Z.$ such that

$$\limsup_{n\to\infty}\limsup_{N\to\infty}\frac{Z_{wn}-ne^{-\lambda N}Z_{w(e^{\lambda N}-1)}}{\sqrt{\{2ne^{-\lambda N}Z_{w(e^{\lambda N}-1)}\log\log n\}}}=+1.$$

But we are now concerned with Q-measure; *i.e.*, with the Poisson process, and the strong law of large numbers therefore tells us that for each fixed w in $(0,\infty)$,

$$e^{-\lambda N}Z_{w(e^{\lambda N}-1)}\to w\ \text{ as }\ N\to\infty,$$

with Q-probability one. Thus in $Q(V_w A)$ we can replace $V_w A$ by the set of mappings $Z.$ for which

$$\limsup_{n\to\infty}\frac{Z_{wn}-wn}{\sqrt{\{2wn\log\log n\}}}=+1.$$

For each fixed w in $(0,\infty)$ the Hartman–Wintner theorem [25] now assures us that $Q(V_w A)=1$, and so $P(A)=1$.

(ii) The next step is to make use of the fact that $Z_N e^{-\lambda N}$ converges to W with P-probability one, and that (P-almost surely) $0<W<\infty$. This enables us to assert that

$$P\left\{\limsup_{n\to\infty}\frac{Z_{\lambda^{-1}\log(1+n)}-nW}{\sqrt{(2nW\log\log n)}}=+1\right\}=1,$$

and an exactly parallel argument shows that the limit inferior is equal to -1, again P-almost surely.

We must now replace n by a positive real number t, and of course one will now have to work not with the original birth process $\{Z_t:t\geqslant 0\}$ but with a separable version $\{\hat{Z}_t:t\geqslant 0\}$, chosen so that $\hat{Z}.$ as a function on $[0,\infty)$ is a monotonic increasing step-function changing only by jumps of amount $+1$. The monotonic-increasing character of $\hat{Z}.$ then shows that in the last formula (and in the corresponding one involving the limit inferior) we can replace $n(\to\infty)$ by $t(\to\infty)$ if we can show that the sequence

$$\frac{\hat{Z}_{\lambda^{-1}\log(1+n)}-\hat{Z}_{\lambda^{-1}\log n}}{\sqrt{(n\log\log n)}}$$

converges P-almost surely to zero.

This however is easily demonstrated by a further appeal to Theorem I. If we write A' for the set of unhatted mappings having the desired property (the convergence to zero of the last expression, minus the hats, as $n \to \infty$), then

$$P(A') = \int_0^\infty Q(V_w A') e^{-w} dw,$$

where now $V_w A'$ consists of those mappings $Z_.$ such that

$$\frac{Z_{nw} - Z_{(n-1)w}}{\sqrt{(n \log \log n)}}$$

tends to zero as $n \to \infty$. If the Poisson process has paths $Z_.$ which almost surely behave like this then we shall be able to conclude that $Q(V_w A') = 1$, and hence that $P(A') = 1$, as required. Let us put $X_n = Z_{nw} - Z_{(n-1)w}$, so that we have to show that

$$X_n / \sqrt{(n \log \log n)}$$

almost surely tends to zero when $n \to \infty$, where the X's are independent random variables with the distribution

$$\mathrm{pr}(X = j) = e^{-w} w^j / j! \quad (j = 0, 1, 2, \dots),$$

and w is a fixed finite positive real number. The strong law of large numbers tells us that

$$\frac{X_1{}^2 + X_2{}^2 + \dots + X_n{}^2}{n}$$

almost surely converges to $E(X^2) = w + w^2 < \infty$, as $n \to \infty$, and so $X_n{}^2 / n$ almost surely tends to zero. This however implies that $X_n / \sqrt{n}$ almost surely tends to zero, which is rather more than we need.

(iii) Let us write

$$\hat{Z}_{\lambda^{-1} \log(1+\tau)} = W\tau + S_\tau \sqrt{(2 W\tau \log \log \tau)},$$

this being the definition of S_τ for $\tau > e$. Then we have shown that, P-almost surely,

$$\lim_{\tau \to \infty} {}^{\sup}_{\inf} S_\tau = \pm 1,$$

and we can quickly infer from this that (P-almost surely) the random function $S_.$ has the segment $[-1, +1]$ as its exact cluster-set when $\tau \to \infty$. This is an elementary consequence of the fact that $S_.$ is the difference between an increasing function and a continuous one, and so cannot for increasing τ pass from the interval $(1 + \epsilon, 1 - \epsilon)$ (which it hits infinitely often) to the interval $(-1 + \epsilon, -1 - \epsilon)$ (which it also hits infinitely often) without intermediately assuming every value in $[1 - \epsilon, -1 + \epsilon]$ (and this is true for all $\epsilon > 0$).

(iv) Finally we have only to observe that

$$\log \hat{Z}_t = \log W + \lambda t + \log(1 - e^{-\lambda t}) + R + \rho R^2,$$

where

$$R = S_{e^{\lambda t}-1} \sqrt{\left\{ \frac{2 \log \log(e^{\lambda t} - 1)}{W(e^{\lambda t} - 1)} \right\}},$$

and where $|\rho| \leqslant 1$ for all sufficiently large t. The assertions of Theorem II then follow easily from what we have proved about $S_.$, so our proof is complete.

From this result we see that if we plot the logarithm of the population size against the standardised time, λt, we will (with probability one) obtain a graph which is very nearly linear for large t. This much is quite well known (see *e.g.*, Harris [24; page 12]), but we now have the additional information that the departure from linearity converges to zero, as $t \to \infty$, in the manner indicated by the final term in (26).

In order to illustrate Theorems I and II the last two slides (Figs. 1 and 2) show (i) ten artificial realisations of the birth process, plotted on a ($\log Z_t$, λt)-diagram, and (ii) one of these realisations together with the " iterated logarithm " curves which have been computed from (26) with $\zeta(t)$ set equal to ± 1.

Further new results about the birth process can similarly be obtained by choosing any other property of the Poisson process (*e.g.*, from [4, 7, or 37]) and transferring it to the birth process by means of Theorem I.

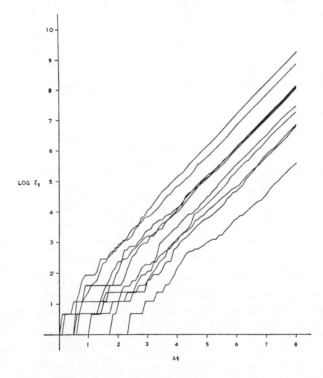

FIG. 1.—Ten independent realisations of the simple birth process $\{Z_t : t \geqslant 0\}$.

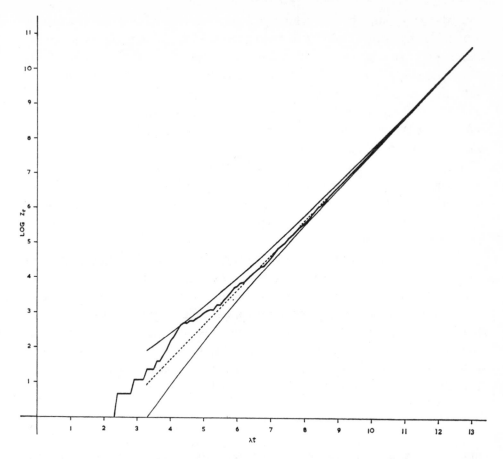

FIG. 2.—A realisation of the simple birth process, and the iterated-logarithm boundaries ($\zeta = \pm 1$) for all realisations having the same initial value $Z_0 = 1$ and the same asymptotic behaviour (the same value of W).

We close this discussion with two final remarks. (1°) It is possible to link up the random variable W appearing in Theorems I and II with the theory of the " ideal boundary " for a Markov process; we have made no use of these ideas here, but some relevant information will be found in reference [**2**]. (2°) The calculations of the present section can be repeated up to a certain point for the birth-and-death process, but now the conditioned process turns out to be essentially non-temporally-homogeneous, so that the situation is much more complicated.

5. *Conclusion and acknowledgements*

I must conclude by thanking not only Professor R. Pyke for valuable comments† on §4, and Miss S. E. Prior for help with the computations,

---

† He has kindly drawn my attention to a paper by D. R. Cox [**3**], especially formula (4 . 12) which, with an unpublished sequel, forms an interesting complement to the work reported here in §4.

but also the many people who most generously supplied the details on which the historical sections of the paper are based. Among those I wish particularly to mention are Professor E. S. Pearson, C.B.E., and Mr. L. J. Gue (the Records and Publications Officer of University College London), for information about the early life of R. B. Hayward; the Master and Fellows of Trinity College, Cambridge, for permission to reproduce the portrait of Watson, and Mr. A. Halcro, who located it; Mr. T. S. Blakeney and Mr. A. K. Rawlinson of the Alpine Club, for information about Watson's climbs in the Alps; Mr. E. C. Rubidge of the staff of the London Mathematical Society, who looked into the possible association of Watson with the early history of the Society†; and Professor Arne Jensen of Danmarks Tekniske Højskole, for information about Erlang's interest in the surname problem.

We cannot have centenaries every year, but I wish it might become more usual for members of the Society to indulge in historical writing. Much may be gleaned from obituaries, but there is also I think a place for brief histories of individual mathematical topics. We may be particularly well placed in time for such activity; our predecessors had perhaps too little to write about, and our successors will certainly have too much. If anyone feels inclined to follow my excursion into Victorian mathematics I think he will find it rewarding.

Note added 1st June 1976. This article must now be regarded as seriously out of date. In a sequel "The genealogy of genealogy; branching processes before (and after) 1873" (Bulletin London Mathematical Society, Vol. 7 (1975), pp. 225–253) I have corrected two things:

(1) The account given here of the Russian work is quite inadequate, and I have therefore redressed this and made clear the place of the Russian work in the whole genealogy of the subject.

(2) I have dealt at some length with the discovery by C. C. Heyde and E. Seneta of the fact that I. J. Bienaymé in 1845 already had the *whole* of the criticality theorem; I have analysed the stimuli which provoked Bienaymé's work and reconstructed (so far as I was able) the details of his lost argument.

---

† He writes: " . . . someone named Watson (no initials, no Revd.) was a founder member of the Society in 1865; the name is lightly pencilled in the book used to record the signatures of newly-elected members. If therefore you have evidence that your Watson was in London at that time there would I suppose be grounds for thinking that he was a founder-member." The Society was founded on 16 January, 1865, and Watson appears to have left Harrow for Berkswell not earlier than January, 1866.

[PLATE 1]

HENRY WILLIAM WATSON, F.R.S.
(1827–1903)

*From a photograph in the Library of
Trinity College, Cambridge, by permission of the Master and Fellows*

References

1. M. S. Bartlett, *An Introduction to Stochastic Processes* (Cambridge, 1955).
2. D. Blackwell and D. G. Kendall, " The Martin-boundary for Pólya's urn scheme, and an application to stochastic population growth ", *J. Applied Probability*, 1 (1964), 284–296.
3. D. R. Cox, " Some statistical methods connected with series of events ", *J. Royal Statist. Soc.*, B, 17 (1955), 137.
4. H. E. Daniels, " The Poisson process with a curved absorbing boundary ", *Proc. Intern. Assoc. Statist. Phys. Sci.*, 1 (1963), 994–1011. (Also available in the *Bulletin* of the I.S.I.)
5. A. de Candolle, *Histoires des Sciences et des Savants depuis deux Siècles* (Geneva, 1873). (See especially p. 388.)
6. J. L. Doob, *Stochastic Processes* (New York, 1953).
7. L. E. Dubins, " Rises and upcrossings of nonnegative martingales ", *Illinois J. of Math.*, 6 (1962), 226–241.
8. E. B. Dynkin, *Markov Processes* (English translation; Berlin, 1965). (See especially Vol. II, pp. 201–202.)
9. A. K. Erlang, " Problem 15 ", *Matem. Tidsskr.* B (1929), 36.
10. C. J. Everett and S. Ulam, " Multiplicative systems in several variables, III ", Los Alamos declassified document 707, 1948.
11. W. Feller, " Die grundlagen der Volterraschen Theorie des Kampfes ums Dasein in wahrscheinlichkeits-theoretischer Behandlung ", *Acta Biotheoretica*, 5 (1939), 11–40.
12. ———, " The birth and death process as diffusion processes ", *J. Math. Pures et Appl.* (9), 38 (1959), 301–345.
13. R. A. Fisher, " On the dominance ratio ", *Proc. Royal Soc. Edinburgh*, 42 (1922), 321–341.
14. ———, " The distribution of gene ratios for rare mutations ", *Proc. Royal Soc. Edinburgh*, 50 (1930), 204–219.
15. ———, *The genetical theory of natural selection* (Oxford, 1930).
16. F. Galton, " Problem 4001 ", *Educational Times*, 1 April, 1873, p. 17.
17. ———, and H. W. Watson, " On the probability of the extinction of families ", *J. Roy. Anthropol. Inst.*, 4 (1874), 138–144.
18. F. Galton, *Natural Inheritance* (London, 1889).
19. ———, *Memories of my life* (London, 1908).
20. J. B. S. Haldane, " A mathematical theory of natural and artificial selection, V ", *Proc. Cambridge Phil. Soc.*, 23 (1927), 838–844.
21. T. E. Harris, " Some theorems on the Bernoullian multiplication process " (Princeton thesis, 1947).
22. ———, " Branching processes ", *Ann. Math. Statist.*, 19 (1948), 474–494.
23. ———, " Some mathematical models for branching processes ", *Proc. 2nd Berkeley Symp.* (1951), 305–328.
24. ———, *The theory of branching processes* (Berlin, 1963).
25. P. Hartman and A. Wintner, " On the law of the iterated logarithm ", *American J. of Math.*, 63 (1941), 169–176.
26. D. Hawkins and S. Ulam, *Theory of multiplicative processes*, Los Alamos declassified document 265, 1944.
27. S. Karlin and J. L. McGregor, " The classification of birth and death processes ", *Trans. American Math. Soc.*, 86 (1957), 366–400.
28. D. G. Kendall, " On the generalised birth-and-death process ", *Ann. Math. Statist.*, 19 (1948), 1–15.
29. ———, " Les processus stochastiques de croissance en biologie ", *Ann. de l'Inst. H. Poincaré*, 13 (1952), 43–108.
30. A. N. Kolmogorov and N. A. Dmitriev, " Branching stochastic processes ", *Doklady Akad. Nauk U.S.S.R.*, 56 (1947), 5–8.

31. N. Levinson, " Limiting theorems for Galton-Watson branching process ", *Illinois J. of Math.*, 3 (1959), 554–565.

32. A. J. Lotka, " The extinction of families, I ", *J. Washington Acad. Sci.*, 21 (1931), 377–380.

33. ——, " The extinction of families, II ", *J. Washington Acad. Sci.*, 21 (1931), 453–459.

34. A. G. McKendrick, " Studies on the theory of continuous probabilities, with special reference to its bearing on natural phenomena of a progressive nature ", *Proc. London Math. Soc.* (2), 13 (1914), 401–416.

35. K. Pearson, *The Life, Letters and Labours of Francis Galton*, Vol. II (Cambridge, 1924).

36. J. H. Pollard, " On the use of the direct matrix product in analysing certain stochastic population models " (to appear in *Biometrika*).

37. R. Pyke, " The supremum and infimum of the Poisson process ", *Ann. Math. Statist.*, 30 (1959), 568–576.

38. G. E. H. Reuter, " Denumerable Markov processes and the associated contraction semi-groups on l ", *Acta Math.*, 97 (1957), 1–46.

39. ——, " Competition processes ", *Proc. 4th Berkeley Symp.*, 2 (1961), 421–430.

40. E. Schröder, " Ober iterirte Funktionen ", *Math. Annalen*, 3 (1871), 296–322.

41. J. G. Semple and G. T. Kneebone, *Algebraic projective geometry* (Oxford, 1952). (See p. 57.)

42. J. F. Steffenson, " On Sandsynligheden for at Afkommet uddør ", *Matem. Tiddskr.* B (1930), 19–23.

43. ——, " Deux problèmes du calcul des probabilités ", *Ann. Inst. Henri Poincaré*, 3 (1933), 319–344.

44. V. Strassen, " An invariance principle for the law of the iterated logarithm ", *Zs.f. Wahrsch.*, 3 (1964), 211–226.

45. H. W. Watson, " Solution to Problem 4001 ", *Educational Times*, 1 August, 1873, pp. 115–116.

46. A. M. Yaglom, " Certain limit theorems of the theory of branching random processes ", *Dokl. Akad. Nauk* (U.S.S.R.), 56 (1947), 795–798.

Churchill College,
    Cambridge.

Studies in the History of Probability and Statistics. XXXI.

The simple branching process, a turning point test and a fundamental inequality: A historical note on I. J. Bienaymé

By C. C. HEYDE and E. SENETA

(1972)

Australian National University

SUMMARY

Some important contributions of I. J. Bienaymé in probability and statistics have been largely forgotten. In this note we discuss three such contributions; on the simple branching process, a (non-parametric) turning point test and the Bienaymé–Chebyshev inequality.

Some key words: History of branching processes; History of probability inequalities; History of non-parametric tests.

1. The Bienaymé (Galton–Watson) process

D. G. Kendall (1966) gives an excellent account of the historical development of the fundamental criticality theorem pertaining to the simple branching process from the time of Galton and Watson's partial contributions to the subject in 1873–4. He deduces that this theorem should be ascribed to Galton (1873), Galton & Watson (1874), Watson (1873), Haldane and Steffensen, Haldane having obtained what is tantamount to a completely correct statement of the theorem for the first time in 1927 and Steffensen giving for the first time a clear statement and detailed proof in 1930.

In fact it would appear necessary, at the very least, to add the name of I. J. Bienaymé to the beginning of this list of contributors. In a communication in (1845), thus anticipating Galton and Watson by some 28 years, Bienaymé shows verbally that the correct statement of the criticality theorem is known to him. No references are given, but the title and the opening paragraph show that his motivation was similar to that of Galton:

On s'est beaucoup occupé de la multiplication possible du nombre des hommes; et récemment diverses observations très curieuses ont été publiées sur la fatalité qui s'attacherait aux corps de noblesse, de bourgeoisie, aux familles des hommes illustres, etc.; fatalité qui, dit-on, ferait disparaître inévitablement ce qu'on a nommé des *familles fermées*.

The following two extracts indicate that not only did Bienaymé have a completely correct statement of the criticality theorem, but had also given some further consideration to the paradoxical critical case.

Si le rapport d'une génération à l'autre, ou la moyenne du nombre des enfants mâles qui remplaceront le nombre des mâles de la génération précédente était mondre que l'unité, on concevrait sans peine que les familles s'étoignissent par la disparition des membres qui les composent. Mais l'analyse montre de plus que quand cette moyenne est égale à l'unité, les familles tendent à disparaître, quoique moins rapidement.

L'analyse montre aussi clairement que le rapport moyen étant supérieur à l'unité, la probabilité de destruction des familles ne peut plus se changer en certitude avec l'aide du temps. Elle ne fait que s'approcher d'une limite finie, assez facile à calculer, et qui offre ce caractère singulier d'être donnée par celle des racines de l'équation (où l'on fait infini le nombre des générations), qui ne convient pas à la question quand le rapport moyen est inferieur à l'unité.

The note ends with the intriguing paragraph:

M. Bienaymé développe diverses autres considérations que les éléments de la question lui ont suggérées, et qu'il se propose de publier bientôt dans un mémoire spécial.

We have not been able to trace the 'mémoire spécial'. Nevertheless, the existence of even the 1845 note appears to justify reference to Bienaymé as possibly the founder of branching process theory. To our knowledge, such a claim is at present absent from the branching process literature.

2. Turning points – a contribution in nonparametric statistics

A very easy test to apply to a series of random observations involves the counting of the number of local maxima and minima, i.e. turning points, in the series. The interval between two turning points is called a phase. Bienaymé (1874) stated that the number of phases, complete and incomplete, in a sequence of N observations is approximately normally distributed about a mean of $\frac{1}{3}(2N-1)$ with variance $(16N-29)/90$. His paper actually gives a slightly different form as a consequence of using a nonstandard normal distribution; we have made the appropriate transformations into modern form. A slightly later paper (Bienaymé, 1875), gives some applications of this result.

The combinatorial aspects of Bienaymé's result have attracted some attention with due acknowledgement: thus Netto (1901, pp. 106–16) summarizes the paper of Bienaymé (1874) and a long series by André published between 1879 and 1896 which includes a recursion formula for the number of phases, including the incomplete phases before the first and after the last turning points, while mention of Bienaymé is made also in the more recent paper of von Schrutka (1941). On the other hand, while the full force of the result as a probability limit theorem is available in the statistical literature, its origins seem to have been largely forgotten. Reference to Bienaymé (1874, 1875) appears in an historical appendix concluding the paper of Wallis & Moore (1941b) who deal in part in their preceding theory with precisely the problem considered by Bienaymé; this paper is a much expanded, and apparently later, version of Wallis & Moore (1941a) which does not refer to Bienaymé. This material is given later by Kendall (1948, pp. 124–5) and Kendall & Stuart (1968, pp. 351–2) but without attribution and a test of randomness based on Bienaymé's result is called a turning point test. Asymptotic normality is inferred from an investigation of skewness and kurtosis, as in Wallis & Moore (1941b). Kendall and Stuart remark that the turning point test is reasonable for a test against cyclicity but poor as a test against trend.

Bienaymé's result represents a remarkable achievement for its time and it would be interesting to know how it was obtained. An early discussion of the result was given by Bertrand (1875). A modern treatment could use indicator variables and the central limit theorem for stationary m-dependent processes. Alternatively, a finite stationary irreducible aperiodic Markov chain can be constructed, and the asymptotic behaviour obtained from the central limit theorem for sums of random variables defined on such a chain.

3. The Bienaymé–Chebyshev inequality

The following result is commonly understood as the Chebyshev inequality and much less commonly as the Bienaymé–Chebyshev inequality. If $E(X^2) < \infty$ and $E(X) = \mu$, then, for any $\epsilon > 0$,

$$\mathrm{pr}\,(|X-\mu| \geqslant \epsilon) \leqslant \mathrm{var}\,(X)/\epsilon^2. \tag{1}$$

In this section we shall give a historical perspective on the attribution for (1).

If X_i ($i = 1, ..., n$) are independent random variables, $E(X_i^2) < \infty$ for each i and $S_n = X_1 + ... + X_n$, then since the Bienaymé equality (Bienaymé, 1853) asserts

$$\text{var}\,(S_n) = \sum_{i=1}^{n} \text{var}\,(X_i),$$

it follows from (1) that

$$\text{pr}\,(|S_n - E(S_n)| \geqslant \epsilon) \leqslant \sum_{i=1}^{n} \text{var}\,(X_i)/\epsilon^2. \tag{2}$$

The result (2) was obtained by Chebyshev (1867) and published simultaneously in Russian and French.

If, in addition, all the X_i are identically distributed, then (2) becomes

$$\text{pr}\,(|S_n - E(S_n)| \geqslant \epsilon) \leqslant n\,\text{var}\,(X_i)/\epsilon^2,$$

and if we put $S_n/n = \bar{X}$ and $\mu = E(X_i) = E(\bar{X})$, then

$$\text{pr}\,(|\bar{X} - \mu| \geqslant \epsilon) \leqslant \text{var}\,(X_i)/(\epsilon^2 n). \tag{3}$$

Clearly, when $n = 1$, (3) also reverts to (1), so that in fact all of (1), (2) and (3) are equivalent.

Now (3) is in essence already contained in the article of Bienaymé (1853, pp. 320–1), albeit in a not particularly obvious form. This article by Bienaymé was reprinted in Liouville's journal immediately following Chebyshev's article of (1867). There is a small editorial comment by Liouville on the reprinted paper but it does not contain any information concerning the inequality.

It should be remarked that Chebyshev himself (1874) gives Bienaymé credit for (3), a tradition continued in several of the writings of Chebyshev's student, A. A. Markov. Chebyshev's claim, however, is in that he has given (3) explicitly (1867) and used it to prove a law of large numbers; Bienaymé gives no hint that he has seen the relevance of (3) in this context.

It is interesting to note that Markov himself seems to have remained a champion of Bienaymé's claims. In his collected works (1951), one finds listed (p. 703) as one of his publications (in Russian) for 1914 a 'Letter to the editor [in connexion with an invalid reference to a memoir of Bienaymé's *Considerations à l'appui de la découverte de Laplace*, mentioned in a note of I. Sh. *In memory of James Bernoulli*]...'.

4. Other contributions

In writing this short historical note on Bienaymé, we do not mean to imply that his contributions discussed herein are his most important contributions in probability and statistics. We have given special attention to the points discussed here since it is clear that the generally accepted attributions have not been entirely accurate. It should be noted, for example, that Bienaymé has made a considerable contribution to the theory of least squares and the development of χ^2 theory, including the Γ distribution, the multivariate central limit theorem and the distribution of linear forms in the multinomial. An interesting historical discussion of this material has been given by Lancaster (1966).

5. Biographical comment

Irénée Jules Bienaymé was born in Paris on 28 August 1796 and died there on 20 October 1878. He joined the Civil Service in 1820 and was appointed general inspector of finance in 1834. After the revolution of 1848 he retired and devoted all his time to scientific work. No extensive biography of Bienaymé has been written but a number of short notes have appeared. The most comprehensive is by de la Gourneria (1878); see also notes by Franceschini (1954) and Dugué (1968). There appears to be a good case for a more detailed treatment, which the authors are undertaking at present.

The authors wish to express their appreciation to Professor H. O. Lancaster for valuable suggestions and for encouragement during the course of this work.

References

BERTRAND, J. (1875). Note relative au théorème de M. Bienaymé. *C.R. hebd. Séance Acad. Sci. Paris* **81**, 458; 491–2.

BIENAYMÉ, I. J. (1845). De la loi de multiplication et de la durée des familles. *Soc. Philomath. Paris Extraits*, Ser. 5, 37–9.

BIENAYMÉ, I. J. (1853). Considérations à l'appui de la decouverte de Laplace sur la loi de probabilité dans la méthode des moindres carrés. *C.R. hebd. Séance Acad. Sci. Paris* **37**, 309–24. Reprinted (1867), *Liouville's J. Math. Pures Appl.* (2) **12**, 158–76.

BIENAYMÉ, I. J. (1874). Sur une question de probabilités. *Bull. Math. Soc. Fr.* **2**, 153–4.

BIENAYMÉ, I. J. (1875). Application d'un théorème nouveau du calcul des probabilités. *C.R. hebd. Séance Acad. Sci. Paris* **81**, 417–23.

CHEBYSHEV, P. L. (1867). Des valeurs moyennes. *Liouville's J. Math. Pures Appl.* (2) **12**, 177–84 (published simultaneously in Russian in *Mat. Sbornik* (2) **2**, 1–9).

CHEBYSHEV, P. L. (1874). Sur les valeurs limites des intégrales. *Liouville's J. Math. Pures Appl.* (2) **19**, 157–60.

DUGUÉ, D. (1968). *International Encyclopaedia of the Social Sciences*, Vol. 2, pp. 73, 74. Chicago: Macmillan and The Free Press.

FRANCESCHINI, E. (1954). *Dictionnaire de Biographie Francaise*, Tome 6, p. 415. Paris: Librairie Letouzey et Ané.

GALTON, F. (1873). Problem 4001. *Educational Times*, 1 April, p. 17.

GALTON, F. & WATSON, H. W. (1874). On the probability of the extinction of families. *J. R. Anthropol. Inst.* **4**, 138–44.

DE LA GOURNERIE, J. (1878). *C. R. hebd. Séance Acad. Sci. Paris* **87**, 617–9.

HALDANE, J. B. S. (1927). A mathematical theory of natural and artificial selection, V. *Proc. Camb. Phil. Soc.* **23**, 838–44.

KENDALL, D. G. (1966). Branching processes since 1873. *J. Lon. Math. Soc.* **41**, 385–406.

KENDALL, M. G. (1948). *The Advanced Theory of Statistics*, vol. 2. London: Griffin.

KENDALL, M. G. & STUART, A. (1968). *The Advanced Theory of Statistics*, 2nd edition, vol. 3. London: Griffin.

LANCASTER, H. O. (1966). Forerunners of the Pearson χ^2. *Aust. J. Statist.* **8**, 1t7–26.

MARKOV, A. A. (1951). *Izbrannie Trudy*. Moscow: Izd. AN SSSR.

NETTO, E. (1901). *Lehrbuch der Combinatorik*. Leipzig.

STEFFENSEN, J. F. (1930). Om Sandsynligheden for at Afkommet uddør. *Matematisk Tidsskrift* B **1**, 19–23.

VON SCHRUTKA, L. (1941). Eine neue Einteilung der Permutationen. *Math. Annalen* **118**, 246–50.

WALLIS, W. A. & MOORE, G. H. (1941 *a*). A significance test for time series analysis. *J. Am. Statist. Ass.* **36**, 401–9.

WALLIS, W. A. & MOORE, G. H. (1941 *b*). A significance test for time series and other ordered observations. Tech. Paper No. 1, *Natl. Bureau Econ. Res.* New York.

WATSON, H. W. (1873). Solution to Problem 4001. *Educational Times*, 1 August, 115–6.

[*Received February* 1972. *Revised March* 1972]

Simon Newcomb, Percy Daniell, and the History of Robust Estimation 1885–1920

STEPHEN M. STIGLER*

(1973)

This article reviews some of the history of robust estimation, empha-sizing the period 1885–1920 and the work of Newcomb, Edgeworth, Shep-pard, and Daniell. Particular attention is paid to lines of development which have excited recent interest, including linear functions of order statistics and mixtures of normal densities as models for heavy-tailed populations.

I. INTRODUCTION

In the eighteenth century, the word "robust" was used to refer to someone who was strong, yet boisterous, crude, and vulgar. By 1953 when Box first gave the word its statistical meaning, the evolution of language had elimi-nated the negative connotation: robust meant simply strong, healthy, sufficiently tough to withstand life's adversities. The subject of robust inference, just like the word "robust," has a long and varied history. It is the aim of this present study to examine a part of this history and its relationship to current work.

The scope of this article will be rather narrow—we shall only be concerned with the *mathematical* background and development of robust estimation up to 1920. Thus we shall be less concerned with the first appearances of estimators such as the median and trimmed mean than with the first mathematical analyses of their behavior and properties. The main emphasis will be on the period 1885–1920, and particular attention will be given to work which is not widely known, yet is relevant to modern lines of thought. Section 3 discusses the contributions of Simon Newcomb to robust estimation, and to the use of normal mixtures as models for heavy-tailed distributions; Section 4 is concerned with the history of the mathe-matical analysis of order statistics in relation to robust estimation, with due attention to the works of Laplace, Sheppard, and Percy Daniell; and Section 5 contains some brief remarks on "*M*-estimators."

The reader may be as surprised as the author was to find to what extent priorities in these areas have been misassigned. While many other points will be touched upon in the article, our major findings are as follows: Laplace [46, 1818] and Sheppard [70, 1899a] seem to

* Stephen M. Stigler is associate professor, Department of Statistics, University of Wisconsin, Madison, Wis. 53706, and is presently on leave (1972–3) at the De-partment of Statistics, University of Chicago. This research was partially supported by Research Grant No. NSF GP 32037x from the Division of Mathematical, Physical and Engineering Sciences of the National Science Foundation at the Department of Statistics, University of Chicago, and by the U.S. Navy through the Office of Naval Research under Contract No. ONR-N00014-67A-0128-0017 at the Department of Statistics, University of Wisconsin. The author wishes to thank William Kruskal, Churchill Eisenhart, and Oscar B. Sheynin for a number of refer-ences and helpful comments.

have been the first to present a large sample theory for one or two order statistics. Simon Newcomb [54, 1886] provided the first sound, modern approach to robust estimation, including the first use of mixtures of normal densities as representing heavy-tailed distributions. Percy Daniell [8, 1920] should be credited with the first mathematical analysis of the class of estimators which are linear functions of order statistics, including the derivation of the optimal weighting functions for estimat-ing scale and location parameters (the so-called "ideal" linear estimators) and the first mathematical treatment of the trimmed mean. Some of Newcomb's work has been commented on recently by Huber [39, 1972], but much of the remainder of the work discussed in this article, including that due to Edgeworth, Galton, Laplace, Sheppard, and Daniell, has been largely ignored in recent years.

We shall begin with a brief overview of the situation prior to 1885.

2. THE SITUATION BEFORE 1885

Scientists have been concerned with what we would call "robustness"—insensitivity of procedures to de-partures from assumptions, particularly the assumption of normality—for as long as they have been employing well-defined procedures, perhaps longer. For example, in the first published work on least squares, Legendre [47, 1805] explicitly provided for the rejection of outliers:

> If among these errors are some which appear too large to be admissible, then those equations which produced these errors will be rejected, as coming from too faulty experiments, and the unknowns will be determined by means of the other equations, which will then give much smaller errors.

Yet most of the early work in *mathematical* statistics was obsessed with "proving" the method of least squares, either starting with the assumption that the sample mean is the best estimate of the mean and deriving the normal distribution, as Gauss did in his first proof in 1809, or starting with the Central Limit Theorem, as did Laplace in 1811. The first mathematical work on robust estimation seems to have been that of Laplace [46, 1818] on the distribution of the median. We shall defer a discussion of Laplace's work until Section 4 where it will

Reprinted from: © **Journal of the American Statistical Association**
December 1973, Volume 68, Number 344
Theory and Methods Section
Pages 872–879

be considered with later work on linear functions of order statistics.

The next statistical problem connected with robust estimation to receive mathematical treatment was the rejection of outliers. In 1852, the first proposal of a criterion for the determination of outliers was published by Benjamin Peirce, the Harvard mathematician-astronomer and father of logician-philosopher C. S. Peirce. Peirce's paper and most others on this subject[1] were not really about robust estimation, as their authors did not concern themselves with the properties of the resulting estimators; rather, they implicitly assumed that after the outlier test was performed the estimation could be done with no thought given to what had gone before, nor what information might be lost. This narrowness of view did not go unnoticed at the time. The first paper proposing an outlier criterion (Peirce [66, 1852]) was soon followed by the first paper criticizing the use of outlier criteria (Airy [1, 1856]). Airy, the Astronomer Royal, wrote:

> And I have, not without surprise to myself, been led to think that the whole theory is defective in its foundation, and illusory in its results; that no rule for the exclusion of observations can be obtained by any process founded purely upon a consideration of the discordance of these observations.

A lively debate ensued, with the participants not always expressing themselves with Airy's restraint. For example, Glaisher [32, 1872] wrote "Professor Pierce's [sic] criterion for the rejection of doubtful observations seems to me to be destitute of scientific precision." At one point an exchange in print between the mathematician Glaisher and the astronomer Stone became so heated that one of Glaisher's papers was itself rejected by the *Monthly Notices of the Royal Astronomical Society* due to the personal nature of his comments (see Glaisher [34, 1874]).

One of the more interesting papers of this time (and one of the most unusual statistical papers of all time) appeared in the *Report of the Superintendent of the U.S. Coast Survey for 1870*. It is by C.S. Peirce, written while he was an Assistant to the Coast Survey (at the time his father was Superintendent of the Survey!). In the paper Peirce presented the then standard material of the theory of errors, but in the language and notation which he had developed for the logic of relations, for which he later became famous. Thus we find, regarding averages,

> Since [m] denotes all men, we may naturally write [m]/m to denote what all men become when that factor is removed which makes [m] refer to *men* rather than to anything else; that is to say, to denote the *number* of men. We may write this for short [m] with heavy brackets. Then t being a relative term ("a tooth of,") by [tl] will be denoted the total number of teeth in the universe. But [t] will be used as equivalent to [tl]/[1], or the average number of teeth that anything has.

Peirce included a sensible—one is tempted to say "logical"—defense of his father's outlier criterion in the

paper (p. 210). By 1885 a number of rejection criteria were in use, often only by the proposer and his employees.[2]

But techniques other than simply "reject outliers, then use the sample mean" were also employed. A variety of weighted means had been used prior to 1885. For example, in 1763 James Short (an English astronomer and noted manufacturer of telescopes) had estimated the sun's parallax based on observations of the transit of Venus of 1761 by averaging three means: the sample mean, the mean of all observations with residuals less than one second, and the mean of those with residuals less than half a second. The median and the midrange had appeared even earlier (Eisenhart [23, 1971]).

By the last half of the nineteenth century, weighted least squares had become a standard topic in the literature of the theory of errors, and it was a frequent practice (at least in astronomical investigations) to weight observations differently, depending upon the scientist's (often subjective) estimate of the "probable error"[3] of the observation. The estimate of the probable error was supposed to be based solely on external evidence: scientists were warned of the possible biases if the magnitude of the observation were allowed to influence its weight (see, for example, Jevons [44, 1874, p. 450]), but it is doubtful that this advice was faithfully adhered to. We shall discuss the use of these weighted means further in the next section, in connection with the contributions of Simon Newcomb.

Other estimators were proposed in this period. In particular, De Morgan [11, 1847, p. 456] had outlined a scheme for discounting the more extreme observations. This method, more fully developed by Glaisher [33, 1873], amounted to starting with the sample mean, then assigning different probable errors to the different observations based on the value of the likelihood function at those observations, and iterating this process. Glaisher's estimate was criticized by both Stone [75, 1873] and Edgeworth [14, 1883], who both (independently) proposed an alternative based on looking for a maximum of the likelihood function (without assuming equal probable errors).[4] Edgeworth later became disenchanted with this alternative [17, 1887a].

At about this time, Francis Galton was making much use of the median [27, 1875], although his motivation was less suspicion of the normal distribution, which he considered a good representation of many real phenomena, than an appreciation of the simplicity, ease of calculation, and ease of interpretation of the median. Many of these same features were also cited in the apparently independent work of Fechner [26, 1878]; see Walker, [76, 1929, p. 84–6]. Also, various formulas for

---

[1] See Anscombe [3, 1960] and Rider [69, 1933] for historical surveys of outlier techniques.

[2] Outlier rejection rules for multivariate data seem to have not been considered prior to 1899, when Sheppard [71, 1899b] proposed a multivariate extension of Chauvenet's [6, 1863] rule.

[3] The probable error of a symmetric distribution is half the interquartile range; for normal distributions p.e. = $(.6745)\sigma$.

[4] Actually, the equations both Stone and Edgeworth derived by differentiation would lead to local minima of the likelihood function. This problem was recently discussed by Mantel [49, 1972].

index numbers were developed during this period; these included weighted averages and geometric means, each designed for a specific purpose.

However, it can still be said that by 1885, the conventional wisdom (but by no means the unanimous view) was that for purposes of estimation, the cautious use of the sample mean was recommended—sometimes weighted, sometimes after discarding outliers, but still the sample mean.

3. SIMON NEWCOMB AND MIXTURES OF NORMAL DENSITIES

1885 can be conveniently taken as the start of one of the most active and innovative periods in the history of mathematical statistics. The story of the development of mathematical statistics into a subject in its own right through the work of such men as Edgeworth, Karl Pearson, Gosset, and Fisher has been told by E.S. Pearson [62, 1967]. Our present, rather narrow purpose is to describe how the modern theory of robust estimation developed over this period. To this end, we shall place particular emphasis on the introduction of mixtures as models for the heavy-tailed distributions which scientists had encountered in practice, and on the use of linear functions of order statistics as robust estimators of location parameters.

Simon Newcomb appears to have been the first to introduce a mixture of normal densities as a model for a heavy-tailed distribution, and to exploit this model to get an estimator of location which was more robust than the sample mean. (Francis Galton and Karl Pearson had modeled measurements of natural populations by normal mixtures about the same time, but with a completely different object in mind, namely to demonstrate how a single population could be broken down into components.) While Newcomb's name may be unfamiliar to present day statisticians, it should not be so to astronomers, applied mathematicians, and economists.

Simon Newcomb (1835–1909) was born in Nova Scotia, attended Harvard, and spent most of his adult life (1861–1897) as a professor of mathematics in the U.S. Navy, working for the U.S. Nautical Almanac Office. He is generally regarded as the greatest American astronomer of the nineteenth century, and was responsible for many of the determinations of astronomical constants which are still accepted today. In addition, he was a powerful applied mathematician, and co-founded and for many years edited the *American Journal of Mathematics*. As an avocation he wrote *Principles of Political Economy* [53, 1885], a book which has established him as a major American economic theorist, and which contains one of the earliest modern mathematical statements of the quantity theory of money.

As was the practice in astronomy at the time, Newcomb made frequent use of weighted means in his estimation of astronomical constants. The relative weights were usually thought of in terms of "probable errors," and

were assigned somewhat subjectively on the basis of Newcomb's judgment of the relative accuracy of the process which produced the observation. For example, after assessing some data on eclipses collected by Ptolemy in the second century A.D., he remarked [51, 1878, p. 41]:

> the [assigned] probable errors are the result of judgment from the terms of [Ptolemy's] description rather than of calculation; they were estimated without any knowledge of the way the comparison with theory would come out, and are printed without subsequent alteration.

With more contemporary data, Newcomb would base his choice of weights upon "the quality of the image and the generally satisfactory way in which the image was kept on the crosswires" [55, 1891a, p. 170] in the case of an experiment he was personally involved with, and upon the number of observers, general opinion of the reporting observatory, and "number and force of the doubtful circumstances "[56, 1891b, p. 383], in cases involving combination of others' measurements. He was apparently aware of criticism of the subjective nature of these assignments, but he maintained that

> Opinions may doubtless differ as to whether a judicious system of weights has always been applied, but it is not likely that any unbiased reassignment would materially affect the result [58, 1898, p. 211].

Newcomb also rejected outliers when necessary, but usually only based on external evidence or really huge deviations.

With this experience in dealing with observations made with differing degrees of precision, it is not surprising that, when faced with a collection of non-normal observations for which there was no satisfactory way to weight them individually, he should consider a mixture of normal densities with different variances as a model. For, having observed that a collection of 684 residuals based on observations of the transits of Mercury had much heavier tails than the corresponding normal distribution (even with excessive deviations ignored), he wrote [52, 1882, p. 382]:

> It is evident that if we have a collection of observations of different degrees of probable error, in which, however, there is no way of distinguishing those of great probable error from those of small probable error, the law of the errors will not be that usually adopted, but there will be a comparative excess of large residuals. It is also evident that in such a case the arithmetical mean does not necessarily give the most probable result. For, in the case of an observation of large residual, there is evidently a preponderance of probability that it belongs to a class with large probable error, and therefore should be assigned least weight. . . . That any general collection of observations of transits of Mercury must be a mixture of observations with different probable errors was made evident to the writer by his observations of the transit of May 6, 1878, which may be here described as an illustration of the subject.

Four years after writing this, Newcomb published a remarkable paper in his own journal, the *American Journal of Mathematics*, in which he used this model to arrive at a more robust estimator of location than the sample mean. In this paper [54, 1886],[5] after criticizing the overuse of outlier criteria and presenting his mixture model, he proceeded to develop an estimator upon the principles of Bayesian decision theory that gave "less weight to the more discordant observations." Adopting squared error as a loss function (Newcomb's word for loss was "evil"), he demonstrated that in general the posterior mean minimizes the expected mean square error, and he suggested the following procedure. 1) Calculate the residuals based on the sample mean, and, using trial and error, fit a mixture of a finite number of normal densities with zero means to these residuals. 2) Take this fitted mixture and, considering the location family it generates, estimate the desired mean by the posterior mean with respect to a uniform prior, given the original observations. Newcomb realized that this procedure presented practical difficulties and gave a number of simplifying approximations to arrive at a usable estimator. He illustrated its use with the data on the transits of Mercury.[6]

As an interesting sidelight, we note that in this article and in a later work, Newcomb made an early use of a simple version of Tukey's sensitivity function (see Andrews et al. [2, 1972, p. 96]). In a paper [60, 1912, p. 212] discussing the unsatisfactory nature of outlier criteria, he wrote that if all observations with large residuals are rejected (and the mean estimated from the remaining observations), then the final result

> becomes a discontinuous function of the residual of the rejected observation, the continuity being broken at the point regarded as the limit of normal error. A simple example will make the case clear. If we have three observed results a, b, c of which the mean is to be taken, and if c be the result which may be abnormal, then so long as c is retained we shall have
>
> $$\text{mean} = \tfrac{1}{3}(a + b + c);$$
>
> the mean will then continuously increase with c. When c passes the normal limit, the mean changes per saltum to
>
> $$\tfrac{1}{2}(a + b).$$

In the same posthumous paper [60, 1912, p. 214], Newcomb also proposed a very simple estimator in the spirit of his 1886 paper: weight the observation X_i by $w_i = c/\max (|X_i - \bar{X}|, c)$, where c is a constant to be specified, stating that when $|X_i - \bar{X}| > c$, "This will lead to practically the same result as if we substituted for the actually observed quantity another quantity corresponding to the residual c." Peter J. Huber has pointed out to me that the fixed point of the iterative version of this estimator, where one replaces $\bar{X}$ by the weighted mean with weights w_i and iterates, is identical to one of Huber's favorite M-estimators (see Section 5), with $\phi(x) = \max [-c, \min (x, c)]$!

4. LAPLACE, SHEPPARD, DANIELL, AND LINEAR FUNCTIONS OF ORDER STATISTICS

With few exceptions, statisticians were quite late in coming to consider any but the simplest linear functions of order statistics as estimators of means. By a linear function of order statistics we shall mean any weighted linear combination of observations where the weights depend *only* on their order, not on their magnitudes or the size of their residuals. The median and the midrange, two members of this class, evidently have a long history (Eisenhart [22, 1964; 23, 1971]), but perhaps the first extensive mathematical analysis to be published involving order statistics was by Laplace. In the second supplement [46, 1818] to his monumental *Théorie Analytique des Probabilités*, Laplace considered the problem we would now call linear regression through the origin: $a_i = p_i y + x_i$, a_i, p_i known, y to be estimated, where the errors x_i were assumed to have an arbitrary continuous, symmetric distribution. By looking for that estimator which minimized the sum of the absolute values of the residuals, he was led to consider an estimator of y which reduces to the median of the a_i's in the case $p_i \equiv 1$. Laplace derived the density of this estimator, showed that this density approaches the normal density as the sample size increases, and gave the necessary and sufficient condition on the error distribution that the median have a smaller asymptotic variance than the sample mean.[7] Laplace's proof is easily adapted to any sample percentile and asymmetrical populations, as was in fact later noted by Edgeworth [15, 1885; 16, 1886]. In addition, Laplace derived the joint asymptotic density of the sample mean and median, and used it to find which linear combination of these estimators has the smallest asymptotic variance. (As the weights depend upon the unknown error distribution, he termed this result "impracticable," but noted that if the error distribution were normal, the best linear combination was the sample mean alone.)[8] Two years before Laplace's investigation, Gauss [31, 1816], considering the problem of estimating the probable error of a normal distribution, had suggested the use of the median of the absolute values of the residuals, and stated (without proof) the asymptotic probable error of the median for this special case. Gauss apparently never published or circulated a proof, for 16 years later Encke [25, 1832], who had corresponded with Gauss, felt it necessary to provide one, attributing it to Dirichlet. It seems likely that Dirichlet's proof for this special case was simply an adaptation of Laplace's, as Dirichlet was quite familiar with Laplace's work, the second supplement in particular (see Dirichlet [12, 1836]).

Later in the nineteenth century, Galton [27, 1875] and particularly Edgeworth [15, 1885; 18, 1887b; 19,

---

[5] Some of his arguments also appear in Newcomb [57, 1895, p. 81–86].

[6] Ogorodnikoff [61, 1928] provided a different simplification of Newcomb's estimator based on a Charlier expansion of the posterior distribution. The relationship between Newcomb's simplified estimator and the maximum likelihood estimator was discussed by Hulme and Symms [40, 1939].

[7] Laplace actually carried through his entire investigation in the more general regression situation, comparing the general estimator with the least squares estimator for this situation. For other views of Laplace's work and its historical context, see Eisenhart [21, 1961] and Stigler [74, 1973].

[8] The possibility of a linear function of two estimators outperforming both has been more fully exploited in the recent Princeton robustness study (see Andrews et al. [2, 1972, p. 132]).

1888], touted the use of the median in situations where heavier tails than the normal could be expected. Specifically, Edgeworth [19, 1888] used Laplace's results to conclude that the median may well be better than the mean when the population distribution is one of Newcomb's mixtures of normal distributions. Also, Edgeworth [16, 1886] seems to have been the first to realize that the median may possess an advantage over the sample mean for serially correlated data.

More complicated linear estimators began to appear in 1889, when Galton (in a footnote on p. 61–62 of *Natural Inheritance*, [28, 1889]) suggested estimating the mean and standard deviation of a normal distribution by what amounts to taking

$$\hat{\mu} = \frac{\xi_p X^{(nq)} - \xi_q X^{(np)}}{\xi_p - \xi_q},$$

$$\hat{\sigma} = \frac{X^{(np)} - X^{(nq)}}{\xi_p - \xi_q},$$

where ξ_p and ξ_q are the p and q percentiles of the standard normal distribution, $X^{(np)}$ and $X^{(nq)}$ are the sample $100p$ and $100q$ percentiles, and p and q are arbitrary but fixed ($0 < p < q < 1$). In 1899 in a long paper on the multivariate normal distribution and its applications, Sheppard proved the joint asymptotic normality of Galton's estimators when the population is normal. He also showed the joint asymptotic normality of $X^{(np)}$ and $X^{(nq)}$, and gave analogues to $\hat{\mu}$ and $\hat{\sigma}$ based on any finite number of sample percentiles [70, 1899, p. 131–2]. Sheppard's (sketchy) proof, which is based on an implicit use of the probability integral transformation, can be easily adapted to any regular distribution.[9]

Sheppard's paper also represented the first attempt since Laplace to optimize performance within a class of linear functions of order statistics. He both showed how the best choice (for normal populations) of p and q can be made [70, 1899, p. 135] and found which linear combination of the three quartiles has the smallest asymptotic variance (again for normal populations) [70, 1899, footnote, p. 134]. Such functions of the three quartiles had been considered earlier by Edgeworth [1893], who neglected the quartiles' correlation and erroneously claimed the estimator with weights in proportions 5:7:5 to be superior to the sample mean for normal populations. Recent work, however, seems to bear out Edgeworth's claim that such an estimator is to be recommended on grounds of robustness (for example, see Gastwirth [30, 1966] and Andrews *et al.* [2, 1972]).

The next mathematical work to appear on order statistics was Karl Pearson's [63, 1902] examination of the Galton difference problem. In this paper, which was inspired by an inquiry of Galton's [29, 1902] as to the most suitable proportion between the values of first and second prizes, Pearson gave the joint density of any two consecutive order statistics and found their expected

difference. He remarked in a footnote that

> I propose on another occasion to consider the application of Galton's problem to a new theory for the rejection of outlying individuals.

This proposal was later carried out by J.O. Irwin [41, 1925].

In 1920, a remarkable paper appeared in the *American Journal of Mathematics* (the journal Simon Newcomb cofounded) by the English mathematician P.J. Daniell. This paper, "Observations Weighted According to Order," has been all but totally overlooked since it's publication. It could in fact be claimed that Daniell was at least thirty years ahead of his time, for it took that long for his major results to be rediscovered. While his paper itself is a model of clarity and rigor, its relevance to modern work is such that it merits a short summary, in his own notation.

The work was apparently inspired by a reading of Poincaré's *Calcul des Probabilités* [68, 1912]. After remarking how Poincaré had suggested discarding extreme observations (when normality is suspect) before taking the mean, Daniell wrote:

> Besides such a discard-average [i.e., the trimmed mean] we might invent others in which weights might be assigned to the measures according to their order. In fact the ordinary average or mean, the median, the discard-average, the numerical deviation (from the median, which makes it a minimum), and the quartile deviation can all be regarded as calculated by a process in which the measures are multiplied by factors which are functions of order. It is the general purpose of this paper to obtain a formula for the mean square deviation of any such expression. This formula may then be used to measure the relative accuracies of all such expressions.

Daniell's analysis proceeded as follows: First he explicitly introduced the probability integral transformation (apparently the first time this was done)[10] and explained how it could be used to find the moments of any function of order statistics. Then, he assumed the population density $p(t)$ was regular (and indefinitely differentiable), and he expanded the inverse of the distribution function in a Taylor series to derive asymptotic expressions for the mean of an order statistic t_r and the mean product of any two. He thus duplicated some of Sheppard's [70, 1899a] results, but in a much more rigorous manner.

Daniell then considered statistics of the form $\bar{t} = \sum_{r=1}^{n} f_r t_r$, where he assumed that the weight f_r associated with the rth order statistic t_r was given by

$$f_r = \frac{1}{n} f\left(\frac{r}{n+1}\right),$$

and put things together to obtain the (now standard)[11] expression for the asymptotic variance of $\bar{t}$,

$$S^2 = \int_{-\infty}^{\infty} \phi^2(t) p(t) dt,$$

---

[9] Twenty years later, Karl Pearson [64, 1920] presented part of Sheppard's proof in more detail, made the obvious step to more general distributions than normal, and much more fully examined the consequences of the result.

[10] The next being Karl Pearson [65, 1931].

[11] See Chernoff, Gastwirth, and Johns [7, 1967].

where $\phi(t)$ is the indefinite integral of $f[x(t)]$,

$$x(t) = \int_{-\infty}^{t} p(u)du.$$

If he was less than specific as to why the remainder terms are uniformly negligible, his standard of rigor was nonetheless far above that of the statistical literature of the time.

In the third section of the paper, Daniell gave the conditions on f under which the asymptotic mean of $\hat{t}$ is the population mean or standard deviation, and defined the "accuracy" of $\hat{t}$ as the ratio of the asymptotic variance of the sample mean (or sample standard deviation, as the case may be) to that of $\hat{t}$. (He also derived the asymptotic variance of the sample standard deviation here.) In the fourth section, Daniell gave the optimal weight function f—that which minimizes S^2—for both the location and scale cases, using standard results from the calculus of variations, and noted that the optimal estimate of σ for the normal case is as accurate as the sample standard deviation in this case. These results were not to appear in print again until Jung [45, 1955], after the papers of Godwin [35, 1949] and Lloyd [48, 1952] had renewed interest in this subject, although they are in Bennett's [4, 1952] unpublished thesis.

The final two sections were concerned with applications. Daniell gave special attention to the "discard-average" (the trimmed mean), presenting the (now standard) expression for its asymptotic variance and evaluating its performance for various Pearson densities, including Student's t. He also gave conditions under which the quartile-discard average is superior to the sample mean. The paper ended with a number of applications to other estimators of location and scale,[12] with numerical results. Daniell did not derive the asymptotic normality of $\hat{t}$, nor did he try to state minimal regularity conditions (indeed, some of his regularity conditions were implied rather than stated). However, taken altogether it is a thoroughly modern paper which almost appears to have been gleaned from the literature of the 1950's and 1960's.

How could such a paper have gone unnoticed for all these years?[13] To see why, we need to learn something of Daniell's life. Percy John Daniell (1889–1946) received a B.A. degree at Cambridge in 1910 (and an M.A. in 1914), where his honors included Senior Wrangler in Mathematics (1909), First Class Physics Tripos (1910), and the Raleigh Prize (1912). His stay at Cambridge would have overlapped R.A. Fisher's, but they were at different colleges and may not have met. After graduation (and brief stays at Göttingen and Liverpool), Daniell went to Rice Institute in Houston, Texas in 1912 as a travelling fellow. He remained at Rice until

1923, becoming a full professor in 1920. It was at Rice that he did his most important work, principally on the theory of integration (including the development of what is now known as the Daniell integral.) In 1924 he returned to England to the University of Sheffield, where he remained until his death at the age of 57. In the latter part of his life he published occasional papers on applied mathematics, on such topics as flame motion, potentials, and quadrature formulas.

The paper [8, 1920], written at Rice, seems to have been Daniell's only related work in statistics. This fact, together with his isolation from active statistical research (both at Rice and Sheffield), was largely responsible for the obscurity of the paper. Daniell's death before his results were rediscovered and widely discussed, and Wilks' overlooking his work in the survey paper of 1948 also served to delay recognition of his priority. As a further irony, these circumstances have helped relegate to obscurity another important paper of Daniell's, "Integral Products and Probability" [9, 1921], in which he presents one of the earliest mathematical treatments of continuous time Markov processes, including the Chapman-Kolmogorov equation (ten years before Kolmogorov) and a short treatment of the Wiener process (two years before Wiener).

5. M-ESTIMATORS

Recently, much attention has been given to a class of robust estimators which Huber calls "M-estimators," M for maximum-likelihood type (see Huber [39, 1972]). T is said to be an M-estimator corresponding to a function ϕ if T is a solution to $\sum \phi(X_i - T) = 0$. Each choice of ϕ determines an estimator; if $\phi = p'/p$, T is the maximum likelihood estimator for the location parameter of the population with density $p(t - \theta)$. As the first appearance of these estimators in the context of robustness seems to be in the work of Jeffreys after 1920 (see Jeffreys [42, 1932] and [43, 1939], in particular), and as this work is outside the scope of this study, we shall not dwell on this subject. However, we cannot resist calling attention to an early reference in which the class of M-estimators is introduced and their consistency claimed.

In a paper examining the various "proofs" of the method of least squares, Ellis [24, 1844] began with Gauss's first proof. Letting x_i's denote observed values, "a" the quantity to be estimated, and $e_i = x_i - a$, Ellis questions Gauss's a priori designation of the arithmetic mean (the solution to $\sum(x_i - a) = 0$) as the most probable value.

> It [the arithmetic mean] is not the only rule to which these considerations might lead us. For not only is $\sum e = 0$ ultimately, but $\sum fe = 0$, where fe is any function such that $fe = -f(-e)$; and therefore we should have
>
> $$\sum f(x - a) = 0,$$
>
> as an equation which ultimately would give the true value of x when the number of observations increases *sine limite*, and which therefore for a finite number of observations may be looked on

---

[12] Including the "discard-deviation," where the *inner* quartiles are discarded.

[13] A fairly complete review of the literature reveals only two published citations, Dodd [13, 1922] and Greenberg [37, 1968], and the descriptions there are superficial and misleading. Daniell's paper came to my attention as the result of a systematic inspection of the *American Journal of Mathematics*.

in precisely the same way as the equation which expresses the rule of the arithmetic mean. There is no discrepancy between these two results. At the limit they coincide: short of the limit both are approximations to the truth. Indeed we might form some idea how far the action of fortuitous causes had disappeared from a given series of observations by assigning different forms of f, and comparing the different values thus found for a.

No satisfactory reason can be assigned why, setting aside mere convenience, the rule of the arithmetic mean should be singled out from other rules which are included in the general equation $\sum f(x - a) = 0$.

Thus Ellis has claimed (without proof or regularity conditions) the consistency of M-estimators,[14] and even suggested the class may be useful for judging to what degree an estimated value depends on the choice of estimator, a stability test. Of course Ellis was not really concerned with robustness, only with illuminating the arbitrary nature of Gauss's proof, but his comments are of interest nonetheless.[15]

[Received January 1973. Revised June 1973.]

REFERENCES

[1] Airy, G.B., (1856), Letter from Professor Airy, Astronomer Royal, to the editor, *Astronomical Journal*, 4, 137–38.

[2] Andrews, D.F., Bickel, P.J., Hampel, F.R., Huber, P.J., Rogers, W.H. and Tukey, J.W., (1972), *Robust Estimates of Location: Survey and Advances*, Princeton: Princeton University Press.

[3] Anscombe, F.J., (1960), "Rejection of Outliers," *Technometrics*, 2, No. 1, 123–47.

[4] Bennett, C.A., (1952), "Asymptotic Properties of Ideal Linear Estimators," Unpublished dissertation, University of Michigan.

[5] Box, G.E.P., (1953), "Non-Normality and Tests on Variances," *Biometrika*, 40, 318–35.

[6] Chauvenet, W., (1863), *A Manual of Spherical and Practical Astronomy, Vol. 2*, Philadelphia: J. B. Lippincott Co., 558–66.

[7] Chernoff, H., Gastwirth, J. and Johns, M.V., (1967), "Asymptotic Distribution of Linear Combinations of Functions of Order Statistics with Applications to Estimation," *Annals of Mathematical Statistics*, 38, No. 1, 52–72.

[8] Daniell, P.J., (1920), "Observations Weighted According to Order," *American Journal of Mathematics*, 42, 222–36.

[9] ———, (1921), "Integral Products and Probability," *American Journal of Mathematics*, 43, 143–62.

[10] David, H.A., (1970), *Order Statistics*, New York: John Wiley and Sons, Inc.

[11] De Morgan, A., (1847), "Theory of Probabilities," *Encyclopedia of Pure Mathematics* (Part of *Encyclopedia Metropolitana*), 393–490.

[12] Dirichlet, G.L., (1836), "Uber die Methode der kleinsten Quadrate," In *G. Lejeune Dirichlet's Werke, Vol. I*, 281–2. Berlin: Reimer, 1889.

[13] Dodd, E.L., (1922), "Functions of Measurements under General Laws of Error," *Skandinavisk Aktuarietidskrift*, 5, 133–58.

[14] Edgeworth, F.Y., (1883), "The Method of Least Squares," *Philosophical Magazine*, 16, Ser. 5, 360–75.

[15] ———, (1885), "Observations and Statistics. An Essay on the Theory of Errors of Observation and the First Principles of Statistics," *Transactions of the Cambridge Philosophical Society*, 14, 138–69.

[16] ———, (1886), "Problems in Probabilities," *Philosophical Magazine*, 22, Ser. 5, 371–84.

[17] ———, (1887a), "On Discordant Observations," *Philosophical Magazine*, 23, Ser. 5, 364–75.

[18] ———, (1887b), "The Choice of Means," *Philosophical Magazine*, 24, Ser. 5; 268–71.

[19] ———, (1888), "On a New Method of Reducing Observations Relating to Several Quantities," *Philosophical Magazine*, 25, Ser. 5, 184–91.

[20] ———, (1893), "Exercises in the Calculation of Errors," *Philosophical Magazine*, 36, Ser. 5, 98–111.

[21] Eisenhart, C., (1961), "Boscovich and the Combination of Observations," Chapter 7, in L.L. Whyte, ed., *R.J. Boscovich Studies of His Life and Work*, London: Allen and Unwin. (Reissued 1963 by Fordham University Press, New York).

[22] ———, (1964), "The Meaning of 'Least' in Least Squares," *Journal of the Washington Academy of Sciences*, 54, 24–33. Reproduced in Precision Measurement and Calibration. Selected NBS Papers on Statistical Concepts and Procedures (ed. H.H. Ku). National Bureau of Standards (US) Special Publication 300, Volume 1. Washington, D.C.: U.S. Government Printing Office, 1969.

[23] ———, (1971), "The Development of the Concept of the Best Mean of a Set of Measurements from Antiquity to the Present Day," 1971 A.S.A. Presidential Address.

[24] Ellis, R.L., (1844), "On the Method of Least Squares," *Transactions of the Cambridge Philosophical Society*, 8, 204–19.

[25] Encke, J.F., (1832), "Über die Methode der Kleinsten Quadrate," *Berliner Astronomisches Jahrbuch für 1834*, 249–312. Translated from the German in R. Taylor, ed., *Scientific Memoirs, Selected from the Transactions of Foreign Academies of Science and Learned Societies and from Foreign Journals, Vol. 2*, 1841, (reissued by Johnson Reprint Corporation, New York and London, 1966).

[26] Fechner, G.T., (1878), "Ueber den Ausgangswerth der Kleinsten Abweichungssumme dessen Bestimmung, Verwendung, und Verallgemeinerung," *Abhandlungen der Mathematisch-Physischen Classe der Königlich Sächsischen Gesellschaft der Wissenschaften, Leipzig*, 11, 1–76.

[27] Galton, F., (1875), "Statistics by Intercomparison, with Remarks on the Law of Frequency of Error," *Philosophical Magazine*, 49, Ser. 4, 33–46.

[28] ———, (1889), *Natural Inheritance*. London: Macmillan and Co., Ltd.

[29] ———, (1902), "The Most Suitable Proportion Between the Values of First and Second Prizes," *Biometrika*, 1, 385–90.

[30] Gastwirth, J., (1966), "On Robust Procedures," *Journal of the American Statistical Association*, 61 (December 1966), 929–48.

[31] Gauss, C.F., (1816), "Bestimmung der Genauigkeit der Beobachtungen," in *Carl Friedrich Gauss Werke*, Band 4, 109–17, Göttingen: Königlichen Gesellschaft der Wissenschaften (1880). Summarized in Section 103 of E.T. Whittaker and G.R. Robinson, *The Calculus of Observations* (London and Glasgow: Blackie and Son, Ltd., 1924; 4th edition (1944) reissued 1967 by Dover Publications, Inc., New York).

[32] Glaisher, J.W.L., (1872), "On the Law of Facility of Errors of Observations, and on the Method of Least Squares," *Memoirs of the Royal Astronomical Society*, 39, Part II, 75–124.

[33] ———, (1873), "On the Rejection of Discordant Observations," *Monthly Notices of the Royal Astronomical Society*, 33, 391–402.

[34] ———, (1874), "Note on a Paper by Mr. Stone, 'On the Rejection of Discordant Observations'," *Monthly Notices of the Royal Astronomical Society*, 34, 251.

[14] Huber proved this in [38, 1964].

[15] In addition to those works cited below, many other works were consulted for references and general information. The information on the life of Simon Newcomb came principally from the *Encyclopedia Britannica*, the *International Encyclopedia of the Social Sciences*, and Newcomb's autobiography [59, 1903]. The information on the life of Percy Daniell came from various editions of *Who's Who* and *American Men of Science*, and Stewart [73, 1947]. Merriman's [50, 1877] bibliography on least squares was quite useful for the period prior to 1877. Good bibliographies of work since 1920 can be found in H.A. David's *Order Statistics* [10, 1970], and in Govindarajulu and Leslie [36, 1972].

[35] Godwin, H.J., (1949), "On the Estimation of Dispersion by Linear Systematic Statistics," *Biometrika*, 36, 92–100.

[36] Govindarajulu, Z. and Leslie, R.T., (1972), *Annotated Bibliography on Robustness Studies of Statistical Procedures*, U.S. Department of H.E.W. DHEW Publication No. (HSN), U.S. Government Printing Office, 72–1051.

[37] Greenberg, B.G., (1968), "Nonparametric Statistics: Order Statistics," *International Encyclopedia of the Social Sciences*, New York: The Macmillan Company and the Free Press.

[38] Huber, P.J., (1964), "Robust Estimation of a Location Parameter," *Annals of Mathematical Statistics*, 35, No. 1, 73–101.

[39] ——, (1972), "Robust Statistics: A Review," *Annals of Mathematical Statistics*, 43, No. 4, 1041–67.

[40] Hulme, H.R. and Symms, L.S.T., (1939), "The Law of Error and the Combination of Observations," *Monthly Notices of the Royal Astronomical Society*, 99, 642–9.

[41] Irwin, J.O., (1925), "On a Criterion for the Rejection of Outlying Observations," *Biometrika*, 17, 238–50.

[42] Jeffreys, H., (1932), "An Alternative to the Rejection of Observations," *Proceedings of the Royal Society*, 137, Ser. A, 78–87.

[43] ——, (1939), "The Law of Error and the Combination of Observations," *Philosophical Transactions of the Royal Society of London*, 237, Ser. A, 231–71.

[44] Jevons, W.S., (1874), *The Principles of Science I*, London: Macmillan, 2nd ed., (1877) reissued 1958 by Dover Publications, Inc., New York.

[45] Jung, J., (1955), "On Linear Estimates Defined by a Continuous Weight Function," *Arkiv For Matematik*, Band 3, nr. 15, 199–209.

[46] Laplace, P.S. de, (1818), *Deuxieme Supplement a la Théorie Analytique des Probabilités*. Paris: Courcier. (pp. 569–623 in *Oeuvres de Laplace*, 7, Paris: Imprimerie Royale (1847); pp. 531–80 in *Oeuvres Complètes de Laplace*, 7, Paris: Gauthier-Villars (1886).)

[47] Legendre, A.M., (1805), "On the Method of Least Squares." Translated from the French in D.E. Smith, ed., *A Source Book in Mathematics*, 576–79, New York: Dover Publications, Inc., 1959.

[48] Lloyd, E.H., (1952), "Least Squares Estimation of Location and Scale Parameters Using Order Statistics," *Biometrika*, 39, 88–95.

[49] Mantel, N., (1972), Letter to the editor, "Another Maximum Likelihood Oddity," *The American Statistician*, 26, No. 5 (December 1972), 45.

[50] Merriman, M., (1877), "A List of Writings Relating to the Method of Least Squares, with Historical and Critical Notes," *Transactions of the Connecticut Academy of Arts and Sciences*, 4, 151–232.

[51] Newcomb, S., (1878), "Researches on the Motion of the Moon, I," *Washington Observations for 1875*—Appendix II (published by the U.S. Naval Observatory, Washington.)

[52] ——, (1882), "Discussion and Results of Observations on Transits of Mercury from 1677 to 1881," *Astronomical Papers 1*, 363–487. Published by the U.S. Nautical Almanac Office.

[53] ——, (1885), *Principles of Political Economy*, New York: Harper and Brothers, Reissued in 1966 by Augustus M. Kelley, New York.

[54] ——, (1886), "A Generalized Theory of the Combination of Observations so as to Obtain the Best Result," *American Journal of Mathematics*, 8, 343–66.

[55] ——, (1891a), "Measures of the Velocity of Light Made under the Direction of the Secretary of the Navy during the Years 1880 to 1882," *Astronomical Papers*, 2, 107–230. Published by the U.S. Nautical Almanac Office.

[56] ——, (1891b), "Discussion of Observations of the Transits of Venus in 1761 and 1769," *Astronomical Papers*, 2, 259–405. Published by the U.S. Nautical Almanac Office.

[57] ——, (1895), *The Elements of the Four Inner Planets and the Fundamental Constants of Astronomy* (Supplement to the American Ephemeris and Nautical Almanac for 1897). Washington: U.S. Government Printing Office.

[58] ——, (1898), "Catalogue of Fundamental Stars for the Epochs 1875 and 1900 Reduced to an Absolute System," *Astronomical Papers*, 8, 77–403. Published by the U.S. Nautical Almanac Office.

[59] ——, (1903), *The Reminiscences of an Astronomer*, Boston: Houghton Mifflin. Reissued in 1972 by Somerset Publishers, New York.

[60] ——, (1912), "Researches on the Motion of the Moon, II," *Astronomical Papers*, 9, 1–249. Published by the U.S. Nautical Almanac Office.

[61] Og[o]rodnikoff, K., (1928), "On the Occurrence of Discordant Observations and a New Method of Treating Them," *Monthly Notices of the Royal Astronomical Society*, 88, 523–32.

[62] Pearson, E.S., (1967), "Studies in the History of Probability and Statistics XVII: Some Reflexions on Continuity in the Development of Mathematical Statistics, 1885–1920," *Biometrika*, 54, No. 2, 341–55.

[63] Pearson, K., (1902), "Note on Francis Galton's Problem," *Biometrika*, 1, 390–99.

[64] ——, (1920), "On the Probable Errors of Frequency Constants, III," *Biometrika*, 13, 113–32.

[65] —— and Pearson, M.V., (1931), "On the Mean Character and Variance of a Ranked Individual, and on the Mean and Variance of the Intervals, between Ranked Individuals, I: Symmetrical Distributions (Normal and Rectangular)," *Biometrika*, 23, 364–97.

[66] Peirce, B., (1852), "Criterion for the Rejection of Doubtful Observations," *Astronomical Journal*, 2, 161–3.

[67] Peirce, C.S., (1873), "On the Theory of Errors of Observations," *Report of the Superintendent of the United States Coast Survey* (1870), 200–24.

[68] Poincaré, H., (1912), *Calcul des Probabilités*, Paris: Gauthier-Villars.

[69] Rider, P.R., (1933), "Criterion for Rejection of Observations," *Washington University Studies—New Series, Science and Technology*, No. 8.

[70] Sheppard, W.F., (1899a), "On the Application of the Theory of Error to Cases of Normal Distribution and Normal Correlation," *Philosophical Transactions of the Royal Society of London*, 192, Ser. A, 101–67.

[71] ——, (1899b), "On the Statistical Rejection of Extreme Variations, Single or Correlated (Normal Variation and Normal Correlation)," *Proceedings of the London Mathematical Society*, 31, 70–99, London, 1900.

[72] Short, J., (1763), "Second Paper Concerning the Parallax of the Sun Determined from the Observations of the Late Transit of Venus; in Which This Subject Is Treated of More at Length, and the Quantity of the Parallax More Fully Ascertained," *Philosophical Transactions of the Royal Society of London*, 53, 300–45.

[73] Stewart, C.A., (1947), "P.J. Daniell," *Journal of the London Mathematical Society*, 22, 75–80.

[74] Stigler, S.M., (1973), "Laplace, Fisher, and the Discovery of the Concept of Sufficiency," *Biometrika*, 60, (December).

[75] Stone, E.J., (1873), "On the Rejection of Discordant Observations," *Monthly Notices of the Royal Astronomical Society*, 34, 9–15.

[76] Walker, H.M., (1929), *Studies in the History of Statistical Method*, Baltimore: Williams and Wilkins.

[77] Wilks, S.S., (1943), "Order Statistics," *Bulletin of the American Mathematical Society*, 5, 6–50.

Biometrika (1971), **58**, 2, *p.* 375

Printed in Great Britain

Studies in the History of Probability and Statistics. XXVII

The hypothesis of elementary errors and the Scandinavian school in statistical theory

By CARL-ERIK SÄRNDAL

University of British Columbia

SUMMARY

This paper follows the development of an aspect of the Theory of Errors known as the Hypothesis of Elementary Errors. This development produced a scientific tradition in statistical inference theory concerned with the causes and effects of nonnormality of statistical data, a tradition sometimes called the Scandinavian school. Its founder was the Swedish astronomer C. V. L. Charlier. Being concerned mainly with the period of Charlier and his followers, this paper first outlines briefly some of the nineteenth-century thinking that formed the essential background for the Scandinavian school. The principal contributions of Charlier and his followers are discussed. Their ideas are compared with simultaneous developments in England, starting from the time of Karl Pearson.

1. SOME NINETEENTH-CENTURY DEVELOPMENTS ON THE CONTINENT

The contribution by Gauss that attached his name to the normal law of error was his famous 'first proof' (1809) of the principle of least squares. Thus the normal law of error was called the Gaussian law of error. Being concerned with astronomical measurements, Gauss's problem was that of the combination and adjustment of observations. The essence of his 'first proof' was the derivation of the normal law of errors through the Principle of the Arithmetic Mean, from which the Principle of Least Squares immediately followed.

On the other hand, Laplace's (1810, 1812) slightly later derivation of the normal law of errors, quite different from Gauss's, essentially consisted in the first demonstration of the Central Limit Theorem. The analytical technique of derivation of the normal law of error used by Laplace, which included extensive use of characteristic functions, became an important tool in Charlier's work on nonnormality. Hence Laplace's work on the normal law, and its subsequent clarification by Poisson (1837), constitutes the point of departure for the trend of thought to be followed in this paper.

While the analytical techniques stemmed from Laplace, the leading theoretical concept in Charlier's work consisted in proliferations of Hagen's (1837) original formulation of the Hypothesis of Elementary Error. Hagen's simple form of the hypothesis stated that the observational error is the algebraic sum of an infinite number of elementary errors, all having the same absolute value and the same probability of being positive as negative. The normal law of error resulted from considering the general term in the expansion of $(\frac{1}{2} + \frac{1}{2})^{2m}$, where m is large.

The legacy from Gauss and Laplace carried a tremendous weight. Critical questioning of the universality of the normal law of errors was little heeded for a large part of the nineteenth century. This impeded critical thought in theoretical and applied statistics and perhaps distorted the progress of mathematical statistics. It is interesting that Charlier (1910) blames this on Gauss rather than on Laplace. Charlier (1910, Preface) states that Gauss made the mistake of believing that any observed deviations from normality in a given set of data was due to an insufficiently large number of observations.

At the end of the nineteenth century a rapid development occurred. The illusion of the normal law as a universal law of errors was shattered. Instead, there emerged conceptions of data distributions that have the characteristics of agnosticism still prevailing today, as shown in our reluctance, or even inability, to state that given data are distributed according to a normal law, or any other narrowly specified law for that matter. If we do adopt a certain distributional law as a working model, we usually acknowledge the fact that this may be just an adequate approximation. Or, as in modern studies of robustness, we may assume only that the true distribution is a member of some family of distributions, or we take yet a further step and assume a nonparametric model. Workers in several fields contributed to the development around the turn of the century. Some of the important names in this connexion were, on the Continent, the economist-demographer Lexis, the astronomers Thiele, Bruhns and Charlier, the psychologist Fechner, and in England, the biologists Galton and K. Pearson, and the economist Edgeworth. The contributions of Lexis and Thiele seem particularly crucial in influencing Charlier.

During the nineteenth century, the analysis of series of demographic data was a main part of the professional responsibilities of mathematical statisticians on the Continent. The work of Lexis (1877) turned out to be an important development towards the understanding of nonnormality of data. In the terminology used around the turn of the century the demographers were concerned with *homograde*, i.e. qualitative, data, as opposed to *heterograde*, i.e. continuously measured, data. Whereas the treatment of heterograde data suffered under the misconception about the universality of the normal law of error, a similarly erroneous nineteenth-century assumption was that homograde data must follow the laws of a so-called Bernoulli series, which is essentially a sample of binomially distributed observations. Thus a histogram of the Bernoulli series should be closely approximated by the normal curve, provided each binomial observation is the result of a large number of Bernoulli trials. However, histograms of actual demographic series were often observed to be far from normal. Searching for an explanation, Lexis constructed several hypothetical models in which nonnormality results under asymptotic conditions, thereby providing further valuable insights into the existence and possible causes of nonnormality.

In modern teminology, a Lexis series is defined in the following way. Perform N sets of s Bernoulli trials each, with the probability p_i $(i = 1, ..., N)$ of success varying from set to set, but being constant within each set. The observed number m_i of successes in the ith set of trials is an observation on a binomial (s, p_i) random variable, and the m_i are independent. The series $m_1, ..., m_N$ is called a Lexis series; the special case where $p_i = p_0$ $(i = 1, ..., N)$ defines the Bernoulli series studied before the time of Lexis. Pick at random one of the N sets. What is the distribution of the random variable M denoting the number of successes in this randomly chosen set? Or, putting the same question in terms that the late nineteenth-century statistician would have been more likely to use: What is the nature of the frequency curve approximating the empirical distribution of the series $m_1, ..., m_N$?

In contrast to the Bernoulli series, the Lexis series will obviously not approach normality. Setting $p_0 = N^{-1}(p_1 + ... + p_N)$ and $q_0 = 1 - p_0$, and, for $r = 2, 3, ...,$

$$\eta_r = N^{-1} \sum_{i=1}^{N} (p_i - p_0)^r,$$

the first two moments of M will be given by

$$E(M) = sp_0, \quad \sigma^2 = \mathrm{var}\,(M) = sp_0 q_0 + s(s-1)\,\eta_2.$$

The ratio of σ^2 to its value under Bernoulli series conditions, i.e. $sp_0 q_0$, is defined by Charlier as the square of the so-called Lexis ratio L, in other words, $L^2 = 1 + (s-1)\,(p_0 q_0)^{-1}\eta_2$. Hence, $L \geqslant 1$ for a Lexis series, a 'super-normal' series, where equality holds only in the special case of the Bernoulli series.

When $s \to \infty$, Charlier (1911) showed that unless all the p_i are equal, the standardized higher cumulants of M tend to nonzero values, for example, $\lambda_3 = \kappa_3/\sigma^3 \to \eta_3 \eta_2^{-\frac{3}{2}}$ and $\lambda_4 = \kappa_4/\sigma^4 \to \eta_4 \eta_2^{-2} - 3$, which emphasizes the failure of the Lexis series to approach normality. Empirical illustrations given by Charlier (1910, Chapter 7) include homograde series formed by the number of births, deaths and marriages in a population observed over a sequence of years. Their observed Lexis ratio tends to be appreciably in excess of the unit value to be expected under Bernoulli series conditions.

The work of Lexis was important in that it accentuated the feeling that something was wrong with the nineteenth-century thinking. Charlier was encouraged thereby to look further into questions of nonnormality of data. Although he kept a clear distinction between homograde and heterograde statistical series, Charlier (1910) was well aware of the similarity between them, both were likely to behave nonnormally. For the homograde series, possible explanations for the nonnormality could be given through probability schemes of the type used by Lexis. For the heterograde series the Hypothesis of Elementary Errors provided Charlier with a possible explanation for the nonnormal variability of observations, just as earlier interpretations of the same hypothesis, such as Hagen's, had provided one of the influential proofs of the normal law of error.

Thiele (1889, 1903) is widely known for his contribution of introducing the concept of cumulants; of great importance also are his early attempts to develop a theory of observations for heterograde data based on a system of nonnormal frequency curves centred around the normal law of error. He was concerned with the representation of laws of error, in which the cumulants of order three and higher could not be regarded as zero or negligible. He was thus led to consider series expansions involving the derivatives of the normal density function such that the coefficients were expressed in terms of the cumulants. Such series expansions had been arrived at earlier by Gram (1879, 1883) as a solution to the problem of how to best approximate, in the sense of least squares, an arbitrary density function by means of a series expansion involving the normal density and its derivatives.

2. Elementary errors and nonnormal frequency curves

It was customary for texts in statistics published during the nineteenth century and early twentieth century to contain some sections of speculation into the causes and nature of 'errors' in general and 'elementary errors' in particular, following and extending Hagen's original idea. In allied fields, e.g. astronomy, such discussion can indeed be found even in modern texts; see, for example, Smart (1958, pp. 35–40). Typically, the Law of Causality was accepted as a matter of belief (Thiele, 1903, Chapter 1), i.e. the assumption that all events that do occur will occur only as a necessary consequence of a previous state of things. This 'previous state of things' must obviously, in any given situation, be represented by a multitude of circumstances, each of which have an 'effect', i.e. each produces a disturbance or error in the making of an observation. Thus one distinguished various kinds of error, such as gross errors, instrumental errors and systematic errors. Certain error sources one might be able to control, others, especially the kind of error thought of as accidental errors, lay beyond control. In any case, it was argued, every observation will be affected by a large number of error sources. In any practical situation it was, of course, easy to speculate about the causes and sources of error, and to mention any number of such sources that seemed plausible enough, and Charlier (1906) provides examples drawn from astronomy and biology.

This theory of the generation of observations seems too vague and hypothetical to be of any substantial usefulness. But, on the other hand, the Central Limit Theorem requires only the well-known very mild conditions in order to be valid. Nevertheless, experience with actual data had indicated that the normal law of errors often failed to give adequate representation of data distributions. Gram (1879) and Thiele (1889) had already contributed theoretical representations, in the form of series expansions, of distributions that could possibly deviate from the normal curve. Charlier (1905a, 1905b, 1906) summarized and completed this trend of thought by showing how the nonnormal frequency curves of Gram and Thiele could be justified through the Hypothesis of Elementary Errors. Realizing that the normal curve itself was only a first approximation to the distribution of a sum of errors, Charlier (1905a) used Laplacean analytical techniques to show how the approximation could be improved by adding correction terms, the importance of which ought to be smaller the larger the number of error sources; for more detail, see §4. Charlier (1906) fitted his nonnormal curves to numerous sets of actual data. His work was sometimes referred to, for example by Wicksell (1917a), by the term 'the genetic theory of frequency', indicating that the curve system was the product of a theory of the generation of observations through the accumulation of errors.

Considering the failure of the normal curve, it was natural that statisticians around the turn of the century were in search of nonnormal curves by which to get improved fit to actual data. The curve systems of Pearson and of Charlier were two outgrowths of this activity. Yet in their theoretical foundations, there two curve systems were fundamentally different. Charlier's system was the product of a 'genetic theory of error', while Pearson's was not. Indeed Charlier (1910, Preface) criticized Karl Pearson for the lack of genetic development in the Pearsonian curve system.

Synonymously with 'fitted curve', the term 'graduated curve' was used to mean a smooth theoretical frequency curve that approximated, and/or could be assumed to have given rise to, the given series of data. A striving for simplicity of models and laws can be seen here; the statistician regarded it as expedient to express the observed data by means of a curve of relatively simple form so as to facilitate the drawing of further conclusions from this data and perhaps other data of similar types. The graduated curves owed their simplicity, both in Pearson's and in Charlier's system, to the fact that a small number of parameters, namely, moments or cumulants, usually not more than four in number, sufficed to give an adequate description of the statistical phenomenon being studied, a fact that appeared to be somewhat surprising to some statisticians at that time. Charlier (1906) offers the following view of a graduated curve:

...experience has shown that this curve really has a certain form, which may be mathematically defined, and what is still more astonishing, that the parameters necessary to mathematically define a certain frequency-curve are generally very few in number. Very often three parameters suffice for representing, with satisfactory approximation, a collection of thousands of individuals. It is the duty of the mathematician to find the equation of this curve.

The existence of a wide ranging system of curves provided a convenient frame of reference when one was to select a curve of graduation for a particular set of data. Pretorius (1930) and Elderton & Johnson (1969) distinguish three main approaches to the fitting of univariate curves: (i) The Pearson system, (ii) series expansion systems, of which Charlier's Type A series is probably the most important example, (iii) translation systems, which involve the transformation of the original variate so as to obtain a normal curve, or possibly a series expansion curve, as a graduated curve for the transformed variable. The latter two systems are intimately connected with the Hypothesis of Elementary Errors, and we dwell on those in more detail in subsequent sections.

Karl Pearson's (1895, 1901) system of skew frequency curves turned out to be an extremely useful and successful frame of reference. It is difficult to find any connexion between the theoretical background for the frequency curves he derived on the one hand, and any theory of errors, on the other, suited to those biological problems to which he applied his curves. The theoretical background of Pearson's system may, of course, be described by saying that the different types of his system are simply continuous functions corresponding to, in the sense of 'smoothing out', the binomial and hypergeometic series, along with certain extensions to account for all the possible solutions corresponding to respective sets of co-efficient setups in the differential equation

$$\frac{1}{y}\frac{dy}{dx} = \frac{-(x-a)}{b_0 + b_1 x + b_2 x^2}.\tag{2.1}$$

In particular, when the variable x is standardized as $z = (x-\mu)/\sigma$, the solutions of (2.1) can be taken to depend only on the standardized third and fourth moments, which therefore took on a particular significance. The curves of the Pearson system can consequently be comprehensively arranged on the basis of the magnitude of the standardized third and fourth moments, as E. S. Pearson has suggested. He hints that Karl Pearson had indeed made some attempt to connect the hypergeometric series underlying his differential equation (2.1) to a system of elementary error components, the customary type of thinking at the time. However, it seems rather difficult because of apparent sparsity of evidence to try to follow Pearson's thinking on this. Thus it seems that we have to accept the basic differential equation as an *ad hoc* tool in deriving a family of frequency curves, one that was not seriously meant to be associated with any Theory of Errors. This does of course not seriously detract from the practical usefulness of the system itself as a means of representing various laws of error. However, it is rather obvious that the probability theory background of the Pearson system can give no clue to the actual situations in which the various curves of the system can be applied.

3. The series expansions of Charlier

Hagen's original Hypothesis of Elementary Errors assumed that each elementary error could take on only the values a and $-a$ with equal probabilities. The following extension of the Hypothesis led Charlier (1905a) to the Type A series. Let there be s sources of error Q_k $(k = 1, ..., s)$, where Q_k produces the error x_k with the law of error $f_k(x)$. The observed variable is $x = x_1 + ... + x_s$, where the x_k are assumed to be independent random variables. The error x_k is assigned the value ma if it happens to fall in the interval $ma \pm \frac{1}{2}a$ ($m = 0, \pm 1$,

$\pm 2, \dots$), where a is fixed. The probability of this event is roughly $af_k(ma)$. We form the corresponding characteristic function

$$\psi_k(\omega) = \sum_{m=-\infty}^{\infty} af_k(ma)\, e^{i\omega ma} \to \int_{-\infty}^{\infty} f_k(x)\, e^{i\omega x}\, dx,$$

letting a tend to zero for $x = ma$ fixed. According to Laplace's addition theorem, the probability of the sum x being in the interval $(x, x+dx)$ could be obtained as

$$f(x)\, dx = \frac{dx}{2\pi} \int_{-\infty}^{\infty} \prod_{k=1}^{s} \psi_k(\omega)\, e^{-i\omega x}\, d\omega.$$

Inserting the expression for $\psi_k(\omega)$, Charlier obtained after some series expansions the Type A series frequency function. It can be written as

$$f(x) = \phi(x) + \sum_{i=3}^{\infty} (-1)^i \frac{A_i}{i!} \frac{d^i \phi(x)}{dx^i}, \tag{3·1}$$

where $\phi(x)$ is the $N(\mu, \sigma)$ density. The parameters are μ, σ, $A_3 = \kappa_3$, $A_4 = \kappa_4$, $A_5 = \kappa_5$ and $A_6 = \kappa_6 + 10\kappa_3^2, \dots$, where κ_i denotes the ith cumulant. Charlier (1905a) also expressed the parameters in terms of functions of the moments of the elementary errors x_k ($k = 1, \dots, s$). Charlier thus arrived at the series expansion of Gram and Thiele by extending to a finer degree of approximation Laplace's analytical treatment of the distribution of a sum of random variables, i.e. he generalized the Central Limit Theorem by going beyond the leading normal density term.

The Type A series fails to be valid if the law of error for the error sources has practically the entire probability mass concentrated at one point. For this situation, Charlier (1905b) formulated a different version of the Hypothesis of Elementary Errors, which led to the series of Type B. Assume that each error source produces only two possible errors, 0 and a ($a > 0$) with probabilities p and q, respectively, where p is close to unity. The frequency function for the Type B series was found to be of the form

$$f(x) = \psi(x) + B_1 \Delta\psi(x) + B_2 \Delta^2\psi(x) + \dots, \tag{3·2}$$

where

$$\psi(x) = \frac{e^{-\lambda}}{\pi} \int_0^{\pi} \cos(\lambda \sin\omega - x\omega) \exp(\lambda \cos\omega)\, d\omega = \frac{e^{-\lambda}\sin(\pi x)}{\pi} \sum_{v=0}^{\infty} \frac{(-1)^v}{v!(x-v)}$$

and $\Delta\psi(x) = \psi(x) - \psi(x-1)$. If x is zero or a positive integer, then $\psi(x)$ reduces to the Poisson probability $\lambda^x e^{-\lambda}/x!$.

Charlier (1908) also considered a slight extension of this idea, in which he assumed that each source can produce three errors, $-a_1$, 0 and a_2 ($a_1, a_2 > 0$), where the probability of error 0 is close to unity. Thereby the sum of the elementary errors can also be negative, and $\psi(x)$ in formula (3·2) has to be replaced by

$$\psi(x) = \frac{e^{-\lambda}}{\pi} \int_0^{\pi} \cos(\kappa \sin\omega - x\omega) \exp(\lambda \cos\omega)\, d\omega.$$

By the nature of Charlier's method of derivation, the parameters μ, σ, A_3, A_4, ... of the Type A curve and $\lambda, \kappa, B_1, B_2, \dots$ of the Type B curve were theoretically dependent on the moments of the elementary errors x_k. For the practical purpose of graduation, however, the parameters were estimated from the sample, and the resulting nonnormal curves could be claimed to be independent of the distributions of the errors. A systematic treatise on the series expansions was published by Jørgensen (1916), which also contains extensive tables helpful in the application of the curves. A Danish actuary, Arne Fisher (1915), was an active proponent, especially in the U.S.A., of the ideas of the Scandinavian school.

Some knowledge of the actual situation generating a certain series of data may help to indicate when a Type A series is preferable to a Type B series. From the underlying formulations of the Hypothesis of Elementary Errors, the Type A series should be of far greater applicability. The Type B curve appears suitable for purposes of graduation in those situations of markedly skew data, where the majority of the observations fall just slightly above a lower limit of the range of variation. In addition to its rather limited practical applicability, the Type B series suffers from drawbacks of a theoretical nature connected with questions of convergence of the series and with possible nonexistence of its moments. The discussion of the Type B series in the literature (Jørgensen, 1916; Steffensen, 1923) is rather limited. Beyond its use for purposes of graduation, there were no other areas where the Type B curve seemed to be tractable enough for making any significant advances of statistical theory. Being in this sense a dead-end approach, we can at this point exclude the Type B series from further discussion.

By contrast, the Type A series, apart from its practical usefulness for graduation purposes, did later prove convenient as a model for treating questions of robustness, especially robustness of validity for normal theory test statistics. Although seemingly cumbersome to handle, the Type A series adapted itself fairly well to the type of analytical treatment needed for studying, for example, the distribution of various statistics under conditions of Type A series nonnormality. In this regard, the series offered an advantage compared with, for example, Pearson curves with matched low moments. A few additional facts about the Type A series are worth observing.

(i) The predominant theory of estimation during the early decades of the twentieth century, before R. A. Fisher's ideas had become widely accepted, was that of the method of moments. The Type A series lent itself admirably to estimation by this method since the parameters, i.e. the coefficients μ, σ, A_3, A_4, ... were moments or simple functions of moments. In practical applications, the terms up to A_4 were usually retained, and the series thus depended on four parameters, the first four cumulants.

(ii) Conditions for the convergence, for every x, of the sum (3·1) to the true density have been given by Cramér (1928). On leaving aside the formal questions of convergence, it is of importance from the practical point of view to know whether two or three terms of (3·1) give good approximation to $f(x)$. When $f(x)$ itself does not deviate very much from a normal curve, it seems likely that the approximation is quite close. Centred around the normal curve as it is, the Type A series therefore seems suited for studying phenomena whose distribution deviates slightly or moderately from the normal law of error.

In this respect the Pearson system is more flexible since it allows without any qualifications even quite extremely nonnormal shapes. In fact, Karl Pearson appears to have been less than enthusiastic about the series expansion approach (Henderson, 1922; Karl Pearson, 1924b). Pearson's curves, as well as Charlier's Type A with three terms, required the first four sample moments for estimating the parameters. Karl Pearson (1924b) was firmly convinced that, due to the high variability of the higher moments, one should never go beyond use of the fourth moment and that 'the mathematician who uses high moments may make interesting contributions to mathematics, but he removes his work from any contact with practical statistics'. In addition, Pearson appears to have thought that little was gained by fitting a series expansion that used only the first few terms of, for example, a Charlier

Type A series; he may have had in mind the cases when the true curve was strongly non-normal and, consequently, the convergence of the series may be slow.

(iii) Possibly the most serious objection against the Type A series is that it sometimes produces negative co-ordinates in the tail of the distribution. It is also possible for the curve to be multimodal. We can, of course, guard against these drawbacks by defining the Type A frequency curves only within those ranges of values of the cumulants for which both of the mentioned anomalous properties are avoided (Elderton & Johnson, 1969, p. 120).

(iv) Considering the 'genetic' foundation of the Type A series afforded by the Hypothesis of Elementary Errors, we would like to know how the magnitudes of the coefficients in (3·1) depend on s, the number of elementary errors. In particular, it would be good if the series (3·1) gave a straightforward expansion in decreasing powers of s. This is, however, not the case; the Edgeworth form of the expansion, on the other hand, meets this requirement (Edgeworth, 1905; Charlier, 1909; Wicksell, 1917b; Cramér, 1928). If all the errors x_i in the sum $x = x_1 + \ldots + x_s$ have identical distributions, then, if $\beta_i = A_i/\sigma^i$, we have $\beta_3 = O(s^{-\frac{1}{2}})$, β_4 and $\beta_6 = O(s)$, while for $i = 5$ and $i \geqslant 7$, β_i is of order $s^{-\frac{3}{2}}$ or smaller. Edgeworth (1905) wrote the differential operators as the exponential of a series wherein the coefficients do diminish straightforwardly. Including terms to order s^{-1}, it takes the form

$$f(x) = \phi(x) - \frac{\kappa_3}{3!}\frac{d^3\phi(x)}{dx^3} + \frac{\kappa_4}{4!}\frac{d^4\phi(x)}{dx^4} + \frac{10\kappa_3^2}{6!}\frac{d^6\phi(x)}{dx^6}. \tag{3·3}$$

The failure of Charlier's series to give a straightforward expansion made Edgeworth (1907) maintain that it was his version rather than Charlier's that provided the true generalization of Laplace's Central Limit Theorem. Only much later did Charlier (1928) meet this criticism by presenting a third type of expansion, called Type C. Its density function is always positive, and it does give a straightforward expansion in decreasing powers of s. The Type C density is of the form

$$f(x) = \exp\left\{\sum_{r=0}^{\infty} \gamma_r H_r(x)\right\},$$

where the γ_r are coefficients, and $H_r(x)$ is the Hermite polynomial of order r. This series has, however, attracted little further attention in the literature.

4. A GENERALIZED HYPOTHESIS OF ELEMENTARY ERRORS

Galton (1879) provided an early example of a nonnormal law of variation. He observed that some sociological and psychological phenomena may be subject to the 'condition of the geometric mean', rather than that of the arithmetic mean, and thereby Galton directed attention to the possible usefulness of a log normal law of variation. As an example, Galton refers to Fechner's law, sensation equals log stimulus.

In § 2, the Method of Translation was mentioned as the third important means of finding a graduated curve to fit to possibly nonnormal data. The idea is due to Edgeworth (1898). A review of translation methods is given by Elderton & Johnson (1969, pp. 122–34). Curve fitting by the Method of Translation can be justified on the basis of a generalized Hypothesis of Elementary Errors. Kapteyn (1903) and Kapteyn & van Uven (1916) had introduced the idea of making the size of the variable dependent not only on the error sources that influence the variable, but also upon the size already attained by the variable. Hence the time element became important. The ideas were formalized by Wicksell (1917a), a student of Charlier's whose argument we now present.

Consider a system of interdependent error sources. These error sources are assumed to be operating one after another in a given sequence. At the time of action of source Q_i, let z_{i-1} be the value already attained by the variable, and let x_i be the error impulse of the error source Q_i. Two assumptions are made.

1. The increment of the variable z produced by the source Q_i is proportional to a certain function θ of the value z_{i-1} attained by the variable at the time of action of source Q_i, where the factor of proportionality is the error impulse, i.e. $z_i = z_{i-1} + x_i \theta(z_{i-1})$.

2. There exists a function $A(z)$ such that $A(z)$ is the sum of the independent error impulses, i.e. $A(z_i) = x_1 + \ldots + x_i$.

We obtain

$$\frac{A(z_i) - A(z_{i-1})}{z_i - z_{i-1}} = \frac{1}{\theta(z_{i-1})},$$

or, if each increment can be considered to be small,

$$A(z) = \int_z^z \frac{dz}{\theta(z)}.$$

Since $A(z)$ is the sum of independent random variables, Charlier's reasoning can now be applied. Thus the distribution of $A(z)$ may successfully be approximated by a Type A series or a Type B series expansion. The first term of the Type A series, i.e. the normal density itself, may give sufficiently good approximation. An important special case is $A(z) = \log z$.

In Hagen's and Charlier's formulations of the Hypothesis of Elementary Errors, the errors are additive, hypothetical and non-observable entities. In the generalized hypothesis dealt with in this section, the role of the elementary errors is played by the error impulses; these latter are additive, hypothetical and non-observable. The generalized hypothesis contains an additional hypothetical element in the form of an assumption about the relation between the increments of z and the value already attained by z; knowledge of this relation is essential in determining the form of the transformation function $A(z)$. The form of this function is thus determined by the biological, economic or sociological background of the problem. Examples that have been quoted (Wicksell, 1917*a*; Quensel, 1945) include the distribution of wealth or income in a given population. It is plausible to assume that the value of an individual's property or wealth is the sum of a large number of increments, where each increment, instigated by a certain growth impulse, is proportional to the size of wealth already attained, i.e. $A(z) = \log z$. This reasoning may also be applied to growth processes for biological organisms.

5. Correlation surfaces and regression problems

On the Continent, two and three variate normal distributions had been considered by writers such as Bravais, Schols and Czuber. The systematic development of the theory of multivariate distributions, or correlation surfaces, as they were called, and multivariate regression was, however, originated in England. Edgeworth (1892) and Karl Pearson (1896) gave the first representations of the general p-variate normal correlation surface. Pearson (1896) demonstrated the fundamental characteristic of a normal correlation surface, namely, that the regression curve of any variable, say, X_p on $X_1, \ldots, X_{p-1}$, is linear.

Karl Pearson also spent considerable time and effort in attempts to create a system of skew correlation surfaces as a continuous system derived from the multivariate hypergeometric frequency surfaces; see K. Pearson (1924*a*) and the references given there. The univariate analogue of this idea had, as we have seen in §3, proved to be successful in providing a very useful system of frequency curves. However, Pearson's attempts to extend the argument to several variables met with severe technical difficulties. Wicksell (1917c) argued that in the multivariate case the scope of Pearson's idea was insufficiently general to meet the desired degree of generality of the correlation surface system that one would like to derive. For consider the bivariate hypergeometric probability

$$\mathrm{pr}\,(S = s,\, S' = s') = \binom{n}{s}\binom{n'}{s'}\binom{N-n-n'}{m-s-s'}\bigg/\binom{N}{m}.$$

The regression curve of S is linear in s', $E(S|S' = s') = n(m-s')/(N-n')$, with an analogous linear expression for the regression curve of S'. Hence, as observed by Wicksell (1917c), a correlation surface system corresponding to the multivariate hypergeometric distribution could only account for those very special cases of multivariate theory where the regression curves are linear.

However, Karl Pearson (1924b) was

convinced of the importance of the double hypergeometric series for describing skew frequency, but the difficulty of integrating the differential equations to the corresponding skew surface has for many years impeded progress.

Only for special cases have solutions been given for the extension to two dimensions of the univariate Pearson differential equation (2·1). For example, van Uven (1947, 1948) derived all the possible bivariate correlation surfaces obtainable as solutions of the pair of partial differential equations

$$\frac{dy}{dx} = \frac{L_i(x_1, x_2)}{Q_i(x_1, x_2)} \quad (i = 1, 2),$$

where L_i $(i = 1, 2)$ are linear and Q_i $(i = 1, 2)$ are quadratic functions in x_1 and x_2. The resulting bivariate frequency surfaces share the common characteristic that the regression curves are linear (Elderton & Johnson, 1969, p. 137).

By contrast, from Charlier's standpoint, it must have seemed fairly obvious how to extend the univariate series expansion idea to correlation surfaces of any dimension. The p-variate correlation surface of Type A has marginals of Type A and was given by Charlier (1914) as

$$f(x_1, \ldots, x_p) = \phi(x_1, \ldots, x_p) + \Sigma\Sigma\ldots\Sigma A_{i_1 \ldots i_p} \frac{\partial^{i_1 + \ldots + i_p}\,\phi(x_1, \ldots, x_p)}{\partial x_1^{i_1}\ldots\partial x_p^{i_p}}, \tag{5·1}$$

where $i_1, \ldots, i_p$ are nonnegative integers such that $i_1 + \ldots + i_p \geqslant 3$, and $\phi(x_1, \ldots, x_p)$ is the p-variate normal correlation surface with parameters $\boldsymbol{\mu}$ and $\boldsymbol{\Lambda}$.

Charlier (1914) also provided a 'genetic' derivation of the bivariate case of (5·1) by reference to the Hypothesis of Elementary Errors. Let w_k be the error produced by the error source Q_k $(k = 1, \ldots, s)$, where the law of error of w_k is a Type A series expansion. Then Charlier showed that

$$x_1 = \sum_{i=1}^{n} a_i w_i, \quad x_2 = \sum_{i=1}^{n} b_i w_i$$

are jointly distributed according to the bivariate Type A correlation surface, where a_k and b_k are constants characteristic of the source Q_k.

Also for the bivariate case of (5·1), Wicksell (1917b) investigated the order of magnitude of the coefficients $\beta_{ij} = A_{ij}/(\sigma_1^i \sigma_2^j)$, where we write i and j in place of i_1, and i_2, in terms of the number s of elementary errors. For $i+j = 3$, $\beta_{ij} = O(s^{-\frac{1}{2}})$, and for $i+j = 4$ or 6, $\beta_{ij} = O(s^{-1})$. For $i+j = 5$ or $\geqslant 7$, β_{ij} is of order $s^{-\frac{3}{2}}$ or smaller.

Edgeworth (1905) also discussed the extension of his type of series expansions to correlation surfaces. Pretorius (1930) illustrated some of the difficulties involved in fitting correlation surfaces, Type A or other, to actual multivariate data. There seems to be rather few good graduations of actual data reported in the literature. For the bivariate Type A surface in particular, one would, for example, like to know more about whether retaining terms up to order $i+j = 4$, which gives 14 parameters, will give a surface that adequately represents real data.

Since his attempts to derive a system of skew correlation surfaces from the multivariate hypergeometric series were less than successful, Karl Pearson did not treat the problem of nonlinear regression curves using a specified correlation surface as a starting point. Instead K. Pearson (1905) approached the problem without making any assumptions as to the form of the correlation surface. By contrast, starting from Type A correlation surface, the conditional distributions could be given explicit form, and the regression curves of various characteristics could be obtained. Using this approach, Wicksell (1917b) derived the regression curves of the first four moments for the Type A bivariate correlation surface. Explicit expressions for these regression curves are obtained easily when the surface is given by the infinite bivariate case of (5·1). Approximate regression curves can then be worked out, depending on the number of terms actually included in the series. If terms to order $i+j = 4$ are retained, the regression curves turn out to be cubic equations whose coefficients are derived from the parameters of the Type A surface. Wicksell (1917b) illustrates his method on several sets of actual bivariate data.

6. ROBUSTNESS STUDIES

The two schools of Pearson and Charlier put heavy emphasis on techniques of curve fitting based on the Method of Moments. Properties of moments, especially in the Pearson school, and of cumulants, especially in the Charlier school, were investigated, for example, their own sampling moments.

The contributions of R. A. Fisher substantially changed the outlook on statistical inference. In particular Fisher made many well known contributions to the stream of results on exact sampling distributions that had started with Student and came to an end in the early 1930's with the results on multivariate normal sampling distributions. The Pearson system came to the fore again, somewhat unexpectedly, in that many of Fisher's results were either exactly or approximately represented by one of the Pearson curves. Furthermore, the Pearson curves came to be used in the early studies of robustness of validity performed in the 1920's. Typically, limited scale Monte Carlo experiments were used, in which small samples were taken from a population represented by one of the Pearson curves, and the performance of standard normal theory test statistics was studied under these conditions of nonnormality. Examples of such studies are Church (1926), Sophister (1928), E. S. Pearson & Adyanthaya (1928, 1929) and E. S. Pearson (1931).

It seems natural that the Scandinavian school would proceed with robustness studies assuming conditions of nonnormality as represented by the Charlier Type A series in its univariate or multivariate form. In this regard, the Charlier Type A series offered certain attractive features as compared to the Pearson curves. Curves from the latter family have been used merely as generators of random nonnormal samples for Monte Carlo type studies. On the other hand, if the population sampled deviates from normality in a way expressed by the Type A series, the distribution of normal theory test statistics could frequently be given explicit expression. For example, although the distribution of $t = (\bar{x} - \mu)\sqrt{n}/s$ is no longer valid if sampling is from a Type A series population, it is not surprising to find that the true distribution of t can be expressed as a series expansion. The leading term of the series is the distribution of t under conditions of exact normality, and the ensuing terms can be considered as corrections for nonnormality (Quensel, 1938). This fact, established for other normal theory test statistics as well, made Charlier's Type A series a convenient tool for studying robustness of validity. Certain questions in this connexion have not been given complete answers. For example, one would like to know how possible anomalies in the tail of the Type A population, like negative ordinates for certain x, might affect the derived distribution of Student's t or the sample correlation coefficient in their respective tails.

Romanowsky (1924) provided a bridge between the Charlier and Pearson approaches by giving series expansion of some of the Pearson curves. These expansions stood in the same relation to the corresponding Pearson curve as the Charlier Type A series stood to the normal curve. For example, the Romanowsky generalization of Pearson's Type III is given by the density

$$f(z) = f_p(z) + \sum_{\nu=1}^{\infty} (-1)^\nu \frac{a_\nu}{\nu!} f_{p+\nu}^{(\nu)}(z),$$

where the leading term
$$f_p(z) = \frac{1}{\Gamma(p)\,\gamma^p} (z-a)^{p-1} \exp\{-(z-a)/\gamma\}$$

is the Pearson Type III density, $f_p^{(\nu)}$ its νth derivative and a_ν are coefficients.

Whereas Fisher had used geometrical views in multidimensional space to derive his results on exact sampling distributions, Romanowsky (1925) demonstrated that characteristic functions provided a convenient tool towards the same end. Using characteristic functions, Quensel (1938) extended the Student–Fisher results on the joint distribution of the sample mean $\bar{x}$ and the sample variance $s^2 = n^{-1}\Sigma(x_i - \bar{x})^2$, assuming a sample of n independent observations from the Type A population $f(x)$ represented by (3·1), where $\phi(x)$ is the $N(0, \sigma)$ density function. The joint density of $\bar{x}$ and s^2 was obtained as

$$f(\bar{x}, s^2) = f_{\frac{1}{2}n-\frac{1}{2}}(s^2)\,\phi_n(\bar{x}) + \Sigma\Sigma(-1)^{i+j} \frac{c_{ij}}{i!\,j!} f_{\frac{1}{2}n-\frac{1}{2}+i}^{(i)}(s^2)\,\phi_n^{(j)}(\bar{x}), \qquad (6·1)$$

where $\phi_n(\bar{x})$ is the $N(0, \sigma/\sqrt{n})$ density and $f_{\frac{1}{2}\nu}(x)$ is the Pearson Type III density with $p = \frac{1}{2}v$, $a = 0$ and $\gamma = 2\sigma^2/n$. The c_{ij} can be expressed in terms of the A_i in (3·1), and $f_k^{(i)}$, ϕ_n^i denote the ith derivatives of f_k and ϕ_n, respectively. Thus the joint density of $\bar{x}$ and s^2 emerges as a sum of products of terms of Charlier's Type A series and of Romanowsky's generalization of Type III. The marginal density of $\bar{x}$ is a Charlier Type A distribution, and that of s^2 is a Romanowsky generalization of Type III. From (6·1) Quensel (1938) derived the distribution of Student's t when the sampled population is the Type A series. The leading term of the resulting series is the density function of Student's t.

To further indicate the general nature of the results obtained (Quensel, 1938), note that if the population is a Charlier Type A bivariate correlation surface, and if the population correlation is zero, then the density of the sample correlation coefficient is represented by Romanowsky's generalization of Pearson's Type II. Its leading term is the Pearson Type II curve which Fisher had obtained for the distribution of the sample correlation coefficient under the assumption of uncorrelated normal samples. Quensel (1952, 1953) has presented further results on the distribution of multidimensional normal theory test statistics under conditions of nonnormality represented by the Type A correlation surface (5·1).

Authors in the British school also have used series expansions for studying robustness, notably Baker (1930, 1935), Geary (1936, 1947) and Gayen (1949, 1950, 1951). Using the four term Edgeworth expansion (3·3), Gayen (1949, 1950) studied the distribution of Student's t and of the variance ratio. For samples from a bivariate, nonzero correlation population of the Edgeworth form, going as far as $i+j = 6$, Gayen (1951) investigated the distribution of the sample correlation coefficient.

7. A BROADER VIEW OF NONNORMALITY

The Type A series expansion with its foundation in the Hypothesis of Elementary Errors offered but one possible model for a possibly nonnormal population. A variety of other mechanisms for producing nonnormal samples were also considered by the Scandinavian school in their research program emphasizing distribution theory and questions of robustness of validity for standard normal theory test statistics. Some contributions along these lines are by Quensel (1938, 1943, 1947, 1952, 1953), Hyrenius (1949, 1950, 1952) and Weibull (1950, 1951, 1953).

The study of robustness of normal theory statistics is conveniently made with reference to a family of frequency curves indexed by one or more parameters, which may indicate the degree of nonnormality. Preferably, the normal curve itself should be included as a member of the family. Tukey (1960) has discussed properties that are desirable in any family of frequency curves used as a frame of reference for robustness studies. The family should be a simple one; convenience for analytical as well as for computational purposes is important. The family should be such that appreciable nonnormality may have large consequences, while being at the same time difficult to discover in a given set of data. It may, however, be argued that analytical tractability needed for deriving the exact distributions under a specified type of nonnormality is today of secondary importance compared to the computational aspects, since quick results from Monte Carlo type sampling experiments are readily obtainable by modern electronic computer equipment. For the purpose of this paper we distinguish three different models of nonnormality in sampling from univariate populations. For each model an exactly normally distributed sample can be obtained as a special case when the parameters of the system take appropriate values. The models are (Hyrenius, 1950; Weibull, 1953):

(i) Sampling of independent observations from a Charlier Type A distribution.

(ii) Sampling of independent observations from a compounded normal distribution.

(iii) Sampling of independent observations, from normal distributions, not necessarily identical.

This scheme can be extended to study the effects of nonnormality in multivariate situations if we substitute Charlier Type A correlation surface, compounded multivariate normal distribution, and non-identical multivariate normal distributions into (i), (ii) and (iii), respectively. These sampling models usually allow exact expressions to be given for the distribution of standard normal theory test statistics, albeit the algebraic work involved may be quite heavy. Often the density function under nonnormality is in terms of a series expansion.

In model (i), the cumulants are conveniently taken as the parameters indexing the family, for example, the skewness and the excess if we use only a three term Type A series expansion. Some of the more important results using the Type A family were mentioned in the previous section.

In model (ii), we assume independent observations $x_1, ..., x_n$ on a random variable X with density

$$f(x) = \sum_{i=1}^{k} \gamma_i \frac{1}{\sigma_i} \phi\left(\frac{x - \mu_i}{\sigma_i}\right). \tag{7.1}$$

This is a general way of representing a compounding of k normal densities. The parameters are $(\mu_i, \sigma_i, \gamma_i)$ $(i = 1, ..., k)$, where the γ_i are nonnegative and add up to unity. Important special cases were (a) $\mu_i = \mu$ $(i = 1, ..., k)$, while the σ_i may differ; (b) $\sigma_i = \sigma$ $(i = 1, ..., k)$, while the μ_i may differ. The concept of compounding of distributions has a long history; see Tukey (1960). The general formulation (7.1) was used by Hyrenius (1950) for investigating the robustness of validity of Student's t. Sampling from a bivariate compound normal population was also considered by Hyrenius (1952). For purposes of robustness studies, $k = 2$ may possibly give a representation that is sufficiently realistic, yet simple enough to elucidate satisfactorily the effects of nonnormality. This is the representation favoured by Tukey (1960), who chose $\sigma_1 = 1$, $\sigma_2 = 3$ and identical means to create unimodal distributions with thicker-than-normal tails. In this form it plays an important role as a tool in the theory of efficiency robust estimation, as seen in the work of Tukey (1960) and in numerous more recent contributions.

In model (iii), the general formulation is in terms of a sample of independent observations $x_1, ..., x_n$, where x_i is normal (μ_ν, σ_ν) for $i \in s_\nu$, where s_ν $(\nu = 1, ..., k)$ are disjoint sets of indices whose union is $\{1, 2, ..., n\}$ and $k \leqslant n$. Important special cases are (a) $\mu_\nu = \mu$ $(\nu = 1, ..., k)$, while the σ_ν may differ; (b) $\sigma_\nu = \sigma$ $(\nu = 1, ..., k)$, while the μ_ν may differ. Important conclusions about robustness may possibly be drawn by restricting the model to $k = 2$ groups.

Quensel (1943), who introduced model (iii), notes that this type of nonnormal sample may often arise in biological, medical and psychological applications. The cause of nonnormality may for example, consist in variations in the accuracy of the measuring device, or to variations over time in the response of individuals undergoing test. Under this model, Quensel (1943, 1947) studied the validity of Student's t and of the variance ratio when the normal populations have identical means but differing variances. One notes that the joint density of $\bar{x}$ and s^2, from which the density of t is derived, turns out again to be in the form of the series expansion (6.1), although the coefficients are different. Student's t turns out to have a lower variance than under conditions of exact normality, so the test should be conservative.

A systematic account of the distributions of traditional normal theory test statistics in one and two dimensions when the individual distributions in model (iii) are normal with different means but identical variances is given by Weibull (1950, 1951, 1953). A mixture of models (ii) and (iii) was considered by Zackrisson (1959).

Models (i), (ii) and (iii) considered above, and possibly combinations of them, cover a wide range of nonnormal conditions. Their usefulness, under simple specifications, as models for studies in efficiency and robust estimation remains unexplored, except for the results for the compound normal model with $k = 2$ referred to above.

The author wishes to express his appreciation to Professor Allan Birnbaum for encouraging this research and for valuable suggestions. Thanks are also due to the two referees for helpful comments.

REFERENCES

BAKER, G. A. (1930). Distribution of the means of samples of n-drawn at random from a population represented by a Gram–Charlier series. *Ann. Math. Statist.* **1**, 199–204.

BAKER, G. A. (1935). Note on the distributions of the standard deviations and second moments of samples from a Gram–Charlier population. *Ann. Math. Statist.* **6**, 127–30.

CHARLIER, C. V. L. (1905a). Über das Fehlergesetz. *Ark. Mat. Astr. Fys.* **2**, No. 8.

CHARLIER, C. V. L. (1905b). Die zweite Form des Fehlergesetzes. *Ark. Mat. Astr. Fys.* **2**, No. 15.

CHARLIER, C. V. L. (1906). Researches into the theory of probability. *Lunds Univ. Arsskrift, N.F. Avd.* 2, **1**, No. 5.

CHARLIER, C. V. L. (1908). Weiteres über das Fehlergesetz. *Ark. Mat. Astr. Fys.* **4**, No. 13.

CHARLIER, C. V. L. (1909). Die strenge Form des Bernoullischen Theorem. *Ark. Mat. Astr. Fys.* **5**, No. 15.

CHARLIER, C. V. L. (1910). *Grunddragen af den Matematiska Statistiken.* Lund. German translation: Vorlesungen über die Grundzüge der Mathematischen Statistik, Hamburg, 1920. English translation: Elements of Mathematical Statistics, Cambridge, Mass., 1947.

CHARLIER, C. V. L. (1911). Researches into the mathematical theory of statistics. 1. *Ark. Mat. Astr. Fys.* **7**, No. 17.

CHARLIER, C. V. L. (1914). Contributions to the mathematical theory of statistics. 6. The correlation function of Type A. *Ark. Mat. Astr. Fys.* **9**, No. 26.

CHARLIER, C. V. L. (1928). A new form of the frequency function. *Lunds Univ. Arsskrift, N.F. Avd.* 2., **24**, No. 8.

CHURCH, A. E. R. (1926). On the means and squared standard deviations of small samples from any population. *Biometrika* **18**, 321–94.

CRAMÉR, H. (1928). On the composition of elementary errors. *Skand. Aktuarietidskr.* **11**, 13–74, 141–80.

EDGEWORTH, F. Y. (1892). Correlated averages. *Phil. Mag.*, Ser 5, **34**, 190–204.

EDGEWORTH, F. Y. (1898). On the representation of statistics by mathematical formulae. *J. R. Statist. Soc.* **61**, 670–700.

EDGEWORTH, F. Y. (1905). The law of error. *Trans. Camb. Phil. Soc.* **20**, 36–65, 113–41.

EDGEWORTH, F. Y. (1907). On the representation of statistical frequency by a series. *J. R. Statist. Soc.* **70**, 102–6.

ELDERTON, W. P. & JOHNSON, N. L. (1969). *Systems of Frequency Curves.* Cambridge University Press.

FISHER, A. (1915). *The Mathematical Theory of Probabilities*, vol. I. New York: MacMillan.

GALTON, F. (1879). The geometric mean in vital and social statistics. *Proc. R. Soc.* **29**, 365–7.

GAUSS, C. F. (1809). *Theoria motus corporum coelestium....* Hamburg.

GAYEN, A. K. (1949). The distribution of 'Student's' t in random samples of any size drawn from non-normal universes. *Biometrika* **36**, 353–69.

GAYEN, A. K. (1950). The distribution of the variance-ratio in random samples of any size drawn from non-normal universes. *Biometrika* **37**, 236–55.

GAYEN, A. K. (1951). The frequency distribution of the product-moment correlation coefficient in random samples of any size drawn from non-normal universes. *Biometrika* **38**, 219–47.

GEARY, R. C. (1936). The distribution of 'Student's' ratio for non-normal samples. *J. R. Statist. Soc. Suppl.*, **3**, 178–84.

GEARY, R. C. (1947). Testing for normality. *Biometrika* **34**, 209–42.

GRAM, J. P. (1879). *On Raekkeudviklinger bestemte ved Hjaelp av de mindste kvadraters Methode.* Copenhagen: Gad.

GRAM, J. P. (1883). Über die Entwicklung reeller Funktionen in Reihen mittelst der Methode der kleinsten Quadrate. *J. Reine Angew. Math.* **94**, 41–73.

HAGEN, G. (1837). *Grundzüge der Wahrscheinlichkeits-Rechnung.* Berlin.

HENDERSON, J. (1922). On expansions in tetrachoric functions. *Biometrika* **14**, 157–85.

HYRENIUS, H. (1949). Sampling distributions from a compound normal parent population. *Skand. Aktuarietidskr.* **32**, 180–7.

HYRENIUS, H. (1950). Distribution of 'Student'–Fisher's t in samples from compound normal functions. *Biometrika* **37**, 429–42.

HYRENIUS, H. (1952). Sampling from bivariate non-normal universes by means of compound normal distributions. *Biometrika* **39**, 238–46.

JØRGENSEN, N. R. (1916). *Undersögelser over frequensflader og korrelation.* Copenhagen: Gad.

KAPTEYN, J. C. (1903). *Skew Frequency Curves in Biology and Statistics.* Groningen: Astronomical Laboratory.

KAPTEYN, J. C. & VAN UVEN, M. J. (1916). *Skew Frequency Curves in Biology and Statistics*. Groningen: Astronomical Laboratory.

LAPLACE, P. S. (1810). Mémoire sur les approximations des formules qui sont fonctions de très-grande nombres, et sur leur application aux probabilités. *Mem. Inst. France for* 1809, pp. 353–415, 559–65.

LAPLACE, P. S. (1812). *Théorie Analytique des Probabilités*. Paris.

LEXIS, W. H. R. A. (1877). *Zur Theorie der Massenerscheinigungen in der menschlichen Gesellschaft*. Freiburg.

PEARSON, E. S. (1931). The analysis of variance in cases of non-normal variation. *Biometrika* **23**, 114–33.

PEARSON, E. S. & ADYANTHAYA, N. K. (1928). The distribution of frequency constants in small samples from non-normal symmetrical and skew populations. *Biometrika* **20**A, 356–60.

PEARSON, E. S. & ADYANTHAYA, N. K. (1929). The distribution of frequency constants in small samples from non-normal symmetrical and skew populations. *Biometrika* **21**, 259–86.

PEARSON, K. (1895). Contributions to the mathematical theory of evolution. II. Skew variation in homogeneous material. *Phil. Trans. R. Soc.* A **186**, 343–414.

PEARSON, K. (1896). Mathematical contributions to the theory of evolution. III. Regression, heredity and panmixia. *Ph." Trans. R. Soc.* A **187**, 187–318.

PEARSON, K. (1901). Mathematical contributions to the theory of evolution. X. Supplement to a memoir on skew variation. *Phil. Trans. R. Soc.* A **197**, 443–59.

PEARSON, K. (1905). Mathematical contributions to the theory of evolution. XIV. On the general theory of skew correlation and non-linear regression. *Drap. Co. Res. Mem.*, *Biom.* Series II, 1–54.

PEARSON, K. (1924a). On a certain double hyper-geometrical series and its representation by continuous frequency surfaces. *Biometrika* **24**, 172–88.

PEARSON, K. (1924b). Note on Professor Romanovsky's generalization of my frequency curves. *Biometrika* **16**, 116–17.

POISSON, S. D. (1837). *Recherches sur la Probabilité des Jugements*, etc. Paris.

PRETORIUS, S. J. (1930). Skew bivariate frequency surfaces examined in the light of numerical illustrations. *Biometrika* **22**, 109–223.

QUENSEL, C. E. (1938). The distributions of the second moment and of the correlation coefficient in samples from populations of Type A. *Lunds Univ. Arsskrift. N.F. Avd. 2*, **34**, No. 4.

QUENSEL, C. E. (1943). An extension of the validity of 'Student'–Fisher's law of distribution. *Skand. Aktuarietidskr.* **26**, 210–9.

QUENSEL, C. E. (1945). On the logarithmico-normal distribution. *Skand. Aktuarietidskr.* **28**, 141–50.

QUENSEL, C. E. (1947). The validity of the z-criterion when the variates are taken from different normal populations. *Skand. Aktuarietidskr.* **30**, 44–55.

QUENSEL, C. E. (1952). The distribution of the second order moments in random samples from non-normal multivariate universes. *Lunds Univ. Arsskrift N.F. Avd. 2*, **48**, No. 4.

QUENSEL, C. E. (1953). The distribution of the partial correlation coefficient in samples from multivariate universes in a special case of non-normally distributed random variables. *Skand. Aktuarietidskr.* **36**, 16–23.

ROMANOWSKY, V. (1924). Generalization of some types of the frequency curves of Professor Pearson. *Biometrika* **16**, 106–116.

ROMANOWSKY, V. (1925). On the moments of standard deviations and of correlation coefficient in samples from normal population. *Metron*, **5**, No. 4, 3–46.

SMART, W. M. (1958). *Combinations of Observations*. Cambridge University Press.

'SOPHISTER' (1928). Discussion of small samples drawn from an infinite skew population. *Biometrika* **20**A, 389–423.

STEFFENSEN, J. F. (1923). *Matematisk Iaktagelseslaere*. Copenhagen: Reitzel.

THIELE, T. N. (1889). *Forelaesninger over almindeling iaktlagelseslaere*. Copenhagen: Gad.

THIELE, T. N. (1903). *Theory of Observations*. London: Layton.

TUKEY, J. W. (1960). A survey of sampling from contaminated populations. In *Essays in Honor of Harold Hotelling*, ed. by I. Olkin. pp. 448–85. Stanford University Press.

VAN UVEN, M. J. (1947, 1948). Extension of Pearson's probability distributions to two variables, I–IV. *Proc. Kon. Akad. Wetens.* **50**, 1063–70, 1252–64; **51**, 41–52, 191–6.

WEIBULL, M. (1950). The distribution of the t and z variables in the case of stratified sample with individuals taken from normal parent populations with varying means. *Skand. Aktuarietidskr.* **33**, 137–67.

WEIBULL, M. (1951). The regression problem involving non-random variates in the case of stratified sample from normal parent population with varying regression coefficients. *Skand. Aktuarietidskr.* **34**, 53–71.

WEIBULL, M. (1953). The distributions of *t*- and *F*-statistics and of correlation and regression co-efficients in stratified samples from normal populations with different means. *Skand. Aktuarietidskr. Suppl.* **36**, 1–106.

WICKSELL, S. D. (1917*a*). On the genetic theory of frequency. *Ark. Mat. Astr. Fys.* **12**, No. 20.

WICKSELL, S. D. (1917*b*). The correlation function of Type A and the regression of its characteristics. *K.Sv. Vetenskapsakad. Handl. N.F.* **58**, No. 3.

WICKSELL, S. D. (1917*c*). The application of solid hypergeometrical series to frequency distributions in space. *Phil. Mag., Series* 6, **33**, 389–94.

ZACKRISSON, U. (1959). The distribution of 'Student's' *t* in samples from individual non-normal populations. *Publ. Statist. Inst.* No. 6, Univ. of Gothenburg.

[*Received August* 1970. *Revised January* 1971]

Some key words: History of Charlier series; Lexis schemes; Asymptotic expansions; Pearson curves; Hypothesis of elementary errors; Central limit theorem; Multivariate distributions; Robust estimation.

Biometrika (1972), **59**, 1, *p.* 205

Printed in Great Britain

Miscellanea

Studies in the History of Probability and Statistics. XXVIII.

On the history of certain expansions used in mathematical statistics

By HARALD CRAMÉR

University of Stockholm

Summary

This note is concerned with the history of the expansions known to probabilitists and statisticians as Charlier's A-series and Edgeworth's series. It has been pointed out (Gnedenko & Kolmogorov, 1968, Chapter 8) that both these types of expansions appear already in the mathematical work of Tchebychev, so that the names usually attached to them are historically incorrect. In the sequel we shall, however, denote them by the names familiar to statisticians. Both expansions are closely connected with the attempts to give a refinement of the classical central limit theorem in probability. The properties which are relevant in this connexion are the asymptotic properties of the sum of a small number of terms. It will be stated here that the first valid proof of these properties was given in two papers by the present author (1925, 1928).

Some key words: History of expansions for distributions; Central limit theorem; Edgeworth series; Charlier series.

In a recent paper, Särndal (1971) gives a historical account of the work connected with the hypothesis of elementary errors, which has been performed by what he calls the Scandinavian school in statistical theory. The work centring about Charlier's A-series occupies an important place in this account. Särndal's presentation of this work does, however, call for some complementary remarks.

Consider the following simple case of the classical central limit theorem. Suppose that we are given a sequence of mutually independent and equidistributed random variables $z_1, z_2, \ldots$ with a common distribution of the continuous type, having zero mean, standard deviation σ and central moments $\mu_3, \mu_4, \ldots$, which need not all be finite. Consider the standardized sum variable

$$y_n = \frac{z_1 + \ldots + z_n}{n^{\frac{1}{2}}\sigma},$$

and let $f_n(x)$ be the density function of y_n. The central limit theorem then asserts that, as $n \to \infty$, under appropriate conditions, $f_n(x)$ tends to $\phi(x)$, the normal $N(0, 1)$ density function.

In order to improve the approximation to $f_n(x)$ supplied by the normal density function, Charlier (1905) introduced his A-series, which gives for $f_n(x)$ an expansion with $\phi(x)$ as its leading term, followed by terms containing the successive derivatives of ϕ, from $\phi^{(3)}$ on:

$$f_n(x) = \phi(x) + \frac{c_3}{3!}\phi^{(3)}(x) + \frac{c_4}{4!}\phi^{(4)}(x) + \ldots. \tag{1a}$$

The coefficients c_i depend on the moments of the common distribution of the z_i, the expressions of c_3 and c_4 being

$$c_3 = -\frac{\mu_3}{n^{\frac{1}{2}}\sigma^3}, \quad c_4 = \frac{1}{n}\left(\frac{\mu_4}{\sigma^4} - 3\right). \tag{2}$$

Thus with increasing n both c_3 and c_4 tend to zero, the order of smallness being $n^{-\frac{1}{2}}$ for c_3, and n^{-1} for c_4. But the simple rule suggested by these relations does not hold, as shown, for example, by c_6, which is of the same order n^{-1} as c_4.

Charlier (1905) asserts that the sum of a small number of terms of the A-series (1 a) gives for large n a better approximation to $f_n(x)$ than does the leading term $\phi(x)$ alone. Särndal (1971, p. 378) writes in this connexion: 'Realizing that the normal curve itself was only a first approximation to the distribution of a sum of errors, Charlier (1905) used Laplacean analytic techniques to show how the approximation could be improved by adding correction terms, the importance of which ought to be smaller the larger the number of error sources.' He further (p. 380) briefly indicates the method of proof used by Charlier

(1905), and then states that Charlier thus arrives at this expansion ' by extending to a finer degree of approximation Laplace's analytical treatment of the distribution of a sum of random variables'. However, Särndal does not mention the fact that Charlier's proof of his statement was entirely false, as it depended essentially on an illegitimate use of what we now call the inversion formula for characteristic functions.

In the course of his discussion of Charlier's 1905 paper, Särndal (1971, p. 380) gives the correct form of the inversion formula which expresses $f_n(x)$ in terms of the corresponding characteristic function. As is well known, this formula contains an integral extended between the limits $\pm \infty$. In Charlier's 1905 paper this integral appears, however, with the limits $\pm \pi$. In a later paper, not quoted by Särndal, Charlier (1914) admits in a brief footnote on p. 2 that the correct limits of the integral should be $\pm \infty$. But he does not mention that his whole proof of 1905 was essentially based on the use of the incorrect limits $\pm \pi$ and becomes invalid when the correct infinite limits are introduced. In fact, the assertion which Charlier claimed to have proved in 1905 is true only in a modified form, and under more restrictive conditions than those given by him. No correct proof of any statement of this character was ever published by Charlier or by any of his students.

Thus the problem of what may be denoted as the asymptotic properties of the A-series remained open for a number of years, until it was taken up and at least partially solved in two papers by the present author (Cramér, 1925, 1928). An account of some of the results obtained in these papers may be found in the book of Gnedenko & Kolmogorov (1968, Chapter 8), as well as in a Cambridge Tract (Cramér, 1970, Chapter 7).

The first paper (Cramér, 1925) is a preliminary publication, while more explicit results were given by Cramér (1928). It is there proved that, under appropriate conditions, asymptotic properties similar to those asserted by Charlier do, in fact, hold for the A-series, but that the corresponding relations appear in a considerably simpler form when the density function $f_n(x)$ is replaced by the corresponding distribution function $F_n(x)$, and the A-series is replaced by the expansion known as Edgeworth's series. The A-series for $F_n(x)$ is obtained by formal integration of $(1a)$, which gives

$$F_n(x) = \Phi(x) + \frac{c_3}{3!} \Phi^{(3)}(x) + \frac{c_4}{4!} \Phi^{(4)}(x) + \cdots, \tag{1b}$$

where $\Phi(x)$ is the normal $N(0, 1)$ distribution function. The series introduced by Edgeworth (1905) assumes for the distribution function $F_n(x)$ the form

$$F_n(x) = \Phi(x) + P_1(x) + P_2(x) + \cdots. \tag{3}$$

Here the $P_i(x)$ are linear aggregates of the derivatives of $\Phi(x)$, the expressions for P_1 and P_2 being

$$P_1(x) = \frac{c_3}{3!} \Phi^{(3)}(x),$$

$$P_2(x) = \frac{c_4}{4!} \Phi^{(4)}(x) + \frac{1}{2} \left(\frac{c_3}{3!}\right)^2 \Phi^{(6)}(x).$$

It thus follows from (2) that P_1 is of the order $n^{-\frac{1}{2}}$, while P_2 is of the order n^{-1}. Edgeworth (1905, 1907) showed that generally, if the requisite moments are finite, P_i is of the order $n^{-\frac{1}{2}i}$.

The asymptotic properties of the Edgeworth expansion (3) for $F_n(x)$ were for the first time given by the present author (1925, 1928). As shown in these papers, the expansion (3) does really have the asymptotic character that Charlier tried to prove for the A-series. It is, in fact, proved (Cramér, 1928, p. 59) that, under fairly general conditions, any partial sum of (3) gives an approximation to $F_n(x)$ with an error of the same order of magnitude for large n as the first neglected term. The Edgeworth expansion for the density function $f_n(x)$, which is obtained by formal differentiation of (3), has similar asymptotic properties, but only under rather more restrictive conditions (Cramér, 1928, p. 63; Gnedenko & Kolmogorov, 1968, Chapter 8). Extensions of the asymptotic properties to cases when the component variables z_i are not equidistributed are also given by Cramér (1928).

We finally observe that the convergence properties of the above expansions, regarded as infinite series, have been discussed by various authors. The results obtained are certainly interesting from a purely mathematical point of view. However, when applying one of the expansions to a probability distribution we cannot discuss the question of convergence or divergence without supposing that all moments have known finite values. Thus it is generally not the convergence theory of the expansions which is statistically relevant, but the question of the asymptotic properties of the sum of a limited number of terms.

REFERENCES

CHARLIER, C. V. L. (1905). Über das Fehlergesetz. *Ark. Mat. Astr. Fys.* **2**, No. 8.

CHARLIER, C. V. L. (1914). Contributions to the mathematical theory of statistics 5. *Ark. Mat. Astr. Fys.* **9**, No. 25.

CRAMÉR, H. (1925). On some classes of series used in mathematical statistics. *Trans. 6th Congr. Scand. Math.* 399–425.

CRAMÉR, H. (1928). On the composition of elementary errors. *Skand. Aktuarietidskrift* **11**, 13–74, 141–80.

CRAMÉR, H. (1970). *Random Variables and Probability Distributions*, 3rd edition. Cambridge University Press.

EDGEWORTH, F. Y. (1905). The law of error. *Trans. Camb. Phil. Soc.* **20**, 36–65, 113–41.

EDGEWORTH, F. Y. (1907). On the representation of statistical frequency by a series. *J. R. Statist. Soc.* **70**, 102–6.

GNEDENKO, B. V. & KOLMOGOROV, A. N. (1968). *Limit Distributions for Sums of Independent Random Variables*, 2nd edition. Cambridge, Mass.: Addison-Wesley.

SÄRNDAL, C.-E. (1971). The hypothesis of elementary errors and the Scandinavian School in statistical theory. *Biometrika* **58**, 375–91.

[*Received October* 1971]

HISTORICAL SURVEY OF THE DEVELOPMENT OF SAMPLING THEORIES AND PRACTICE*

By You Poh Seng
(1951)

THE use of sampling in statistical surveys is a relatively modern development. Before the end of the last century sampling had been rarely employed, and even then not in a completely scientific manner. Chiefly owing to the fact that the statisticians responsible for the employment of the method never gave an account of how they took the samples, of what difficulties they encountered, or of what steps they took to counter these difficulties, and did not discuss the accuracy of their results, it is impossible for us to decide whether properly to call them sampling surveys, and how to evaluate their experience. Thus, in 1861, Dr. William Farr used a partial sample for the collection of statistical data in the English Census of population: all he mentioned of the procedure was that he took 14 "sub-districts" with a total population of 264,327, and examined the number of families and persons in relation to houses, special emphasis being laid on family composition (Census, 1861). In the same report we find detailed observations in ten chosen counties on the number of landlords distributed according to number of employees and size of farms. All we can assume is that the samples were taken randomly, and certainly if he used any controls he did not mention them.

It is also interesting to note that the so-called sample investigations were usually taken in conjunction with a census. No sample survey as a separate investigation had ever been undertaken. But owing to reticence concerning the sample design, sampling could not really be regarded as systematically developed.

It was not until A. N. Kiaer took office as Director of the Bureau of Statistics in Oslo (then Kristiania) that the practice, and to a lesser extent the theory, of sampling began to be systematically developed: not until he took the platform at the International Institute of Statistics in 1895 was sampling method in statistical investigations first introduced to and debated by an international body of statisticians.

The work of Kiaer can be regarded as a turning-point in the history of Statistics. Born in 1838, he was made head of the Norwegian Bureau of Statistics when it became an independent body for the collection and interpretation of social and demographic facts. As head of the Bureau of Statistics, he was responsible for the decennial censuses of population and agriculture during the last quarter of the century, the measurement of movement of population, and many other official investigations.

He was the first to use the sampling method in collecting data independently of the census. He systematically built up the case for using this method, and carried out several purely sampling investigations for the Bureau of which he was head.

The main part of his work on the use of the method in the collection of official statistics appeared in various publications of the International Institute of Statistics (Kiaer 1895, 1899, 1901). Here he brought the method to the serious attention and discussion of other statisticians.

In the session of the Institute in Berne in 1895 he attempted to clarify the meaning of "dé-nombrements representatifs" (representative investigations). He stressed that by "dénombre-ments" he did not mean a census or any ordinary enumeration, nor did he intend that it could take the place of a census; it was used to signify a special type of enumeration aimed at collecting detailed information which could not be obtained in a census.

Thus he intended to introduce a word to express neither a haphazard enumeration nor a full inquiry but rather an investigation based on information collected from individuals who had been first selected according to a particular representative method.

Modern usage can express this more clearly. What he was in search of was a "sampling

* This article is based on the first chapter of the author's Ph.D. thesis: "Techniques of Sampling
. . .", presented in May, 1949.

investigation" based on information collected from sampling units which have been first selected according to a sampling design.

We shall have occasion later on to elaborate on his meaning with regard to his definition of "dénombrement representatif". Meanwhile it is of interest to follow him in the detailed description of his use of the method in two particular surveys carried out under his direction in Norway.

The first was an investigation carried out in 1894 throughout Norway on a proposed retirement pension and sickness insurance scheme. The most notable point was that, while the decennial census did not, and could not, give the details required for the purpose of this investigation, he had no need to carry out a full survey corresponding to a census to collect these details. Here, in fact, is an instance of the vast potentiality of sampling survey, and of its advantage for an independent survey of conditions touching very closely on the social and economic life of a large proportion of the population.

There were over 60 questions asked in this survey, which was to be made by special investigators. They were to interview more than 120,000 adults, of whom 80,000 were to constitute the investigation proper, and 40,000 a parallel investigation to be made in localities inhabited by the working classes.

For the investigation proper, the total number of interviews were allocated proportionately between the towns and country districts, the proportions being determined by the Population Census taken in 1891, so that about 20,000 persons were interviewed in the towns and 60,000 in the country districts.

The method of selection of the sampling units in the town areas differed from that in the country districts.

Towns.—Thirteen towns were chosen, including all the five large towns with populations of 20,000 or over. As there were 61 towns in Norway at that time, this meant that about one-fifth of them were investigated. The number of interviews were not distributed among the larger and smaller towns in proportion to the total population of each. Since the total population of all the smaller towns is relatively greater than that of the larger towns, it follows that more interviews should be allocated to the smaller towns chosen. So that, while Kristiania had to return 6,350 schedules (or an equivalent of one-sixteenth of its adult population in 1891), some other towns chosen had to return a total number of schedules corresponding to about one-ninth of their adult population, and some a third. Kiaer claimed that this method of distribution of interviews was not only rational, but also advantageous in that the final results could be derived more readily.

In Kristiania, where there was an annual enumeration of the population, information in this respect was rather detailed and therefore useful. Thus in 1892, according to the enumeration taken that year, there were 400 streets in the town divided into four categories, viz.:

(a) 100 streets each with population 100 or over;
(b) 187 streets with 101–500;
(c)   80 streets with 501–1,000; and
(d)   33 streets with 1,001 and over.

The interviews were to be undertaken as follows:

(a) The entire population of one-twentieth of the least populated streets.

(b) One-tenth of the streets of the second category, but only half the houses in each to be enumerated.

(c) One-quarter of the streets of the third category, one-fifth of the houses in each to be enumerated.

(d) One-half of the streets of the fourth category, one-tenth of the houses in each.

In the selection for enumeration care was taken to ensure an even spread throughout the town.

The same procedure was to be adopted in the other large towns and the medium towns, while in the small towns only the adult population of three or four houses were interviewed. It is to be mentioned that Kiaer did not go into the details of the sampling procedure in these towns, but it should be clear that the general procedure was the same.

Country districts.—A different procedure was adopted in the sampling in country districts. The interviews were to be distributed among the 18 rural prefectures in Norway according to their rural population in 1891. To obtain as correct representation as possible, the communes in each prefecture were classified according to predominantly agricultural, stock-farming, forestry, industrial, shipbuilding, and fishing. A number of representative communes were chosen from each category. The interviews in each prefecture were distributed so that the total selected for examination in each category of commune was proportional to the total population for that category in the entire prefecture.

The total number of communes represented was 109, an average of 6 per prefecture, while the total number of communes in the whole country was 498 in 1891.

In each category in each prefecture, however, the number of interviews allocated to each chosen commune was not proportional to its population, owing to the fact that some communes chosen represented, by virtue of their principal occupations, a population too large or too small compared to the total communes in the category; e.g. certain communes, while representative geographically, were, owing to the nature of their special industries, too predominant in their representation of those particular industries. This difficulty was overcome by distributing to these communes a relatively smaller number of interviews than to others, and *vice versa*. Thus in some prefectures there were some communes with half their adult population enumerated, some with one-third, some with one-sixth, and some with even less.

In the chosen communes themselves an effort was made to distribute, as far as possible, the interviews among the different parishes of the communes proportionately to their populations, distinguishing those with sparse and those with dense populations. Distribution in each sub-division of each commune was left to the discretion of enumerators, who were instructed to choose the routes of investigation to conform with the purpose of representativeness of the study, care being taken to enumerate not only socially "average" houses, but also in general houses representing different social and economic conditions in the commune.

A special precaution was introduced in the study. To guard against possible imperfect representation, a certain number of interviews were reserved for each prefecture for correction of differences or errors. For this purpose, for each prefecture the number of interviews that should have been returned for each occupation was calculated according to the census of 1891. These were then compared with the numbers actually returned, and adjustments were made by supplementary reports to make up for the gaps in the two figures, or, where the numbers actually returned were larger than those expected, the excess was eliminated. If the difference were not too large, it could be ignored, partly because it could be compensated for in the sum total for the whole prefecture, and partly because it is neither possible nor necessary that a representative investigation should correspond to a full census in all its details.

The second survey described by Kiaer in the same report was mentioned to show that a representative investigation could be made in more ways than one. This survey had for its purpose the ascertainment of distribution of incomes of adult males in Norway by occupations, ages, and civil status.

For this survey, 23 representative towns and 127 representative rural communes were chosen. In each chosen area the individual returns in the General Census of 1891 for males aged 17, 22, 27, 32, 37, . . ., up to 97 were extracted from all the returns, so that only one-fifth of the adult males for each representative town or commune were included in the survey. To reduce still further the resulting large number of returns, only those returns of persons whose names began with the letters A, B, C, L, M and N were used (for Kristiania and other populated towns, only persons whose names began with L, M and N were included). To the information given on the returns for these selected persons were added details regarding their incomes collected in a special interview-survey. This survey comprised an average of only about 3·3 per cent. of the male population of Norway living in country districts and 1·6 per cent. of that living in towns (or a total number of only about 10,800 individuals).

Comparisons of census and survey results regarding distribution by professions (or occupations) showed favourable representation, the only large difference, that for the marine population, being explained by the fact that the census figures applied to the "actual" population, while the survey figures applied to the population "de droit", which in Norway comprised a large number of foreign sailors.

Apart from all his other achievements, his clear presentation of a method of drawing the "representative" samples was sufficient to win for Kiaer a place among the pioneers of sampling investigations. Of course, there was considerable vagueness in the presentation of his technique, and he did not analyse his survey results as we, in the present state of our knowledge, would do. He did, however, by his persistent efforts, bring about the adoption by the Institute of a recommendation urging the specification of the conditions under which the selection of observation units is made. To this we shall return.

By scientific insight Kiaer realized that the more or less recognized method of completely random selection is not the only, nor even the best, method of drawing a sample. Thus, in the discussion following the presentation of his report, he maintained that the conditions of a country could be thoroughly studied by taking a large enough number of small geographic units or localities spread over the country. These units could serve to "represent" the country if the selection of the units had been carried out *rationally*. Here we notice two very important conditions of a successful sampling investigation: proper representation and rational selection of units.

The method he used for proper representation is, to us, a well-worked-out method of stratification, the stratification factors being geographic, social and economic. In the first of the two investigations mentioned above, geographic stratification was achieved by the division of the universe into town areas and country districts, and by further subdivision of these. For stratification according to social conditions he used, in the towns, the division into the four categories of streets, it being assumed that the least populous streets represent the wealthier stratum of the town and the most populous streets the poorer stratum, while the intermediate streets would therefore represent average or medium social and economic conditions.

In the country districts the stratification used was according to main occupations, and further, definite instructions were issued to interviewers to select, not only socially "average" houses for investigation, but also houses representing different social and economic conditions. Of course, bias due to human choice was inevitable, but at least the method of stratification was clear. Further, Kiaer introduced proportional selection in each chosen stratum based on the population details of the previous census.

The complete theory of stratification was not developed until the time of Neyman, but even so, Kiaer, by sheer common sense, had already adumbrated a method of stratification which even today would be a useful model for developing a sampling design requiring multiple factor stratification.

To summarize the method, we can do no better than to quote Kiaer: "It is fundamental to observe that the accuracy of the results of a survey depends, not on the larger or smaller number of observations made, but on the method of obtaining correct representation". And this is fundamental indeed. Certainly before Kiaer's work there was a tendency to distrust any sample survey which included only a small fraction of the universe. Nor, even to-day, is this tendency moribund. Kiaer showed that this was no valid criticism, provided the method of selecting sampling units ensured correct representation. Later developments provided the justification for his viewpoint.

In the general discussion at the same session of the Institute there was strong opposition to the representative method of investigation. Professor G. von Mayr (of Munich University), L. Bodio, the secretary-general of the Institute and President of the Supreme Council of Statistics of Italy, and G. E. Milliet were opposed to the method, all their arguments depending on the alleged sanctity of the census method.

On the other hand, Cheysson pleaded that there should be no prejudice against the new method, which, however, he wrongly designated monography. Levasseur, the Vice-President of the Institute, stressed that it was necessary to deal with three distinct cases: general statistics, which was statistics properly speaking and concerned itself with the totality of phenomena of the same order throughout one region, one state or province; monography, which was concerned with detailed description of an object or phenomenon, or of some aspects of particular units of the universe; and finally, statistical explorations. The first two were complementary to each other; but there should be room for the third, which dealt with the study of conditions by means of statistics not of the totality of phenomena, but of a determinate and restricted part of the phenomena. Schmoller was of the opinion that the method was useful where the materials for investigation were complex or more numerous than could be handled by a general statistical investigation.

Summing up the case for the opposition, Mayr stated that, though there were a large number of social facts which could not be enumerated, or measured, or weighed, but could be studied by monography, the "representative" investigations of Kiaer were not concerned with such facts. Rather they were concerned with issues where measurements or counts of the whole could be made, but where only parts of this whole were deliberately ("de propos délibéré") drawn and studied. While he agreed with Levasseur that these studies had their value as "investigations on a fixed point", that is, that they had their special value, he maintained that it was not correct to regard the results of such investigations as giving sufficient information in default of a complete inquiry of the whole. Such a complete inquiry, he insisted, could never be replaced by a partial "representative" investigation.

In view of such opposition, it is interesting to follow up Kiaer's defence of the method in the next session held four years later in St. Petersburg. His long discourse in this second session is of importance also for his clear distinction between representative and typological investigations.

Kiaer's definition of representative investigation was that it was a partial inquiry in which observation was made on a large number of units distributed throughout a country or territory so that their totality would form a miniature of the whole country or territory. These units were not chosen arbitrarily, but according to a rational scheme based on the general results of some previous statistical investigations. The distribution of the observations was so arranged that the results could be controlled in many respects by these general statistics. As to investigations by "types" (i.e. average cases), while these were useful, they would seem to abound with disadvantages compared with representative investigations. Thus, even if one knew the proportions which the individuals represented by the types made of the whole field of inquiry, the types were far from giving plausible results for the totality, for the totality comprised not only the types, but also a variety of extreme and non-typical cases found in reality. It was therefore necessary, in order that the investigation should present a true miniature of the whole, to observe not only the "types", but also all kinds of phenomena. This was the kind of investigation which could be made, if not exactly, at least approximately, by the representative method which neglected neither the "types" nor the variations from the "types".

To the objection that representative investigations had only a special value relative to the parts observed and that they could not therefore be interpreted for the whole, he replied that this was applicable only to investigations which were not representative. In so far as these investigations were representative, they constituted in the totality of observed units a "photograph which reproduces the details of the original in its true relative proportions".

He admitted that he could not understand the logic of the argument that the monographic method was concerned with objects which could not be counted, or weighed, or measured, while statistics was concerned with objects which could in their totality, but which "de propos délibéré", were only counted, or weighed, or measured in parts. This argument would seem to confine the use of monographs to descriptions or other non-numerical matters, while numerical matters should be reserved exclusively for the field of general statistics. Such an argument showed ignorance of the work of many statisticians who had applied numerical measurements in monographic studies—for example, the work of Theysson on the comparative budgets of ten monographs or types of families. Once it was admitted that it was necessary in certain cases to use partial investigations, it was surely desirable to improve those investigations by precise counting, weighing and measurement.

But those opposing it either could not accept this argument, or did not desire to do so. The dissension had at least revealed the area of agreement and disagreement. All agreed that a partial investigation could never have the same value as a complete inquiry, and that there were some cases where a partial investigation was inappropriate even if it was not possible to make a complete inquiry. They differed on Kiaer's assertion that there were many cases where complete statistics were impossible to obtain, but where partial investigations could be used to advantage. For this it was sufficient to refer to the numerous publications of the Departments of Labour in America and many European countries recounting many partial investigations. One could apply to representative investigations the words which Bodio pronounced in the session at Berne, in connection with the work of Engel on budgets of working families: "Monography and enumeration are two ways of investigating social facts which are complementary to each other. The latter, by itself, gives only the general profiles of the phenomena, the silhouette, so to speak, of the figures. Monography (and Kiaer added, partial investigations in general) permits us to push our analysis into all the details of the economic and moral life of the people: it supplies blood, flesh and nerves to the skeleton built up by general statistics. Enumeration in turn completes the results obtained by monography". Here we have a lucid statement of the reciprocal relationship between general statistics and partial investigations in general.

According to Kiaer, it was important to stress that the scientific value of partial investigations depended more on the representative character than on the number of the observations. There were, he asserted, numerous ways of obtaining representativeness. If comparisons showed conformity between the results of the partial investigations and those of the general census, valid conclusions could be derived, it being assumed that to the extent that any partial investigation could be shown to be correct in the controlled factors, it was probably correct also in the factors which could not be controlled.

Kiaer pointed out that the results of a partial investigation could be controlled to a certain extent even in default of general statistics. Leaving aside regularity of phenomena, which was itself a control, the results of a partial investigation could be controlled by making use of the results of other partial investigations obtained by different representative schemes, and it was clear that if we could obtain the same (or approximately the same) results by various methods, greater faith could be placed on these results.

Such was the force of the arguments put forth by Kiaer in favour of representative investigations that, although the Institute was not prepared to recommend the use of the method, it felt compelled to nominate a sub-committee to discuss more fully every aspect of the problem. It was arranged that this sub-committee should report at the next session.

In the same year in an article in the *Allgemeines Statistisches Archiv* Kiaer reiterated the case he had stated at the Institute, and outlined the representative investigations carried out under his aegis in Norway. One can summarize the salient points of this article:

(i) The representative method of investigation is applicable not only in the field of social and economic inquiries, but also in that of agriculture and forestry.

(ii) To obtain an exact representative selection it is necessary to group the different communities under investigation. Thus in social inquiries, the towns and country communes are differentiated, and are further divided into large, medium, small, coastal, inland, industrial and rural. This principle of grouping (or stratifying) homogeneous parts of a country must be applied with care to obtain a really representative sample.

(iii) The questions propounded should as far as possible follow the lines of the general census, so that the results can be controlled, and so that the statistics can be thoroughly analysed.

(iv) If possible two or more different systems of obtaining a representative sample should be used so that greater faith can be placed on the results of the inquiry, and proof of the usefulness of the method can be obtained.

(v) It is important to study and develop the practical and theoretical aspects of the method, so that proper limits can be set to representative statistics.

In the following session of the Institute in Budapest, in 1901, Kiaer was especially eloquent in his plea for the representative method of investigations. He reiterated that a detailed investigation of a town, or of a certain quarter of a town, could hardly be called a representative investigation. If the town, or quarter of the town, could be regarded as a type, then we had a typological investigation; but one could not generalize those results. If, however, one were to gather the information from a number of units distributed at random, so that in some parts of the universe one observed a large number, and in others a small number of units, one would have partial investigation, not generally representative. To attain representativeness one should have a large number of units distributed so that localities of different characteristics are represented in the same proportion as in the universe.

For practical proofs of the representative character of the method, Kiaer could cite numerous experiments he had carried out in Norway which showed that certain statistical phenomena required more numerous and more carefully selected observations than did others. Thus population growth and migration statistics could not be investigated by the representative method as easily as the relative figures concerning births and deaths..

Where an investigation dealt with matters which had not previously been investigated, or which could not be controlled by comparison with the census, Kiaer suggested the division of the investigation into two or three distinct parts, using for each a different representative method of selection. If these several parts gave similar results, accurate representativeness could be guaranteed.

To conclude, Kiaer re-affirmed his opinion that the representative method could be recommended in many investigations as being preferable to partial investigations made at random and without regard to the rule that results collected for the different parts should be distributed proportionately.

The discussion following Kiaer's general report was of particular importance. There was, of course, the general objection that the method was beset with difficulties and could be dangerous to apply, particularly where the population studied contained dynamic characteristics caused, for example, by migrations between countries, or between provinces within countries.

A very interesting contribution was made by Bortkiewicz on the "Method of Controls". His view was that, while the method constituted a most important advance, the form in which it was presented by Kiaer was subject to a grave objection, largely because of what Bortkiewicz called the "coincidence of approximation". If, for instance, one found 1·8 per cent. of one's partial observations belonged to a certain profession, and actually 2 per cent. was found for the universe, one would declare oneself satisfied with the results of the investigation. But, in so far as these two were not strictly equal (and never could be), our conclusion was to that extent subjective. The one way to overcome this difficulty would be to appeal to the Calculus of Probability, to determine if the difference between the two figures (0·2 per cent. in our example) could or could not in fact spring from chance; and the probability could be calculated according to Poisson's Formulae. If the difference was larger than could be regarded as fortuitous, we should have to conclude that our investigation was not representative of the whole. According to this criterion, Bortkiewicz had put Kiaer's results as presented in the *Allgemeines Statistisches Archiv* to the test, and had found them to be generally not satisfactory.

Bortkiewicz thus interjected into Kiaer's argument a more scientific test of representativeness. Bortkiewicz did not formulate the tests to be applied in a stratified sample, but the fact remained that he did bring the possibility of the application of such an objective test to the notice of statisticians in general. Although he was not the first statistician to employ this method in connection with sampling, he was the first to express the basic idea.

In the ensuing session of the Institute in Berlin, in 1903, Kiaer's purpose was fulfilled. In this session the Sub-Committee which had previously been appointed by the Institute to study the representative method proposed the following resolution:

"The Committee, considering that the correct application of the representative method, in a certain number of cases, can furnish exact and detailed observations from which the results can be generalized, within certain limits, recommends its use, provided that in the publication of the results the conditions under which the selection of the observation units is made are completely specified. The question will be kept on the agenda, so that a report may be presented in the next session on the application of the method in practice and on the value of the results arrived at".

Thus sampling, in whose genesis Kiaer had played such a predominant role, received the official imprimatur of the world's statisticians. Every outstanding figure in the further development of sampling has concentrated on some particular aspects of the theory: Kiaer was the first, and to date the last, to take a catholic view of the field.

It is true that Kiaer discussed the method of random sampling; it is true that his own method of selection was akin to what was later called the stratification method; but the basis of his method was more intuitive than scientific. Enlightened intuition preceded scientific validation. Kiaer showed the possibilities of the sampling method independent of and complementary to the census. He cleared the ground for others to cultivate. The crops that have so far come to fruition represent the particular interests of his successors; the varieties that remain to be evolved are limitless.

Kiaer had shown, and the International Institute of Statistics had recommended, that sampling could be used to advantage in the study of social and economic problems. A critical evaluation of the results of the sample techniques was the next step in the development. If, following Kiaer, one admits the Method of Controls, then Bortkiewicz's caveat becomes relevant; what confidence can one place on the representative character of the sample if, for the controlled factors, there is a difference between the constant calculated from the partial observations and the constant expected from the general statistics used as the basis for control? And thus, what confidence can one place on the analyses of the sample observations? Next, how can one estimate the population constants from the results of the samples? And what confidence should be placed on these estimates? An understanding of, and an improvement in, the techniques of sampling were clearly necessary.

This was Professor Bowley's task and his achievement. Bortkiewicz, as we noted, had raised the first problem but did not adumbrate a reasoned solution to it. However, he did suggest the line of attack upon it. It was not until 1906 that Bowley, in his Presidential address to the Economic Science and Statistics Section of the British Association for the Advancement of Science, presented the problem in its crystallized form and suggested a systematic solution. Owing largely to the work of Professors Karl Pearson and Edgeworth, statistical theory had developed extensively during the last quarter of the nineteenth century, but its application to practical statistics had been scant. As Bowley wrote: "In recent years progress . . . of theory has, indeed, been rapid, and a great number of important and thoroughly criticized methods are ready for use, and are, in fact, in constant use by biologists and botanists; but there has been remarkably little application to practical statistical problems. In the thirty years following the publication of Quetelet's *Lettres*, attention was mainly given to establishing the constancy of great numbers and averages based thereon, an important but limited work, while the relation of the frequency of deviations to the law of error was regarded rather as a statistical curiosity. Professor Edgeworth's illustrations in 1885 of the importance of mathematical methods in testing the truth of practical deductions have as yet borne singularly little fruit . . . it is time that it [mathematical statistics] was brought to bear on the criticism and analysis of existing industrial statistics . . . most of our statistics remain untested and their significance not analysed. The simple method of samples, . . ., for which all the materials have existed for at least twenty years, has (so far as I know) been completely ignored" (Bowley 1906).

Bowley's proposal to make use of the theory of probability was of great significance; it instituted a new epoch in the theory of sampling, supplementing Kiaer's original work.

To quote Bowley again: "The region to which I am devoting particular attention is that where the theory of probability is involved, not because there are not many other directions in which mathematical methods are useful, but because this is of the greatest importance and the least generally understood. All depends on a complete grasp of the nature of the measurement when we say, for example, that from certain data the most probable estimate of average wages is 24s.; it is as likely as not, however, to be as much as 4d. from this value: the standard deviation is 6d.; the chances are 10 to 1 against the average being over 24s.·8d., 100,000 to 1 against it being over 26s. This is the kind of statement to which calculations lead. The result may be briefly indicated as 24s.·±6d., when the 'standard deviation' is adopted as the measure of accuracy. In a normal curve of frequency about two-thirds of the area is within the standard deviation: the chance that a given observation should be within this distance of the true average is 2:1. The unit of measurement thus devised is most subtle and most complex. When it is applicable it gives the only complete measure of precision. When the initial difficulty of appreciating the nature of mathematical probability is overcome, a difficulty which rather grows than diminishes as one works at it, there still remains the greater task of deciding in what cases it can properly be applied and on the method of calculation. It has, in my opinion, often been used where it is not appropriate, where the chances of deviation are not those indicated by a normal curve . . . Thus it has sometimes been argued that if *pn* cases of a particular kind are found in *n* instances, then (without further analysis of the relation of the cases to the whole group) the 'statistical coefficient' for the class is $p \pm \sqrt{\dfrac{p(1-p)}{n}}$, a deduction not based on sound theory; if in fact (here I follow Lexis) the deviation found from this formula is compared with that actually found from several observed values of *p*, the two do not in general coincide. In general, two lines of analysis are possible: we may find an empirical formula (with Professor Karl Pearson) which fits this class of observations, and by evaluating the constants determine an appropriate curve of frequency, and hence allot the chances of possible differences between our observation and the unknown true value; or we may accept Professor Edgeworth's analysis of the causes which would produce his generalized law of great numbers, and determine *a priori* or by experiment whether this universal law may be expected or is to be found in the case in question".

His solution of the problem followed the analysis of Edgeworth, based on the "Central Limit Theorem". Thus: "If quantities are distributed according to almost any curve of frequency, satisfying simple and common conditions, the average of successive groups of, say, 10, 20, 100, . . ., *n* of these conform to a normal curve (the more and more closely as *n* is increased) whose standard deviation diminishes in inverse ratio to the number in each sample. . . . Take, first, a number of small samples (say, 4 or of 10 in each) and observe the curve of frequency for these; if there is a reasonable indication of the shape of the normal curve appearing, I calculate the 'standard deviation' for this grouping, say σ, and proceed with confidence to deduce that the average of a much larger, say, of *n*, will have a normal curve of frequency, with deviation nearly

$\sigma \cdot \sqrt{\dfrac{10}{n}}$, where 10 was the number in the first group of samples. If we can apply this method . . .,

we are able to give not only a numerical average, but a reasoned estimate for the real physical quantity of which the average is a local or temporary instance".

Here, then, is the foundation of the theory which not only supplies a justification of the sampling method, but also enables us to estimate the true value of our statistical constants. Also it indicates to what degree of confidence we can accept these estimates. It need not be stressed that the theory as evolved by Bowley is applicable solely to random samples. ". . . the chances are the same for all the items of the groups to be sampled, and that the way they are taken is absolutely independent of their magnitude".

As we shall see, different sampling types require modification to this theory as applied to the random sampling type. The theory of probability was not new, but in its application to sampling statistics Bowley attained his pre-eminent position, as he says: "The method of sampling is, of course, only one of many instances of the application of the theory of probability to statistics . . . when it is used the test of precision is ignored. We are thus throwing aside a very powerful weapon of research. It is frequently impossible to cover a whole area, as the census does, or as Rowntree here [York, where the address was delivered] and Mr. Booth in London successfully accomplished, but it is not necessary. We can obtain as good results as we please by sampling, and very often quite small samples are enough; the only difficulty is to ensure that every person or thing has the same chance of inclusion in the investigation".

We notice that while, only a few years previously, Kiaer had to use all the eloquence at his command to plead for the use of the sampling method, Bowley, by his new "powerful weapon of research", could boldly declare that a full census or inquiry was not necessary, and further, that often small samples sufficed for the survey. He had the authority of a theoretician applying himself to the practice of sampling statistics; Kiaer had only his own intuition and courage, with no theoretical support.

In the ensuing twenty years Bowley and those he gathered around him completed a series of sampling surveys into the social and economic conditions of many towns of England. He contributed significantly to the "New Survey of London Life and Labour", and at the same time developed the theory of sampling in his monograph, "The Measurement of The Precision attained in Sampling" (Bowley, 1926).

Meanwhile, the International Institute of Statistics, after a lapse of nearly two decades, returned to the problem with new vigour. At the conclusion of the discussion at its Berlin session in 1903 it recommended that the representative method be used, subject to the reservation which we have noted above, and further suggested that a report be drawn up in the following session on the application of the method. Such a report was not compiled in its next session, and the question was shelved until, in 1924, a Committee was appointed to study "The Application of the Representative Method in Statistics". The Committee consisted of Professor A. L. Bowley, Professor Corrado Gini, Mr. Adolph Jensen, M. Lucien March, Professor Verrijn Stuart and Professor Franz Zizek.

Their report was presented at the Rome session of the Institute in 1926. In it Jensen dealt with the practical application of sampling, whilst Bowley (1926) treated it theoretically.

A wealth of interesting detail was presented by Jensen: his criticisms of insufficiently representative investigations, his analyses of the results of those investigations which could properly be called representative, and much also that was to prove of great significance.

The most important point concerning the theoretical aspect was the introduction of a new design of sampling, that of purposive selection. Till then the random type and the stratified type of sampling were the only two in general use*. Experiments on the purposive method of selection had already been carried out†: the novel aspects involved were the sampling by groups instead of individual units and the intentional dependence on correlation (between the quantity sought and one or more known quantities). That the intra-class correlation would also be involved was not recognized till much later.

In the discussion on these reports the objections made over two decades previously to the representative method as first proposed by Kiaer were not advanced, partly because of the theoretical background so carefully prepared by Bowley, and partly because in the intervening years constant use of the method had established its worth. The recommendation accepted by the Institute shows the advance that had been made. It reads:

"The International Institute of Statistics . . . I. . . . calls attention to the very considerable advantages which can be obtained by applying the Representative Method under the following conditions:—

"The results of a partial investigation should only be generalized provided that the sample used is in its nature sufficiently representative of the totality. In such respects the sample may be selected in different ways; the following two main cases, however, are to be distinguished:

"A. Random Selection: A number of units are selected in such a way that exact equality of chance of inclusion is the dominant rule . . . ;

"B. Purposive Selection: A number of groups of units are selected which together yield nearly the same characteristics as the totality. In order to have any knowledge of the precision of the estimates, it is necessary that sufficient groups should be included to allow the variations between the characteristics of the groups to be measured . . . ;

"II. Recommends that the investigation should be so arranged . . . as to allow of a mathematical treatment of the precision of the results, and that with these results should be given an indication of the extent of the error to which they are liable;

"III. Repeats the wish . . . that in the reports on the results of every representative investigation an explicit account in detail of the method of selecting the sample adopted should be given."

It is a curious fact that though the stratified type of sampling has been in use for nearly as long as the random type, it was not recognized in this report as a sampling type on the same plane as the random or purposive selection. There is no denying that it should be so considered for, by it, the selected sample can be made to approximate a "representative miniature" of the universe.

However, with this report, we are at the stage when the representative method has at last been recognized as of use in statistical investigations; when theory has at last started to be systematically developed; when one can confidently employ the sampling method in the collection of statistics and regard the results so obtained as properly "genuine statistics"; and finally, in deciding on the sampling design to use in any particular investigation, one can improve the accuracy of the results of such investigation based on theoretical and, therefore, purely objective considerations. This last fact leads to the next equally important stage of developing alternative sampling types or designs, where the previously accepted types or designs are found to be inaccurate, or to be difficult or impossible to apply, because of imposed limitation.

Professor J. Neyman, in the *Journal of the Royal Statistical Society* (1934), gave a theoretical

---

* Although it wanted another eight years before the theory of the stratified type was fully developed by Professor J. Neyman.

† Jensen's inclusion of Kiaer's investigations in the category of "purposive selection" is not strictly correct. It is true that Kiaer employed the "Method of Controls"; but while for purposive selection controls are used in the selection of groups for sampling and the selected groups are then completely observed or interviewed, Kiaer's method did not impose these restrictions. He used these controls merely to stratify his groups (towns or country communes) in the first place, and the selected groups (presumably selected by random selection) were further stratified. His controls were used *after* the surveys for testing the representative character of the selection observation units. To illustrate this we have only to refer to the safeguard he mentioned for the survey on the health insurance scheme in the Berne session of the Institute in 1895. Further, his selected groups were not completely interviewed, but were sampled in proportion to the population according to the general census of 1891.

criticism of the method of Purposive Selection. In doing so he placed the methods of stratified and group sampling on a sound theoretical basis.

But that was not all that Neyman achieved. Until then the theory of estimation was mainly that of "point estimation" based on Bayes' Theorem of Inverse Probability. This requires "the knowledge of probabilities *a priori* attached to different admissible hypotheses concerning the values of the collective characters of the population . . .", so that "we are met with conclusions based, *inter alia*, on some quite arbitrary hypotheses concerning the probabilities *a priori*, and Professor Bowley accompanies his results with the following remark: 'it is to be emphasized that the inference thus formulated is based on assumptions that are difficult to verify and which are not applicable in all cases' ". Proceeding along a different track, Neyman elaborated on and refined the theory of "interval estimation", as suggested by Professor R. A. Fisher "to remove the difficulties involved in the lack of knowledge of the *a priori* probability law". This new method of estimation, later further investigated by E. S. Pearson, S. S. Wilks and many others, became the recognized theory of estimation.

R. A. Fisher took another important step in the development of sampling theory. His work was concentrated largely on biological and agricultural research, but was later found to be of great importance to general sampling problems. The researches of R. A. Fisher and his contemporaries were concerned with agricultural field experiments, where the repetition of each treatment only a few times rendered the procedure of estimating the error of each treatment mean from deviations of the yields of the individual plots from that mean incompatible with theoretical requirements for precision, and there was the additional difficulty due to the fact that each replicate of the experiment was arranged in a compact block of plots on the ground to eliminate fertility differences as far as possible.

Fisher's technique of the analysis of variance enabled error estimates from different treatments to be pooled, while eliminating variation due to blocks or other features of the layout of the experiment. To ensure validity of the error-estimates the principle of randomization was introduced.

Various experiments based on the new technique were carried out in the Rothamsted Experimental Station and other experimental centres in England by such authorities as Fisher, Mackenzie, Clapham, Irwin, Cochran, Wishart, Yates, Zacopanay, and others. Their results were published in various specialist journals and summarized by Yates in 1946 (Yates, 1946, Irwin, 1929, 1938). Fisher, (1925) presented a fully systematized theory of experimental designs and analysis

The first application of the new technique to sampling problems was made in 1929 by Clapham, who estimated the yields of experimental plots of cereals from a number of small sampling units cut from each plot, and used the analysis of variance to calculate the sampling errors to which sampling units of various types were subject. The method was further tested by Clapham in 1931, and its efficiency was thoroughly examined in 1935 by Yates and Zacopanay. Thus the way was clear for the development of sampling techniques applicable to many agricultural and biological problems, in which the sampling problems were approached from a new angle, namely, that of estimating sampling errors from the results of the observations. By means of this new approach it was possible not only to ascertain the adequacy of the sampling actually undertaken, but also to increase the efficiency of future sampling of the same type of material.

Yates (6) gave the principles on which this new work was based:

"(1) If bias is to be avoided, the selection of the samples must be determined by some process uninfluenced by the qualities of the objects sampled and free from any element of choice on the part of the observer.

"(2) If a valid estimate of sampling error is to be available each batch of material must be so sampled that two or more sampling units are obtained from it. These sampling units must be a random selection from the whole aggregate of sampling units that can be taken from the batch of material, and all the sampling units in the aggregate must be of approximately the same size and pattern and must together comprise the whole of the batch of material".

The first condition was satisfied by the statisticians who drew up the report to the International Institute of Statistics in 1926. The second condition embodied the really new advance. As Yates wrote: "Realization of the functions of strict processes of randomization in agricultural field experiments had led to a corresponding realization of its importance in providing a valid estimate of error in sampling. 'At random' no longer meant 'haphazard'. Again, the analysis of variance, by making possible the pooling of estimates of error and the separation of components of error which were not homogeneous, enabled the number of independent sampling units taken from each batch of the material to be reduced to a small number, and so permitted the use of relatively complicated sampling schemes often involving sampling in two or more stages".

Applying this to sampling in social and economic research, the first point to make is that experiments cannot be replicated as can treatments in agriculture, mainly because in most cases conditions cannot be controlled, and where they can the expense would be prohibitive. Nevertheless, the new technique makes possible really valid estimation of sampling errors in stratified sampling to determine the efficiency of different types of sampling units, in sub-sampling, line sampling, and other sampling designs. It thus facilitates the selection of the most appropriate design for various conditions and limitations.

The new technique has the added advantage that, where the natural units of the population being surveyed vary widely in size and must be taken as they are, it enables the efficiency of such units to be determined.

The development of the t and z tests enabled error-estimates as calculated by the analysis of variance technique to be used as a basis for exact tests of significance. The publication by Fisher in 1935 of "The Design of Experiments" to a certain extent completed this phase of the development as applied to agricultural and biological experiments. It has yet to be satisfactorily applied to social economic problems.

While the analysis of variance technique was being developed, another important phase of sampling theory was initiated. Combinatory analysis was applied by Professor Carver (1930) to the estimation of errors in sampling He was the founder of the *Annals of Mathematical Statistics*, and in the first volume of that journal he outlined his new technique. By means of the new theory, error-estimates could easily be calculated for almost any type or design of sampling, and, in conjunction with the analysis of variance technique, the general outline of sampling theory reached its final form about 1935. The present form of combinatory analysis as applied to sampling theory is the work of Professor Paul S. Dwyer, whose work in the *Annals of Mathematical Statistics* in 1938 could be regarded as a continuation and completion of the work started by Carver (Dwyer, 1938).

Just before the outbreak of the Second World War the development of sampling theory and its application to social and economic research shifted from Europe to America.* This does not mean that since 1938 sampling theory was not developed or that sampling surveys were not undertaken in England and elsewhere in Europe: it does mean that where there was development of the theory it was perhaps less systematic or co-ordinated elsewhere than in the United States This last fact can be attributed mainly to the lack of a Central Statistical Agency (corresponding, for example, to the Bureau of the Census in the U.S.A.) to undertake full-scale sample surveys to test the validity of any sampling design as theoretically developed by its technical staff, or to decide on any survey that it deemed should be undertaken. Even before this, sampling theory had not been neglected in the United States. Carroll D. Wright, working in Massachusetts, was indeed a pioneer in sampling, holding a parallel position to that of Kiaer at the end of the last century. The main reason for regarding Kiaer as the true pioneer is that, while both of them had the foresight to recognize the importance of sampling in social and economic research and the courage to apply it in obtaining official statistics for the different statistical bureaux of which they were chiefs, Kiaer was the first to point the way to a systematic development of the problem of sampling and to present it to the world for further development. Kiaer was the better-trained theoretical statistician, applying his theory, crude as it was, to the collection of sample statistics, while Wright was the practical statistician *par excellence*.

Wright's first large-scale statistical investigation was the Massachusetts Census of 1875, taken under his direction as Chief of the then newly-organized Bureau of Labor. This Bureau was the first of its kind in the United States, and indeed in the world, for, though statistical bureaux then existed in many European countries, they were concerned with all types of statistics—demographic, social, economic and so on. The Massachusetts Bureau of Labor, owing largely to

---

* In agricultural and biological experimentation, the work of the Rothamsted Experimental Station, and other experimental stations in England under such eminent statisticians as Professor Fisher, Clapham and later on, Dr. F. Yates, enabled the development and application of sampling theory to be kept on a high plane throughout the War.

Wright's inspiration and guidance, began a new field in governmental statistics, that of industry and labour. Since then, 34 similar State Bureaux have been inaugurated in the United States.

In 1884 Wright was chosen to organize the National Bureau of Labor and to extend his work from one state to the entire U.S.A. From then until his retirement from the position of National Commissioner of Labor in 1904, to become President of the newly created Clark College at Worcester, Mass., he succeeded in making the National Bureau a positive force in the industrial development of the nation's planning. He published a long series of reports covering the entire field of labour questions.

Unfortunately this long series of reports reveals very little of the actual method employed by Wright to obtain "representative" samples. That he did use the representative method, as understood at that time, and not any random, haphazard method of selection, is clear from his communication to Kiaer, which Kiaer read at the Budapest session of the International Institute of Statistics in 1901: "The experience of the U.S. Department of Labour has continually strengthened my own views as to the value of representative statistics. The first annual report (Industrial Depressions), the fourth (Working Women in Large Cities), . . . are all emphatic evidences of the values of this method, while nearly all the special reports published in our bi-monthly *Bulletin* add to the weight of the evidence. In fact, offices like this and the State Bureaux of Labour Statistics must use the representative method; and it is best that they should, for on most of the topics on which they deal representative facts are quite sufficient at least when we consider the vast cost of securing statements of aggregates.

"So, taking all things into consideration, I am of the opinion that the conclusions given in your St. Petersburg report and your more detailed studies in the *Allgemeines statistisches Archiv* are eminently wise and sound".

As Wright did not make clear the methods of selection of his samples, the credit for the systematic development of the theory of sampling rests among European statisticians. But the bold and farseeing step of applying sampling theories to the collection of government statistics on a vast scale, independent of the census, secured Wright's place among the pioneers of sampling. State statistical departments were organized, modelled on the Massachusetts Labor Bureau; Federal Bureaux of Statistics were started, not to control the statistical departments of the various States, but to co-ordinate their work, and simultaneously to collect nation-wide statistics independently of the State departments. These Federal Bureaux followed the pattern he had developed.

We are here concerned only with the development of governmental statistical departments in so far as it advanced the theory and practice of sampling. Here, too, statisticians in the United States of America were the first in the field. True, the British Ministry of Labour had in the twenties taken samples of "insured work-people unemployed" of several categories (Hilton, 1924, 1928), and sampling was also used in the calculation of the first British Cost of Living Index Number. Several European countries, for example, Norway, Denmark, Sweden, Hungary, and Italy, had also employed the sampling method in the collection of their government statistics. But their scale was by no means vast, and any application of the sampling method was spasmodic. Thus there was not in any country outside the U.S.A. any annual, or regular, sample surveys, while in the State of Massachusetts an annual sample survey of business was started at the beginning of the century, followed thereafter by the biennial Statistics of Business for the whole of the United States. Both these steps were taken under the direction or at the suggestion of Wright.

Not only in the governmental departments has this development been rapid, but also in private institutions and universities, as well as in the co-operative efforts of the public and private agencies. In this regard there are significant differences between the U.S.A. and other countries. For example, in England during the first decade of the present century, under the guidance of Bowley, researches had been carried out into the social and economic conditions of several towns—Reading, Northampton, Warrington, Stanley, Bolton, and others. These were sponsored by institutions such as the Ratan Tata Foundations. The great investigation into the London life and labour in the late twenties was carried out largely at the instance of the London School of Economics. But such endeavours were spasmodic, and though intensive in themselves did not coalesce into any nation-wide projects—further, they relied on sampling methods previously used by Bowley, and did not carry out new researches into sampling methods; the resources to do so were inadequate.

In the U.S.A. the picture was altogether different. There the efforts of the charity organization movement, the study of economic and social problems by the universities and the development of market research all helped towards the development of the techniques of social investigations.

The Pittsburgh Survey of 1907 was the first complete social survey—not merely a sample survey (Kellogg, 1914). Directed by Paul U. Kellogg, financed by the Russell Sage Foundation, it purported to be "an appraisal . . . of how human engineering had kept pace with mechanical in the American steel district . . . an attempt to throw light on these and kindred economic forces, not by theoretical discussion of them, but by spreading both the objective facts of life and labour . . .". Its subject matter included the study of wages, hours of work, work accidents, questions of industrial relations and conditions, family budgets and home conditions of steel workers, and a host of other questions. It made use of the case-work method developed by B. S. Rowntree in his York Inquiry in England, and by Charles Booth in his inquiry into the "Life and Labour of the People in London". But while in England the case-method survey was, after the findings of Bowley, not used at all, or used, as in the second survey of the "London Life and Labour", in conjunction with the sampling method,* in the United States its use increased rather than diminished. Thus the next large-scale survey, that of Springfield, Ill., in 1914, had as its "method of investigation . . . the study of the records, published and unpublished, compiled and uncompiled, of organizations and institutions in the community and of outside agencies which had data on Springfield; personal visits to and observation of Springfield organizations and institutions in operations; the gathering of facts through intensive studies or tests planned fcr certain sections of the city, or of the population; special study of the activities of particular agencies or groups of agencies and interviews with officers in charge; first-hand observation of conditions throughout the city; written inquiries and personal interviews with individuals in possession of experience or information pertaining to the problem in hand; and study of legislation relating to local conditions and procedures". So that the survey, besides applying the case-method, used the experience of the civic and social workers in finding facts relating to current social situations, the experience of the engineer in understanding the structural relations of different types of community conditions, of the surveyor in relating his work and study to a definite geographic area, of the social research worker, of the physician, city planner and social worker in bringing the problem down to human terms, of the journalist and publicity worker in interpreting facts in terms of human experience. In short, such social survey is a co-operative undertaking which applies scientific method to the study and treatment of current related social problems and conditions having definite geographic limits and bearings. It endeavours to publicize its facts, conclusions and recommendations so that they shall be the common knowledge of the community and a force for coordinated action.

At the same time the specialized survey was being developed. For example in Cleveland there was a series of investigations each dealing specifically with one subject—education, recreation, criminal justice, hospitals, and health. In these it was neither easy nor desirable to employ the sampling method. Intensive investigation is the only sure way of bringing out the required information; moreover, the limited scale of the survey does not warrant the use of sampling to select a certain proportion of units for observation. General and special surveys undertaken within a definite geographic area, such as a city, do not necessarily involve sampling.

Since then social surveys have flourished in the United States. Several thousand projects on special studies and several hundred general surveys have been completed. There were many institutions sponsoring and participating in surveys; university and college organizations; agricultural experiment stations; health and medical associations; family welfare and charity societies; councils of social agencies; child labour committees; consumers' leagues; bureaux of municipal research; commissions of efficiency and economy; housing committees; playground and recreation committees; committees on industrial relations; market research associations; and many others. Practically every type of private organization interested in improving the conditions under which people live and work, and a large number of municipal, state and

---

* In general one can say that the case-method was used to obtain information which is not easily measurable, as in the second London Survey, the "local distribution of poverty, street by street, and . . . direct information as to poverty due to personal habits as distinct from that due to deficient income"; while the sample method was used "to obtain detailed and precise information on the composition of the working-class families, housing accommodation and rent, number and ages of earners, total income, etc.", i.e. information which could be measured.

federal bodies, have made the study of social conditions an important feature of their regular work, not only to give citizens the information necessary to form intelligent opinions upon matters of public concern, but also to develop plans for their current work.

In the first decade of the century there was in the U.S.A. a new development in social surveys. Federal agencies began to co-operate with private agencies. The first work of this kind was the survey of individuals and schools in the city of Richmond, Virginia, carried out, in co-operation, by U.S. Bureau of Education, the U.S. Bureau of Labor Statistics, the National Society for the Promotion of Individual Education, the Russell Sage Foundation of New York City and the city of Richmond.

Yet another development was the institution of more or less permanent survey or research bureaux in different localities.

There were also agencies which studied, not the social welfare or conditions of the people, but the consumption and expenditure habits of some part of the country, thus trying to determine whether a particular industry could with advantage be established, or whether a particular commodity could be sold easily, and various other phases of the problem of market research. Projects were undertaken by national newspapers to fathom public opinion on national affairs, such as Prohibition, Presidential elections, and many other matters.

It is therefore not surprising that sampling developed with such rapidity in U.S.A. Where the surveys are on general lines and national in character, or where market research schemes have to cover, not one state or one district, but the whole country, case-method, or typological, or intensive methods of investigation are impossible, or if possible, are too expensive.

As early as the beginning of the century there had been sampling surveys on social conditions, an example of which was provided by the study of living conditions and migration of the Upper East Side of New York (Mark, 1907). Houses were the units of investigation. A number were chosen as representative of conditions in their particular blocks, or as consisting of families typical of a class more widely distributed throughout the district. Of course, no account of the method of selection and justification for the selection was given, as was also true of nearly all sampling surveys undertaken at the time.

At about the same time the case-method investigation of the American charity organizations, of C. Booth in his survey into the "Life and Labour or London", and of Professor Lindsay in his report to the National Conference of Charities (in 1899), was criticized by Kleene (1908), chiefly in connection with its use in surveys for the study of causes of pauperism. He adduced two reasons for the inapplicability of the case-method—the relative paucity of recorded information, and the difficulty amongst what there was of interpreting the causative contribution of particular social ills. He thus cast doubt on its use, for measuring the effects of social conditions—merely counting cases—assuming that they are of equal value, and not utilizing proper methods of selection vitiated the use of the method unless it were applied conjointly with an intensive survey, as had been done in some of the specialized fields we have mentioned above.

We need not mention every survey that does employ the sampling method. It is sufficient to say that by the thirties the development of sampling for general social surveys, and sometimes for special surveys, had advanced so far that hardly a survey failed to use it for the selection of units for observation. Thus sampling was used in surveys into living conditions, population migration, standard of living, family incomes, expenditures and savings, education problems, farm conditions, farm acreages, relief family conditions, relief and rehabilitation problems, racial problems, health surveys, housing surveys, effects of rural credit, psychological research, occupational characteristics of workers, fertility questions, and many other problems of social and economic significance. Marketing research schemes, consumer opinion and consumer purchases surveys, too, make considerable use of this new method. Public opinion polls, straw votes, newspaper surveys, all make use, sometimes unwisely no doubt, of sampling. And lastly, sampling is now being increasingly used in industry for research on quality control of standardized products.

In the theory of sampling, too, there had been important advances, so that the criticism can hardly be raised that theory does not keep pace with practice. Indeed it is more the other way round; for it is difficult, if not impossible, to develop social studies to such an extent as to endow them with the status of experimental research. Each survey costs so much time, money and labour; social conditions are difficult to control, and are so complex that although theories for "ideal" sampling designs have been developed by mathematical statisticians ("ideal", that is, in various circumstances and under various limitations), no "experiments" are possible to prove their usefulness, and consequently they have to wait until some agencies have made use of them. The U S. government bureaux are, of course, an exception.   They have professional experts to develop new theories for particular circumstances, and at the same time have the resources and boldness to put these into actual use.   Thus the double sampling design developed by Neyman was employed by the Bureau of Home Economics and that of Agricultural Economics in the study of rural and urban family expenditures and savings; the sub-sampling design in connection with block and area sampling and the study of the relative efficiencies of various sizes of sampling units for estimating the characteristics of a finite population, were developed by the Bureau of the Census (notably by such statisticians as Hansen and Hurwitz), and employed by it for the survey of the Labour Force; the theory of area sampling was developed and used by the Department of Agriculture in its plan for a master sample of agriculture; the theory of questionnaire-interview-follow-up and its employment was perfected by the Bureau of the Census in its survey of business sales; the theory of grid or "systematic" sampling for survey of lumber resources and of lumber produced, by the Bureau of the Census; the theory of matching for regularly repeated surveys was designed and put into practice by Jessen for the Department of Agriculture.

While the U.S.A., chiefly, is to be credited with such rapid advances in the development of sampling theory and its application, other countries have not been idle in both these aspects. The most notable example is India, where since the establishment of the Indian Statistical Institute on December 17th, 1931, for research work and for the training of statisticians, both the theoretical and practical aspects of sampling have developed rapidly and successfully, especially in the field of large-scale crop surveys, the field which has been aptly designated statistical engineering by Professor Mahalanobis, the guiding spirit of the Institute for the greater part of its existence to the present day.

Two types of surveys were developed by this Institute as contract works for the Government, from the grants for which it obtained the greater part of its income.   The first type is the large-scale statistical engineering project to obtain reliable estimates of acreage, rate of yield per acre, and total production of food and fibre crops, such as rice, wheat, jute, etc., or of economic or demographic factors relating to indebtedness, unemployment, destitution, paddy land, plough cattle, birth and death rates of rural families.   The second is more localized, relating to cost and level of living, housing, food-consumption, consumer surveys, public opinion, and so on.   For the first type of surveys Professor Mahalanobis (1944) has developed a comprehensive theory of area or grid method of sampling, and has further introduced the idea of cost and variance functions for the determination of an optimum or most economical design for sampling in any particular instance.

Also, in connection with large-scale surveys, where the work has to be continued at suitable intervals (such as yearly, quarterly, etc.), the exploratory method of sampling has been developed and used by the Institute.   A survey is first carried out on a very small scale with the primary object of collecting basic information required for preparing an efficient design for later surveys.   Sometimes such preliminary surveys have to be carried out more than once, and the scale of operations is gradually increased until finally the whole area or universe is fully covered.   For example, in the Bengal crop survey, a pilot survey was first carried out in 1937 covering only 124 square miles at a total cost of about £1,100.   In 1938 a second survey was organized, in the light of the experience gained in the previous season, on a larger scale covering about 400 square miles, at a cost of £2,500; next year the total area covered was nearly 2,600 square miles, at a cost of £6,000.   In 1940 the area covered was nearly 20,600 square miles, at a cost of £8,100; and finally in 1941 the whole jute tract of Bengal, measuring about 60,000 square miles, was surveyed at a cost of £10,100 (Mahalanobis, 1946).   These "pilot" surveys provided sufficient experience to expedite the exploratory phase of future surveys, especially where conditions are expected to be similar, as, for example, the Bihar crop survey, where the exploratory work was done in only two districts, covering about 8,000 square miles, from February to April, 1944.   Furthermore, the exploratory phase is useful in providing the necessary training to the field staff and the computational and technical staff.

In the second type of surveys work has been carried out during the war on the study of family budgets, housing and other economic conditions of factory workers in an industrial area north of Calcutta, and another inquiry on labour conditions carried out in Nagpur at about the same time. The design of the surveys enabled the analysis of variance to be carried out, and it might be mentioned that these provide practically the first instances of the application of the Fisherian analysis of variance in socio-economic surveys, especially in respect of personal equations or bias of the investigating staff, thus enabling this type of error to be separated and eliminated.

India can therefore safely claim to rank with the United States as amongst the foremost users of the sampling method in social and economic research. And it is a very happy combination, for in the United States we have the typical example of an industrial and highly developed country, while in India the conditions approximate more nearly to those of a country not so highly developed, or, more specifically, to the conditions of those countries, which, like China, have no genuine statistics, and where such statistics, if they are to be obtained at all, have to be obtained mainly by sample surveys, for which the experience of India will serve as a guide and as an example worthy of imitating.

Thus, half a century after Kiaer and Wright started the use of sampling in large-scale social surveys, and four decades after Bowley developed the statistical theory of sampling, we are now in a position which many other sciences and scientific theories have taken centuries to attain. To-day nearly every social survey worthy of the name is making use of the sampling method, correctly or otherwise, depending on the understanding on the part of the planners of the theory, of the circumstances under which they have to carry out the surveys, of the materials, human and otherwise, at their command. Any lack of understanding of the sampling theory is due largely to the fact that, with the exception of a chapter or so in nearly every text-book on Statistics, mainly on random and stratified sampling, no real effort has been made to systematize all the sampling types and designs which have been separately developed in practically every learned journal of statistics, agriculture, eugenics, marketing research, public opinion, social research, economic research, etc.—with the exception, that is, of the recent work of Dr. F. Yates (1949). For there are, in actual fact, two very important related aspects of sampling which every sampler must take into consideration:

(*a*) The static aspect which *treats* every sample survey for its own immediate worth. Under it every sampling plan has for its aim the attainment of the maximum precision, and therefore information, possible at a given cost, time, labour and other available materials, or, alternatively, a given precision or information at the minimum cost, time, labour.

(*b*) The dynamic or sequential aspect which does not, of course, exclude the static aspect. Under this aspect a sample survey, besides its immediate purpose, as stated under (*a*), can serve as a quasi-laboratory experiment for future surveys. It should enable a thorough analysis of the particular sampling plan, of the particular sampling unit used, the particular sampling ratio, the particular method of inquiry, etc., to be carried our so that future surveys, carried out under more or less similar conditions, can be planned with greater assurance of success, or so that modifications can be introduced to obtain greater precision or information.

References

Bowley, A. L. (1906), *J. R. Statist. Soc.*, **69**, 548.
—— (1926), *Bull. Inst. Int. Statist.*, **22**, Liv. I, [6].
Carver, H. C. (1930), "Fundamentals of the theory of sampling", *Ann. Math. Statist.*, **1**, 10.
Census of England and Wales, 1861, Vol. III, General report, pp. 93–99. [Tables 32–42.]
Dwyer, P. S. (1938), *Ann. Math. Statist.*, **9**, 1, 97.
Fisher, R. A. (1925), *Statistical Methods for Research Workers.* London: Oliver & Boyd.
Hilton, J. (1924), "Inquiry by sample", *J.R. Statist. Soc.*, **87**, 562.
—— (1928), "Some further inquiries by sample", *ibid.*, **91**, 519.
Irwin, J. O. (1929), "Crop forecasting and the use of meteorological data in its improvement", *Conference of Empire Meteorologists, Agric. Section*, **2**, 220.
—— Cochran, W., and Wishart, J. (1938), "Crop estimation and its relation to agricultural meteorology", *Supp. J. R. Statist. Soc.*, **5**, 1.
Kellogg, P. U. (1914), *The Pittsburg District: Civic Frontage.* N.Y.: Russell Sage Foundation.
Kiaer, A. N. (1895), "Observations et expériences concernant les dénombrements représentatifs", *Bull. Inst. Int. Statist.*, **9**, Liv. I, 176.
—— (1899), "Sur les méthodes représentatives ou typologiques appliquées à la statistique", *ibid.*, **11**, Liv. I, 180.
—— (1901), "Sur les méthodes représentatives ou typologiques", *ibid.*, **13**, Liv. I, 66.
—— (1899), "Die repräsentative Untersuchungsmethode", *Allg. Statist. Arch.*, **5**, 1.

Kleene, G. (1908), "Statistical study of causes of destitution", *J. Amer. Statist. Ass.*, **11**, 273.

Mahalanobis, P. C. (1944), "On large-scale sample surveys", *Philos. Trans.*, B, **231**, 329.

—— (1946), "Recent experiments in statistical sampling in the Indian Statistical Institute", *J. R. Statist. Soc.*, **109**, 325.

Mark, M. L. (1907), "The Upper East Side: a study in living conditions and migration", *J. Amer. Statist. Ass.*, **10**, 345.

Neyman, J. (1934), "On the two different aspects of the representative method", *J. R. Statist. Soc.*, **97**, 558.

Yates, F. (1946), "A review of recent statistical developments in sampling and sampling surveys", *ibid.*, **109**, 12.

—— (1949), *Sampling Methods for Censuses and Surveys*. London: Griffin.

SIR ARTHUR LYON BOWLEY   (1869–1957)
by
W. F. MAUNDER   (1972)

IN the summer of 1895, George Bernard Shaw penned an intentionally venomous letter to Beatrice Webb giving free vent to his indignation about the 'atrocious malversion' of the Hutchinson bequest; this latter, intended for the support of the Fabian cause, was the windfall which Sydney Webb was then using to found the London School of Economics. Shaw's professed view was simple:

> 'Any pretence about having no bias at all, about "pure" or "abstract" research, or the like evasions and unrealities must be kept for the enemy.'

The Webb's vision of an academic institution devoted to impartial research and teaching in the Social Sciences could be only a manifestation of temporary (as he hoped) insanity. Not surprisingly, the first lecturers to be appointed he found dismally unsuited for the task of Collectivist propaganda. The eminent among them received scathing comment. That Arthur Bowley escaped castigation may have been due either to the fact that he was then an unknown young man or, if known to him, that he met with Shavian approval. Certainly, by 1901 Shaw was sending friendly advice to him about 'vaccination and serumpathy' outlining a preposterous scheme whereby he could immortalise himself 'as the founder of Scientific Pathological Statistics'.

Bowley, indeed, was in for the start of the School's first session, which opened in October of 1895, but as a part-time lecturer on his free Wednesday afternoons, cycling up from Leatherhead where he was teaching Mathematics at a boy's school. This was the beginning of a lifetime's connection which progressed through a Readership to a School Professorship and, in 1919, to a full-time University Chair.

The exact confluence of events which led Bowley into his career can hardly be unravelled in retrospect but at least some precipitating causes may be detected. In 1888 he had entered Trinity College, Cambridge, to read Mathematics, the victim of high expectations. His Mathematics master at Christ's Hospital had cherished the ambition of educating a Senior Wrangler and when Bowley obtained the first place in the Trinity scholarship examination it looked as though it was an aspiration which needed only time for fulfilment. Bowley recorded: 'the programme was that I should devote myself to Mathematics, become Senior Wrangler, obtain a College Fellowship and promote knowledge and my own income.' The apparent cause of the plan's miscarriage was ill health which developed in his second year and appears to have remained with him for most of the remainder of his undergraduate years despite a recuperative voyage to Egypt, paid for by his tutor and other dons, and a lengthy stay of several months at Bournemouth. He sat his examinations in 1891 none the less (being listed as tenth Wrangler must have been a considerable relief in the circumstances) but, presumably to make good the time he had lost, he could not take his degree without a further period of residence.

Arthur Bowley has left a note *In Praise of Mathematics* (partly reproduced in the *Family Memoirs*) which recounts his attitude at this time; he has evident sympathy with the view that mathematics is an end in itself and that 'it exceeds all other sciences in that it is based upon the essential nature of thought and is the one thing that is permanent, while physicists and chemists continually change their theories'. He comments:

> 'There is indeed a satisfaction in the development of a mathematical proof, with its rigid hypothesis, clear-cut, terse and logical argument, and far reaching result. It is analogous to the artisan's pleasure in a fine piece of work. Again it is almost an aesthetic joy to discover some general theorem, which is found to contain very many earlier theorems as special cases.'

But he comes down finally on the other side:

> 'It is not, however, that it is by the pleasure of the practitioner that the importance of mathematics is to be measured, but by its use to the engineer, the physicist, the chemist and others, whether to construct buildings, to examine invalids by X-rays, or to burst the atom.'

Its use to the economist is not mentioned but his final year at Cambridge was devoted to reading economics as well as chemistry and physics. His tutor arranged an introduction to Alfred Marshall and here we may be tolerably certain was the effective influence that transformed him from a mathematician into a statistician. A personal contact was established and it was this which seems to have led later to the LSE appointment; a more immediate consequence, however, was Marshall's suggestion that Bowley should compete for the Cobden Essay Prize. The title set for that year (1892) was *Changes in the Volume, Character and Geographical Distribution of England's Foreign Trade in the XIXth Century and their Causes*. It not only won the prize but the work was also successful as a publication, continuing to be used as a text for a number of years. The subject does not make an immediate appeal as being one of absorbing interest but, in fact, it reads very well even today; the analysis is lucidly presented and the more remote effects of trade are strikingly brought out. The advantages of Free Trade did not blind him to its possible dangers and less pleasant consequences. There was a warning which echoes today in the fears of some about Common Market entry: 'In the process (of increasing international specialization) much must be sacrificed . . . . a nation may lose its greatest branch of trade and obtain no other; and the flow of population will take place in exactly the same way as from town to town . . . . many countries have thus fallen in power and population.' On the technical side, one notices particularly that in his discussion of the balance of trade he does, in fact, develop all the essentials of what we distinguish as the balance of payments. Indeed it is to be regretted that a fuller treatment, apparently envisaged originally, was not actually included; for, in a letter to Cannan long after, Bowley wrote in reference to estimates of earnings on foreign investments:

'By an elaborate series of equations, I got certain estimates which Hobson approved twenty years later. I showed the mathematics to Giffen, who very truly said that the method was too refined for the data, and the beautiful analysis was boiled down to a foot-note.'

A conclusion about wages which he drew from the study was, one might think, little to the liking of the times; the results he adduced were to the effect that highly paid labour need not be an unfavourable condition for profitable production nor need the competition of badly paid foreign labour be necessarily harmful. Half a century later, Leontief's analysis of American foreign trade produced what was regarded as the paradoxical finding that the country renowned above all others for its high wage rates had a comparative international advantage not in capital-intensive goods but in labour-intensive goods. Bowley, of course, was not calling for indiscriminate wage increases and, with our post-war history in mind, one reads somewhat ruefully his warning that if the increase in wages relative to productivity is greater in one country than in another, then the former will be at a disadvantage.

The work was far removed from an undergraduate project and it is not surprising that it was thought worth reprinting thirty years later.

International trade was not a field which Bowley was to make a major interest and indeed he published only one other full scale work on the subject; this was an analysis of the country's trading position in the First World War (*The Effect of the War on the External Trade of the United Kingdom*) which limited its perspective to the problems of the period and, unlike the earlier essay, is now only likely to interest the historian.

The precipitating events which led Bowley to become a statistician are straightforward enough but this is only watching the movement of surface water. He also worked in the Cavendish Laboratory in his final year at Cambridge—why did he not become a natural scientist? There must have been predisposing causes which are not so easily discovered but there is enough evidence of a kind sufficient to make a convincing picture of his approach to life.

He was the son of the vicarage (born in 1869 in a Bristol parish) but one cannot hope to evaluate the influence of this factor; indeed, outwardly no visible signs persisted and the formalities of religion seem to have been little to his taste. His school teaching career was effectively ended when in 1899 he departed from his post at Leatherhead on his declaration to his headmaster that 'the boy's religion was only perfunctory attendance at chapel, that the services were vain repetition and that I refuse to attend chapel in the future' (*Family Memoirs*). Bearing in mind the conventions of the period, this was strong stuff and testimony to no mean force of character; from where he derived his willingness to publicly maintain his convictions at cost to himself is another matter. His chosen concern was to be, rather literally, man's bread but one senses that his dedication and commitment to his work were seated in something more profound than a comfortable humanitarianism.

Before going up to Cambridge he had been actively involved in the social work of his father's parish and he had had first-hand experience of slum conditions at an early age. At Trinity at least two incidents occurred which reveal his attitude. Together with some other students, he tried to start a custom of entertaining the dining hall waiters as a gesture to the principle of social equality; one suspects that probably they turned out to be rather uncomfortable affairs but it makes a point. On another occasion he played some part in organizing a boycott of Commemoration Dinner which was regarded as a wasteful use of resources. Both incidents are reminiscent of student activities with which we are now more than familiar; one may reflect though that a 'sit-out' is not only a rather more dignified protest than a 'sit-in' but also more persuasive to those of a contrary view.

Bowley did not become a social or political agitator but all his work was indirectly concerned with welfare in the broad sense and a large part was directly devoted to an investigation of social conditions. Indeed, it has been said that he regarded his part in the *New Survey of London Life and Labour*, which occupied his last years at LSE, as the culmination of his professional career. It is not unreasonable to speculate that he rejected the idea of any simple panacea in favour of slow, patient work in which any true improvement to man's lot had to be accomplished a brick at a time. This places him on firmly recognizable ground; neither retreat into 'ivory tower' academia nor the passionate espousal of any particular political creed was to be his path.

After winning the Cobden Prize, Bowley went on to consolidate his reputation by entering for the Adam Smith Prize in 1894 and repeating his previous success. The subject was *Changes in Average Wages in the United Kingdom between* 1860 *and* 1891 and it is a paper which, at first sight from our stance, it is hard to appreciate with true justice. Marshall described it as brilliant and it is a measure of its quite basic importance that, precisely because it was written, we now regard the methods as commonplace. The basic principle, like most ideas so useful in practice, was very simple; it was merely to study wages by ratios (or index numbers) displaying their movement instead of attempting to look at actual levels. There was nothing new in the idea of index numbers at this time, of course, but nobody before had seen their potential for use in this particular way. The problem was that while there was no dearth of data on wages, the figures related to different occupations, in different places at different times collected on every variety of definition. Any attempt at direct aggregation or averaging was meaningless but if only two figures could be found collected on the same basis for the same occupation in the same town but at different dates, then a reliable measure of change could be derived. Then, given enough links of this kind, they could be forged into a continuous chain covering the whole period on a consistent basis. By this means it was possible to extract a coherent picture from 'the great mass of wage statistics which had hitherto been almost useless because of its fragmentary character' (as Marshall put it). The work was a highly ingenious and painstakingly delicate exercise in linking and combining disparate series for different trades, places and times. Even a casual inspection immediately impresses the reader with the huge volume of material which is digested and the sheer physical labour of computation which is implied. Presumably this was accomplished without a research assistant and without a calculating machine. The essential principle was that each individual indicator obtained from any given system of collection could be expected to be approximately invariant with respect to the particular system used. An example he quoted, which illustrates and supports his theoretical analysis in a striking way, was for the Building Trades, where two distinct sets of data were available: results derived from figures collected by Chambers of Commerce could be compared with those coming from Trades Unions. With 'no point of resemblance' between the original figures (and every reason for suspecting quite contrary biases) the resulting index series were almost identical.

A most important aspect of this paper was the scrupulous attention which was paid to error, with an attempt to distinguish those which might be brought within the theory of random errors from those arising from exceptional circumstances with a different order of magnitude. An interesting side-light on his approach is shed by the following quotation in which he compared the ideal data to observations made with a chronometer and the actual material to measurements with a sundial:

'If the time of sunset on many days scattered at random throughout the year is written down from a chronometer and from a sundial, and the averages taken for each method, the averages will be practically the same. Trying this for twenty-four days taken at random throughout the year, I found 6 hours 8 minutes and 6 hours 10½ minutes as the averages. The more days taken, the nearer both would be to 6 hours.'

That this investigation of wages was not a matter of idle curiosity to him is very evident for he started by asking 'Who are benefiting most by the development of industry: those who obtain profits or interest, or those who receive wages?' and ended by declaring that he would wish to see the share of wage earners increase at least until 'the sources of comfort, pleasure and relaxation now out of reach of any but the most prosperous of them has become the lot of the ordinary labourer, and till poverty is no longer possible for anyone engaged in ordinary work'. This was a question that haunted him throughout his life and again and again he returned to the same theme.

One conjectures that winning the Cobden and Adam Smith Prizes, although as far as is known neither was handed in at the stroke of midnight on the closing date, may have been a similar solace to Bowley as the Lothian and Matthew Arnold Prizes had been to Curzon in somewhat similar circumstances a few years earlier.

In 1895 the wages essay, with some revisions, was read as a paper at a meeting of the Royal Statistical Society and earned its author the Guy Medal in Silver. The subject remained a lifelong interest and, indeed, Bowley's last published paper was a communication to the Royal Statistical Society in 1952 on the same topic (he was then 82). In his later work his concepts became more refined: thus, while in his first essay the differences between the movements in wage rates, earnings and the total wages bill are somewhat blurred, by 1921 when he wrote the account of wages under wartime conditions, (*Prices and Wages in the United Kingdom* 1914-1920) he introduced clear-cut distinctions and, indeed, would appear to have been the first person to draw attention to the important phenomenon of 'wage drift' (although he did not use that term) with a detailed examination of why the movement of actual earnings should diverge from that of wage rates. His attempts to obtain theoretical distributions to fit the observed skew character of wages distributions are probably no longer of great interest although his speculation in 1900 that for a sufficiently homogeneous group, in which wages depended only on skill or ability, the distribution might be nearly normal, is certainly an idea which has been used by later workers.

465

He clearly saw the need for a theory in terms of a probability process which would satisfactorily explain the generation of income distributions but was unable to offer one; indeed, it is still awaited.

His interest in wages led him naturally into the closely allied fields of unemployment and prices, particularly of what was then termed the 'cost of living'. The latter term is still far from dead in popular use, but Bowley fought hard against its inherent looseness of thought and at least the official title of the (then) Ministry of Labour's index was eventually changed to 'Retail Prices'. Incidentally, it may also be remarked, that his influence must have contributed considerably to such a series being officially provided; his early papers refer continually to the unsatisfactory state of affairs in which an indifferent wholesale price index had to be used as a proxy for a retail price index.

Although, by the nature of things of no lasting interest since it has been overtaken by the vast growth in data available, there was a most ingenious method which he developed for the measurement of unemployment and which rewards attention if only as a curiosity. This work dates from the pre-national insurance period and the only direct data then available derived from the limited experience of trade unions. Bowley experimented with the use of both a wider class of numerical data indirectly related to employment and what he described as 'measurement by adjectives'. He noted that what was then the Labour Department *Gazette* gave a verbal assessment of the state of employment in various industries which compared the current month with both the previous month and the corresponding month a year previous; these descriptions had used a consistent five point scale ranging from 'very good' to 'very bad' over a number of years. The statements were first checked for internal consistency, with satisfactory results, by linking the monthly changes back to the position a year earlier to see if that result tallied with the direct comparison that was given. A graphical analysis followed in which it was assumed that the four changes could be assigned equal numerical significance with an arbitrary scale; this was done for over 20 industries separately and the results combined into a general measure. The agreement of the latter, where it could be checked, with the limited Labour Department index suggested that the method gave reliable results and hence could be used both to extend an index over industries and years for which numerical data were lacking and also as an adjustment factor for indirect numerical data of broader coverage. One is suspicious that some circularity is involved in the testing of this method but it attracts attention for its ingenuity rather than its practical uses.

All of Bowley's work in the field of employment, wages and prices continually brought him up against the problems in the use of index numbers. The series of papers in which he tackled their theory and practice would have left him with a substantial reputation if that had been his sole interest. The upheavals of the First World War brought to the fore the divergence between base weighted and current weighted measures of retail prices, since the typical basket of goods purchased by the average family in 1918 had changed radically from that of 1914. Bowley's studies of the effects had demonstrated that the official index being base weighted, would over-estimate the price rise (at least in respect of the items covered) by not allowing for substitution effects, but the point was not readily taken; a Joint Committee of the Parliamentary Committee of the Trades Union Congress and other labour associations undertook an investigation, based on a new family budget collection of their own in 1920, and came up with an increase based on current weights greatly in excess of the official figure. Bowley's rejoinder, one feels, does him great credit for he had been handed a splendid opportunity for cleverness; the surprising result was due to arithmetical blunders. The mistake was pointed out with gentleness and the basic material welcomed as valuable evidence. Indeed, the latter was of some value in convincing the sceptics of the reliability of sampling since the correctly worked results, although based on only some 600 budgets, were closely in accord with other data.

Great attention was being given at this time to the inclusion of sliding scale 'cost of living' adjustments in wage agreements and Bowley was inclined to favour it on the grounds that it made for better industrial relations. 'Neither employers nor workmen can afford to neglect any aid to the peaceful settlement of wages', he wrote. It is a method now very much out of fashion, of course, but I am not convinced we shall not see another turn of the wheel.

It was from this period that Bowley began to urge the continuing collection of family budgets although with slow effect; indeed, it seems quite likely that his successful defence of the official 'cost of living' series, within its restricted field, militated heavily against his proposal being adopted. A warning with which he accompanied the proposal is still relevant today: his fear was that households willing and able to supply budgets were unlikely to be properly representative in 'intelligence and the exercise of thrift'. It is still an open question, I believe, as to how far the well-known deficiencies in the estimates of tobacco and alcohol consumption obtained from the present *Family Expenditure Surveys* might be attributable to the 30 per cent or so of the sample who do not co-operate.

Among the other topics, related to this field, in which Bowley made some of the earliest contributions were the treatment of quality changes, the use of adult-equivalent units in measuring household size, the economies of scale in household size and the relation between wholesale and retail prices. In the last, his use of lagged relationships and variate differences are both methods which have now become very popular.

On the theoretical side, his most important contribution was in the study of error in index numbers and in two early papers (*Journal of the Royal Statistical Society*, 1897 and 1911) he provided formulae for the standard error of a weighted average, the ratio of two unweighted averages and the ratio of two weighted averages. In the most important cases of the more common index number formulae he gave the classical formulation of the effects of errors in weights as opposed to errors in the variables being averaged. The rule of thumb commonly expressed in a form such as 'small variations in weights may be neglected' can be compared with his own statement to some point: 'the effect of errors of weights is small compared with that of errors of quantities, when there are many quantities whose dispersion is small, no preponderent weights, and little correlation between weights and quantities.' It is noticeable that Bowley's attention to error in index numbers has singularly failed to influence subsequent practice in this respect; presumably the reason is that in so far as his results related to probability statements they depended on the assumption that the observations at hand behaved as though generated by a natural random process whereas, in practice, items are purposefully selected and also there is a whole range of other imponderables. On the former objection it is to be observed that later Bowley appears to have mooted the idea that proper random sampling techniques could be applied to the problem but it still remains to make any progress in practice; on the latter point, his own conclusion was that the measurable error of an index number 'when properly defined and related to an objective' was more important than the wrong choice of formula but this has not taken us much further forward since indices are rarely, if ever, related to a single objective.

The most widely known of his contributions on index numbers is perhaps the formula to which his name was given but this has been a source of puzzlement to some since it is known that both Edgeworth and Marshall had made a similar suggestion previously. The essence of it—the averaging of base and current weights—is a fairly obvious intuitive idea, but Bowley arrived at it in a quite different way. He postulated an unchanged utility function and then sought to find the change in expenditure that would be necessary between the two dates in order to leave an individual's utility position unchanged, given that it was maximized subject to prevailing prices at both periods. The formula then emerges from the leading term in a Taylor's series expansion and it was this development which Ragnar Frisch christened as the 'Bowley Index'. It might seem somewhat surprising that, with such an eminently reasonable property, the formula has not been widely used but, in practice, it presents almost insurmountable difficulties of data collection to achieve an up-to-date measurement. It appears likely though that the form currently used for the Retail Price Index gives a result which approximates quite closely to that which a Bowley Index figure would give.

His interest and reputation in the field of prices and wages naturally came to involve Bowley in many official enquiries. In 1920 he was called as an official witness before the Court of Inquiry set up to regulate the wages and conditions of employment of dock labour (the first to be established under the Industrial Courts Act of 1919). The proceedings were rather muddle-headed but a large part centred round an attempt to determine what wage was necessary to support a family 'not at a mere poverty line standard but a standard of living which would recognise the citizenship of the workers' (as the Court's President, Lord Shaw of Dunfermline, expressed it). The docker's case was presented by Ernest Bevin, then their national organiser, and he had not taken kindly to an optimal food budget which Bowley had produced for a family of five (two adults and three children) on a total expenditure of 40/-

per week, In fact, Bevin had made the whole thing look rather ridiculous by arranging a 'cookery' exhibition in which every item was split into exactly equal portions; thus each person's 11 ozs. of cheese was divided into so many identical cubes for each meal of the week, and so on. Bowley defended the budget along fairly obvious lines and pointed out that 'existing supplies do not provide more for the population as a whole than some slight excess upon what is given in that budget'. At which Bevin brought in some references to ship owners' meals at the Savoy and, perhaps anticipating he was facing one of the same ilk, enquired: 'Do you live on 40/-, Professor?' The Professor had evidently done his homework rather thoroughly; after pointing out that his family was larger than the standard, he continued: 'My expenditure in the last three weeks on food alone was 47/- per week. Add a certain amount of garden produce and it would be 52/-. Reduce the family and it works out at 46/-. The 6/- is accounted for through a larger allowance of milk. There has been no butter in my house for three months.' Bevin could only console himself with the thought that the middle classes had better opportunities of education and knew more of food values, to which Bowley's response leaves one with some uncertainty: 'The education of the middle class woman does not teach her to cook.'

Bowley, in fact, welcomed the eventual award (the claim was met in full) with the hope that it would be a great and notable achievement in the improvement of dock labour conditions but his fear was that insufficient attention was being paid to the limitations on the height of wages in general imposed by the size of the national product. At this time he was frequently re-iterating his view that the practical problem in raising wages and reducing hours was to obtain '54 hours' product for 48 hours' labour'.

His concern with this general issue of the limit to the possible improvement in wages seems to have been, indeed, the main factor in his extensive work on national income estimation in the inter-war years. It is fascinating to observe concepts and the refinement of procedures growing under his hand, for together with Josiah Stamp with whom he collaborated, theirs was the definitive influence of the period. When he first turned his attention to the subject, estimates were rudimentary in both principle and execution; the aggregate as usually calculated, and as given in some of his early papers, consisted of just three items—income above the tax limit, wages and 'intermediate income' (i.e. income which was not included in the first two categories). By the end of the period, when the torch had passed to Colin Clark, their efforts had brought the whole subject into its recognizable modern form and, indeed, it is clear that the official estimates, first undertaken as the basis of planning in the Second World War, would have been impossible without this work.

He dealt, at one point or another, with nearly all the issues which today's student of the subject encounters; not surprisingly, he was not able to suggest answers to all the problems he raised and many such matters remain still unresolved. Typical of his development of the subject is his treatment of taxation; in 1913 he was pondering the effects of changes between direct and indirect taxation in relation to both the valuation of Government services and the final aggregate. The arguments seem sound enough as far as they went, but the upshot was confused with no clear conclusion emerging. In 1922 he returned to the issue and, even if it is too much to say that 'all was light', one certainly feels that his lucid exposition cut a broad path through the jungle. Although he did not use the terms, he clearly propounded the difference between the 'market prices' and 'factor cost' bases of valuation. Indeed, this paper on "The Definition of National Income" (in *The Economic Journal*) is a very definite landmark; perhaps its greatest contribution is the all important distinction between income as the reward for current productive effort and transfer incomes (a term explicitly introduced) not corresponding to any such effort during the period of account. Thus the services of Civil Servants and Police are current income, but pensions and payments of interest to National Debt holders are not, although they are paid out of the nation's current income. Hence, national income is no longer the simple sum of individual incomes and we are left with two different aggregates: the one, which excludes transfer incomes and includes collective incomes and is identical with total product, and the other, which includes transfer incomes and excludes collective incomes. Bowley's suggestion that both were needed as serving separate purposes has become, of course, standard practice.

The application of the principles set out in this paper were worked out in, what was later republished as, *Three Studies in the National Income;* the last of these was methodologically the most sophisticated, but the first is of particular interest because it gave a definitive answer to a vital problem, the answer to which was previously a matter, at best, of speculation, and, more generally, of violent prejudice. This is the question, that Bowley took very seriously, of to what extent wages could be raised by a re-distribution of income. Even the Webb's were unduly partisan on this issue and had declared in a Labour Party statement of policy in 1918 that 'two-thirds of (the population), that is to say, the manual working class, obtain for all their needs only one-third of the produce of each year's work' with the implication that wages could be raised dramatically and immediately. Bowley's sober analysis should have summarily disposed of this naive view for his general conclusion, founded on a scrupulously impartial and thorough investigation, was that 'the wealth of the country, however divided, was insufficient . . . for a general high standard'. His calculation was, in fact, that an equal distribution might just have sufficed to raise everybody to the Rowntree minimum standard and this, of course, assumed no loss of product in the process. The exercise was not intended as an attack on socialism but simply as an abstract study of the maximum possible effects of redistribution of the existing product under the most favourable assumptions. Indeed, we are left in no doubt of both an intense sympathy with the objective of raising the standards of the less well-off and a strong disapproval of ostentatious luxury; on the latter, he makes the very fair point that 'when it is realized that the whole income of the nation was only sufficient for reasonable needs if equally divided, luxurious expenditure is seen to be more unjustifiable even than has been commonly supposed'. Nor did he wish an interpretation to be given to his results as crude as the proposition he had demolished, that is to leave well alone since there was no hope for improvement. He advocated new or increased public expenditure on old age pensions, widowed families, health, education, housing subsidies, income supplements for the poorest, and 'the supply of milk and other goods below cost'. Some of these he recognized as simple transfers, others he argued could be regarded as capital investment which could lead to increased productivity in the future. As always, it was the last factor which he urged as the basic source of hope and he detailed a series of measures, of which some have since been adopted and others are still being advocated.

An essential part of the development of national accounting was the provision of better data and here, as in most other applications, research in economic statistics depends heavily on official sources. At the beginning of the century, the official dictum appears to have been that no figure should be released to the public which could not be sustained as exactly correct in a court of law; the pursuit of accuracy was not unpraiseworthy but an extreme pedantry was discouraging to both the collection and publication of new data. Moreover, the preoccupation with arithmetical accuracy led to neglecting consideration of what kind of figure it was useful to produce. The scientific enquirer, Bowley said (in his Presidential Address to Section F of the British Association in 1906) 'is left in the position of a man who inquires a distance in France and is told that it is 8·543 kilometres along the high-road, and then some way along a path'. On top of this, a great deal of what was produced with such meticulous care and unstinted labour, was then unintelligible through lack of any accompanying explanation and, even if it had been understandable, of no great interest to anybody.

This whole problem engaged Bowley's attention from an early date and the theme occurs both in a number of papers specifically devoted to the problem as well as incidentally in many other of his writings. In addition, one may suppose that he was also influencing the course of development both in personal contact with official statisticians and indirectly through the climate of opinion that he created. A great deal of this work is concerned with the fundamentals of measurement in the social sciences—questions such as the definition of the unit, the universe of coverage, comparability over time and place, and the limits of accuracy—which are now firmly embedded in our discipline. Nobody appears to have made such an explicit and systematic treatment of these issues before and, of course, once attention is directed to them they may appear almost trivial; in fact, it is very easy for them to escape attention buried under implicit assumptions and when they are brought to light they are matters of very considerable complexity.

On the more general issue of official policy, the needs he constantly emphasized were for: data directly related to the magnitude of interest; Government responsibility for statistics other than those for departmental administrative requirements; detailed guides to the method of collection and processing of data; co-ordination between different departments to obtain uniformity of treatment; timely publication; and an assessment of error margins in place of a futile attempt at unattainable exactness. The means of the reform was to be a Central Thinking Office of Statistics, coupled with the creation of a trained statistical officer class in the Civil Service.

Many of these ideas were embodied in a formal petition made to His Majesty's Government in 1919 by the Royal Statistical Society, but it met with a cool reception from the officials who took it as an attack on them instead of, as Bowley put it, 'on the system under which it is their misfortune to find their efforts frustrated'. It was not accepted that the Government had responsibility for providing statistics on all economic and social aspects of the community and the establishment of a Central Statistical Office was deemed impracticable. None the less, the cause progressed slowly in the inter-war years, received a decisive impetus from the Second World War and since then, whatever complaints remain, has effected a radical revolution. The process has been the result of many people's labour, of course, and one can scarcely hope to apportion credit, but, on any showing, it is fair to claim that Bowley played a key role if only by virtue of the particular historical period in which he lived. One imagines, for example, that a Petty or a King born a couple of centuries later would have played the same role.

Switching to the other end of the statistical spectrum, Bowley's contributions to mathematical statistics appear so overshadowed by those of, for example, the Pearson father and son, of Neyman and of Fisher, to take only the obvious names, that one might be inclined to dismiss them as of little significance. However, he not only maintained a deep interest in the theoretical side of the subject but also played a very considerable part in its interpretation and dissemination. Indeed, the remarkable feature is the broad range of his work. A number of his books, even if intended as 'text-books', were rather more than that for his period. The *Elements of Statistics* went through six editions between 1901 and 1937 and complemented rather than competed with the well-known Yule text when that appeared ten years later. On its publication, Foxwell wrote in a personal letter 'in my opinion it will be *the* textbook of statistics for the next generation'. His interpretation of Edgeworth in *F. Y. Edgeworth's Contributions to Mathematical Statistics* rescued much of what otherwise might have been lost. He wrote the first text on mathematical economics (*The Mathematical Groundwork of Economics*, 1924) and also probably the first text on pure mathematics intended for social science students (*A General Course of Pure Mothematics*, 1913). The latter met with Whittaker's warm approval but some rather heavy handed censure from Hardy; in a private letter to Bowley he remarked 'what I felt about the book was that it was an attempt to do what is simply not possible for anyone but a professional mathematician in the strict sense'. In a further letter, after Bowley had successfully challenged at least one of his criticisms, he admitted somewhat grudgingly 'Your proof of the multiplication theorem (of infinite series), therefore, is, I think, substantially correct'.

There might well be a case for thinking that towards the end of his active career Bowley's attitude towards the new developments in theory was unhelpful and negative. In moving a vote of thanks to Fisher for his paper on inductive inference at a Royal Statistical Society meeting in 1935, he expressed a genuine regard in the highest terms for his work in general but then gave reign to a wit, not usually so caustic, over part of the paper under discussion:

> 'I found the treatment to be very obscure. I took it as a weekend problem, and first tried it as an acrostic, but I found that I could not satisfy all the 'lights'. I tried it then as a cross-word puzzle, but I have not the facility of Sir Josiah Stamp for solving such conundrums. Next I took it as an anagram, remembering that Hook stated his law of elasticity in that form, but when I found that there were only two vowels to eleven consonants, some of which were Greek capitals, I came to the conclusion that it might be Polish or Russian, and therefore best left to Dr. Neyman or Dr. Isserlis. Finally, I thought it must be a cypher, and after a great deal of investigation, decided that Professor Fisher had hidden the key in former papers, as is his custom, and I gave it up.'

Many, no doubt, have had similar bouts of despair in attempting to understand Fisher.

In the same way Bowley had been unconvinced by Neyman's theory of confidence limits and, in a similar role at a meeting a year earlier, had expressed a doubt that a 'confidence interval' was not a 'confidence trick'. The trouble was, of course, that throughout his life he had been seeking the answer to the wrong question (or, rather quite correctly knew that there was no answer but still thought one was wanted) and could not adjust readily to those who produced an answer to another, more useful question. One might reasonably suppose that by this time his faculties had passed their zenith and that he was not so readily perceptive of the merits of new ideas but, even so, it should be remarked that at this very time he was collaborating with R. G. D. Allen in their pathbreaking work in econometrics; their book *Family Expenditure*, which appeared in 1935, is rightly regarded as one of the great foundation stones of the modern study of the subject.

In this selective review of Arthur Bowley's contributions to economic and social statistics, I have left until the last that field which seems to me to have provided his single greatest achievement and which he developed virtually single handed. I refer to the technique of social surveys by sampling or, as it used to be called, 'the representative method'. The possibilities of the method occurred to him early and, in fact, he drew attention to them in his British Association Address of 1906. He gave his audience a full demonstration of the technique and, in particular, emphasized the great virtue of probability sampling in that not only can an estimate of any desired precision be obtained but that a measure of the precision can be obtained from the sample itself. It was a subject which occupied him through the greater part of his active career; it almost became a King Charles's head, and no opportunity was lost to draw attention to its potential whatever the topic on which he was writing.

To be an enthusiast is one thing but to be a successful practitioner is another and his penetration of the practical problems is as remarkable as his vision of the possibilities. His insistence on meticulous attention to detail at all stages of a project, in planning, in field work and in analysis of results, set a standard which has not been surpassed. On occasions he was thought unreasonable; in the early twenties a criticism he made of a Ministry of Labour sample design for not allowing an exactly equal chance of selection for every member of the universe was evidently thought to be rather tiresomely pedantic. Three years later, Hilton, then Director of Statistics at the Ministry, sent him a handsome apology admitting that a curious fault in their estimates which had been evident for some time had at last been tracked down to exactly the point that Bowley had picked on.

Looking at his first survey of Reading in 1912, the most impressive feature is that the infant emerges virtually fully grown. That the report still reads as though it might have been referring to a survey conducted last year, is a measure of its influence on subsequent practice. The town directory carefully scrutinized, was used as a sampling frame of houses; the interviewers had strict instructions never to substitute another house for that selected, in the event of failure to obtain response; all households were included in the case of multiple occupation of houses; check questions were included for comparison of results with independent data.

This first experiment was quickly followed by four others in Northampton, Warrington, Stanley and Bolton, constituting the famous Five Towns Survey completed just before the First World War. A primary objective in all of them was to determine the incidence of poverty and it is worth recalling his conclusion that 'to raise the wages of the worst paid workers is the most pressing social task with which the country is confronted today'. These same towns were again surveyed in a second round in 1924, revealing a considerable improvement due to substantial gains in earnings of unskilled workers and to smaller family sizes but marred by increased unemployment.

Bowley's final and most massive undertaking was the House Sample Inquiry of the *New Survey of London Life and Labour* which covered some 30,000 households. Again a primary object of the enquiry was to answer the 'insistent questions which are in all men's minds; in what direction are we moving? Is poverty diminishing or increasing?' (as the Director of the project, Sir Hubert Llewellyn Smith, put it). In this, only a few years before he was to become emeritus, there is a loud echo of the undergraduate ponderings which had led him into statistics.

It is, I think, in the practice of sampling that Bowley must be recognized as pre-eminent, and there is ample testimony from his contemporaries as to that, but his influence on the theory of sampling was also of prime importance. He wrote a fundamental review of its existing state for the International Statistical Institute in 1925 and dealt with various aspects at greater or lesser length in a number of other publications. This, though, all appears very dated now and, in contrast to the practice, one is sharply reminded of the extent of later developments. That so many of these developments are implicit in his work is though not the least part of his influence.

His theory was confined, in the main, to simple random sampling where the sample units are also the units of interest but in practice he used both cluster sampling and stratified sampling (the latter of which terms he introduced). These were the growth points for further theoretical work by others.

Bowley lived for over twenty years after his retirement in 1936 from his Chair at the London School of Economics and continued to take a productive interest in many branches of his subject. When well into his seventies, he returned to active work to become the wartime head of the Oxford Institute of Statistics and surprised his colleagues by his soon realized intention of being no mere figurehead. He left with them a warm impression of an alarmingly active mind, humanity and dedication. His later years appear to have been contented ones although idleness was never to his liking; he had books, music and weaving while, no doubt what he would have put above all, he was blessed with the well matched companionship of a talented wife until his end.

One feels that, in a sense, Bowley lived just at the right time. His was the first Chair devoted exclusively to statistics in the social sciences and he established what was in fact the first department (although the School had no formal departmental organization at that time). Of course, he built on the work of others (he freely recognized his debt to Giffen and Edgeworth, in particular) but he gave form and direction to what had been previously a somewhat amorphous discipline. It was then not too large for a man of his gifts to put his imprint upon it; that imprint, as I see it, was a scrupulous scientific approach motivated by a genuine concern for people.

A NOTE ON SOURCES

Both Professor Marion Bowley and Professor Sir Roy Allen were most generous in their help in locating material and in providing answers to a variety of enquiries: neither, of course, is in any way responsible for the accuracy of my statements or views. *A Memoir of Professor Sir Arthur Bowley* (1869-1957) *and His Family* by Dr. Agatha H. Bowley (privately published in 1972) is a fascinating source of personal detail to which I am greatly indebted. The Royal Statistical Society Librarian, Miss S. I. Kearsey, was put to a lot of trouble in locating correspondence and was most helpful in the arrangements she made. Mr. G. C. Allen, of the British Library of Political and Economic Science, is to be thanked-likewise for tracing many other items of interest.

The bibliography given in the obituary by R. G. D. Allen and R. F. George (*Journal of the Royal Statistical Society*, Vol. 120, 1957, pp. 236-241) is substantially complete except for very minor publications. As far as the latter have been traced in the preparation of this lecture, they consist mostly of newspaper articles and reports of public lectures. A list has been retained by me but is not reproduced here.

A list of the correspondence which was traced is given below (excluding letters of a purely routine nature):

1895	20th May	Lord Farrer to A.L.B.	(Guy Medal)		RSS
1901	19th Jan	George Bernard Shaw to A.L.B.	(Vaccination statistics)		RSS
	7th Feb.	H. S. Foxwell to A.L.B.	(*Elements of Statistics*)		RSS
	7th Feb.	Alfred Marshall to A.L.B.	(*Elements of Statistics*)		RSS
	21st Feb.	Alfred Marshall to A.L.B.	(Further comments)		RSS
	3rd Mar.	Alfred Marshall to A.L.B.	(Methods of economic investigation)		RSS
	30th Dec.	Alfred Marshall to A.L.B.	(Retail prices)		RSS
(?)	8th Nov.	F. Y. Edgeworth to A.L.B.	(*Elements of Statistics*)		RSS
1906	7th Aug.	H. S. Foxwell to A.L.B.	(B.A. Presidential Address)		RSS
	6th Sept.	A.L.B. to Edwin Cannan	(Infinitesimal texation)		BL
	12th Sept.	A.L.B. to Edwin Cannan	(Infinitesimal taxation)		BL
	9th Dec.	A.L.B. to Edwin Cannan	(Diminishing returns)		BL
1908	7th Dec.	Edwin Cannan to A.L.B.	(Diminishing returns)		BL
1914	25th Mar.	E. T. Whittaker to A.L.B.	(*A General Course of Pure Mathematics*)		RSS
	8th May	G. H. Hardy to A.L.B.	(*A General Course of Pure Mathematics*)		RSS
(?)	(?)	G. H. Hardy to A.L.B.	(Further Comments)		RSS
1918	4th Mar.	A. Bonar Law to A.L.B.	(Cost of living Committee)		RSS
1919	6th Mar.	A.L.B. to Sydney Webb	(Coal Commission)		BL
1921	28th Aug.	A.L.B. to Edwin Cannan	(Currency issue)		BL
1924	27th Mar.	Edwin Cannan to A.L.B.	(Consumer's surplus)		BL
	31st Mar.	A.L.B. to Edwin Cannan	(Consumer's surplus)		BL
	7th Apr.	Edwin Cannan to A.L.B.	(Consumer's surplus)		BL
1927	1st June	John Hilton to A.L.B.	(Sampling)		*
1932	1st June	A.L.B. to Edwin Cannan	(Balance of Trade)		BL
1933	20th Oct.	India Office to A.L.B.	(Mission to India)		RSS
1935	17th Oct.	W. A. Appleton to A.L.B.	(Guy Gold Medal)		RSS

1953	19th Oct.	R. F. George to A.L.B.	(Manuscript exhibition)	RSS
	25th Oct.	A.L.B. to R. F. George	(Essay on socialism)	RSS
	28th Oct.	R. F. George to A.L.B.	(Continuation)	RSS
	1st Nov.	A.L.B. to R. F. George	(Bicycling habits)	RSS
	3rd Nov.	R. F. George to A.L.B.	(Reply)	RSS

BL : British Library of Political and Economic Science
RSS : Royal Statistical Society Library
* : Copy provided by Sir Roy Allen

Note on the History of Sampling Methods in Russia

By S. S. Žarković
(1956)

Various articles have been published on the history of the theory and application of sampling methods. They include: "A review of recent statistical developments in sampling surveys", by F. Yates, published in this *Journal*, Vol. CIX, 1946; "History of the uses of modern sampling procedures", by F. F. Stephen, published in the *Journal of the American Statistical Association*, Vol. 43, 1948; and the "Historical survey of the development of sampling theories and practice", by You Poh Seng, published in this *Journal*, Vol. CXIV, 1951.

Some of these articles have considerable importance because of their systematic approach to the subject. None, however, contains any reference to Russian contributions to the development of sampling methods during the First World War and in the period immediately following it, when Lenin was still alive. I should like to call attention to this fact because, at least as far as I know, it may be said that during those years it was precisely in Russia that the most intensive work in the field of sampling was being carried on.

In this connection the mathematical traditions of Russia should not be overlooked. At the time in question Russian statisticians were aware of the results of Tchebyshev's work. Markov, whose interest in statistical problems is well known, was still living. A. A. Tchuprow was also working in Russia during the First World War, and his contributions to statistical theory are known throughout the world. In addition, a number of highly competent statisticians were active, among them A. Kaufmann, V. Romanovsky, E. Slutsky, B. Iastremsky, Boiarsky, Tchetwerikov, Obukhov, etc. Many of them were young and subsequent events prevented them from becoming known outside Russian borders.

To understand Russian statistics at that time one must remember that the general trend in statistics during the first years after the revolution differed markedly from that apparent later. Judging from publications, it appears that in the earlier period theoretical work was completely free. The problems which Russian post-revolution statistical theory dealt with were those common to many countries at that time and were thus being widely studied. The methods used for solving practical problems were likewise those common to the current statistical methodology.

Another feature of Russian statistics in those years was the interest theoreticians took in practical questions and the effect it had on their intensive application of theory to the solution of practical problems. These Russian statisticians watched the development of statistical theory all over the world, they published translations of the most important foreign contributions and they reviewed for the benefit of their readers all important results, whatever country provided them. This keen activity supplied the base from which they sought solutions to their own practical problems. I would go so far as to say that, in consequence, Russian statistics in the early thirties were on a par with the best in other countries.

A very detailed picture of all aspects of this statistical activity may be obtained by reading *Vestnik Statistiki*, the organ of the Central Statistical Administration, which started publication in 1919. One section of the *Vestnik* was devoted to theory, another to reviews of statistical literature and accounts of the work on which different statistical agencies were engaged. There were also notes on statistically important events and facts, etc. The main source of the information given in this article is, in fact, *Vestnik*.

A study of that publication and of other available sources reveals that sampling methods were in practical use during the First World War and that their application was extensive thereafter. In his book *The Application of Mathematical Statistics to Experimental Work*, Moscow, 1946, V. Romanovsky states that A. A. Tchuprow had designed some sampling surveys during the First World War. In *Vestnik*, Vol. III, 1919, S. S. Kon wrote "On the question of the application of sampling methods in the processing of the agricultural census". In Vol. IV, 1919, there is an article by N. Tchetwerikov, director of the Methodological Division of the Central Statistical Administration, "On sampling methods". In this interesting article, important from the historical point of view, Tchetwerikov says that A. A. Tchuprow read his report "On sampling methods" to the meeting of the Russian research workers in natural sciences and medicine, held in 1900. According to Tchetwerikov, in this report only probability samples are dealt with and the basic theory of surveys is developed.

As far as Tchetwerikov himself is concerned the theory of sampling represents in his opinion an objective procedure based on random selection of sampling units with the known probabilities of selection of each particular unit. The basic feature of sample surveys is not the size of the sample, says Tchetwerikov, as was believed at that time by some Russian statisticians, but the way in which the sampling units are drawn into the sample.   The aim of this article is to present the logical bases of sampling methods;   with regard to formulae, Tchetwerikov says they can be found in Tchuprow's works.   In his attempts to make clear the meaning of sampling methods, Tchetwerikov speaks on simple random samples selected with or without replacement of units, on stratified samples, on cluster sampling, and on systematic selection of sampling units.   In his presentation of sampling methods he mentions some problems important even to-day.   Thus he says that the precision of estimates does not depend only upon the size of the sample but also on how the units are selected, on the magnitude of the variation of the phenomenon under study and the shape of the distribution.   In connection with the latter he calls attention to the need for the new theoretical developments in order to show how the precision of estimates is influenced by distributions deviating from the normal type.   This is the field in which new results are expected even to-day.   In addition, the possibility of using supplementary information in sample surveys is described, and so the article covers the basic chapters of the modern theory of sampling.

In Vol. VI, 1920, there is another article by Tchetwerikov entitled "An experiment on the processing of mass data by means of sampling methods".   Vol. VII, 1920, contains a study on "Cattle breeding according to the results of 10 per cent. sample census of agricultural holdings in 1919".   In Vol. VIII, 1921, there is an article on "The origin of sampling methods and the first results obtained on the basis of their application in Russia", by A. Gurev.   This article is important because it shows that the ideas popularized by Kiaer were used in Russian theory and practice in the second half of the nineteenth century, i.e. much earlier than when Kiaer started his work in this field.   Gurev presents many examples to illustrate this statement.   In the same volume A. Gozulov reviews the "Results of the sample survey of children aged 0–18 years taken in Rostov and Nakhichevani".   Vol. IX contains a report on the results of the 10 per cent. tabulation of the federal agricultural census in the Province of Pskov.

From the above it is clear that before 1920 not only were theoretical articles on sampling methods available for study but practical applications were being made.   It is also evident that the use of such methods was not confined to the central authority;   they were also employed in different surveys and tabulation programmes in the provincial statistical offices.   Already, by that time, sampling had obviously become popular.

In Vol. X, 1922, pp. 179–180, reference is made to the preparation of a pamphlet *Review of the Data on the Tenure of Land*, on the basis of a sample taken in each republic.   On p. 181 the same problem is discussed in connection with a sample from agricultural holdings, drawn from other agricultural statistics, data being provided on the allocation of the sample among counties, communes and individual households.   On p. 33 of the Appendix, A. I. Krashcheva deals with a sample census of agricultural holdings in 1919 and some information is given about the sample tabulation of the 1920 census.   These data form part of a paper which, when read before the federal statistical conference, was followed by an animated discussion regarding the size of the sample needed for the same tabulation.   On p. 76 information is given concerning research into the needs of the population for different commodities, the study having been planned as a sample survey.

In Vol. XI, 1922, pp. 151–2, 162, there is another note on the use of sampling methods in agricultural statistics.

Vol. XII, pp. 181–188, mentions, among others, a paper read at a meeting of statisticians in October, 1922, by I. A. Poplavsky, on "The use of sampling methods in transport statistics" in which he suggested that data be collected by means of registration of traffic at certain points of the transport routes.   On p. 191 it is stated that the Methodological Division of the Central Statistical Administration has been made responsible for the study of the precision of the results obtained by "mechanical" (i.e. systematic) selection of the sample, and of the question whether the units should be households or clusters of households.   In the Appendix, p. 29, A. I. Krashcheva's name appears again.   She was probably the first woman to take a deep interest in the theory and practice of sampling methods and is often mentioned in *Vestnik* in connection with them.   On this occasion, she had read to the All-Russia Statisticians' Conference, in November, 1922, a report on "Methods of acreage statistics", in which she spoke on the application of sampling methods and drew attention to some attendant theoretical problems requiring study.   A. I. Kogen also spoke to that meeting on sampling methods.   A resolution (p. 31–32) was passed suggesting the need for further study of a number of problems in the field of sampling methods.   The large number reported as speaking on the problem shows in itself how popular such methods then were in Russia.

Vol. XIII gives, on p. 189, an account of a report made to the same meeting by N. L. Dubenetsky on current agricultural statistics advocating the application of sampling methods in that field.

A resolution is also recorded to the effect that "the Methodological Division, in collaboration with the Division of Current Agricultural Statistics, be directed to study the question of the percentage of households that should be drawn into the sample from the whole mass of households . . . and the question of how to select the sample". There is also a reference, on p. 191, to a decision of the Central Statistical Administration whereby the Methodological Division was directed to undertake some experiments in order to clarify the problem of cluster sampling. Some fresh data on sampling methods are to be found on pp. 208–9, 298 and again on p. 300, where a review of Lebedev's report is presented. The title of this report is "The problem of the methodology of sample surveys of agricultural holdings".

In Vol. XVI, A. Krashcheva, on pp. 88–89 again deals with the application of sampling methods, this time in the study of the dynamics of peasant households. The possibility of using such methods in connection with some other problems of agricultural statistics is discussed on pp. 172–175, 178–179 and 184, and again in Vol. XVII, pp. 165–168 and 263–267.

One more item remains to be mentioned, the book *Basic Theory of Sampling Methods*, by A. G. Kowalsky, published in 1924 in Saratov and reviewed in Vol. XX, 1925. I have had no opportunity to see the book itself; from the review it appears to have been very similar to modern text-books on the subject. The author declares that he has confined himself to probability samples because only that type of sample makes it possible to develop an objective and scientific theory of surveys. Another feature of the book is its attempt to give a full mathematical treatment of the topic, and in connection with this many formulae and mathematical derivations are provided. The reviewer also states that some new and unknown theories are presented. One of these (according to the review) is the optimum allocation of the sample among different strata in a stratified population. This theory was introduced into the statistics of western countries by J. Neyman in 1934. Later it was found that Tchuprow had developed the principles of the theory in *Metron* in 1923. Apparently, Kowalsky developed the same theory independently of Tchuprow. He had a formula showing that the best allocation of the total size of the sample is not attained when distribution among the strata is proportional to the number of their units, but, instead, when the stratum sample sizes are proportional to the stratum numbers of units and "the degree of the variation of the phenomenon under study in a given stratum" (p. 260). The reviewer adds that the book includes a new mathematical treatment of the systematically selected samples.

The above exposition is sufficient to prove that Russian statistics should not be neglected in a historical review of the development of the theory and practice of sampling methods. The same holds good for some other fields in which Russian statistics probably then led the rest of the world.

In subsequent years political considerations became an increasingly pronounced factor in the development of Russia's statistics. This brought about the gradual disappearance of the use of theory in the practical activity of the statistical administration. In the late thirties *Vestnik Statistiki* began to close its pages to papers in which statistical problems were dealt with mathematically. At the end of that period they disappeared completely and have not appeared since. The result of this new trend was that statisticians abandoned practice to continue their work at the universities and other scientific institutions where they pursued statistics under the name of some other subject. Officially, Romanovsky, Kolmogorov, Smirnov and many others are mathematicians divorced from statistics. A very interesting example is E. Slutsky, who enjoys world wide renown as one of the forerunners of econometrics. He gave up statistics to embark on a new career in astronomy. Some distinguished names such as Tchetwerikov, Obukhov and a number of others appeared in meteorology. Many younger statisticians are working now as engineers, technicians, and so on. According to official views, statistics became an instrument for planning the national economy. Consequently, its basis is the Marx-Lenin political economy; it represents a social science or, in other words, a class science. The law of large numbers, the idea of random deviations, and everything else belonging to the mathematical theory of statistics were swept away as the constituent elements of the false universal theory of statistical science (see Pisarev, Methodological questions of Soviet Statistics, *Planovoe Hoziaistvo*, 1940; *Vestnik Statistiki*, 1950, which contains a report on the famous conference on methodology, held at the Central Statistical Administration). Nevertheless, the relatively long tradition of sampling methods made it difficult to eliminate them. Thus, S. V. Sholtz, in his text book on agricultural statistics, *Kurs selskohoziaistvenoi statistiki*, Moscow, 1945, claims that such methods represent the best and most objective way of obtaining data about yields. However, sampling methods are now being discussed without any formulae.

It is to be hoped that some day one of the living Russian statisticians who was scientifically active during and after the First World War may write a historical review of Russian statistics, and thus of the development of the theory and application of sampling methods in that country.

A Supplement to "Note on the History of Sampling Methods in Russia"

By S. S. ZARKOVICH
(1962)
Food and Agriculture Organization, U.N., Rome

SOME years ago I published a "Note on the History of Sampling Methods in Russia" (this *Journal*, **119**, 1956, pp. 336–338). In that note a reference was made to the *Basic Theory of Sampling Methods* by A. J. Kowalsky, a book published in 1924 in Saratov. I had not had a chance of seeing the book and mention of it was made on the basis of a review published in *Vestnik Statistiki*, **20**, 1925. According to what was said in the review it was possible to consider this book as a treatment of the subject along modern lines and very much in advance of the ideas and practice of sample surveys of that time. Recently, however, I was able to study the book in the Lenin Library in Moscow and the purpose of this Supplement is to give some additional information about it. This information might be of interest, considering the content and the year of publication of the work.

The book was published in the *Scientific Papers* of the State University of Saratov, **2**, Part 4, 1924, pp. 60–138. In the introduction Kowalsky says that in deriving certain formulae he had the help of the mathematician B. K. Riesenkampf, who had formulated an independent development of the mathematical theory of sampling. In Part I the author presents sample surveys as an objective scientific procedure, based on the theory of probability and containing no subjective elements, to which at that time recourse was often made in survey practice. In this respect Kowalsky says that he is adopting the attitude formulated by Tchuprov in the paper presented to the statistical sub-section of the session of the all-Russia congress of research workers in natural sciences and medicine in 1894. The same attitude was expressed in a number of other papers published by Tchuprov and others in subsequent years. Kowalsky proceeds to a review of the many uses of sampling methods in Russian statistics and points out the improvements needed if sample surveys are to be made into an objective and scientific procedure. To this purpose, he proceeds in the subsequent chapters to present the mathematical foundations of the theory of sampling.

In the establishment of the theory Kowalsky does not follow the authors who deal with sampling with replacement or with infinite populations, as such cases are of little practical importance in his area of interests. Instead he presents the theory of sampling without replacement from finite populations and is aware, together with Tchuprov, that sampling with replacement is a special case of this more general approach.

Part II of the book deals with the estimation of the proportions and the number of units having a given attribute. Formulae for unbiased estimates from simple random samples are shown together with their variances. A section follows on the difference between sampling with and sampling without replacement. Next comes the theory of stratified sampling presented in a general way, viz. with formulae which make it possible to use any sampling fraction from individual strata. This leads to the problem of the "most convenient proportion of sampling" from various strata which Kowalsky solves by introducing what is now called optimum allocation.

Using the method of the Lagrange undetermined multipliers, Kowalsky minimizes the variance of the estimated mean from a stratified population and obtains (p. 93) the size of the sample in the ith stratum

$$n_i = n \frac{N_i \sqrt{(P_i Q_i)}}{\Sigma N_i \sqrt{(P_i Q_i)}}$$

which is the same formulation now used by most authors in arriving at the optimum allocation.

The same result was published by Tchuprov in 1923 and by Neyman in 1934. Although Kowalsky associates himself proudly with the line of thought adopted by Tchuprov, it is difficult to say whether he benefited from Tchuprov in the establishment of the optimum allocation formula. Kowalsky gave much credit to Tchuprov throughout his book and it is hard to believe that he would have failed to do so with respect to this result if it were suggested by Tchuprov. It is even less likely that Kowalsky saw the paper by Tchuprov mentioned above. Kowalsky lived in a provincial city where the circulation of Western literature in the post-Revolution years was more than restricted. In addition, there are indications that Kowalsky had to wait a long time before he was able to find a publisher for his work.

The next section of Part II deals with the problem of the efficiency of stratified sampling compared to simple random sampling. This question was not clear at that time and Kowalsky solved it elegantly (p. 95) for the case of proportional allocation by showing that the variance of the estimated proportion from such a sample could be expressed approximately as

$$\sigma_p^2 \doteq \frac{P.Q}{n} \left(1 - \frac{n}{N}\right) - \frac{\Sigma N_i (P_i - P)^2}{N.n} \left(1 - \frac{n}{N}\right).$$

This formula shows that Kowalsky saw clearly the circumstances in which proportional allocation leads to improved efficiency.

Part III deals with the case of the continuous variable and presents the same theory as in Part II. Here there is a reference to Tchuprov as the first to develop (in 1917) the variance of the mean in sampling without replacement, although the result was not published. Kowalsky also says that this formula was developed independently by Riesenkampf. Regarding the gain due to stratification Kowalsky again restricts his discussion to proportional allocation and points out in what way this gain depends upon the differences between strata means.

Another interesting section in this part is the treatment of systematic samples, which were at that time often considered as equivalent to simple random samples. Kowalsky shows that systematic sampling could be conceived of as sampling of one unit from a population divided into n equal strata, n being the size of the sample. Hence a formula of the variance in the case of stratified sampling could be used to derive the precision of systematic samples compared with simple random samples. Kowalsky is aware that such a formula cannot be used in estimating the variance since strata means remain unknown. However, it helped him to introduce the concept of homogeneity as the essential element of the precision of systematic samples. Although in a somewhat different form, this concept is now generally used in presenting the efficiency of systematic samples. Kowalsky breaks down the total variation into its components and shows that systematic sampling may in some cases have a high relative efficiency with respect to simple random sampling. He is also aware of the possibility that the opposite may take place.

The last part of the book deals with numerical examples and explains the computation of estimates and their variances.

Although Kowalsky's book does not contain any treatment of multi-stage and multi-phase sampling nor the use of the supplementary information, it is a pleasure for any statistician interested in this field to see that a text published some 30 years before the main books on sampling was in fact written from the modern standpoint on the subject.

REFERENCES

NEYMAN, J. (1934), "On the two different aspects of the representative method: the method of stratified sampling and the method of purposive selection", *J. R. statist. Soc.*, **109**, 558–606.
TCHUPROV, A. A. (1923), "On the mathematical expectation of the moments of frequency distributions in the case of correlated observations", *Metron*, **2**, 646–680.

QUEEN MARY
COLLEGE
LIBRARY